Immunology: Functions and Disorders of the Immune System

Immunology: Functions and Disorders of the Immune System

Editor: Avery Steele

STATES
ACADEMIC PRESS
www.statesacademicpress.com

States Academic Press,
109 South 5th Street,
Brooklyn, NY 11249, USA

Visit us on the World Wide Web at:
www.statesacademicpress.com

ISBN: 978-1-63989-283-9 (Hardback)

Cataloging-in-Publication Data

Immunology : functions and disorders of the immune system / edited by Avery Steele.
p. cm.
Includes bibliographical references and index.
ISBN 978-1-63989-283-9
1. Immunology. 2. Immune system. 3. Immunologic diseases. 4. Serology. I. Steele, Avery.
QR181 .I46 2022
616.079--dc23

Table of Contents

Preface

Over the recent decade, advancements and applications have progressed exponentially. This has led to the increased interest in this field and projects are being conducted to enhance knowledge. The main objective of this book is to present some of the critical challenges and provide insights into possible solutions. This book will answer the varied questions that arise in the field and also provide an increased scope for furthering studies.

The network of biological processes which protect an organism from different types of diseases is termed as the immune system. It deals with a wide variety of pathogens ranging from parasites to viruses, as well as potentially infectious objects like wood splinters. The branch of biology which focuses on the immune systems of various organisms is known as immunology. It makes use of immunology charts to contextualize the physiological functions of immune system, and detect different types of immunological disorders. Inflammatory disease, cancer, hypersensitivity, autoimmune disease, and immune deficiency are some of the common examples of disorders related to the immune system. The discipline of immunology can be further classified into developmental immunology, classical immunology, reproductive immunology, diagnostic immunology and theoretical immunology. The book aims to shed light on some of the unexplored aspects of the functions and disorders of the immune system, and the recent researches in this field. It strives to provide a fair idea about this discipline and to help develop a better understanding of the latest advances within this field. This book will help the readers in keeping pace with the rapid changes in this field.

I hope that this book, with its visionary approach, will be a valuable addition and will promote interest among readers. Each of the authors has provided their extraordinary competence in their specific fields by providing different perspectives as they come from diverse nations and regions. I thank them for their contributions.

Editor

1

The Immunology of Cancer Cells

Gregory Lee[1,2]*, Cheng-Yuan Huang[1], Suefay Liu[1] and Hao Zhang[1,2]

[1]*UBC Center for Reproductive Health, Vancouver, Canada*

[2]*Department of Pathology, Shantou University Medical College, Shantou, China*

***Corresponding author:** Gregory Lee, UBC Center for Reproductive Health, Vancouver, Canada, Tel: 778-322-4651; E-mail: cyglee@yahoo.com*

Abstract

RP215 is a monoclonal antibody discovered in 1987 which was later found to react specifically with a carbohydrate-associated epitope detected mainly among immunoglobulin heavy chains from cultured cancer cells. With RP215 as the unique probe for cancerous immunoglobulins, the experimental evidences seem to suggest that the immunology of cancer cells plays dramatic roles in the growth/ proliferation and immune protection of cancer cells. RP215-specific cancerous immunoglobulins serve in part to capture or neutralize circulating antibodies or antigen hostile to cancer cells in human body. In the conventional immune system, B and T lymphocytes and their associated antigen receptors play key roles in the adaptive immune response against pathogens and/or cancer cells through established mechanisms. In contrast, cancer cells co-express both B and T lymphocyte-associated antigen receptors and immune protection is exercised by completely different mechanisms. For example, these cancer cell-expressed antigen receptors demonstrate a lack of class switching, or limited hyper mutation. Furthermore, toll-like receptors involved in the innate immune system in cancer cells are strongly affected by these cancer cell-expressed antigen receptors. Therefore, both normal and cancerous immune systems can co-exist and can be operated simultaneously within the human body. A potential therapeutic strategy may be developed by using RP215 as a drug candidate to target cancer cells based on these observations.

Keywords: Cancerous Immunoglobins; RP215 Monoclonal Antibody; Cancer Immunology; Antigen Receptors.

Abbreviations

Fuc: Fucose; Gal: Galactose; GalNAc: N-acetylgalactosamine; GlcNAc: N-acetylglucosamine; Hex: Hexose; IgG: Immunoglobulin G; IgM: Immunoglobulin M; Man: Mannose; NeuAc: N-acetylneuraminic acid; NeuGc: N-glycolylneuraminic acid; siRNA: Small interfering ribonucleic acid

Introduction

The principles and mechanisms of adaptive immunity in the normal immune system have been well established [1]. The antigen receptors of B and T cells origins are two key players in fighting off infections/inflammation arising from the invasions of foreign pathogens and/or cancer initiated within the human body environment. B and T cells in our normal immune system may respond to cancer-associated antigens by inducing specific humoral and cellular immune response [1]. The end results may be neutralization or cytotoxic killing of cancer cells within our body environment. However, this process may be only one side of the known actions and has been well accepted in the field of immunology.

Under our normal immune environment, the cancer cells initiated within our body may have also evolved to protect themselves immunologically [2]. Despite many years of studies during the last two decades, little progress has been made regarding the molecular mechanisms of immune protection among cancer cells. The etiology the expression of immunoglobulin in cancer cells remains unknown due to the lack of adequate probes for such investigations [3].

As early as three decades ago, monoclonal antibodies were generated against OC-3-VGH ovarian cancer cells [4]. RP215 was among one of the three thousand selected monoclonal antibodies produced and was later shown to react specifically with a carbohydrate-associated epitope detected mainly on cancer cell-derived immunoglobulin heavy chains but not those of B-cell derived human immunoglobulins (IgG) [4, 5]. Through years of biological and immunological studies, RP215 was demonstrated to be a suitable alternative for antibodies against antigen receptors or anti-immunoglobulins for mechanistic investigations in the immunology of cancer cells [4 -8].

Through these investigations, it has become apparent that cancerous antigen receptors should, in principle, play certain roles in the growth/ proliferation as well as immune protection by cancer cells [7]. However, the mechanisms of action for the immune protection by cancer cells may not be the same as those established in the normal immune system [1].

Based on these considerations, results of relevant studies are summarized to highlight the possible roles of cancerous immunoglobulins in the immunity of cancer cells. The experimental observations accumulated so far may be sufficient for us to draw a conclusion regarding the mechanisms of immune protection in cancer cells. Therefore, an alternative cancer immune system may be proposed and compared with that of the traditional one. Specially, these cancerous immunoglobulins may play differential roles in immune protection as compared to those of immune surveillance by normal immunoglobulins. Through such comprehensive analyses, a rational approach for anti-

***Corresponding author email:** cyglee@yahoo.com*

cancer drug development may be designed for the therapeutic treatment of human cancer in the future [9, 10].

Immunoglobulins expressed by normal B-lymphocytes and cancer cells

Differential Expressions: Expressions of immunoglobulins by normal B lymphocytes are distinctly different from those by cancer cells [11-13]. The characteristic comparisons of these two independent immune systems are summarized in Table 1 [14-18]. Generally speaking, immunoglobulins are expressed by B cells through a long series of interactions with various types of immune cells, such as antigen-presenting cells and/or T cells [1]. These interactions may involve activation, differentiation, antibody expression, and maturation. The molecular mechanisms of antibody productions by B lymphocytes have been well established and include gene conversion, somatic hyper mutation, class switching and maturation [1].

In contrast, cancerous antigen receptors, including immunoglobulins and T cell receptors, are expressed in a less complicated manner [19,12,13]. The incidence of somatic hyper mutation in the variable regions of cancerous immunoglobulins is greatly reduced in comparison to that of normal B lymphocytes [15,20]. The number of different immunoglobulins expressed on the surface of cancer cells and the different selected forms are also limited [12]. Immunoglobulins of different classes or subclasses can be expressed simultaneously by cancer cells derived from a single clone [18]. This is in contrast with to normal B lymphocytes. In addition, identical immunoglobulins can sometimes be detected in different cancer tissues among unrelated cancer patients [12].

Table 1: Comparison of Normal and Cancer Immune Systems.

Structure/ Functions	Normal Immune Cells	Cancer Cells
Expression of Antigen receptors [19]	a. Immunoglobullins expressed by B-lymphocytes b. T cell receptors expressed by T-lymphocytes	Both antigen receptors expressed by cancer cells of one single clone
Class switching of immunoglobulins [14,15]	Class switching of immunoglobulins; One single type B cell can express only one type of immunoglobin class and subclass	No class switching; immunoglobulins of different classes or subclasses can be expressed by cancer cells from one single clone
Hypermutation in Variable Regions of Immunoglobulins [14-16]	High frequency of hypermutation in the Fab domains of immunoglobulins leading to unlimited diversity of B lymphocytes	Limited mutations or diversity in the variable regions of immunoglobulins (less than 100 detected)
Glycosylation Patterns [17]	No O-linked glycans, and only one N-glycosylation at N297 position of IgG heavy chains; terminal NeuAc only	Both O-linked and N-linked glycans are detected in cancerous IgG heavy chains; terminal NeuAc and NeuGc
Interactions with Toll-like Receptors (Innate Immunity) [8]	No known interactions with toll-like receptors	Strong interaction with toll-like receptors within cancer cells
Relative Immunoactivity [18]	Normal immunoactivity	Weak immunoactivity (less than 1-5%) due to aberrant glycosylations

Differential Glycosylation Patterns of Cancer:

- **N-glycosylation of Cancerous IgG:** Glycosylation patterns are quite different between normal and cancerous immunoglobulins [17,21]. For example, normal human IgG is generally N-glycosylated at N-297 position in the heavy chain subunit. O-linked-glycosylation among the heavy chain regions of normal IgG is rarely found [17,21]. Typically, branched N-linked glycan structures were detected in both normal and cancerous IgG's [17,21]. However, additional N-glycan structures, including those of multi-mannose or multi-branched structures, were also identified in cancerous IgG [17]. The results of such comparative analysis are presented in Table 2.
- **O-glycosylation of cancerous IgG:** Compared to normal human IgG, O-linked glycans can only be detected in cancerous IgG. O-linked glycan analysis and glycopeptide mapping of cancerous IgG were performed with the proposed O-linked glycan structure/compositions presented in Table 3. These O-glycans were obtained from CA215 and/or cancerous IgG isolated independently from OC-3-VGH (ovary) and C-33A (cervix) cancer cell lines [17]. Enzyme immunoassay kit for normal human IgG was employed to determine the relative immunoactivity of normal human IgG and affinity-purified cancerous IgG. Unexpectedly, cancerous IgG exhibited much lower immunoactivity when compared with that of normal IgG. This phenomenon can only be explained by the aberrant glycosylation. Apparently, the aberrant glycosylations can result in significant alterations in the protein structures or conformations and thus reduce immunobinding activity [18].

Functional Analyses of Cancerous Immunoglobulins

The biological and immunological properties of normal human immunoglobulin have been known for decades [1]. These immunoglobulins may serve as one of the antigen receptors or as antibodies to neutralize foreign pathogens and cancer cells which are initiated within the body environment [5,22,23]. In contrast, cancerous immunoglobulins may be required for immune protection, and serve in part to defend against immune surveillance by the body's immune system [2, 3].

Detection of Circulating "Antigens" Recognized by Cancerous Immunoglobulins:

- Since RP215 reacts mainly with the heavy chains of cancerous immunoglobulins, the corresponding antigen designated as CA215 can be isolated in sufficient quantity [18]. CA215 can serve as an adequate ligand to capture circulating "antigen" which can be recognized by cancerous

Table 2: Profile of N-linked Glycans Unique to CA215 or cancer IgG, but not in normal human IgG.

Observed Mass m/z	Charge state	Proposed Structure
[M+Na]+ +	[M+2Na]2+	
1169	Double	$GlcNAc_5Man_3Hex_2$
1172	Single	$GlcNAc_2Man_3$
1330	Double	$GlcNAc_4Man_3Hex_2Fuc_1NeuGc_1$ [a]
1366[a]	Double	$GlcNAc_5Man_3Hex_2NeuGc_1$
1438[a]	Double	$GlcNAc_4Man_3Hex_2NeuGc_2$
1467[a]	Double	$GlcNAc_5Man_3Hex_3NeuGc_1$
1498	Double	$GlcNAc_6Man_3Hex_4$
1525[a]	Double	$GlcNAc_4Man_3Hex_2Fuc_1NeuGc_2$
1580	Single	$GlcNAc_2Man_5$ [b,c]
1621	Single	$GlcNAc_3Man_3Hex_1$
1785[b]	Single	$GlcNAc_2Man_6$

[a]OM with NeuGc as the terminal sialic acid.
[b]OM with high mannose structure.
[c]Lots CA215A and CA215B (acid-eluted) were used for analysis.
Obtained from [17] with permission.

Table 3: Comparative Profiles of Permethylated O-linked Glycans of Five Different CA215 Samples Which Consist Mainly of Cancerous IgG.

Sample ID	Observed Mass m/z [M+Na]+ Structure	Proposed Structure
CA215 (lots: A, B, and C)[a]	534	GalNAc1Gal1
CA215 (lots: A and B)	708	GalNAc1Gal1Fuc1
CA215 (lots: A, B, D, C, E, and F)[b]	896	GalNAc1Gal1NeuAc1
CA215 (lots: C, E, and F)	926	GalNAc1Gal1NeuGc1
CA215C	940[c]	GalNAc1GlcNAc1NeuAc1
CA215(lots: A, B, and C)	1140	GalNAc1GlcNAc1Gal1NeuAc
CA215 (lots: C, D, E, and F)	1257	GalNAc1Gal1NeuAc2
CA215 (lots: C, E, and F)	1317	GalNAc1Gal1NeuGc2
CA215 (lots: A, B, C, E, and F)	1345	GalNAc1GlcNAc1Gal2NeuAc
CA215 (lots: C, E, and F)	1375	GalNAc1GlcNAc1Gal2NeuGc

[a]CA215 lots A, B and C were from OC-3-VGH ovarian cancer cells (CA215-OC-3) lots A and B were obtained through acid elution, whereas lots C, D, E and F were obtained through elution with 3M urea.
[b]Lot CA215D was obtained by an additional purification of urea-eluted CA215 (S15K-100425) with goat anti-human IgG affinity column followed by the same analysis (CA215D is designated as affinity-purified cancerous IgG). CA215 lots E and F were from C-33A cervical cancer cells (CA215-C33A).
[c]Detected by MALDI-TOF MS method but not found by NSI-MS method.
[d]N-acetylgalactosamine (□), N-acetylglucosamine (■), Fucose (▲), Galactose (●), N- acetylneuraminic acid (♦) and N-glycolylneuraminic acid (♦).
Obtained from [17] with permission.

immunoglobulins through specific immunobinding. By using CA215 affinity chromatography, relevant antigen can be isolated from pooled human specimens. The isolated "antigen" can then be characterized through biochemical and immunological analysis. Attempts are being made to elucidate molecular mechanisms of immune protection by cancer cells [6-8]. Based on results our preliminary analysis, pooled normal human serum specimens were found to show affinity to CA215 and/or cancerous immunoglobulins oGu & Lee, unpublished observations). It remains to be demonstrated if these CA215 (or cancerous immunoglobulins) cross-reacting proteins or antibodies are those recognizing cancer-associated antigens identified in serum samples of general population in humans.

- **Roles of Cancerous Immunoglobulins in the Growth/proliferation of Cancer Cells**: Based on previous experimental observation by several laboratories, it has been generally accepted that the expressions of cancerous immunoglobulins on the surface of cancer cells may be required for growth/proliferation of cancer cells [2,3,8]. Knock down of cancerous IgG expression by transfection with IgG-related siRNA or siRNA plasmids could result in retarded or inhibited growth of cancer cells *in vitro* and *in vivo*[3,24]. Blocking of surface bound immunoglobulins with specific antibodies such as RP215, as well as anti-human IgG or anti-T cell receptors, was also found to result in induced apoptosis of cancer cells *in vitro* [6-8]. Complement-dependent cytotoxicity can also be induced in cancer cells with antibodies against the surface-bound receptors. Furthermore, RP215 was also demonstrated to cause tumor volume reductions in nude mouse models [22].

Effects of RP215 and antibodies against antigen receptors on the gene expression of cancer cells have been investigated [8]. It was generally observed that selected genes involved in growth/regulation of cancer cells are strongly influenced by antibody binding to surface antigen receptors. Their effects on the gene regulations of cancer cells are highly correlated, indicating the biosimilarity of these three ligands [8]. They also have strong effects on the regulations of the toll-like receptors in the innate immune system of cancer cells. All of these experimental observations are consistent with the protective roles of antigen receptors in cancer cells, especially their growth and proliferation within the normal immune environment.

Compared to the unique functional properties of cancerous immunoglobulins, none of these factors are found to operate similarly in the normal immune system [1]. This phenomenon makes the cancer immunity uniquely different from that of traditional immune system established so far (Table 1).

Conclusion

In this mini-review, scientific information has been presented to highlight the fundamental differences between the traditional normal immune system and the cancer immune system. These differences may stem from the evolution of different organisms. We believe that in higher organisms such as mammals, two separate immune systems should exist independently to perform differential functions of immune surveillance and immune protection in our body environment. The growth/ proliferation of cancer cells depends highly on the balance of these two opposite factors inside the body. Only when the detailed molecular interactions between cancerous immunoglobulins and the complicated body immune system are completely resolved, immunotherapy of human cancer may have a better chance of success.

Acknowledgements

This research project was supported in parts by NRC-IRAP program (#794354) of Canada and the NSERC summer student scholarship to Suefay Liu of McGill University. The editorial assistance of Dr. Chichi Liu and technical support of Yiting Tang is acknowledged. Initial observations of circulating " antigen" in human serum were made in Dr. Jiang Gu's lab at Shantou University, Shantou, China. Helpful discussion on this topic with Dr. Gu is acknowledged.

Conflict of interest

Gregory Lee is co-founder of Vancouver BioTech Ltd. The others have nothing to declare.

References

1. Murphy K, Travers P, & Walport M. (2008). The Adaptive Immune Response. In: *Janeway's Immunobiology* (7th Ed.) Part IV. (7th ed., pp. *323-496*). Garland Science, Taylor & Francis Group.
2. Chen Z , & Gu J (2007) Immunoglobulin G expression in carcinomas and cancer cell lines. *FASEB J, 21(11)*: 2931-2938.
3. Qiu X, Zhu X, Zhang L, Mao Y, Zhang J. et al. (2003) Human epithelial cancers secrete immunoglobulin G with unidentified specificity to promote growth and survival of tumor cells. Cancer Research, 63(19): 6488-6495.
4. Lee CY, Chen KW, Sheu FS et al. (1992) Studies of a tumor-associated antigen, COX-1, recognized by a monoclonal antibody. Cancer Immunology, Immunotherapy, 35(1): 19-26.
5. 5. Lee G, Laflamme E , Chien CH, & Ting HH (2008) Molecular identity of a pan cancer marker, CA215. Cancer biology & therapy, 7(12): 2007-2014.
6. Lee G (2012a) Cancerous immunoglobulins and CA215: implications in cancer immunology. American Journal of Immunology, 8(4):101-116.
7. Lee G, Liu S (2013) Roles of antigen receptors and CA215 in the innate immunity of cancer cells. Open Journal of Immunology,3(3): 127-138.
8. Tang Y, Zhang H, & Lee G (2013) Similar gene regulation patterns for growth inhibition of cancer cells by RP215 or anti-antigen receptors. Journal of Cancer Science & Therapy, 5: 200-208.
9. Lee G, Zhu M, & Ge B (2012b) Potential Monoclonal Antibody Therapy for the Treatment of Ovarian Cancer. In Farghaly, S. A (Ed.) ,(*pp. 385-406). Ovarian Cancer-Basic Science Perspective*. Vancouver: InTech.
10. Lee G, Cheung A, Ge B, Zhu M. et al. (2012c) CA215 and GnRH receptor as targets for cancer therapy. Cancer Immunology, Immunotherapy, 61(10): 1805-1817.
11. Babbage G, Ottensmeier CH, Blaydes J, Stevenson FK, & Sahota S (2006) Immunoglobulin heavy chain locus events and expression of activation-induced cytidine deaminase in epithelial breast cancer cell lines. Cancer Research, 66(8): 3996-4000.
12. Zheng J, Huang J, MaoY, Liu S, Sun X, et al. (2009) Immunoglobulin gene transcripts have distinct V_HDJ_H recombination characteristics in human epithelial cancer cells. Journal of Biological Chemistry, 284(20): 13610-13619.
13. Zhu X, Wu L, Zhang L, Hao P, Zhang S, et al. (2010) Distinct regulatory mechanism of immunoglobulin gene transcription in epithelial cancer cells. Cellular and Molecular Immunology, 7(4): 279-286.
14. Early P, Huang H, Davis M, Calame K,& Hood L (1980) An immunoglobulin heavy chain variable region gene is generated from three segments of DNA: V_H, D and J_H. *Cell, 19(4):* 981-992.
15. Honjo T, Kinoshita K, Muramatsu M (2002) Molecular mechanism of class switch recombination: Linkage with somatic hypermutation. Annual Review of Immunology, 20: 165-196.
16. Jung D, Giallourakis C, Mostoslavsky R, & Alt F (2006) Mechanism and control of V(D)J recombination at the immunoglobulin heavy chain locus. Annual Review of Immunology, 24: 541-570.
17. Lee G, Azadi P (2012d) Peptide mapping and glycoanalysis of cancer cell-expressed glycoproteins CA215 recognized by RP215 monoclonal antibody. International Journal of Carbohydrate Chemistry, 31(1): 10-30.
18. Lee G, Cheung AP, Li B, Ge B, & Chow PM, (2012e) Molecular and immuno-characteristics of immunoglobulin-like glycoproteins in cancer cell-expressed biomarker, CA215. *Immunological Investigations,41(4):* 429-446.
19. Lee G, Zhu M, Ge B, Potzold S (2012f) Wildspread expressions of immunoglobulin super family proteins in cancer cells. Cancer Immunology, Immunotherapy,61(1): 89-99.
20. Dudley D D, Chaudhuri J, Bassing CH, & Alt FW (2005) Mechanism and control of V(D)J recombination versus class switch recombination: Similarities and differences. Advances in Immunology, 86: 43-112.
21. Arnold J N, Wormald MR, Sim RB, Rudd PM, & Dwek RA (2007) The impact of glycosylation on the biological function and structure of human immunoglobulins. Annual Review of Immunology, 25: 21-50.
22. Lee G, Chu RA, & Ting HH (2009) Preclinical assessment of anti-cancer drugs by using RP215 monoclonal antibody. Cancer Biology Therapy, 8(2):161-166.
23. Lee G, & Ge B (2010) Inhibition of in vitro tumor cell growth by RP215 monoclonal antibody and antibodies raised against its anti-idiotype antibodies. Cancer Immunology, Immunotherapy, 59(9): 1347-1356.
24. Hu D, Duan Z, Li M, Jiang Y, Liu H, et al. (2011) Heterogeneity of aberrant immunoglobulin expression in cancer cells. Cellular & Molecular Immunology, 8(6): 479-485.

2

Regulation of Murine Dendritic Cell Functions by Calcium Channels

Gabriela Mellado-Sánchez, Héctor Vivanco-Cid, and Adriana Sumoza-Toledo*

Multidisciplinary Laboratory of Biomedical Sciences, Institute of Medico-Biological, Research Campus Veracruz Universidad Veracruzana, Iturbide S/N. CP. 91700, Veracruz, Mexico

***Corresponding author:** Adriana Sumoza-Toledo, Multidisciplinary Laboratory of Biomedical Sciences, Institute of Medico-Biological, Research Campus Veracruz Universidad Veracruzana, Iturbide S/N. CP. 91700, Veracruz, Mexico, E-mail: asumoza@uv.mx*

Abstract

Dendritic Cells (DCs) are highly potent Antigen-Presenting Cells (APCs) that have a key role in mediating tolerance or immunity to self and non-self antigens. In their immature stage DCs are highly phagocytic and undergo a maturation process after taking up an antigen. DC maturation is characterized by activation of mechanisms of antigen presentation, increased expression of Major Histocompatibility Complex (MHC) class II and co-stimulatory molecules in the plasma membrane, and secretion of cytokines and chemokines. Despite the fact that the role of calcium (Ca^{2+}) in DC function has been clearly established, regulation of Ca^{2+} signals in these cells is not well known. However, recently it has been demonstrated that functional capacitative Ca^{2+} release-activated Ca^{2+} (CRAC), Transient Receptor Potential Melastatin-2(TRPM2) and TRP Vanilloid- 1 (TRPV1) channels are critical for mouse DC maturation and migration. Also, Ryanodine Receptor-1 (RyR1) signaling activated by L-type Ca^{2+} channel CaV1.2 cause rapid MHC-II expression in the plasma membrane of DCs. The understanding of the regulation of Ca^{2+} signals in DCs is essential, to potentially modulate DC functions in disease processes. Therefore, in this review, we discuss recent studies on the expression and roles of Ca^{2+} channels in DC biology and function.

Keywords: Calcium; Dendritic cells; Ion channels; CRAC; TRPM2; TRPV1; RyR

Introduction

Dendritic cells (DCs) are Antigen-Presenting Cells (APCs) that play a critical role in the regulation of both: innate and adaptive immune responses. Initially, DCs were described by Ralph Steinman in 1973 [1], as a different immune cell population in the spleen and lymph nodes of mice. DCs are the only APCs that have the ability to induce a primary immune response in naïve T lymphocytes, and therefore they are considered the most potent APCs, influencing the type and quality of the response [2]. DCs can exist in two main states: In a steady state immature Dendritic Cells (iDCs) and fully mature DCs (mDCs). The distinction between immature and mature DCs is based on phenotypic markers and biological functions [3]. iDCs lack or have low levels of several important accessory molecules that mediate binding and stimulation of T cells, such as CD40, CD54, CD58, CD80, CD83 and CD86. They also express high levels of intracellular Major Histocompatibility Complex (MHC) class II molecules. On the cell surface, iDCs express high levels of chemokine receptors such as CCR1, CCR5, and CCR6. Functionally, iDCs are characterized by high endocytic activity and low T-cell stimulation potential [4-7]. Phenotypic maturation is characterized by down-regulation of the capacity to capture antigens and up regulation of antigen processing and presentation functions. The mDCs phenotype is characterized by expression of high levels on the surface of MHC II, CCR7, CD40, CD54, CD80, CD83, CD86, CD58 and low expression of CCR1, CCR5, CCR6 [4-7]. DCs are also able to interact with other cells besides T cells, such as Natural Killer (NK) cells, neutrophils, and epithelial cells [8-10]. Other critical roles of DCs in immunity are the maintenance of B cell function, the establishment of immunological memory, and the maintenance of peripheral tolerance [11].

DC Subsets

DCs are widely distributed in all tissues, especially in those that provide an environmental interface, such as the skin and mucosal tissues. Similar to other immune cells, DCs are divided in subsets, which have been shown to possess a differential ontogeny, morphology, phenotype, transcriptional programs and functions [12]. In mice, DCs can be subdivided into $CD8^+$ $CD11b^+$ and $CD8^-$ $CD11b^+$ conventional DCs (cDCs), a lineage originated from a myeloid progenitor in the bone marrow. cDCs characteristically express high levels of MHC class II and the integrin CD11c, but not B220 marker [13-16]. cDC subsets are activated by microbial products through cell surface Toll- like Receptors (TLRs) to produce inflammatory cytokines such as interleukin (IL)-1, IL-6, IL-12, and tumor necrosis factor-alpha (TNF-α) and they are specialized in the activation of $CD8^+$ and $CD4^+$ T cells [17]. They have a predominant role in MHC- II presentation and immunological tolerance, inducing clonal deletion of auto reactive T cell or Treg differentiation [18,19]. $CD8^+$ $CD11b^+$ DCs are specialized in the induction of $CD8^+$ T cell immunity. They are the main source of IL-12 and IL- 15 [18], two cytokines involved in the differentiation of cytotoxic $CD8^+$ T cells and have the ability to prime $CD8^+$ T cell responses in a cross-presentation dependent mechanism [20]. $CD8^-$ $CD11b^+$cDCs can sense pathogens and migrate from non-lymphoid tissues to regional lymph nodes charged with self and foreign antigens.

***Corresponding author email:** asumoza@uv.mx*

Other cDCs subsets include migratory CD103$^+$ CD11b$^-$ DCs, CD103$^-$ CD11b$^+$ DCs, and Langerhans Cells (LCs), which are abundant in the intestinal mucosa and skin [21-24]. DCs can also be originated from a lymphoid progenitor. Plasmacytoid DCs (pDCs) are the prominent subset of this group, which phenotypically express CD8α$^+$ CD11b$^-$ B220$^+$ DC SING$^+$ [25,26]. The other pDCs specific surface marker is the murine Siglec H [27]. pDCs are a specialized population that have the ability to produce very large amounts of interferon alpha/beta (IFN-α/β) upon activation and a limited ability to prime naïve CD4$^+$ and CD8$^+$ T cells. They are an important DCs subset in viral and anti-tumoral immunity [26]. Other DCs subsets include *in vitro* or *in vivo* inflammatory or infection-derived DCs, which develop from monocytes in response to stimulation such as Granulocyte-Macrophage Colony-Stimulating Factor (GM-CSF), IL-4 and TNF-α [28]. A summary of DC subsets is showed in Table 1.

Early studies using Ca^{2+} ionophores and Ca2$^+$ chelators have shown that Ca^{2+} signals may trigger maturation and functional properties of DCs [29-31]. Intracellular Ca^{2+} ions are crucial second messengers to initiate signaling pathways for fundamental cellular functions, such as cell cycle, survival, apoptosis, migration, and gene expression[32,33]. Regulation of intracellular Ca^{2+} concentrations ($[Ca^{2+}]_i$; ~100 nM) involves both Ca^{2+} entry from the extracellular space and Ca^{2+} release from intracellular stores, such as calciosomes, Endoplasmic Reticulum (ER), lysosomes, or mitochondria, by specialized pumps and ion channels [32-34]. Although $[Ca^{2+}]_i$ increase triggers signaling pathways in the cell, the exquisite spatial and temporal organization of Ca^{2+} oscillations, waves and sparks might also provide a code for selective activation of signaling pathways and their duration. For example, a short $[Ca^{2+}]_i$ increase is observed in lymphocytes during immunological synapse, release of lytic granules, and cytotoxicity. In contrast, prolonged $[Ca^{2+}]_i$ increase regulates cytokine production, cell differentiation, effector

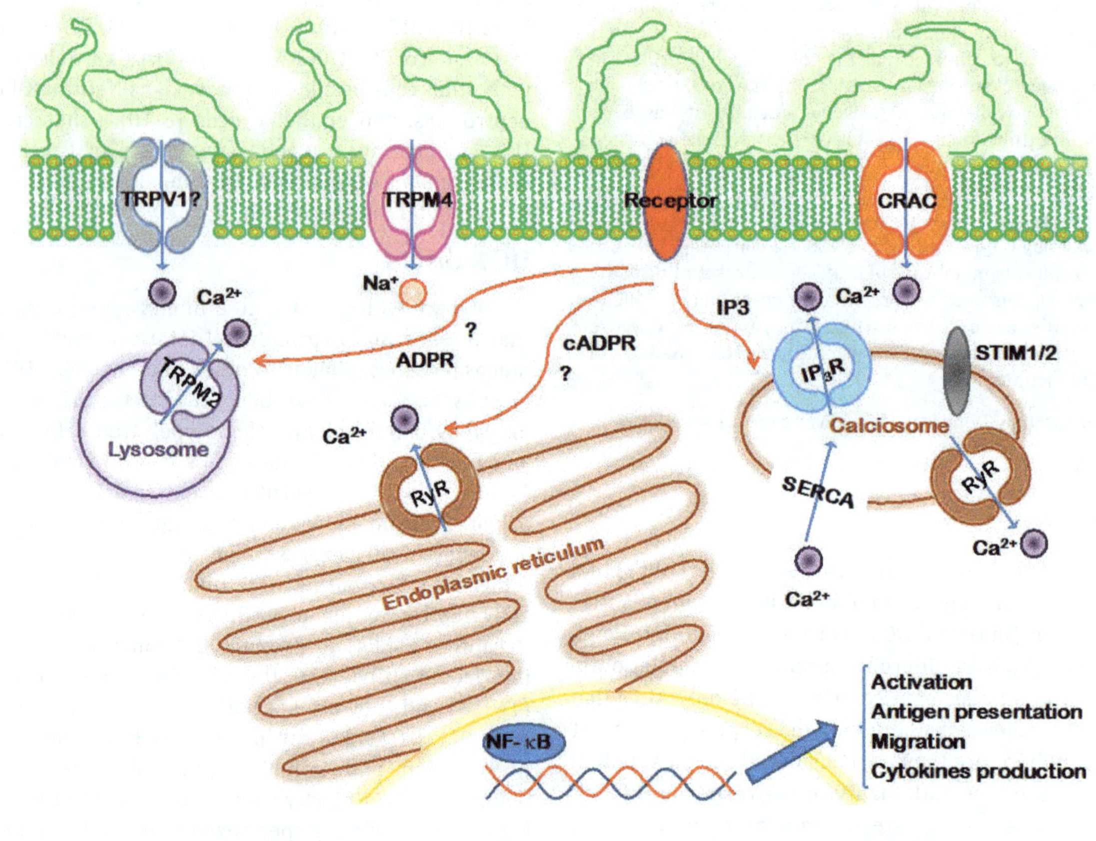

Figure 1: Calcium channels in DCs: Extracellular signals (chemokines, cytokines, microbial peptides, etc) are recognized by DCs by means of G protein-coupled receptors or receptor protein tyrosine kinases, activating the formation of 1,4,5-triphosphate (IP3) that in turn binds to IP3 receptors in the ER and calciosomes, causing Ca^{2+} release. Decrease in the luminal Ca^{2+} in the ER is detected by the Stromal Interaction Molecule 1/2 (STIM1/2) resulting in the activation of Capacitative Ca^{2+} Release-Activated Ca^{2+} (CRAC) channels, allowing Ca^{2+} influx across the plasma membrane. Chemokines also activate Transient Receptor Potential Melastatin- 2(TRPM2) channels in DC lysosomes. The Ca^{2+} signals activate transcription factors such as Nuclear Factor of Activated T cells (NFAT) or Nuclear Factor-κB (NF-κB) for gene expression. TRPM4, a Ca^{2+} activated TRP channel that allows Na^+ into the cell is expressed in the plasma membrane of DCs and indirectly regulates DC functions by decreasing the driving force for Ca^{2+} entry through CRAC channels. DCs also express TRP Vanilloid-1 (TRPV1) and Ryanodine receptor (RyR) channels but their functions are still not clear.

Table 1: Dendritic cell subsets.

DC subsets	CD8α	CD103	CD205	CD11b	B220 or CD45RA	DC-SING	Langerin (CD207)	MHC class II	CD11c
pDCs	+	-	-	-	+	++	-	+	+
CD8α+ DCs	+	low	+	+	-	-	+/-	++	+++
CD8α-CD11b+ DCs	-	+/-	+	+	-	ND	-	++	+++
CD103+ DCs	-	+	++	-	-	-	+	++	++
Lung Intestine	-	+	-	+	-	-	-	+	++
Langerhans cells	-	-	++	+	-	-	++	++	++
Monocyte-derived inflammatory DCs	-	-	-	+	-	+	-	++	++

functions, etc. The present review addresses the role of Ca^{2+} channels in DC functions [33,35].

Ca^{2+} Release-Activated Ca^{2+} Channels (CRAC) in DC

The main mechanism for Ca^{2+} entry in immune cells, including DCs, is the Store-Operated Ca^{2+} Entry (SOCE). SOCE is activated by Ca^{2+} release from the intracellular stores and involves the activation of Capacitative Ca^{2+} Release-Activated Ca^{2+} (CRAC) channels in the plasma membrane (Figure 1) [34,36,37]. SOCE-mediated Ca^{2+} influx provides ions not only for signaling purposes, but also for ER and calciosomes store refilling. SOCE activation can be initiated by stimulation of G protein-coupled receptors or stimulation of receptor protein tyrosine kinases by external signals (cytokines, chemokines, bacterial peptides, etc), leading to activation of Phospholipase C (PLC) that in turn hydrolyzes phosphatidylinositol- 4,5- bisphosphate (PIP2) to release Inositol-1,4,5-triphosphate (IP3) and Diacylglycerol (DAG) [33,34].The subsequent binding of IP3 to IP3 receptors in the ER and calciosomes causes a rapid and transient Ca^{2+} release, raising the $[Ca^{2+}]_i$ (Figure 1). On the other hand, the decrease in the luminal Ca^{2+} in the ER is detected by the stromal interaction molecule 1/2 (STIM1, STIM2; Ca^{2+} sensors; Figure 1), resulting in its conformational change (oligomerization and aggregation) and activation of CRAC channels [33,34]. CRAC channels, which pore is formed by CRACM/Orai 1-3 proteins, then allow influx of extracellular Ca^{2+} across the plasma membrane (Figure 1). CRAC are highly Ca^{2+}-selective, low conductance channels with a characteristic inwardly rectifying current-voltage relationship [33,34]. Interestingly, Orai and STIM proteins may have different tissue distribution, selectivity and conductivity for Ca^{2+}.

As a result of $[Ca^{2+}]_i$ increase several signaling pathways and transcription factors are activated, such as the calmodulin-calcineurin pathway that activate the Nuclear Factor of Activated T cells (NFAT), the Ca^{2+}- dependent kinase-calmodulin (CaMK) pathway which activate the Cyclic-adenosine monophosphate-Responsive Element Binding protein (CREB), and the nuclear factor B (NFκB) pathway. Moreover, the DAG formed from PIP2 hydrolysis can activate the Protein kinase C pathway (PKC), and Ras-mitogen-activated protein kinase, which ultimately activate transcription factors such as Activating Protein-2 (AP-2) and NFκB [33,34].

Although the presence of CRAC currents and its role in DC maturation have previously been demonstrated in mouse DCs [36], it has only recently been shown that Orai2 and STIM2 are most abundant in DCs [38]. Furthermore, recruitment of Orai2 and STIM2 towards the immunological synapse has been observed during antigen presentation of DC to T lymphocytes [38]. Likewise, studies using CRAC blockers have shown that this channel plays an important role in DC maturation, cytokine production (TNF-α and IL-6) and chemotaxis [37]. DC maturation can be triggered *in vitro* by increasing $[Ca^{2+}]_i$ by stimulating them with peptidoglycan (PGN), CpG DNA, microbial products like Lipopolysaccharide (LPS) [39,40], or ionophores [29-31]. It has also been suggested that LPS, PGN and CpG induced activation of PLC γ2 [39], which in turn acts on PIP2 to produce IP3 that leads to Ca^{2+} release from intracellular stores; followed by CRAC channel activation (reviewed in [34]) causing the nuclear translocation of calcineurin-dependent NFAT factor and cytokine production, such as IL-2 [33,41]. On other hand, DC maturation with Ca^{2+} ionophoresis associated with NFκB activation, likely by activating Calcium/Calmodulin-dependent Kinase II (CaMKII), which inactivates NFκB-inhibiting molecule IkB similar to what has been shown in T cells [42].

In addition, DC chemotaxis depends on Ca^{2+} influx. DC chemotactic response to chemokines, including (C-X-C motif) ligand 12 (CXCL12) and (C-C motif) ligand 21 (CCL21), results in PLC activation, IP3 production, Ca^{2+} release from intracellular stores, and subsequent activation of CRAC channels and Ca^{2+} influx [31,40,43,44].

Transient Receptor Potential (TRP) Channel in DC

Our previous study has shown that lysosomal Ca^{2+} release through TRP Melastatin-2 (TRPM2) channel, the second member of the TRP melastatin-related channel family, plays an important role in DC maturation and chemotaxis (Figure 1) [40]. TRPM2 channel is expressed in DC only in lysosomes [40]. This channel is synergically activated by Adenosine Diphosphate Ribose (ADPR) and Ca^{2+}, and allows entry of sodium (Na^+), Ca^{2+}, potassium (K^+) and caesium (Cs^+) into the cytosol. In addition to Ca^{2+}, cyclic ADPR (cADPR), hydrogen peroxide (H_2O_2) and Nicotinic acid Adenine Dinucleotide Phosphate (NAADP) may directly or indirectly facilitate TRPM2 gating by ADPR [45]. DCs may produce ADPR by means of CD38 activity, an ectoenzyme that use β-Nicotinamide Adenine Dinucleotide (β-NAD$^+$) as a substrate to catalyse the production of ADPR, cADPR, and NAADP, and by activation of the Poly(ADPR)-Polymerase/Poly(ADP-ribose) Glycohydrolase (PARP/PARG) pathway during DNA repair, replication and transcription [45]. DCs lacking TRPM2 channels express reduced levels of co stimulatory molecules, such as CD80, CD86, MHC-II and CD83, in the plasma membrane when they are stimulated with TNF-α and CpG DNA, than TRPM2 expressing-DCs [40]. They also show reduced Ca^{2+} signals in response to CXCL12 and CCL21, affecting the chemotaxis response towards these chemokines [40]. However, the mechanisms that link CD38 and PARP/PARG pathways to TRPM2 and to chemokine receptors are still not clearly understood.

DCs also express TRP Vanilloid-1 (TRPV1) protein in their plasma membrane, another non-selective Ca^{2+} channel of the TRP family, which is activated by capsaicin. But, there is controversial data on the expression and function of this channel in DCs. Earlier studies from Basu and Srivastava showed that extracellular Ca^{2+} influx via TRPV1 activation induces mouse DC maturation and provokes increase in the expression level of MHC class II and CD86 on the surface [46]. Conversely, O'Connell PJ et al. [47] did not detect TRPV1 transcripts and TRPV1 currents in bone marrow derived-mouse DCs. A recent study by Tóth BI et al. [48] shows molecular and functional expression of TRPV1 channels in monocyte derived-human DCs. Although DC stimulation with capsaicin induces Ca^{2+} mobilization, this reduces the expression level of maturation markers in DCs, such as CD83 and CCR7 [48]. On the other hand, TRPM4, a Ca^{2+}-activated TRP channel that allows Na^+ into the cell, indirectly regulates DC migration but not maturation by decreasing the driving force for Ca^{2+} entry through CRAC channels [43].

Ryanodine and Purinergic Receptors in DC

Ryanodine Receptor-1 (RyR1), a channel expressed in intracellular Ca^{2+} stores, is also expressed in DCs [49,50]. RyR1 signaling coupled with L-type Ca^{2+} channel CaV1.2, which has been also detected in DCs, cause rapid MHC class II expression in the plasma membrane of DCs [50]. Interestingly, RyRs are also activated by cADPR and NAADP$^+$, and might contribute through these pathways to DC maturation [44,51,52]. Finally, DCs express Purinergic Receptors (P2Rs), P2X (ligand-gated ion channels) and P2Y (G-protein coupled receptors) on their surface, such as P2X1, P2X4, and P2X7, and P2Y1, P2Y2, P2Y4, P2Y6, P2Y11, and P2Y14, respectively. DC stimulation with Adenosine Triphosphate (ATP), a Damage-Associated Molecular Pattern (DAMP) molecule released by injured cells during inflammation and necrosis, or UTP results in characteristic Ca^{2+} signaling associated to P2X or P2Y, mainly P2X7 [53-57].

Concluding Remarks

Not much is known about Ca^{2+} channel expression and Ca^{2+} regulation in DCs. Recent studies have addressed the role of CRAC, TRPV1, TRPM2, RyR1 and CaV1.2 channels in DC maturation and migration. However, the mechanisms that lead to activation of these channels during DC function are not well understood. Moreover, future studies still need to address which channels regulate Ca^{2+} signals during antigen presentation, immune synapse, apoptosis, and other DC functions. The meaning of Ca^{2+} oscillations, frequency and patterns are unknown, which might play an important role in establishing and/or maintaining immunological tolerance or immunity to self and non-self antigens.

References

1. Steinman RM, Cohn ZA. Pillars Article: Identification of a novel cell type in peripheral lymphoid organs of mice. I. Morphology, quantitation, tissue distribution. J Immunol. 2007; 178(1): 5-25.
2. Guermonprez P, Valladeau J, Zitvogel L, Théry C, Amigorena S. Antigen presentation and T cell stimulation by dendritic cells. Annu Rev Immunol. 2002; 20: 621-67.
3. Merad M, Sathe P, Helft J, Miller J, Mortha A. The dendritic cell lineage: ontogeny and function of dendritic cells and their subsets in the steady state and the inflamed setting. Annu Rev Immunol. 2013; 31: 563-604. doi: 10.1146/annurev-immunol-020711-074950.
4. Dalod M, Chelbi R, Malissen B, Lawrence T. Dendritic cell maturation: functional specialization through signaling specificity and transcriptional programming. EMBO J. 2014; 33(10): 1104-1116. doi: 10.1002/embj.201488027.
5. Dieu MC, Vanbervliet B, Vicari A, Bridon JM, Oldham E, Aït-Yahia S, et al. Selective recruitment of immature and mature dendritic cells by distinct chemokines expressed in different anatomic sites. J Exp Med. 1998; 188(2): 373-386.
6. Sallusto F, Schaerli P, Loetscher P, Schaniel C, Lenig D, Mackay CR, et al. Rapid and coordinated switch in chemokine receptor expression during dendritic cell maturation. Eur J Immunol. 1998; 28(9): 2760-2769.
7. Yanagihara S, Komura E, Nagafune J, Watarai H, Yamaguchi Y. EBI1/CCR7 is a new member of dendritic cell chemokine receptor that is up-regulated upon maturation. J Immunol. 1998; 161(6): 3096-3102.
8. Ferlazzo G, Morandi B. Cross-Talks between Natural Killer Cells and Distinct Subsets of Dendritic Cells. Front Immunol. 2014; 5: 159. doi: 10.3389/fimmu.2014.00159.
9. Bennouna S, Bliss SK, Curiel TJ, Denkers EY. Cross-talk in the innate immune system: neutrophils instruct recruitment and activation of dendritic cells during microbial infection. J Immunol. 2003; 171(11): 6052-6058.
10. Rescigno M. Dendritic cell-epithelial cell crosstalk in the gut. Immunol Rev. 2014; 260(1): 118-128. doi: 10.1111/imr.12181.
11. Liu YJ. Dendritic cell subsets and lineages, and their functions in innate and adaptive immunity. Cell. 2001; 106(3): 259-262.
12. Liu K. Nussenzweig MC. Origin and development of dendritic cells. Immunol Rev. 2010; 234(1): 45-54. doi: 10.1111/j.0105-2896.2009.00879.x.
13. Vremec D, Zorbas M, Scollay R, Saunders DJ, Ardavin CF, Wu L, et al. The surface phenotype of dendritic cells purified from mouse thymus and spleen: investigation of the CD8 expression by a subpopulation of dendritic cells. J Exp Med. 1992; 176(1): 47-58.
14. Vremec D, Pooley J, Hochrein H, Wu L, Shortman K. CD4 and CD8 expression by dendritic cell subtypes in mouse thymus and spleen. J Immunol. 2000; 164(6): 2978-2986.

15. den Haan JM, Lehar SM, Bevan MJ. CD8(+) but not CD8(-) dendritic cells cross-prime cytotoxic T cells in vivo. J Exp Med. 2000; 192(2): 1685-1696.

16. Belz GT, Nutt SL. Transcriptional programming of the dendritic cell network. Nat Rev Immunol. 2012; 12(2): 101-113. doi: 10.1038/nri3149.

17. Kaisho T. Pathogen sensors and chemokine receptors in dendritic cell subsets. Vaccine. 2012; 30(52): 7652-7657. doi: 10.1016/j.vaccine.2012.10.043.

18. Maldonado RA, von Andrian UH. How tolerogenic dendritic cells induce regulatory T cells. Adv Immunol. 2010; 108: 111-165. doi: 10.1016/B978-0-12-380995-7.00004-5.

19. Steinman RM, Hawiger D, Nussenzweig MC. Tolerogenic dendritic cells. Annu Rev Immunol. 2003; 21: 685-711.

20. Morelli AE, Zahorchak AF, Larregina AT, Colvin BL, Logar AJ, Takayama T, et al. Cytokine production by mouse myeloid dendritic cells in relation to differentiation and terminal maturation induced by lipopolysaccharide or CD40 ligation. Blood. 2001; 98(5): 1512-1523.

21. Bedoui S, Whitney PG, Waithman J, Eidsmo L, Wakim L, Caminschi I, et al. Cross-presentation of viral and self antigens by skin-derived CD103+ dendritic cells. Nat Immunol. 2009; 10(5): 488-495. doi: 10.1038/ni.1724.

22. Ng SC, Kamm MA, Stagg AJ, Knight SC. Intestinal dendritic cells: their role in bacterial recognition, lymphocyte homing, and intestinal inflammation. Inflamm Bowel Dis. 2010; 16(10): 1787-1807. doi: 10.1002/ibd.21247.

23. del Rio ML, Bernhardt G, Rodriguez-Barbosa JI, Förster R. Development and functional specialization of CD103+ dendritic cells. Immunol Rev. 2010; 234(1): 268-281. doi: 10.1111/j.0105-2896.2009.00874.x.

24. Milling S, Yrlid U, Cerovic V, MacPherson G. Subsets of migrating intestinal dendritic cells. Immunol Rev. 2010; 234(1): 259-267.

25. Asselin-Paturel C, Boonstra A, Dalod M, Durand I, Yessaad N, Dezutter-Dambuyant C, et al. Mouse type I IFN-producing cells are immature APCs with plasmacytoid morphology. Nat Immunol. 2001; 2(12): 1144-1150.

26. Reizis B, Bunin A, Ghosh HS, Lewis KL, Sisirak V. Plasmacytoid dendritic cells: recent progress and open questions. Annu Rev Immunol. 2011; 29: 163-183. doi: 10.1146/annurev-immunol-031210-101345.

27. Zhang J, Raper A, Sugita N, Hingorani R, Salio M, Palmowski MJ, et al. Characterization of Siglec-H as a novel endocytic receptor expressed on murine plasmacytoid dendritic cell precursors. Blood. 2006; 107(9): 3600-3608.

28. Segura E, Amigorena S. Inflammatory dendritic cells in mice and humans. Trends Immunol. 2013; 34(9): 440-445. doi: 10.1016/j.it.2013.06.001.

29. Bagley KC, Abdelwahab SF, Tuskan RG, Lewis GK. Calcium signaling through phospholipase C activates dendritic cells to mature and is necessary for the activation and maturation of dendritic cells induced by diverse agonists. Clin Diagn Lab Immunol. 2004; 11(1): 77-82.

30. Czerniecki BJ, Carter C, Rivoltini L, Koski GK, Kim HI, Weng DE, et al. Calcium ionophore-treated peripheral blood monocytes and dendritic cells rapidly display characteristics of activated dendritic cells. J Immunol. 1997; 159(8): 3823-3837.

31. Koski GK, Schwartz GN, Weng DE, Czerniecki BJ, Carter C, Gress RE, et al. Calcium mobilization in human myeloid cells results in acquisition of individual dendritic cell-like characteristics through discrete signaling pathways. J Immunol. 1999; 163(1): 82-92.

32. Chen YF, Chen YT, Chiu WT, Shen MR. Remodeling of calcium signaling in tumor progression. J Biomed Sci. 2013; 20: 23. doi: 10.1186/1423-0127-20-23.

33. Izquierdo JH, Bonilla-Abadía F, Cañas CA, Tobón GJ. Calcium, channels, intracellular signaling and autoimmunity. Reumatol Clin. 2014; 10(1): 43-47. doi: 10.1016/j.reuma.2013.05.008.

34. Shumilina E, Huber SM, Lang F. Ca2+ signaling in the regulation of dendritic cell functions. Am J Physiol Cell Physiol. 2011; 300(6): C1205-1214. doi: 10.1152/ajpcell.00039.2011.

35. Feske S. Calcium signalling in lymphocyte activation and disease. Nat Rev Immunol. 2007; 7(9): 690-702.

36. Hsu Sf, O'Connell PJ, Klyachko VA, Badminton MN, Thomson AW, Jackson MB, et al. Fundamental Ca2+ signaling mechanisms in mouse dendritic cells: CRAC is the major Ca2+ entry pathway. J Immunol. 2001; 166(10): 6126-6133.

37. Matzner N, Zemtsova IM, Nguyen TX, Duszenko M, Shumilina E, Lang F. Ion channels modulating mouse dendritic cell functions. J Immunol. 2008; 181(10): 6803-6809.

38. Bandyopadhyay BC, Pingle SC, Ahern GP. Store-operated Ca^{2}+ signaling in dendritic cells occurs independently of STIM1. J Leukoc Biol. 2011; 89(1): 57-62. doi: 10.1189/jlb.0610381.

39. Aki D, Minoda Y, Yoshida H, Watanabe S, Yoshida R, Takaesu G, et al. Peptidoglycan and lipopolysaccharide activate PLCgamma2, leading to enhanced cytokine production in macrophages and dendritic cells. Genes Cells. 2008; 13(2): 199-208. doi: 10.1111/j.1365-2443.2007.01159.x.

40. Sumoza-Toledo A, Lange I, Cortado H, Bhagat H, Mori Y, Fleig A, et al. Dendritic cell maturation and chemotaxis is regulated by TRPM2-mediated lysosomal Ca2+ release. FASEB J. 2011; 25(10): 3529-3542. doi: 10.1096/fj.10-178483.

41. Zanoni I , Ostuni R, Capuano G, Collini M, Caccia M, Ronchi AE, et al. CD14 regulates the dendritic cell life cycle after LPS exposure through NFAT activation. Nature. 2009; 460(7252): 264-268. doi: 10.1038/nature08118.

42. Li Q, Verma IM. NF-kappaB regulation in the immune system. Nat Rev Immunol. 2002; 2(10): 725-734.

43. Barbet G, Demion M, Moura IC, Serafini N, Léger T, Vrtovsnik F, et al. The calcium-activated nonselective cation channel TRPM4 is essential for the migration but not the maturation of dendritic cells. Nat Immunol. 2008; 9(10): 1148-1156. doi: 10.1038/ni.1648.

44. Partida-Sanchez S, Gasser A, Fliegert R, Siebrands CC, Dammermann W, Shi G, et al. Chemotaxis of mouse bone marrow neutrophils and dendritic cells is controlled by adp-ribose, the major product generated by the CD38 enzyme reaction. J Immunol. 2007; 179(11): 7827-7839.

45. Sumoza-Toledo A, Penner R. TRPM2: a multifunctional ion channel for calcium signalling. J Physiol. 2011; 589(7): 1515-1525. doi: 10.1113/jphysiol.2010.201855.

46. Basu S, Srivastava P. Immunological role of neuronal receptor vanilloid receptor 1 expressed on dendritic cells. Proc Natl Acad Sci U S A. 2005; 102(14): 5120-5125.

47. O'Connell PJ, Pingle SC, Ahern GP. Dendritic cells do not transduce inflammatory stimuli via the capsaicin receptor TRPV1. FEBS Lett. 2005; 579(23): 5135-5139.

48. Tóth BI, Benko S, Szöllosi AG, Kovács L, Rajnavölgyi E, Bíró T. Transient receptor potential vanilloid-1 signaling inhibits differentiation and activation of human dendritic cells. FEBS Lett. 2009; 583(10): 1619-1624. doi: 10.1016/j.febslet.2009.04.031.

49. O'Connell PJ, Klyachko VA, Ahern GP. Identification of functional type 1 ryanodine receptors in mouse dendritic cells. FEBS Lett. 2002; 512(1-3): 67-70.

50. Vukcevic M, Spagnoli GC, Iezzi G, Zorzato F, Treves S. Ryanodine receptor activation by Ca v 1.2 is involved in dendritic cell major histocompatibility complex class II surface expression. J Biol Chem. 2008(50); 283: 34913-34922. doi: 10.1074/jbc.M804472200.

51. Lee HC. Multiplicity of Ca2+ messengers and Ca2+ stores: a perspective from cyclic ADP-ribose and NAADP. Curr Mol Med. 2004; 4(3): 227-237.

52. Partida-Sánchez S, Goodrich S, Kusser K, Oppenheimer N, Randall TD, Lund FE. Regulation of dendritic cell trafficking by the ADP-ribosyl cyclase CD38: impact on the development of humoral immunity. Immunity. 2004; 20(3): 279-291.

53. Di Virgilio F. Purinergic mechanism in the immune system: A signal of danger for dendritic cells. Purinergic Signal. 2005; 1(3): 205-209. doi: 10.1007/s11302-005-6312-z.

54. Ferrari D, La Sala A, Chiozzi P, Morelli A, Falzoni S, Girolomoni G, et al. The P2 purinergic receptors of human dendritic cells: identification and coupling to cytokine release. FASEB J. 2000; 14(15): 2466-2476.

55. Mutini C, Falzoni S, Ferrari D, Chiozzi P, Morelli A, Baricordi OR, et al. Mouse dendritic cells express the P2X7 purinergic receptor: characterization and possible participation in antigen presentation. J Immunol. 1999; 163(4): 1958-1965.

56. Schnurr M, Toy T, Stoitzner P, Cameron P, Shin A, Beecroft T, et al. ATP gradients inhibit the migratory capacity of specific human dendritic cell types: implications for P2Y11 receptor signaling. Blood. 2003; 102(2): 613-620.

57. Shin A, Toy T, Rothenfusser S, Robson N, Vorac J, Dauer M, et al. P2Y receptor signaling regulates phenotype and IFN-alpha secretion of human plasmacytoid dendritic cells. Blood. 2008; 111(6): 3062-3069.

3

Pseudomonas Quinolone Signal Modulates Cystic Fibrosis Epithelial Cell Response through the Toll-Like Receptor 4

Patricia R. Wardwell[1,2], David R. Wilson[1,2], Dewayne L. Garner[1,2] and Rebecca A. Bader[1,2*]

[1]*Department of Biomedical & Chemical Engineering, Syracuse Biomaterials Institute, Syracuse University, Syracuse, NY, USA*

[2]*Onondaga Community College, Syracuse, NY, USA*

***Corresponding author:** Rebecca A. Bader, Department of Biomedical & Chemical Engineering, Syracuse Biomaterials Institute, Syracuse University, Syracuse, NY, USA, E-mail: babader@syr.edu*

Abstract

Pseudomonas aeruginosa is an opportunistic human pathogen that causes severe infection in compromised individuals, including persons with Cystic Fibrosis (CF). The expression of many virulence factors in *P. aeruginosa* is controlled by Quorum Sensing Molecules (QSMs) that are synthesized and secreted by this bacterium. Recent studies suggest that QSMs are also capable of interspecies communication, with exposure of mammalian cells to Acyl-Homoserine Lactones (AHLs) and Pseudomonas Quinolone Signal (PQS) resulting in an immunomodulatory response. Although the initial immune response is intended to clear/contain the infection, this process is ineffective in CF lungs, and the persistent, excessive inflammation eventually leads to structural damage to the tissue. The goal of this study was to examine the response of IB3-1 CF airway epithelial cell line to PQS, alone and in conjunction with Pseudomonas-Derived Lipopolysaccharides (pLPS). In contrast to results obtained with other cell types suggesting that PQS is anti-inflammatory, PQS induced inflammation in the IB3-1 CF cell line and exacerbated inflammation when administered simultaneously with LPS, as determined by ELISA for two markers of inflammation, interleukin-6 and interleukin-8. In addition, PQS was shown, for the first time, to act through toll-like receptor 4, a receptor that traditionally has only been associated with LPS. A better understanding of the role that QSMs play in the inflammatory response can potentially lead to new strategies to minimize airway destruction.

Keywords: *Pseudomonas aeruginosa*; Pseudomonas lipopolysaccharides; Pseudomonas quinolone signal; Cystic fibrosis; Epithelial cells

Introduction

Pseudomonas aeruginosa is an opportunistic human pathogen that causes a wide range of infections in compromised individuals [1]. In Cystic Fibrosis (CF) patients, *P. aeruginosa* is the most prevalent bacteria to colonize the lungs and is significantly correlated with increase in morbidity and mortality [2]. The treatment of CF is complicated by the resistance of *P. aeruginosa* to a number of antibiotic treatments, due in part to the formation of thick, mucoid biofilms [2]. Additional difficulties in the treatment of the *P. aeruginosa* infection in CF lungs arise from chronic inflammation resultant from a dysregulated innate immune response. Early stages of infection are hallmarked by the presense of neutrophils intended to clear the infection; however, the process is not adequate in CF lungs, and the persistent inflammation leads to severe, structural tissue damage [3]. In healthy lung tissue, the resident epithelial cells serve as a protective barrier; in contrast, with CF, these epithelial cells are thought to serve as the originators for the proinflammatory signaling that is perpetuated by bacteria-derived stimuli.

Indirect and direct interactions between the bacteria and host cells serve to promote infection and inflammation. For example, bacterial Quorum Sensing (QS) molecules, which regulate the expression of genes associated with virulence factor production, biofilm formation, and antibotic resistance, have been shown to play a critical role in inter-kingdom signaling. Two QS systems in *P. aeruginosa* are based upon the interaction of N-acyl-homoserine lactones (AHLs), specifically N-(3-oxododecanoyl) Homoserine Lactone (3-oxo-C_{12}-HSL) and N-butyryl Homoserine Lactone (C_4-HSL), with inducible transcription factors, LasR and RhIR, respectively, that activate gene expression [4]. Another, lesser studied, QS system relies upon the association of 2-heptyl-3-hydroxy-4-quinolone, referred to as Pseudomonas Quinolone Signal (PQS), with PqsR [5]. Although the primary role of QS molecules is to control gene expression by *P. aeruginosa*, these molecules also elicit a response from host cells [6,7].

In general, AHLs are thought to modulate the immune response in a concentration dependent manner with low concentrations (< 10 μM) reducing the production of inflammatory mediators and higher concentrations activating pro-inflammatory signaling pathways, such as nuclear factor kappa B (NF-κB) [8]. However, the exact nature of the immunomodulatory impact further depends on the structure of the AHL [9], the presence of additional bacteria-derived stimuli (especially lipopolysaccharides, LPS) [10], and the type of mammalian cell [11]. The structurally distinct PQS molecule has likewise given rise to a mixture of seemingly conflicting results. At non-cytotoxic concentrations (1-5 μM), administration of PQS to monocytes and bronchoalveolar cells

***Corresponding author email:** babader@syr.edu*

resulted in a reduction in the secretion of two pro-inflammatory mediators, tumor necrosis factor-α (TNF-α and interleukin-6 (IL-6), purportedly via suppression of the NF-κB pathway. However, administration of high concentrations of PQS (≥ 10 μM) resulted in enhanced TNF-α secretion [6]. Feasibly, the QSMs act in concert to ensure that the innate immune response required for removing bacteria is impaired while maintaining the pro-inflammatory environment that yields structural damage.

The gap in knowledge regarding the precise role of QSMs, particularly PQS, in the dysregulation of the host inflammatory response highlights the need for further research on the complexity of inter-kingdom signaling. Towards this end, our lab has been working towards understanding of the role that PQS plays in modulating the inflammatory response, particularly as pertains to the profoundly altered, CF airway epithelial cells. In this study, ELISA was used to assess changes in the secretion of two proinflammatory mediators, Interleukin-6 (IL-6) and Interleukin-8 (IL-8,) by the IB3-1 CF epithelial cell line resultant from treatment with combinations of LPS derived from *P. aerugnosa*, pLPS, and PQS. The selected proteins are known to be elevated in patients with CF [12]. In addition, pleiotropic IL-6 serves as a key player in the overall immune response [13], while IL-8 is active in the recruitment of neutrophils that have been directly linked to reduced pulmonary function and increased lung damage [14,15]. To further understand how PQS interacts with the host cell, identical experiments were conducted following treatment of cells with toll-like receptor 4 antibody (TLR4-pAB). Toll like receptors are a family of transmembrane cell receptors that recognize bacteria and other microorganisms, with TLR4 serving as a receptor for Gram-negative lipopolysaccharides [16]. Although AHLs have been shown to induce a cellular response through a mechanism that does not involve the TLRs, the interaction of PQS with TLRs has not previously been studied.

Materials and Methods

Cell culture and cytotoxicity

IB3-1 CF epithelial cells were obtained from Professor Pam Zeitlin at Johns Hopkins University and cultured at 37°C, 5% CO_2 in LHC-8 complete growth media (Invitrogen), supplemented with 5% fetal bovine serum (Atlanta Biologics) on 25 cm^2 collagen-coated tissue culture flasks. pLPS and PQS were procured from Sigma Aldrich. Cytotoxicity of PQS (0- 200 μg/ ml) and pLPS (0-1000 μg/ ml) towards the IB3-1 cells was assessed using a WST-8 cell proliferation assay (Cayman Chemical) following the manufacturer's instructions.

Immunomodulation by PQS and pLPS

Media for stimulation of the IB3-1 cells was prepared via dilution of pLPS and PQS stock solutions in DMSO at 5 mg/ mL and 10 mg/ mL, respectively. The DMSO did not have an impact on cellular proliferation at such low concentrations within the media. Upon reaching confluence within the 25 cm^2 tissue culture plates, the cells were seeded into coated 24 well plates at a density of 25,000 cells per well in 500 μL of media. Plated cells were allowed to adhere for 24 h in unstimulated media, at which point the media was replaced with 500 μL of stimulated media containing various combinations of pLPS and PQS. After 24 h, the supernatant was collected and stored at -80°C until analysis.

Toll-like receptor 4 blocking

LPS is known to induce inflammation through complexation with TLR4 [16]. To identify if TLR4 also plays a role in the host cell response to PQS, cells were seeded into a 24 well plate, as described above. At 23 h after seeding, the media was replaced with LHC-8 media with 5% FBS and 1 μg/ mL of TLR4- pAB (Santa Cruz Biotechnology). After 1h of exposure to the TLR4-pAB, the media was removed and replaced with media containing combinations of pLPS and/or PQS with 1 μg/ mL TLR4-pAB. After 24 h, the supernatant was collected and stored at -80°C until analysis. In additional studies to confirm the role of TLR4 in PQS interactions, cells were treated with 1 μg/ mL of IgG isotype control antibody prior to stimulation with either PQS alone (10 μg/ mL) or in combination with pLPS (100 μg/ mL).

Quantitative analysis of inflammatory cytokines

IL-6 and IL-8 ELISA kits (Peprotech) were run according to the manufacturer's instructions to determine the levels of IL-6 and IL-8 in the collected supernatant. Samples were run in duplicate, and each experiment was repeated independently at least three times.

Statistical analysis

All data is expressed as the mean ± the standard deviation. IL-6 and IL-8 levels were expressed relative to cells that remained untreated. To compare cytokine profiles obtained from exposure of the cells to various stimulant combinations, with or without TLR4-pAB, one-way ANOVA with Fisher's LSD post-hoc test was performed. For each combination of stimulants, the Student's t-test was used to compare IL-6 and IL-8 concentrations with and without TLR4-pAB. Student's t-test was likewise used to compare IL-6 and IL-8 concentrations for cells treated with TLR4-pAB and isotype control antibody. Differences were considered to be significant when $p \leq 0.05$.

Results and Discussion

Cytotoxicity of PQS and pLPS

Both pLPS and PQS were found to be non-cytotoxic towards the IB3-1 cells over the majority of concentrations observed (Figure 1). The lack of PQS cytotoxicity is consistent with results obtained previously by other investigators [17,18]. In accord with the observed outcome in this study, LPS derived from *P. aeruginosa* is less cytotoxic than that derived from *Escherichia coli* and other enterobacterial species, purportedly due to differences in the structure of the bisphosphorylated lipid (lipid A) component of the molecule. Structural changes in Lipid A may likewise cause an enhanced inflammatory response. In general, an increase in the degree of acylation from a penta-acylated form to hexa- or hepta-acylated form is directly correlated to the production of pro-inflammatory mediators [19,20].

Immunomodulation by PQS and pLPS through toll-like receptor 4

As anticipated, stimulation with pLPS alone enhanced the

secretion of pro-inflammatory mediators by the IB3-1 epithelial cells, although the increase in IL-6 was not significant (Figure 2). Cellular activation by pLPS is thought to occur by binding to CD14, followed by interaction with TLR4 and co-factor MD-2 [16]. Interaction with TLR4 in turn leads to the recruitment of Myeloid Differentiation Primary Response Gene 88 (MyD88) adapter protein through interaction with the TIR (Toll-Interleukin-1 Receptor) domain, which subsequently stimulates the production of proinflammatory mediators through activation of the NF-kB and Mitogen-Activated Protein Kinase (MAPK) pathways [21-23]. A number of seemingly contradictory reports exist pertaining to the location of TLR4 on lung epithelial cells [24]. In the current study, the addition of TLR4-pAB resulted in reduced secretion of IL-6 and IL-8 when the cells were stimulated by pLPS, regardless of the presence of PQS. The latter result strongly indicates that pLPS-induced activation occurs at TLR4 on the cell surface.

Stimulation with PQS alone yielded a trend towards increased secretion of IL-6 and IL-8. These results contrast reports that PQS possesses anti-inflammatory properties [6] (Figure 2). However, to our knowledge, all prior studies were conducted with lymphocytes or phagocytes, suggesting that impact of PQS may be cell-type dependent, as has been previously observed for AHLs. For example, at 10 μM, 3-oxo-C_{12}-HSLreduced the secretion of Interleukin-12 (IL-12), a pro-inflammatory cytokine, by LPS-stimulated macrophages [10], while exposure of lung fibroblasts and epithelial cells to the same concentration, without an additional stimulant, enhanced the secretion of IL-8 [11]. Thus, PQS may effectively mirror AHLs in regards to inter-kingdom signaling.

Together, PQS and LPS acted in a synergistic manner, as indicated by increased IL-6 and IL-8 relative to levels obtained with either mediator alone. Surprisingly, blocking the TLR4

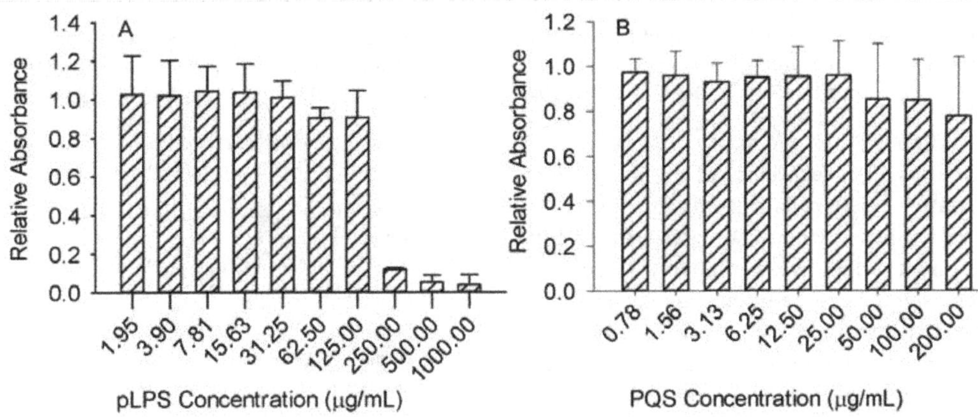

Figure 1: Increasing pLPS concentrations were non-cytotoxic towards the IB3-1 cells up to 125 μg/ ml (A), while increasing PQS concentrations had no significant impact on cellular viability over the range of concentrations observed (B).

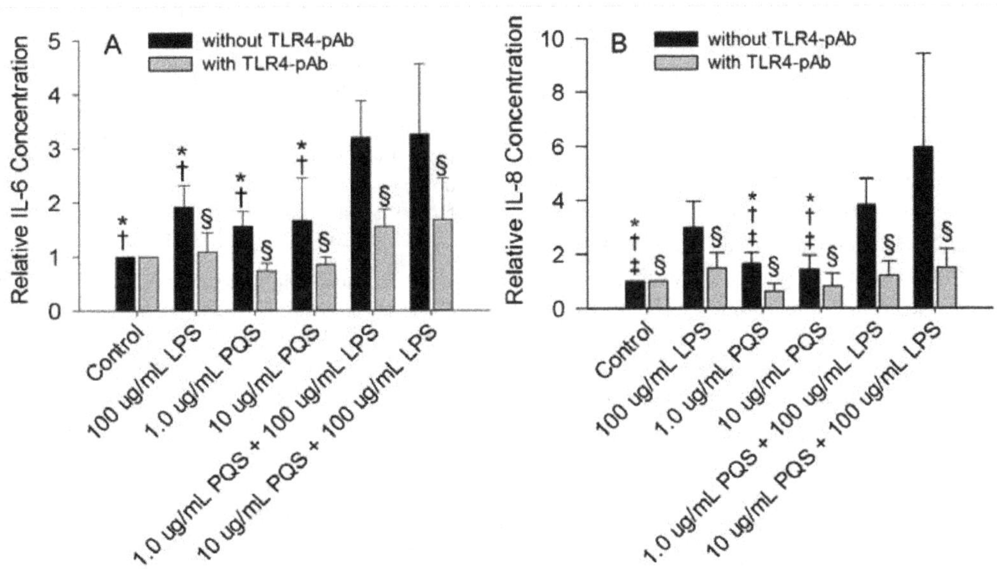

Figure 2: ELISA was performed to determine the levels of IL-8 (A) and IL-6 (B) secretion after stimulation with pLPS and/or PQS for 24 hours in the presence (□) or absence (■) of TLR4-pAB. Results are expressed as fold changes relative to a non-stimulated control. All data is presented as mean ± SD (N = 3). *, $p < 0.05$ versus 1.0 μg/ ml PQS + 100 μg/ml LPS; ‡, $p < 0.05$ versus 100 μg/ml LPS; †, $p < 0.05$ versus 10 μg/ ml PQS + 100 μg/ ml LPS, and §, $p < 0.05$ versus the corresponding stimulation condition without TLR4 blocking.

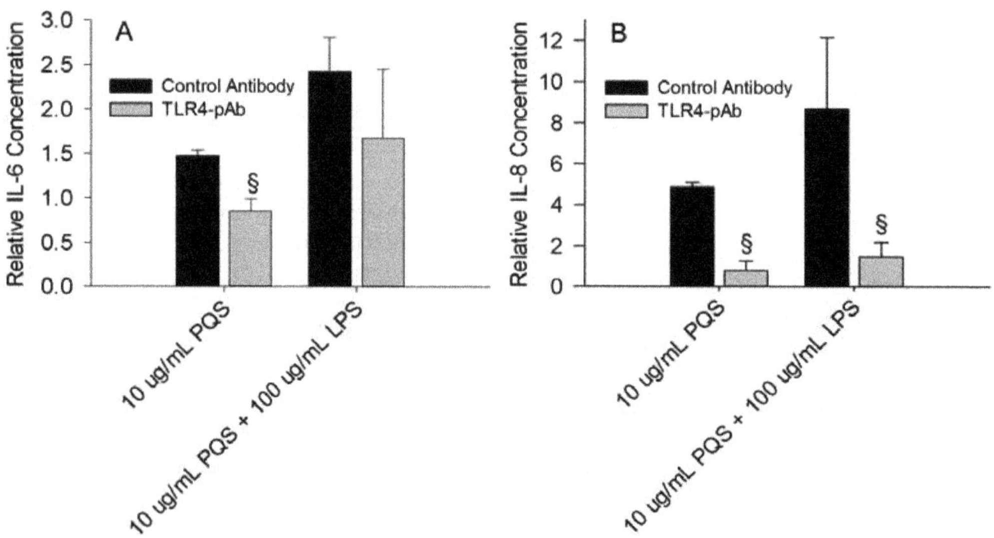

Figure 3: ELISA was performed to determine the levels of IL-8 (A) and IL-6 (B) secretion after stimulation with PQS alone or in combination with LPS for 24 hours in the presence of TLR4-pAB (□) or isotype control antibody (■). Results are expressed as fold changes relative to a non-stimulated control. All data is presented as mean ± SD (N = 3). §, $p < 0.05$ versus the corresponding stimulation condition with isotype control antibody blocking.

receptor reduced secretion of IL-6 and IL-8 when cells were subsequently stimulated by PQS, regardless of the presence of pLPS. To confirm the latter result, cells treated with the TLR4-pAB or isotype control antibody prior to stimulation with PQS alone or in combination with pLPS were compared (Figure 3). The TLR4-pAB antibody significantly reduced cytokine secretion relative to any changes that resulted from non-specific antibody binding; thereby, supporting the notion that PQS interacts with TLR4. A similar technique has been used previously to validate that *P. aeruginosa* induces a cellular response through interaction with TLR5 [25]. Although LPS is known to act through TLR4 [16], this is the first report of PQS interacting with a toll-like receptor. Subsequent studies will explore (1) whether PQS interacts directly with TLR4 or requires prior complexation with other membrane proteins and (2) if PQS activates the NF-kB and MAPK proinflammatory signaling pathways through recruitment of MyD88, as occurs with pLPS.

In conclusion, through improved understanding of inter-kingdom signaling and the factors involved in Immunomodulation, more effective treatment strategies for diseases associated with infection and inflammation can be designed. For the first time, we demonstrated that PQS acts through TLR4 and can act synergistically with pLPS to enhance the inflammatory response. Future studies will aim towards garnering further understanding of the mechanism by which PQS interacts with TLR4 and activates proinflammatory pathways. In addition, this study paves the way for future studies on inter-kingdom signaling linked to other structurally similar Quinolone, such as the PQS precursor 4-Hydroxy-2-Heptyl quinoline (HHQ) that are released by *P. aeruginosa* [26,27].

References

1. Bjarnsholt T, Jensen PO, Jakobsen TH, Phipps R, Nielsen AK, Rybtke MT, et al. Quorum sensing and virulence of Pseudomonas aeruginosa during lung infection of cystic fibrosis patients. PLoS One. 2010; 5(4): e10115. doi: 10.1371/journal.pone.0010115.
2. Deretic V, Govan JR, Konyecsni WM, Martin DW. Mucoid Pseudomonas aeruginosa in cystic fibrosis: mutations in the muc loci affect transcription of the algR and algD genes in response to environmental stimuli. Mol Microbial. 1990; 4(2): 189-196.
3. Rieber N, Hector A, Carevic M, Hartl D. Current concepts of immune dysregulation in cystic fibrosis. Int J Biochem Cell Biol. 2014; 52: 108-112. doi: 10.1016/j.biocel.2014.01.017.
4. Passador L, Cook JM, Gambello MJ, Rust L, Iglewski BH. Expression of Pseudomonas aeruginosa virulence genes requires cell-to-cell communication. Science. 1993; 260(5111): 1127-1130.
5. Pesci EC, Milbank JB, Pearson JP, McKnight S, Kende AS, Greenberg EP, et al. Quinolone signaling in the cell-to-cell communication system of Pseudomonas aeruginosa. Proc Natl Acad Sci USA. 1999; 96(20): 11229-11234.
6. Kim K, Kim YU, Koh BH, Hwang SS, Kim SH, Lépine F, et al. HHQ and PQS, two Pseudomonas aeruginosa quorum-sensing molecules, down-regulate the innate immune responses through the nuclear factor-kappa B pathway. Immunology. 2010; 129(4): 578-588. doi: 10.1111/j.1365-2567.2009.03160.x.
7. Reen FJ, Mooij MJ, Holcombe LJ, McSweeney CM, McGlacken GP, Morrissey JP, et al. The Pseudomonas quinolone signal (PQS), and its precursor HHQ, modulate interspecies and interkingdom behaviour. FEMS Microbiol Ecol, 2011; 77(2): 413-428. doi: 10.1111/j.1574-6941.2011.01121.x.
8. Cooley M, Chhabra SR, Williams P. N-Acylhomoserine lactone-mediated quorum sensing: a twist in the tail and a blow for host immunity. Chem Biol. 2008; 15(11): 1141-1147. doi: 10.1016/j.chembiol.2008.10.010.
9. Davis BM, Jensen R, Williams P, O'Shea P. (2010). The interaction of N-acylhomoserine lactone quorum sensing signaling molecules with biological membranes: implications for inter-kingdom signaling. PLoS One. 2010; 5(10): e13522. doi: 10.1371/journal.pone.0013522.
10. Telford G, Wheeler D, Williams P, Tomkins PT, Appleby P, Sewell H,

et al. The Pseudomonas aeruginosa quorum-sensing signal molecule N-(3-oxododecanoyl)-L-homoserine lactone has immunomodulatory activity. Infect Immun. 1998; 66(1): 36-42.

11. Smith RS, Fedyk ER, Springer TA, Mukaida N, Iglewski BH, Phipps RP. IL-8 production in human lung fibroblasts and epithelial cells activated by the Pseudomonas autoinducer N-3-oxododecanoyl homoserine lactone is transcriptionally regulated by NF-kappa B and activator protein-2. J Immunol. 2001; 167(1); 366-374.

12. Elizur A, Cannon CL, Ferkol TW. Airway inflammation in cystic fibrosis. Chest. 2008; 133(2): 489-495. doi: 10.1378/chest.07-1631.

13. Rincon M, Irvin CG. Role of IL-6 in asthma and other inflammatory pulmonary diseases. Int J Biol Sci. 2012; 8(9): 1281-1290. doi: 10.7150/ijbs.4874.

14. Bonfield TL, Panuska JR, Konstan MW, Hilliard KA, Hilliard JB, Ghnaim H, et al. Inflammatory cytokines in cystic fibrosis lungs. Am J Respir Crit Care Med. 1995; 152(6): 2111-2118.

15. Downey DG, Bell SC, Elborn JS. Neutrophils in cystic fibrosis. Thorax. 2009; 64(1): 81-88.

16. Miller SI, Ernst RK, Bader MW. LPS, TLR4 and infectious disease diversity. Nat Rev Microbiol. 3(1): 36-46.

17. Skindersoe ME, Zeuthen LH, Brix S, Fink LN, Lazenby J, Whittall C, et al. Pseudomonas aeruginosa quorum-sensing signal molecules interfere with dendritic cell-induced T-cell proliferation. FEMS Immunol Med Microbiol. 2009; 55(3): 335-345. doi: 10.1111/j.1574-695X.2008.00533.x.

18. Reen FJ, Clarke SL, Legendre C, McSweeney CM, Eccles KS, Lawrence SE, et al. Structure-function analysis of the C-3 position in analogues of microbial behavioural modulators HHQ and PQS. Org Biomol Chem. 2012; 10(44): 8903-8910. doi: 10.1039/c2ob26823j.

19. Hajjar AM, Ernst RK, Tsai JH, Wilson CB, Miller SI. Human Toll-like receptor 4 recognizes host-specific LPS modifications. Nat immunol. 2002; 3(4): 354-359.

20. Ernst RK, Hajjar AM, Tsai JH, Moskowitz SM, Wilson CB, Miller SI. Pseudomonas aeruginosa lipid A diversity and its recognition by Toll-like receptor 4. J Endotoxin Res. 2003; 9(6): 395-400.

21. Chow JC, Young DW, Golenbock DT, Christ WJ, Gusovsky F. Toll-like receptor-4 mediates lipopolysaccharide-induced signal transduction. J Biol Chem. 1999; 274(16): 10689-10692.

22. Lu YC, Yeh WC, Ohashi PS. LPS/ TLR4 signal transduction pathway. Cytokine. 2008; 42(2): 145-151. doi: 10.1016/j.cyto.2008.01.006.

23. Warner N, Nunez G. MyD88: a critical adaptor protein in innate immunity signal transduction. J Immunol. 2013; 190(1): 3-4. doi: 10.4049/jimmunol.1203103.

24. Guillot L, Medjane S, Le-Barillec K, Balloy V, Danel C, Chignard M, et al. Response of human pulmonary epithelial cells to lipopolysaccharide involves Toll-like receptor 4 (TLR4)-dependent signaling pathways: evidence for an intracellular compartmentalization of TLR4. J Biol Chem. 2004; 279(4): 2712-2718.

25. Blohmke CJ, Victor RE, Hirschfeld AF, Elias IM, Hancock DG, Lane CR, et al. Innate immunity mediated by TLR5 as a novel antiinflammatory target for cystic fibrosis lung disease. J Immunol. 2008; 180(11): 7764-7773.

26. Lépine F, Milot S, Déziel E, He J, Rahme LG. Electrospray/ mass spectrometric identification and analysis of 4-hydroxy-2-alkylquinolines (HAQs) produced by Pseudomonas aeruginosa. J Am Soc Mass Spectrom. 2004; 15(6): 862-869.

27. Mashburn LM, Whiteley M. Membrane vesicles traffic signals and facilitate group activities in a prokaryote. Nature. 2005; 437(7057): 422-425.

4

mTOR Signaling in Regulatory T Cell Differentiation and Expansion

Cristin Coquillard, Valery Vilchez, Francesc Marti, Roberto Gedaly*

Department of Surgery, Division of Transplantation, University of Kentucky - College of Medicine, 800, Rose Street, C453 Lexington, KY, 40536, USA

***Corresponding author:** Roberto Gedaly, Department of Surgery, Division of Transplantation, University of Kentucky, College of Medicine, 800, Rose Street, C453 Lexington, KY, 40536, USA, E-mail: rgeda2@uky.edu*

Abstract

mTOR signaling is gaining recognition for its role in dictating T cell fate through the interaction and balance between two complexes, mTORC1 and mTORC2. naive T cells do not require mTOR for initial T Cell Receptor (TCR)-mediated cytokine production. However, mTOR plays a critical role in determining antigen-induced effector and regulatory T cell (Treg) fate decisions. mTOR activation is essential for T cell commitment to Th1, Th2 and Th17 effector lineages. In the absence of mTOR, naive T cells preferentially differentiate into Treg cells. Previous studies have placed mTOR as a critical mediator of Treg development and function, suggesting that specific targeting could lead to new therapeutic opportunities in generating tolerance. This review focuses on the recent advances in mTOR signaling and downstream events in T cells, emphasizing the contrasting role of mTOR in effector versus Treg cells and its clinical implications.

Keywords: Regulatory T cells; mTOR signaling; mTOR inhibition; Rapamycin

Abbreviations

APC: Antigen Presenting Cells; AMPK: AMP-Activated Protein Kinase; CTLA4: Cytotoxic T Lymphocyte Antigen 4; eIF4E: Eukaryotic Translation Initiation Factor 4E; FKBP12: FK506-Binding Protein; FRB: FKBP12-Rapamycin Binding; FOXO1: Forkhead Box O1; FOXO3: Forkhead Box O3; GITR: Glucocorticoid-Induced Necrosis Factor Receptor; IL-2: Interleukin 2; IL-2R: Interleukein-2 Receptor; mTOR: Mammalian Target of Rapamycin; mTORC1: Mammalian Target of Rapamycin Complex 1; mTORC2: Mammalian Target of Rapamycin Complex 2; PI3K: Phosphatidylinositol-3-OH Kinase; PPAR-g: Peroxisome Proliferator-Activated Receptor-Gamma; PRAS40: Proline-Rich AKT/ PKB Substrate; PTEN: phosphatase and Tensin Homolog; Raptor: Regulatory Protein associated with mTOR; Rictor: Rapamycin-Insensitive Companion of mTOR; SGK1: Serum Glucocorticoid-Regulated Kinase 1; S6K: p70 Ribosomal S6 Kinase; SREBP1: Sterol Regulatory Binding Protein 1; TCR: T Cell Receptor; TSC1: Tuberous Sclerosis Complex 1; TSC2: Tuberous Sclerosis Complex 2; ULK1: UNC-51-like Kinase

Introduction

Appropriate T cell Activation Requires Three Major Signals: T Cell Receptor (TCR) engagement, co-stimulatory signals, and the presence of inflammatory cytokines. These signals are sensed by mTOR. Once activated, mTOR interacts with and influences T cell signaling pathways essential in regulating cell differentiation, proliferation, survival and metabolism. Multiple agonists (TCR engagement and co-stimulation, growths factors, and cytokines) regulate T cell fate by acting on cell surface plasma membrane receptors and stimulating mTOR through the activation of PI3K-AKT and the Ras-ERK-RSK signaling pathways [1].

The mTOR signaling pathway plays a crucial role in dictating T cell fate through the interaction and balance of two mTOR-containing complexes, mTOR Complex 1 (mTORC1) and mTOR Complex 2 (mTORC2). mTORC1 and mTORC2 are involved in a distinct set of cellular signals and cause separate downstream effects [2,3].

PI3K/mTOR signaling has been shown to promote effector T cell activation and differentiation, and our group and others have demonstrated the reorganization of this pathway during the differentiation of naive T cells into functional suppressor Tregs [4,5]. Indeed, T cell specific loss of mTOR correlates with a diminished generation of effector Th1, Th2 and Th17 cells, and enhanced generation of Tregs. Tregs play a major role in the prevention of autoimmunity by suppressing T cell responses to self-antigens and by limiting the response to foreign antigens. These cells consistently express high levels of the cell surface IL-2 Receptor (IL-2R) α-chain (CD25) and the forkhead family transcription regulator, Foxp3. Rapamycin and its analogs (rapalogs) promote tolerance in experimental models by favoring Treg-dependent suppression in human T cells, and are currently being used to prevent rejection in solid organ transplantation [6,7].

As mTOR is emerging as a critical regulator of the balance between regulatory and effector T cells, there is increasing interest in the development of molecular-targeted therapies that control upstream and downstream events of mTOR activities. The focus of our manuscript is to describe the contrasting role of mTOR in effector versus Treg cells and the clinical implications of its inhibition.

***Corresponding author email:** rgeda2@uky.edu*

mTOR structure and complexes

The TORs are large proteins (~ 250kDa) belonging to the phosphatidylinositol kinase-related kinase family (PIKK), which are characterized by a carboxy-terminal serine/threonine kinase domain similar in sequence to that found in the phosphatidylinositol 3-kinases [8]. At its amino-terminal, mTOR contains 20 HEAT (Huntington, EF3, A subunit of PP2A, TOR1) repeats, which function as protein interaction sites, followed by a large central FAT (FRAP, ATM, TRAP) domain. Toward the carboxy-terminal, mTOR includes a FRB (FKB12-rapamycin binding) domain, a kinase domain and a second FAT domain called the FATC domain (FAT C-terminus) [9] [Figure 1A].

In mammalian cells, mTOR exists as the catalytic subunit of two structurally and functionally different complexes, mTORC1 and mTORC2. Each is distinguished by a unique protein essential for assembly of the complex, binding substrate, and regulation [2]. In mTORC1, that protein is Raptor (regulatory protein associated with mTOR) while in mTORC2 it is Rictor (Rapamycin-insensitive companion of mTOR) [3,10]. Other components are also unique to each complex: mTORC1 contains PRAS40 (Proline-rich AKT/PKB substrate 40 kDa), which acts as a negative regulator of mTOR [11] while mTORC2 contains Protor and mSIN1, both of which appear to play a role in downstream events [12,15]. In addition to these unique proteins, both complexes share mLST8 and Deptor [2] [Figure 1B].

mTOR signaling

mTORC1 integrates information from multiple pathways, including growth factors, nutrients, oxygen, and energy status. The interplay of these complex signals converges to promote cell growth and proliferation by increasing anabolic processes like protein and lipid synthesis and decreasing catabolic processes like autophagy [14]. Much of what is known about mTORC1 arose from studies using rapamycin, which binds to the FRB domain of mTOR along with FKBP12 (FK506-binding protein of 12 kDa) and inhibits its function[15]. mTORC1 is thus known as the rapamycin-sensitive complex. Compared to mTORC1, relatively little is known about the regulation of mTORC2. To categorize mTORC2 as a rapamycin-insensitive complex is a slight misnomer due to rapamycin's ability to inhibit mTORC2 at high dose and/or after prolonged use [16]. When activated, mTORC2 phosphorylates the kinase AKT at Serine 473 and induces its full activation [17]. Since AKT is also upstream activator of mTORC1 [18], AKT activation has an important role in mTORC1-mTORC2 crosstalk as one of the multiple intertwined bioloops that regulate mTOR activity. The general consensus suggests that mTORC2 is activated by growth factors, and that this complex plays a role in rearrangement of the cytoskeleton and in cell cycle progression,

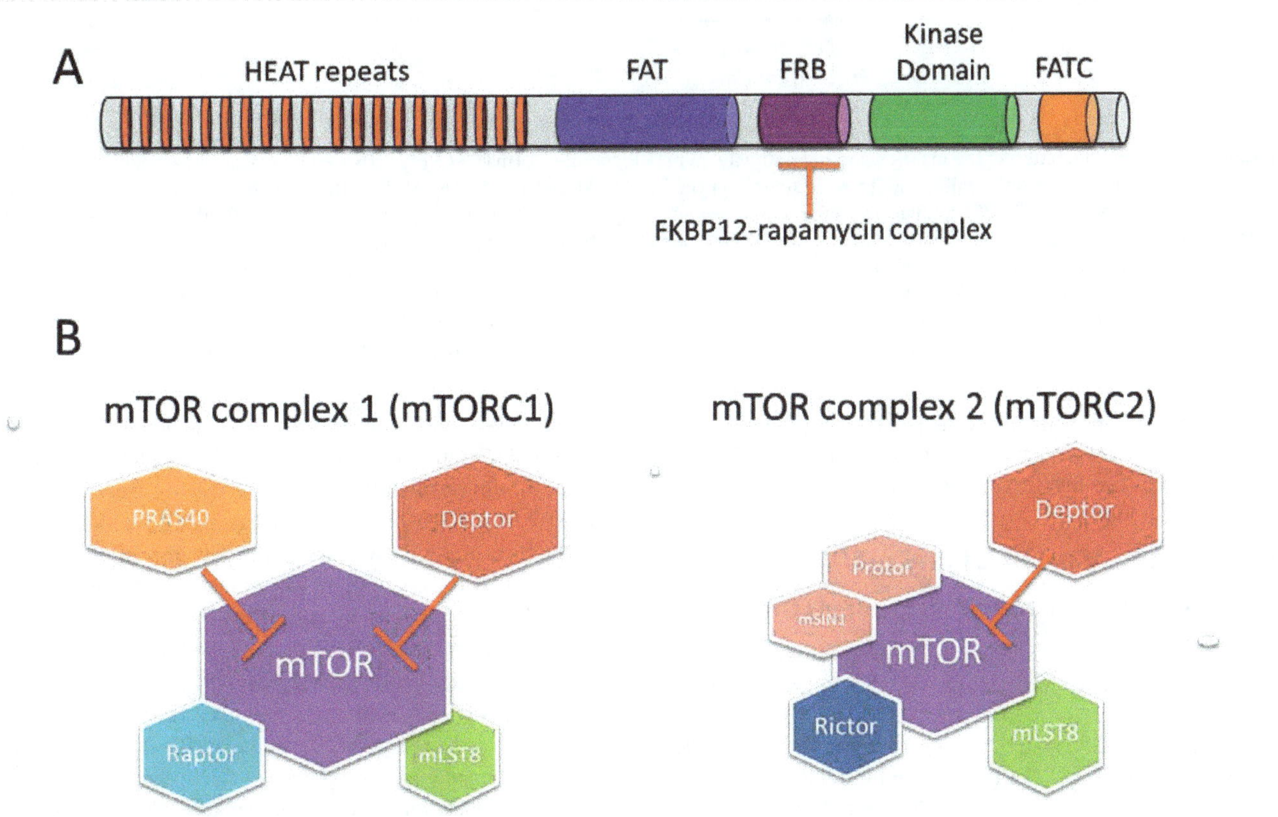

Figure 1: The primary structure of mTOR including HEAT repeats, FAT, FRB, Kinase, and FATC domains. B. The protein subunits of the mTOR complexes, both complexes include, in addition to mTOR, mLST8 and Deptor, which acts as an inhibitor of both complexes. mTORC1 additionally includes PRAS40, an inhibitor of complex 1, and Raptor. mTORC2 also includes Rictor, Protor, and mSIN1.

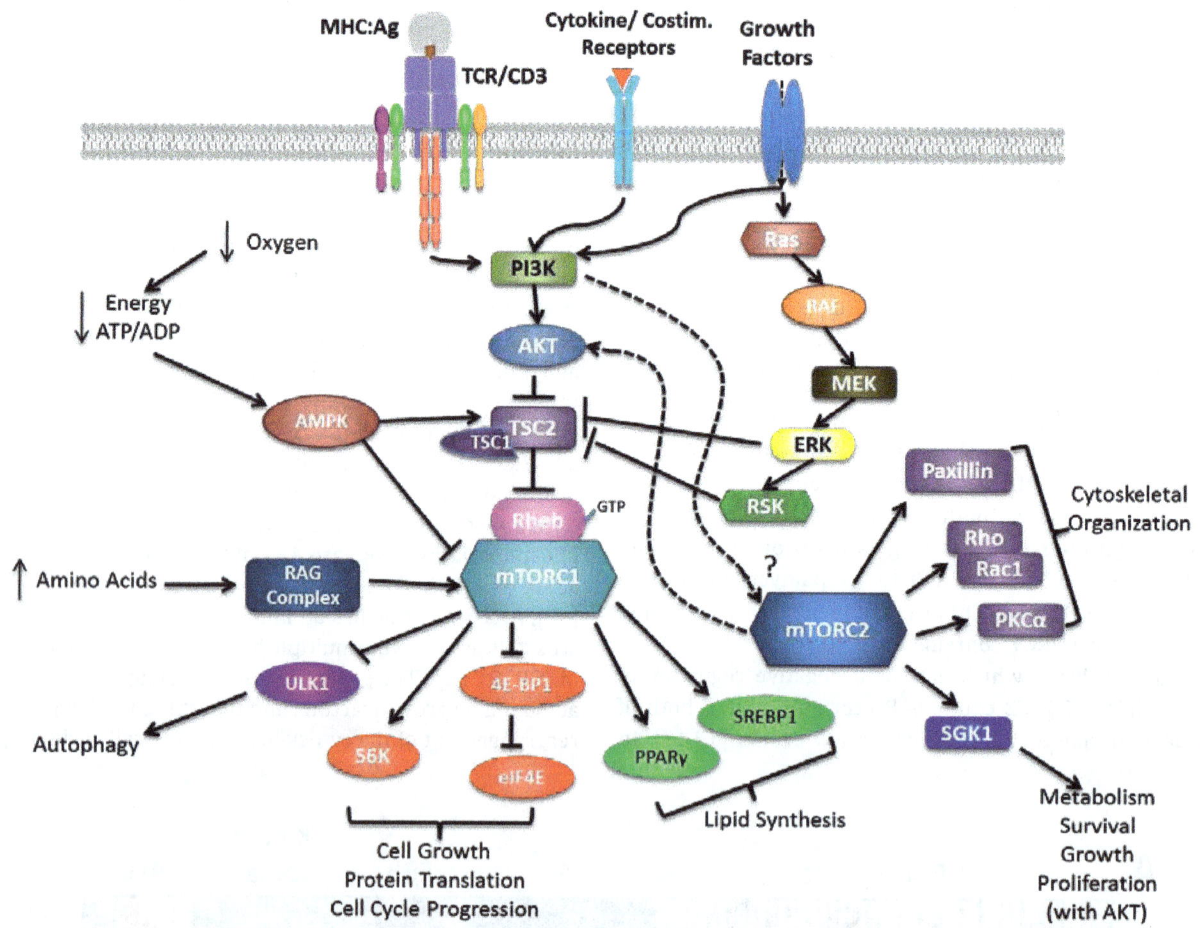

Figure 2: Simplified depiction of the mTOR signaling pathway. m mTOR complex 1 (mTORC1) is activated by extracellular signals like TCR engagement and CD28 co stimulation, insulin, and growth factors via the PI3K/ Akt and Ras/ RAF/ MED/ ERK/ RSK pathways. Intracellularly, sufficient amino acids necessary for mTORC1 activation where as low oxygen and energy inhibit mTORC1. Downstream, mTORC1 activates cell growth and protein translation through S6K and elF4E as well as lipid synthesis through PPAR-gamma and SREBP1, while inhibiting autophagy through ULK1. Other signaling pathways also affect mTORC1 that are not depicted here, including Wnt and TNF-alpha. The signaling pathways are not as well elucidated for mTOR complex 2 (mTORC2). It is thought to be activated by growth factors through a PI3K pathway that differs from that of mTORC1. Downstream, mTORC2 activates Akt, putting it upstream of mTORC1 signaling, and SGK1, both of which influence metabolism, survival, growth, and proliferation. mTORC2 also influences cytoskeletal organization through activation of Paxillin, PKC- alpha, Rho, and Rac1

cell survival, and anabolism by phosphorylating the ACG kinases in a pathway that intersects with the activation of mTORC1 [2]. An overview of the mTOR complexes and signaling pathway is summarized in Figure 2.

The Yin-Yang of mTORC1 and mTORC2

Upstream and downstream events of mTORC1

The tuberous sclerosis complex, a heterodimer comprised of TSC1 and TSC2, is the point at which signals from several different cellular pathways are integrated in the regulation of mTORC1. TSC1/2 acts as a GTPase-activating protein (GAP) towards Rheb, promoting the hydrolysis of Rheb bound to GTP, and converting it to an inactive, GDP bound state. TSC1/2 therefore functions as a negative regulator of mTORC1 [19,20]. When Rheb is in an active GTP bound state, its translocation to the cell membrane stimulates mTORC1-mediated signaling [19].

Regulation of the TSC complex is primarily mediated by two major signaling pathways: the PI3K-AKT and the RAS-ERK axis. Binding of insulin or other growth factors or cytokines to their cell-surface receptors activates PI3K, which results in the activation of AKT. AKT then directly phosphorylates TSC2 on several residues, there by inhibiting its GAP activity for Rheb and allowing activation of mTORC1 [21]. AKT also directly phosphorylates PRAS40, relieving its inhibition of mTORC1 [22]. The tumor suppressor PTEN (phosphatase and tensin homolog deleted on chromosome 10) opposes the activity of PI3K and negatively regulates this pathway [22]. Alternatively, growth factor-induced RAS-ERK pathway can activate mTORC1 through ERK and RSK-mediated phosphorylation of TSC2 and/or Raptor [23]. Other pathways including TNF and Wnt/β-catenin signaling have also been described as extracellular upstream regulators of mTORC1 [24]. Likewise, AMPK signaling inhibits mTORC1 when intracellular conditions like hypoxia and low cellular energy are detected [14,25,26].

Activated mTORC1 acts on a broad range of downstream substrates [15]. One mechanism by which mTORC1 induces protein synthesis is through inhibition of the eukaryotic initiation factor 4E binding protein 1 (4E-BP1), which itself acts as a translation inhibitor [27]. Another major downstream target of mTORC1 is the p70 ribosomal S6 kinase (S6K). When activated, S6K upregulates protein translation and elongation, mRNA synthesis, and ribosome biogenesis [2,27]. In a regulatory feedback loop, S6K also acts upstream of mTORC1 by inhibiting PI3K activation and rebalancing mTOR activity [15,28,29]. In this context, the preferential short-term inhibition of mTORC1 by rapamycin and the subsequent loss of the S6K negative feedback loop may lead to unexpected secondary effects such as hyper-activation of AKT and induction of mTORC2 downstream effects.

Upstream and downstream of mTORC2

The activation of mTORC2 is significantly less well understood than that of mTORC1. Early studies demonstrated that mTORC2 was activated by insulin via the PI3K pathway, but signaling steps beyond PI3K are distinct from mTORC1 activation and have mostly remained unclear. Recent studies have implicated ribosomal binding of mTORC2 and ER stress in mTORC2 signaling, but further studies are needed to fully understand the signaling inputs to mTORC2 [30,31].

Similarly, not much is known regarding the downstream effects of mTORC2 and its role in the regulation of cellular functions. mTORC2 controls the actin cytoskeleton by promoting the activation of the ACG kinase protein kinase C-alpha (PKCα), inducing GTP loading of the GTPases Rho and Rac1, and the phosphorylation and relocalization of paxillin [10]. Along with PKCα, mTORC2 also activates two other members of the AGC family of kinases-SGK1 and AKT. Some of the main downstream effects of mTORC2 are caused by its inhibition of the transcription repressors FOXO1 and FOXO3 (Forkhead box O1/O3).

mTORC1 and mTORC2 activities are affected by cross-regulatory loops promoted by downstream events of both complexes. As previously mentioned, one important mechanism of cross-talk involves mTOR regulation of AKT. mTORC2 phosphorylates AKT at serine 473, which, together with the phosphorylation at threonine 308 (PI3K-dependent activation of PDK1), is needed for complete AKT activity [17]. Physiological or pharmacological inhibition of mTORC1 (by Deptor in cancer cells or by rapamycin) may indirectly promote the activation of AKT by relieving the negative feedback of S6K on PI3K [32]. The complexities of the mTORC1 and mTORC2 signaling and crosstalk have been excellently and thoroughly reviewed in recent publications [2,15,33,34].

mTOR signaling on T cell fate decisions

Naive mature T cells circulate in the periphery in a quiescent state characterized by small size and exit from active division (G0). Regular engagement of the TCR with self-peptide presented by MHC, and the antiapoptotic interaction of IL-7 with IL-7R, promote survival of these cells in this resting state [35]. In quiescent T cells, homeostasis and the initial TCR-mediated cytokine production are surprisingly independent of mTOR signaling [36]. Deficiency of TSC1 is highly disruptive of naive T cell homeostasis, which indicates the important role of TSC1 and the requirement of low functional mTORC1 during T cell quiescence [37,38].

Recognition of foreign antigen is necessary to activate effector T cells, but engagement of the TCR with the antigen: MHC complex alone (signal 1) is not sufficient for activation. Rather, co-stimulation with a second signal (signal 2) like CD28 is required for full activation, which induces the production of IL-2 necessary for T cell activation and proliferation (signal 3) [31]. In the absence of signal 2, signal 1 alone induces a T cell anergic state, characterized by hypo responsiveness to secondary TCR stimulation [39]. Anergy is likely among the most prominent mechanisms in controlling peripheral tolerance to self-antigen. Engagement of the TCR can induce low-level activation of mTOR through the PI3K pathway, but this is amplified by signal 2, which strongly activates the PI3K/ AKT/ mTOR axis [40].

Previous studies indicated that mTOR inhibition with rapamycin induces T cell anergy even in the presence of co-stimulation, thus highlighting the importance of mTOR signaling in effective T cell activation [35]. In contrast, recent studies support that the absence of mTOR does not alter the TCR-induced signaling cascade and the ability of naive T cells to produce IL-2. The same naive T cells fail to differentiate into an effector cells under skewing conditions both *in vitro* and *in vivo* [37].

The role of mTOR in effector T cells

The activation status of the cell and the integration of multiple signals from the local environment determine the intensity, duration and type of mTOR signaling [41]. During TCR activation, naive $CD4^+$ T cells can differentiate into effector Th1, Th2, or Th17 cells or suppressor iTregs depending on the particular cytokine milieu where the activation occurs. An environment rich in IL-12 and IFN-γ will likely produce Th1 cells, while the presence of IL-6, IL-21, and TGF-β will skew differentiation towards the production of Th17cells [42]. Though quite different in their functions, Th1 and Th17 cells share a common intracellular signaling pathway; and the differentiation of each of them requires signaling via the mTORC1 complex [43]. In contrast, there are conflicting reports on the role of mTORC2 in Th1 differentiation. A study by Lee et al. [44] reported that naive T cells are unable to differentiate into the Th1 lineage when deficient in Rictor or Rheb, although later studies indicate otherwise [45]. There is consensus that Th2 cells fail to differentiate in the absence of Rictor but develop normally in the absence of Rheb, pointing to mTORC2 rather than mTORC1 as key regulator of Th2 differentiation [43, 44].

The role of mTOR in treg cells

Treg cells are critical in toning down the immune response to prevent a pathological inflammatory response or autoimmunity [46]. Though it became clear they existed decades ago, it was not until 2003 that the phenotype of suppressor Treg cells was characterized as $CD4^+CD25^+Foxp3^+$, although none of these markers are exclusively expressed in Tregs. Tregs express, among others, cytotoxic T lymphocyte antigen 4 (CTLA4) and glucocorticoid-induced necrosis factor receptor (GITR), but do not express the IL-7 receptor (CD127) [47]. Tregs comprise only a small percentage of the T cells in the body (5-10%), but they are absolutely necessary for immune homeostasis.

Treg cells exist in two phenotypically indistinguishable yet distinct subtypes that differ in their origin: natural Tregs (nTregs) and induced Tregs (iTregs). nTregs develop in the thymus, mainly in response to self-antigens and require high affinity interactions between the TCR and peptide: MHC with CD28 co-stimulation. iTregs develop from naive CD4+CD25- T cells in the periphery [48], and their repertoire is more specific for foreign antigens [49]. In contrast to nTregs, iTregs have a lower activation threshold for TCR stimulation and have no requirement for CD28 co-stimulation, although they require IL-2 and TGF-β signals. The differences in antigen-recognition between nTregs (self-antigen) and iTregs (foreign) may partially account for their distinct regulatory activities, with nTregs preferentially involved in preserving self-tolerance and functionally active iTregs present at inflammatory sites [49,50]. Unlike effector T cells, Tregs demonstrate a low level of mTOR activity throughout their lives [51]. In fact, their development occurs in conditions that promote minimal mTOR activity, such as low affinity or brief TCR binding and stimulation [52,53].

Emerging data suggest that the number and function of Tregs in humans are maintained even in senescence [54], supporting the relevance of those peripherally developed iTregs. Recent evidence emphasizes the functional plasticity of T cells. In this context, iTregs can be generated from Th1, Th2, and Th17 effector cells as well as from naive cells [55,56]. Our group has demonstrated that human primary naive $CD4^+$ T cells differentiate into Tregs only after multiple rounds of cell division and re-configuration of the PI3K/ AKT/ mTOR pathway [2,3]. This is similar to the "two-step differentiation model" proposed by Guo et al. [57] according to which the initial TCR-dependent conditioning step induces the expression of CD25 and "primes" the cell to efficiently respond to the second "cytokine-dependent" phase that results in the acquisition of the Treg phenotype.

Several signals have been described to govern the generation, function and stability of Treg cells. Smad3 plays a critical role in promoting regulatory T cell differentiation. TGF-β signaling activates Smad3 which, along with mTOR-induced NF-AT, contributes to the induction of Foxp3 by promoting acetylation of the FOXP3 enhancer. mTOR plays simultaneous roles of inducing specific effector T cell differentiation by enhancing STAT activation and inhibiting Smad3 phosphorylation. Delgoffe et al. [36] suggested that in the presence of active mTOR, TCR engaged T cells display normal IL-2 production and differentiate towards Th1, Th2 and Th17. However, the absence of mTOR leads to induction of functionally stable Treg cells. The authors propose that in the absence of mTOR, mTORC2-mediated activation of AKT is absent, resulting in hyper phosphorylation of Smad3. Our laboratory has shown that the expression of AKT protein is substantially higher in the Treg subpopulation when compared to conventional T cells, likely due to the increased stability of non-active AKT. Moreover, this low-active AKT in Tregs co-immunoprecipitates with Smad3, suggesting a new level of cross-talk between the TFG-β and TCR pathways in iTregs (data not published).

Critical among the signals that guide the development and function of Tregs are also those mediated by IL-2R, as an increase in this pathway has been associated with an accumulation of Treg cells *in vivo* [58]. Among the signals activated via IL-2R, the phosphorylation of the transcription factor STAT5 appears to play a key role in the generation and expansion of Treg cells. IL-2 also initiates signaling through PI3K. Tregs express high levels of PTEN, the main negative regulator of PI3K suggesting a restrained activation of PI3K signaling. Huynh et al. [59] demonstrated that PTEN-mediated control of PI3K activity is critical for the stability of mouse Tregs. Using a PTEN-deficient mouse model, specifically in the $Foxp3^+$ population, they found that decreased control of PI3K activity led to the accumulation of transient $Foxp3^+CD25^-$ cells and subsequent loss of expression of Foxp3.

Other mTOR-dependent pathways may negatively regulate Foxp3 expression. The FOXO transcription factors directly bind to the Foxp3 promoter, and they are inactivated by mTORC2-dependent, AKT-mediated phosphorylation. Moreover, the expression of the constitutively active form of AKT in Treg cells has been associated with a diminished Treg cell gene signature, including reduced expression of Foxp3 and IL2ra (IL-2R α subunit CD25) [60]. The fact that an increase in Tregs was observed following deletion of mTOR but not in the Rheb and Rictor T cell conditional knockouts suggests that loss of both of these pathways is necessary to enhance the generation of Tregs [43]. In the same context, the rapamycin-induced blockage of mTOR signaling also generates Tregs through induction of Foxp3 expression regardless of the presence of effector T cell skewing cytokines and strong TCR engagement.

The above-mentioned balance of the feedback loops between mTORC1 and mTORC2 may impact the effects of rapamycin-based therapy, which was initially described as a selective allosteric inhibitor of mTORC1. Blocking the mTORC1-dependent S6K activation stops the negative feedback loop over PI3K, which leads to hyper activation of AKT. Therefore, the single targeting of mTOR has to be carefully addressed in a cell type-specific, time and concentration dependent-manner.

mTOR as a mediator of the metabolism of proliferating T cells

In contrast to the rather well characterized intracellular signaling cascades, the role of cytokine-induced metabolic changes in T cell fate and function have been recognized only recently [61]. mTOR is known to play a key role in integrating signals that respond to nutrient availability and growth factors from the extracellular environment and to adjust the cell's metabolism accordingly.

In the quiescent metabolic state of naive T cells, the low activity of mTOR correlates with the mitochondrial fatty acid oxidation as the primary energy source of the cell [62]. Upon effective antigen recognition, the clonal expansion of antigen-primed T cells demands the sudden increase in aerobic glycolysis to support their new energetic and anabolic demands. This preferential metabolic circuitry in functionally active T cells is similar to the aerobic glycolysis of cancer cells, a phenomenon known as the Warburg effect. Recent studies have identified specific signaling events and transcription factors involved in the coordination of cell proliferation and metabolism in activated T cells that might dictate the outcome of adaptive immune responses. Likewise, there is a growing body of evidence coupling the plasticity of cell metabolism and bioenergetic pathways with T cell fate lineage decisions. PI3K/ mTOR signaling plays a central role in the metabolic regulation of effector T cell activation and differentiation.

The activation of antigen-engaged naive T cells requires a rapid mTORC1-dependent readjustment of the metabolic machinery towards glycolysis [63]. The glycolytic switch in antigen-experienced cells is mTORC2-dependent [64]. These findings depict a model of evolving mTOR-complex prevalence in the regulation of effector T cell responses and emphasize the key role of mTOR activity in the metabolic reprogramming of T cells. The integration of T cell metabolism and function is especially relevant for the maintenance of the immune homeostasis between effector and regulatory T cell function. While pro-inflammatory effector Th1, Th2 and Th17 activities rely mostly on glycolysis to meet the cell energy demands, both natural and iTreg cells and CD8^{+} memory T cells do not primarily use glycolysis for energy generation and instead rely on fatty acid metabolism [65]. Compared to effector T cells, memory T cells have enhanced mitochondrial spare respiratory capacity and greater mitochondrial mass than naive T cells, which provides the extra energy necessary to promote survival [65-67]. Targeting cell metabolism is envisioned as a novel therapeutic strategy in cancer but is also emerging as an attractive immunotherapeutic approach [Figure 3].

Perspectives of pharmacological targeting of mTor signaling pathway on treg development

The physiologic relevance and potential involvement of the mTOR pathway as the facilitator of T cell lineage commitment in several pathological conditions have attracted major interest as a pharmacological target. Inhibition or deficiency in this pathway may lead to acquisition or increase of suppressor Treg function, and attracted interest as pharmacological target in autoimmune disease, transplantation and graft versus host disease (GVHD).

Rapamycin promotes *de novo* expression of Foxp3 in naive T cells, and it has demonstrable beneficial effects on the survival and proliferation of Tregs *in vivo* and *in vitro* [6,7]. As a result,

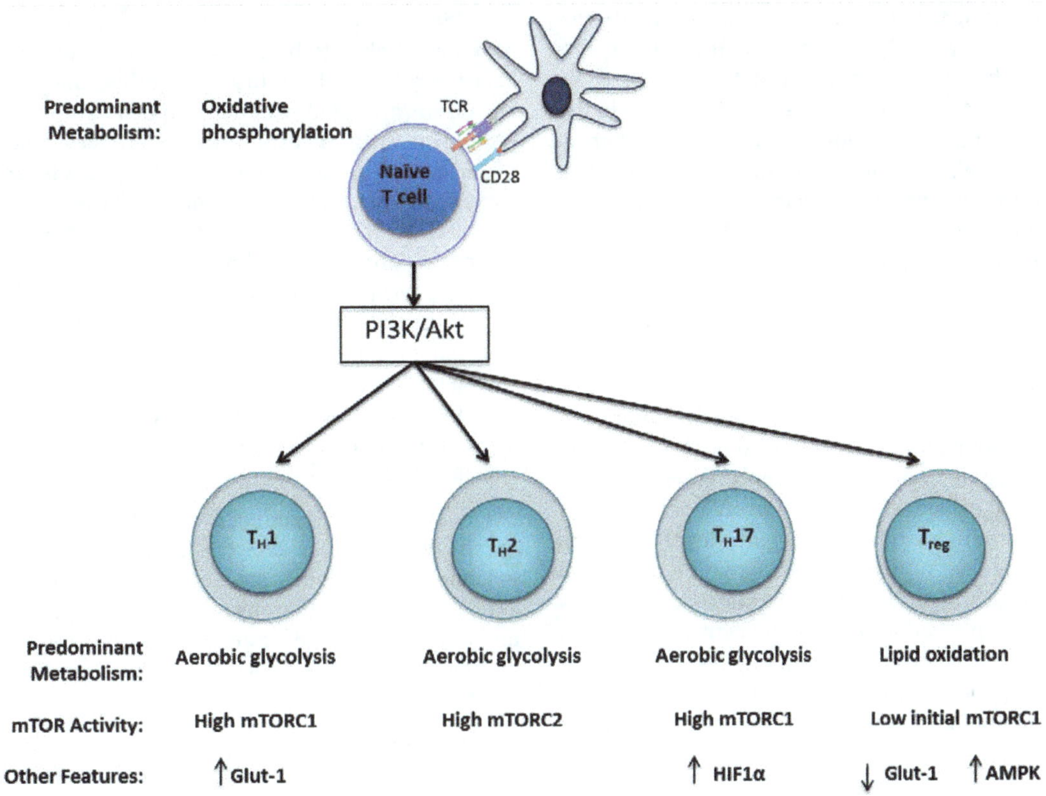

Figure 3: The differentiation of CD4+ T cells requires changes in metabolism in which mTOR activity plays in important role. While pro-inflammatory effector Th1, Th2 and Th17 activities rely mostly on glycolysis to meet the cell energy demands, Tregs exhibit higher dependency on mitochondrial oxidative phosphorylation.

Modified from Coe, et al. Metabolic regulation of regulatory T cell development and function. Frontiers in Immunology. 2014; 5: 1-6.

rapamycin inhibits effector T cell proliferation but promotes Treg accumulation. As previously mentioned, the preferential inhibition of mTORC1 by rapamycin may lead to non-desirable effects such as losing the inhibitory effect of S6K over PI3K-induced AKT activation. Alternative therapeutic strategies have been proposed that combine mTOR inhibition and upstream PI3K/AKT inactivation in cancer [68].

One of the limitations in developing Treg-based therapies is the low frequency of circulating Tregs, especially antigen-specific Tregs, and the unfavorable *ex vivo* expansion properties. However, recent technical advances in *ex vivo* expansion regimens of nTregs or generation of iTregs have made it feasible to reconsider Treg therapies [69-71]. *Ex vivo* conversion of $CD4^{+}CD25^{-}$ naive T cells into iTregs with suppressor function represents an alternative strategy to *ex vivo* nTreg isolation and expansion. Our group [4] and others have shown that Treg culture media with relative low dose of IL-2 in the presence of rapamycin resulted in a steady and consistent growth of Treg cells essentially free of contaminating non-Tregs. The high degree of plasticity and insufficient number of antigen-specific Tregs that can be obtained with *ex-vivo* expansion still represents major challenges associated with the infusion of human Tregs in clinical practice.

Several phase I trials have been conducted to assess their effect on GVHD following hematopoietic stem cell transplantation (HSCT) and in type I diabetes [72]. The first human trial employing adoptive transfer of nTregs to suppress GVHD was reported by Edinger et al. [73] treating five HSCT recipients with fresh, bead-purified donor nTregs. The authors reported no infusional toxicity, increase in infection rate or GVHD. A similar study using purified nTregs was performed in 2008 by Di Ianni et al. [74] to evaluate the effect of these cells on GVHD prevention in 28 patients. nTregs were infused into patients three days prior to $CD34^{+}$ cells supplemented with frozen/thawed mature donor T cells in the absence of any post-transplant immunosuppression. The authors showed that adoptive transfer of Tregs prevented GVHD, promoted lymphoid reconstitution and enhanced immunity to opportunistic infections. Trzonkowski et al. [75] used sort-purified nTregs ($CD4^{+}25^{+}127^{-}$) expanded *in vitro* to treat two patients with GVHD with overall positive outcomes. More recently, Brunstein et al. [76] established a method of $CD4^{+}CD25^{+}FoxP3^{+}$ Treg enrichment from cryopreserved umbilical cord blood (UCB), and evaluated the safety profile of these cells in 23 patients. The authors reported no infusional toxicities with a reduced incidence of acute GVHD when compared to control subjects. These initial clinical trials evaluating the safety and efficacy of Tregs in treating GVHD demonstrate promising safety and potentially efficacy profiles and increase the interest to consider wider applications of Treg-based therapy in other diseases including autoimmunity and solid organ transplantation [77,78].

Beta-cell specific auto reactive T cells can be found in patients with Type I Diabetes (T1D) and in healthy controls, and are usually controlled by a network of regulatory mechanisms including Treg cells. Increasing the number of Tregs by adoptive transfer can be used to prevent and treat even established T1D [79]. Putnam et al. [80] isolated Tregs from recent-onset T1D patients and healthy control subjects by fluorescence-activated cell sorting and compared their capacity to expand *in vitro*. The authors found that expansion of $CD4^{+}CD127^{lo/-}$ cells required the addition of rapamycin to maintain lineage purity. In contrast, expansion of $CD4^{+}CD127^{lo/-}CD25^{+}$ T-cells, resulted in high yield, functional Tregs that maintained higher Foxp3 expression in the absence of rapamycin. Tregs from T1D patients and control subjects expanded similarly and were equally capable of suppressing T-cell proliferation. Pre-clinical data support that low dose IL-2 drives tolerance via Treg activation and that the IL-2/IL-2R axis is impaired in T1D [81]. Rapamycin/ IL-2 combination treatment of NOD mice effectively treats autoimmune diabetes [82]. Based on these findings, Long et al. [83] performed a phase 1 clinical trial to test the safety and immunologic effects of rapamycin/ IL-2 combination therapy in T1D patients. Nine subjects were treated with rapamycin for three months and IL-2 for one month. The authors found that Treg cells transiently increased within the first month of therapy, although the response of these cells to IL-2 (as measured by STAT5 phosphorylation) improved and persisted after treatment. Notably, clinical and metabolic data demonstrated a transient worsening in all subjects.

In solid organ transplantation, a growing body of evidence recognizes the balance between graft-reactive effector cells and graft-protective suppressor Tregs as the ultimate determinant of long-term allograft survival, in which mTOR pharmacological inhibition may play a role [84]. Higher levels of circulating Tregs have been linked to better outcomes and higher percentages of Tregs within T cell infiltrates in transplant biopsies and have been also correlated with better function of transplanted organs [85,86]. In 2013, Yamashita et al. [87] reported the administration of *in vitro* generated iTregs to 10 patients undergoing liver transplantation. Administration of these cells was safe and appeared to facilitate early weaning of immunesuppression in half of the patients. These findings are promising but preliminary, and long-term outcomes in large cohorts should be evaluated to determine the real benefit of Treg-based therapy in solid organ transplantation. Funded by the European Commission's Seventh Framework, the ONE Study, currently underway, is a Phase I/ IIa clinical trial designed to test the safety and practicality of seven different regulatory cell populations in living donor kidney transplantation [88].

A deeper understanding of the activity of mTOR in regulatory T cells will help elucidate the uniqueness of this signaling and will open novel doors toward therapeutic manipulation of Tregs. Treg-based cellular therapy has not caused any infusional toxicity and it has established a limited safety record with regard to risk of infection, relapse or early mortality. However, efficacy data is still currently lacking.

In summary, the mTOR signaling pathway is critical in T cell fate decisions that promote a reduction of effector T cells and differentiation and expansion of regulatory T cells. A better understanding of the complexities of the PI3K/ AKT/ mTOR pathway in the regulation of T cells will be essential in designing novel therapeutic strategies for future applications in cancer, immunology/ allergy and clinical transplantation.

References

1. Shaw R, Cantley LC. Ras PI, (3)K and mTOR signaling controls tumour cell growth. Nature. 2006; 441(7092): 424-430.
2. Zoncu R, Efeyan A, Sabatini DM. mTOR: from growth signal integration to cancer, diabetes and ageing. Nat Rev Mol Cell Biol. 2011; 12(1): 21-35. doi: 10.1038/nrm3025.
3. Kim DH, Sarbassov DD, Ali SM, King JE, Latek RR, Erdjument-Bromage H, et al. mTOR interacts with raptor to form a nutrient-sensitive complex that signals to the cell growth machinery. Cell. 2002; 110(2): 163-175.
4. Ellis GI, Reneer MC, Vélez-Ortega AC, McCool A, Martí F. Generation of induced regulatory T cells from primary human naive and memory T cells. J Vis Exp. 2012; (62): 3738. doi: 10.3791/3738.
5. Reneer MC, Estes DJ, Vélez-Ortega AC, Norris A, Mayer M, Marti F. Peripherally induced human regulatory T cells uncouple Kv1.3 activation from TCR-associated signaling. Eur J Immunol. 2011; 41(11):3170-3175. doi: 10.1002/eji.201141492.
6. McMahon G, Weir MR, Li XC, Mandelbrot DA. The Evolving Role of mTOR inhibition in Transplantation Tolerance. J Am SocNephrol. 2011; 22(3): 408-415. doi: 10.1681/ASN.2010040351.
7. Battaglia M, Stabilini A, Roncarolo MG. Rapamycin selectively expands $CD4^+CD25^+FoxP3^+$ regulatory T cells. Blood. 2005; 105(12): 4743-4748.
8. Dobashi Y, Watanabe Y, Miwa C, Suzuki S, Koyama S. Mammalian target of rapamycin: a central node of complex signaling cascades. Int J Clin Exp Pathol. 2011; 4(5): 476-495.
9. Ma XM, Blenis J. Molecular mechanisms of mTOR-mediated translational control. Nat Rev Mol Cell Biol. 2009; 10(5): 307-318. doi: 10.1038/nrm2672.
10. Sarbassov DD, Ali SM, Kim DH, Guertin DA, Latek RR, Erdjument-Bromage H, et al. Rictor, a novel binding partner of mTOR, defines a rapamycin-insensitive and raptor-independent pathway that regulates the cytoskeleton. Curr Biol. 2004; 14(14): 1296-1302.
11. Vander Haar E, Lee SI, Bandhakavi S, Griffin TJ, Kim DH. Insulin signalling to mTOR mediated by the Akt/PKB substrate PRAS40. Nat Cell Biol. 2007; 9(3): 316-323.
12. Lu M, Wang J, Ives HE, Pearce D. mSIN1 protein mediates SGK1 protein interaction with mTORC2 protein complex and is required for selective activation of the epithelial sodium channel. J Biol Chem. 2011; 286(35): 30647-30654. doi: 10.1074/jbc.M111.257592.
13. Pearce LR, Sommer EM, Sakamoto K, Wullschleger S, Alessi DR. Protor-1 is required for efficient mTORC2-mediated activation of SGK1 in the kidney. Biochem J. 2011; 436(1): 169-179. doi: 10.1042/BJ20102103.
14. Laplante M, Sabatini DM. mTOR signaling at a glance. J Cell Sci. 2009; 122(10): 3589-3594. doi: 10.1242/jcs.051011.
15. Shimobayashi M, Hall MN. Making new contacts: the mTOR network in metabolism and signalling crosstalk. Nat Rev Mol Cell Biol. 2014; 15(3): 155-162. doi: 10.1038/nrm3757.
16. Sarbassov DD, Ali SM, Sengupta S, Sheen JH, Hsu PP, Bagley AF, et al. Prolonged rapamycin treatment inhibits mTORC2 assembly and Akt/PKB. Mol Cell. 2006; 22(2): 159-168.
17. Bracho-Valdes I, Moreno-Alvarez P, Valencia-Martines I, Robles-Molina E, Chavez-Vargas L, Vazquez-Prado J. mTORC1- and mTORC2-interacting proteins keep their multifunctional partners focused. Life. 2011; 63(10): 896-914. doi: 10.1002/iub.558.
18. Vanhaesebroeck B, Guillermet-Guibert J, Graupera M, Bilanges B. The emerging mechanisms of isoform-specific PI3K signaling. Nat Rev Mol Cell Biol. 2010; 11(5): 329-341. doi: 10.1038/nrm2882.
19. Long X, Lin Y, Ortiz-Vega S, Yonezawa K, Avruch J. Rheb binds and regulates the mTOR kinase. Curr Biol. 2005; 15(8): 702-713.
20. Tee AR, Manning BD, Roux PP, Cantley LC, Blenis J. Tuberous Sclerosis Complex Gene Products, Tuberin and Hamartin, Control mTOR Signaling by Acting as a GTPase-Activating Protein Complex toward Rheb. Current Biology. 2003; 13(15): 1259-1268.
21. Cai SL, Tee AR, Short JD, Bergeron JM, Kim J, Shen K, et al. Activity of TSC2 is inhibited by AKT-mediated phosphorylation and membrane partitioning. J Cell Biol. 2006; 173(2): 279-289.
22. Carracedo A, Pandolfi PP. The PTEN-PI3K pathway: of feedbacks and cross-talks. Oncogene. 2008; 27(41): 5527-5541. doi: 10.1038/onc.
23. Carriere A, Romeo Y, Acosta-Jaquez HA, Moreau J, Bonneil E, Thibault P, et al. ERK1/2 phosphorylate Raptor to promote Ras-dependent activation of mTOR complex 1 (mTORC1). J Biol Chem. 2011; 286(1): 567-577. doi: 10.1074/jbc.M110.159046.
24. Dan HC, Baldwin AS. Differential involvement of IkappaB kinases alpha and beta in cytokine- and insulin-induced mammalian target of rapamycin activation determined by Akt. J Immunol. 2008; 180(11): 7582-7589.
25. Guertin DA, Sabatini DM. Defining the role of mTOR in cancer. Cancer Cell. 2007; 12(1): 9-22.
26. Gwinn DM, Shackelford DB, Egan DF, Mihaylova MM, Mery A, Vasquez DS, et al. AMPK phosphorylation of raptor mediates a metabolic checkpoint. Mol Cell. 2008; 30(2): 214-226. doi: 10.1016/j.molcel.2008.03.003.
27. Chauvin C, Koka V, Nouschi A, Mieulet V, Hoareau-Aveilla C, Dreazen A, et al. Ribosomal protein S6 kinase activity controls the ribosome biogenesis transcriptional program. Oncogene. 2014; 33(4): 474-483. doi: 10.1038/onc.2012.606.
28. Harrington LS, Findlay GM, Gray A, Tolkacheva T, Wigfield S, Rebholz H, et al. The TSC1-2 tumor suppressor controls insulin–PI3K signaling via regulation of IRS proteins. The Journal of Cell Biology. 2004; 166(2): 213–223.
29. Zhang H, Bajraszewski N, Wu E, Wang H, Moseman AP, Dabora SL, et al. PDGFRs are critical for PI3K/Akt activation and negatively regulated by mTOR. J Clin Invest. 2007; 117(3):730-738.
30. Zinzalla V, Stracka D, Oppliger W, Hall MN. Activation of mTORC2 by association with the ribosome. Cell. 2011; 144(5): 757-768. doi: 10.1016/j.cell.2011.02.014.
31. Appenzeller-Herzog C, Hall MN. Bidirectional crosstalk between endoplasmic reticulum stress and mTOR signaling. Trends Cell Biol. 2012; 22(5): 274-282. doi: 10.1016/j.tcb.2012.02.006.
32. Efeyan A, Sabatini DM. mTOR and cancer: many loops in one pathway. Current Opinion in Cell Biology. 2010; 22: 169-176. doi: 10.1016/j.ceb.2009.10.007.
33. Laplante M, Sabatini DM. mTOR signaling in growth control and disease. Cell. 2012, 13; 149(2): 274-93. doi: 10.1016/j.cell.2012.03.017.
34. Xie J, Proud CG. Crosstalk between mTOR complexes. Nat Cell Biol. 2013; 15(11):1263-1265. doi: 10.1038/ncb2877.
35. Sprent J, Surh CD. Normal T cell homeostasis: the conversion of naive cells into memory-phenotype cells. Nat Immunol. 2011; 131(6): 478-

484.

36. Delgoffe GM, Kole TP, Zheng Y, Zarek PE, Matthews KL, Xiao B, et al. The mTOR kinase differentially regulates effector and regulatory T cell lineage commitment. Immunity. 2009; 30(6): 832-844. doi: 10.1016/j.immuni.2009.04.014.

37. O'Brien TF, Gorentla BK, Xie D, Srivatsan S, McLeod IX, He YW, et al. Regulation of T-cell survival and mitochondrial homeostasis by TSC1. Eur J Immunol. 2011; 41(11): 3361-3370. doi: 10.1002/eji.201141411.

38. Wu Q, Liu Y, Chen C, Ikenoue T, Qiao Y, Li CS, et al. The tuberous sclerosis complex-mammalian target of rapamycin pathway maintains the quiescence and survival of naive T cells. J Immunol. 2011; 187(3): 1106-1112. doi: 10.4049/jimmunol.1003968.

39. Schwartz RH. T cell anergy. Annu Rev Immunol. 2003; 21: 305-334.

40. Koch U, Radtke F. Mechanisms of T cell development and transformation. Annu Rev Cell Dev Biol. 2011; 27: 539-562. doi: 10.1146/annurev-cellbio-092910-154008.

41. Waickman AT, Powell JD. mTOR, metabolism, and the regulation of T-cell differentiation and function. Immunol Rev. 2012; 249(1): 43-58. doi: 10.1111/j.1600-065X.2012.01152.x.

42. Zhu J, Yamane H, Paul WE. Differentiation of effector CD4 T cell populations (*). Annu Rev Immunol. 2010; 28: 445-89. doi: 10.1146/annurev-immunol-030409-101212.

43. Delgoffe GM, Pollizzi KN, Waickman AT, Heikamp E, Meyers DJ, Horton MR, et al. The kinase mTOR regulates the differentiation of helper T cells through the selective activation of signaling by mTORC1 and mTORC2. Nat Immunol. 2011; 12(4): 295-303. doi: 10.1038/ni.2005.

44. Lee K, Gudapati P, Dragovic S, Spencer C, Joyce S, Killeen N, et al. Mammalian target of rapamycin protein complex 2 regulates differentiation of Th1 and Th2 cell subsets via distinct signaling pathways. Immunity. 2010; 32(6): 743-53. doi: 10.1016/j.immuni.2010.06.002.

45. Heikamp EB, Patel CH, Collins S, Waickman A, Oh MH, Sun IH, et al. The AGC kinase SGK1 regulates TH1 and TH2 differentiation downstream of the mTORC2 complex. Nat Immunol. 2014; 15(5): 457-64. doi: 10.1038/ni.2867.

46. Vignali DA, Collison LW, Workman CJ. How regulatory T cells work. Nat Rev Immunol. 2008; 8(7): 523-32. doi: 10.1038/nri2343.

47. Lastovicka J. The Phenotypic Markers of CD4+CD25+ T Regulatory Lymphocytes. Research in Immunology: An International Journal. 2013; 2013.

48. Horwitz DA, Zheng SG, Gray JD. Natural and TGF-beta-induced Foxp3(+)CD4(+) CD25(+) regulatory T cells are not mirror images of each other. Trends Immunol. 2008; 29(9): 429-35. doi: 10.1016/j.it.2008.06.005.

49. Bluestone JA, Abbas AK. Natural versus adaptive regulatory T cells. Nat Rev Immunol. 2003; 3(3):253-7.

50. Reneer MC, Marti F. The balancing act of AKT in T cells. Front Biol. 2013; 8(2): 160–74.

51. Zeiser R, Leveson-Gower DB, Zambricki EA, Kambham N, Beilhack A, Loh J, et al. Differential impact of mammalian target of rapamycin inhibition on CD4+CD25+Foxp3+ regulatory T cells compared with conventional CD4+ T cells. Blood. 2008; 111(1): 453-62.

52. Gottschalk RA, Corse E, Allison JP. TCR ligand density and affinity determine peripheral induction of Foxp3 in vivo. J Exp Med. 2010; 207(8): 1701-11. doi: 10.1084/jem.20091999.

53. Katzman SD, O'Gorman WE, Villarino AV, Gallo E, Friedman RS, Krummel MF, et al. Duration of antigen receptor signaling determines T-cell tolerance or activation. Proc Natl Acad Sci. 2010; 107(42): 18085-90. doi: 10.1073/pnas.1010560107.

54. Jagger A, Shimojima Y, Goronzy JJ, Weyand CM. Regulatory T cells and the Immune Aging Process: A mini-review. Gerontology 2014; 60(2):130-7. doi: 10.1159/000355303.

55. Hori S. Developmetal plasticity of Foxp3+ regulatory T cells. Curr Opin Immunol. 2010; 22(5): 575-82. doi: 10.1016/j.coi.2010.08.004.

56. Zhou L, Lopes JE, Chong MM, Ivanov II, Min R, Victora GD, et al. TGF-beta-induced Foxp3 inhibits Th17 cell differentiation by antagonizing ROR gamma function. Nature. 2008; 453(7192): 236-40.

57. Guo Z, Khattar M, Schroder PM, Miyahara Y, Wang G, He X, et al. A dynamic dual role of IL-2 signaling in the two-step differentiation process of adaptive regulatory T cells. J Immunol. 2013; 190(7): 3153–62. doi: 10.4049/jimmunol.1200751.

58. Webster KE, Walters S, Kohler RE, Mrkvan T, Boyman O, Surh CD, et al. In vivo expansion of T reg cells with IL-2-mAb complexes: induction of resistance to EAE and long-term acceptance of islet allografts without immunosuppression. J Exp Med. 2009; 206(4): 751-60. doi: 10.1084/jem.20082824.

59. Huynh A, DuPage M, Priyadharshini B, Sage PT, Quiros J, Borges CM, et al. Control of PI(3) kinase in Treg cels maintains homeostasis and lineage stability. Nat Immunol. 2015; 16(2): 188-96. doi: 10.1038/ni.3077.

60. Haxhinasto S, Mathis D, Benoist C. The AKT-mTOR axis regulates de novo differentiation of CD4+Foxp3+ cells. J Exp Med. 2008; 205(3): 565-74. doi: 10.1084/jem.20071477.

61. Wang R, Green DR. Metabolic checkpoints in activated T cells. Nat Immunol. 2012; 13(10): 907-15. doi: 10.1038/ni.2386.

62. Wang R, Dillon CP, Shi LZ, Milasta S, Carter R, Finkelstein D, et al. The transcription factor Myc controls metabolic reprogramming upon T lymphocyte activation. Immunity. 2011; 35(6): 871-82. doi: 10.1016/j.immuni.2011.09.021.

63. Pearce EL, Poffenberger MC, Chang CH, Jones RG. Fueling immunity: insights into metabolism and lymphocyte function. Science. 2013; 342(6155): 1242454. doi: 10.1126/science.1242454.

64. Gubser PM, Bantug GR, Razik L, Fischer M, Dimeloe S, Hoenger G, Durovic B, et al. Rapid effector function of memory CD8+ T cells requires an immediate-early glycolytic switch. Nat Immunol. 2013; 14(10): 1064-72. doi: 10.1038/ni.2687.

65. Pollizzi KN, Powell JD. Integrating canonical and metabolic signalling programmes in the regulation of T cell responses. Nat Rev Immunol. 2014 ;14(7): 435-46. doi: 10.1038/nri3701.

66. Van der Windt GJ, O'Sullivan D, Everts B, Huang SC, Buck MD, Curtis JD, et al. CD8 memory T cells have a bioenergetic advantage that underlies their rapid recall ability. Proc Natl Acad Sci U S A. 2013;110(35): 14336-41. doi: 10.1073/pnas.1221740110.

67. Pearce EL, Walsh MC, Cejas PJ, Harms GM, Shen H, Wang LS, et al. Enhancing CD8 T-cell memory by modulating fatty acid metabolism. Nature. 2009; 460(7251): 103-7. doi: 10.1038/nature08097.

68. Maira SM, Stauffer F, Brueggen J, Furet P, Schnell C, Fritsch C, et al. Identification and characterization of NVP-BEZ235, a new orally available dual phosphatidylinositol 3-kinase/mammalian target of rapamycin inhibitor with potent in vivo antitumor activity. Mol Cancer

Ther. 2008; 7(7): 1851-63. doi: 10.1158/1535-7163.MCT-08-0017.

69. Levings MK, Sangregorio R, Roncarolo MG. Human CD25(+)CD4(+) T regulatory cells suppress naive and memory T cell proliferation and can be expanded in vitro without loss of function. J Exp Med. 2001; 193(11): 1295-302.

70. Earle KE, Tang Q, Zhou X, Liu W, Zhu S, Bonyhadi ML, et al. In vitro expanded human CD4+CD25+ regulatory T cells suppress effector T cell proliferation. Clin Immunol. 2005; 115(1): 3-9.

71. Tang Q, Henriksen KJ, Bi M, Finger EB, Szot G, Ye J, et al. In vitro-expanded antigen-specific regulatory T cells suppress autoimmune diabetes. J Exp Med. 2004; 199(11): 1455-65.

72. Hippen KL, Riley JL, June CH, Blazar BR. Clinical perspectives for regulatory T cells in transplantation tolerance. SeminImmunol. 2011; 23(6): 462-8. doi: 10.1016/j.smim.2011.07.008.

73. Hoffmann P, Boeld TJ, Eder R, Albrecht J, Doser K, Piseshka B, et al. Isolation of CD4+CD25+ regulatory T cells for clinical trials. Biol Blood Marrow Transplant. 2006; 12(3): 267-74.

74. Di Ianni M, Falzetti F, Carotti A, Terenzi A, Castellino F, Bonifacio E, et al. Tregs prevent GVHD and promote immune reconstitution in HLA-haploidentical transplantation. Blood. 2011; 117(14): 3921-3928. doi: 10.1182/blood-2010-10-311894.

75. Trzonkowski P, Bieniaszewska M, Juscinska J, Dobyszuk A, Krzystyniak A, Marek N, et al. First-in-man clinical results of the treatment of patients with graft versus host disease with human ex vivo expanded CD4+CD25+CD127- T regulatory cells. Clin Immunol. 2009; 133(1): 22-26. doi: 10.1016/j.clim.2009.06.001.

76. Brunstein CG, Miller JS, Cao Q, McKenna DH, Hippen KL, Curtsinger J, et al. Infusion of ex vivo expanded T regulatory cells in adults transplanted with umbilical cord blood: safety profile and detection kinetics. Blood. 2011; 117(3): 1061-1070. doi: 10.1182/blood-2010-07-293795.

77. Tang O. T-reg: The Ups and Downs. ESOT+AST Meeting 2014 - Personalized Transplantation. Madrid (Spain).

78. Veerapathran A, Pidala J, Beato F, Yu XZ, Anasetti C. Ex vivo expansion of human Tregs specific for alloantigens presented directly or indirectly. Blood. 2011; 118(20): 5671-5680. doi: 10.1182/blood-2011-02-337097.

79. Jaeckel E, Mpofu N, Saal N, Manns MP. Role of regulatory T cells for the treatment of type 1 diabetes mellitus. Horm Metab Res. 2008; 40(2): 126-136. doi: 10.1055/s-2008-1042427.

80. Putnam AL, BruskoTM, Lee MR, Liu W, Szot GL, Ghosh T, et al. Expansion of Human Regulatory T-Cells From Patients With Type 1 Diabetes. Diabetes. 2009; 58(3): 652-662. doi: 10.2337/db08-1168.

81. Bayer AL, Pugliese A. The IL-2/IL-2R system: from basic science to therapeutic applications to enhance immune regulation. Immunol Res. 2013; 57(1-3): 197-209. doi: 10.1007/s12026-013-8452-5.

82. Long SA, Buckner JH, Greenbaum CJ. IL-2 therapy in type 1 diabetes: "Trials" and tribulations. Clinical Immunology. 2013; 149(3): 324-331. doi: 10.1016/j.clim.2013.02.005.

83. Long SA, Rieck M, Sanda S, Bollyky JB, Samuels PL, Goland R, et al. Rapamycin/IL-2 combination therapy in patients with type 1 diabetes augments Tregs yet transiently impairs beta-cell function. Diabetes. 2012; 61(9): 2340–2348. doi: 10.2337/db12-0049.

84. Tang Q, Bluestone JA, Kang SM. CD4(+)Foxp3(+) regulatory T cell therapy in transplantation. J Mol Cell Biol 2011; 4(1): 11–21. doi: 10.1093/jmcb/mjr047.

85. Demirkiran A, Kok A, Kwekkeboom J, Kusters JG, Metselaar HJ, Tilanus HW, et al. Low circulatory regulatory T-cell levels after acute rejection in liver transplantation. Liver Transpl 2006; 12(2): 277-284.

86. Bestard O, Cruzado JM, Rama I, Torras J, Goma M, Seron D, et al. Presence of Foxp3+ regulatory T cells predicts outcome of subclinical rejection of renal allografts. J Am Soc Nephrol 2008; 19(10): 2020-2026. doi: 10.1681/ASN.2007111174.

87. Yamashita K, Zaitsu M, Nagatsu A, et al. A clinical trial of regulatory T cell-based immunotherapy for tolerance induction in living donor liver transplantation. Vienna: European Society for Organ Transplantation, 2013.

88. Geissler EK. The ONE study compares cell therapy products in organ transplantation: Introduction to a review series on suppressive monocyte-derived cells. Transplant Res 2012; 1:11-13.

5

A Positive Pre-Transplant Endothelial Precursor Cell Crossmatch does not Imply Reduced Long-Term Kidney Graft Function

Markus Gäbel[1], Ayeda AlMahri[2,3], Lennart Rydberg[2], Jan Holgersson[2], and Michael E. Breimer[1]*

[1]Department of Surgery, Institute of Clinical Sciences, Sahlgrenska University Hospital, Bruna stråket 5, S-413 45 Gothenburg, Sweden
[2]Department of Clinical Chemistry and Transfusion Medicine, Institute of Biomedicine,
Sahlgrenska Academy at the University of Gothenburg, Sahlgrenska University Hospital, Gothenburg, Sweden
[3]Sheikh Khalifa Medical City (SKMC), Abu Dhabi, United Arab Emirates

***Corresponding author:** Michael Breimer, Department of Surgery, Institute of Clinical Sciences, Sahlgrenska University Hospital, Bruna stråket 5, S-41345 Gothenburg, Sweden, E-mail: michael.breimer@clinchem.gu.se*

Absract

A flow cytometric crossmatch test detecting antibodies specific for donor Endothelial Precursor Cells (EPC) was evaluated in a multicenter study in 2005-06. A Positive Pre-Transplant EPC Crossmatch (EPCXM) was associated with a higher frequency of early rejections and reduced renal function at three and six months. The long-term follow-up of all patients (n = 53/147) recruited at our center is reported. Patients were retrospectively evaluated regarding rejections, patient/ graft survival and renal function over a four-year follow-up.

As for the whole multicenter study patient population, significantly more early rejections occurred in EPCXM positive compared to EPCXM negative patients (5/7 vs. 5/46, $p = 0.002$). The EPCXM positive group had higher SCr at three (183 vs. 118 μmol/l, $p = 0.01$) and six (172 vs. 124 μmol/l, $p = 0.02$) months compared to the EPCXM negative group, and measured Glomerular Filtration Rate (mGFR) was decreased in the EPCXM positive group at 6 months (50 vs. 29 ml/min, $p = 0.01$). SCr decreased and mGFR increased over time in the EPCXM positive group, while SCr increased slightly and mGFR decreased slightly in the EPCXM negative group eliminating the difference in renal function between the groups.

A positive EPCXM pre-transplantation is associated with higher frequency of early graft rejections, but does not influence long (4 year) term renal function.

Keywords: Antibody-mediated rejection; Anti-endothelial cell antibodies; Crossmatch; Kidney graft function; Non-HLA

Abbreviations

ABS: Antibodies; ACR: Acute Cellular Rejection; AMR: Antibody Mediated Rejection; CAN: Chronic Allograft Nephropathy; CDC: Complement Dependent Cytotoxicity; EPC: Endothelial Precursor Cell; EPCXM: Endothelial Precursor Cell Crossmatch; LXM: Lymphocyte Crossmatch; mGFR: measured Glomerular Filtration Rate; PRA: Panel Reactive Antibodies; SCr: Serum Creatinine; XM: Crossmatch

Introduction

With the introduction of solid-phase techniques for detection and specificity-determination of HLA Antibodies (Abs) [1,2], their importance for both acute and more chronic forms of Antibody-Mediated Rejection (AMR) has been established (reviewed in [3-5]). It is also clear that non-HLA-specific Abs contribute to AMR [6-9]. This notion is reinforced by the fact that patients receiving HLA identical grafts may be lost in AMR [10]. Graft losses due to AMR may also be seen in patients having negative Lymphocyte Crossmatch (LXM) tests and no donor-specific HLA Abs [11,12]. The degree of sensitization has been shown to reduce long-term graft survival in recipients of HLA identical grafts, suggesting that Abs against other targets than HLA may contribute to poor graft survival [13].

Despite the fact that there are several reports on the significance of non-HLA Abs for graft survival, this clinical problem remains poorly defined. In large, this can be explained by lack of suitable assays for detection of this population of Abs, which in addition can be expected to be heterogeneous with regard to the antigens recognized. Thus finding an assay detecting all of the potential specificities may be difficult. Since Endothelial Cells (EC) are likely to be the most prominent target cells for non-HLA Abs causing AMR, many tests used in the past have utilized various cultured EC [12,14,15]. Problems with this strategy include that it is difficult for a clinical routine laboratory to keep cells in culture for EC Crossmatch (XM) testing and that cultured cell lines usually prohibit donor-specific XM testing.

Recently, a novel flow cytometric XM test was evaluated in a multicenter kidney transplantation trial [16]. This XM test utilizes as target cells donor-derived Endothelial Precursor Cells (EPC) defined by expression of the angiopoietin receptor, Tie-2 [16,17]. Patients with a positive EPCXM had a significantly increased frequency of rejections as well as higher Serum Creatinine (SCr) levels at three and six months post-transplantation [16]. This communication reports the four-year follow-up of all patients

***Corresponding author email:** michael.breimer@clinchem.gu.se*

recruited at our center (n = 53) and reveals that long term graft survival and renal function are not significantly different between the anti-EPC positive and negative patient groups beyond six months and during the four-year follow-up.

Materials and Methods

Patients

Fifty-three patients at our institution previously reported in the Multicenter Trial (MCT) [16] were retrospectively reviewed in this study (Table 1). Patients were accepted for transplantation based on negative T- and B-cell Complement-Dependent Cytotoxic XM (CDCXM) tests. Sixty-one patients were originally recruited between November 2005 and October 2006. Eight patients were excluded; two because they were not transplanted, three because there were not enough cells to perform an EPCXM, two patients lost their grafts early from surgical complications and one patient had a negative control outside the specified range in the EPCXM. Twenty-eight patients received kidneys from living donors and 25 received kidneys from deceased donors. Two patients received ABO incompatible grafts from living donors, one had a positive EPCXM and one had a negative EPCXM.

All patients were retrospectively reviewed up to four years post-Tx and no patient was lost to follow up. Relevant clinical data including SCr, measured Glomerular Filtration Rate (mGFR) and rejection episodes were recorded. GFR was measured by the ^{51}Cr -EDTA or Inulin clearance techniques depending on the local hospital practice. Three patients died during the follow-up period; one patient died of septicemia nine months after transplantation and two patients died of cardiac failure 14 months after transplantation. All three patients had negative EPCXM and died with functioning grafts. Three grafts were lost during the follow-up period. One graft in an EPCXM positive patient was lost due to hepatorenal syndrome and two grafts were lost in EPCXM negative patients, one to Chronic Allograft Nephropathy (CAN) and one to recurrence of IgA nephropathy. The MCT was approved by the Stockholm regional human ethics committee (docket no. 2005/222-31/1).

Table 1: Patient Demographics.

	All patients (n = 53)	**EPCXM positive patients (n = 7)**	**EPCXM negative patients (n = 46)**	**MCT (n = 147)**
Age (years)	48 ± 13	51 ± 12	47 ± 13	46 ± 14.5
Male	36	4	32	87
Female	17	3	14	60
Living donors	28	3	25	122
Deceased donors	25	4	21	25
HLA-sensitization				
NS (PRA > 10%)	40(76%)	3(43%)	37(80%)	113(77.5%)
S (PRA 10 - 80%)	8(15%)	2(28.5%)	6(13%)	25(17%)
HS(PR>80%)	5(9%)	2(28.5%)	3(7%)	8(5.5%)
LXM[a]				
T-cell	32(4+)	5(3+)	27(1+)	
B-cell	32(3+)	5(2+)	27(1+)	

[a]Number of B- and T-cell flow cytometric lymphocyte crossmatch-tests performed and number of tests with Positive results in parenthesis.
EPCXM: Endothelial Precursor Cell Crossmatch; LXM: Lymphocyte Crossmatch; NS: Non-Sensitized; S: Sensitized;
HS: Highly Sensitized; MCT: Multicenter Trial

Table 2: Rejections.

	< 3 Months after transplantation		**> 3 Months after transplantation**	
Rejection type[a]	EPCXM positive patients	EPCXM negative patients	EPCXM positive patients	EPCXM negative patients
Antibody mediated	0	0	0	0
Borderline	1	1	0	0
Type IA	0	2	0	2
Type IB	0	0	0	0
Type IIA	4	1	0	1
Type IIB	0	1	0	0
Time to rejection (mean)	6 days	20 days	-	27 months

[a]Biopsy-proven rejections according to the 2003 upgraded Banff 97 classification of renal allograft rejection.
EPCXM: Endothelial Precursor Cell Crossmatch

Immunosuppression

Induction therapy was given in 22/53 patients and included Anti-Thymocyte Globulin (ATG) (n = 7), rituximab (n = 10) and IL-2 receptor antagonists (n = 6). One patient received both rituximab and an IL-2 receptor antagonist. Initial maintenance immunosuppression included tacrolimus and Mycophenolate Mofetil (MMF; n = 26), cyclosporine and MMF (n = 21), rapamycin and MMF (n = 3), tacrolimus and rapamycin (n = 1), tacrolimus, rapamycin and MMF (n = 1) and tacrolimus and azathioprine (n = 1). All patients received steroids.

Often acute rejection episodes occurring < 3 months after transplantation, five episodes were treated with methylprednisolone and five with Anti-Thymocyte Globulin (ATG). One patient who received an ABO-incompatible graft underwent plasmapheresis in addition to ATG. In three acute rejections that occurred late, > 3 months after transplantation, two patients were treated with methylprednisolone and one patient was not treated at all because the kidney was considered too marginal. The two patients that received ABO-incompatible grafts were pretreated with blood group-specific immunoadsorption (GlycoSorb-ABO®, Glycorex Transplantation AB, Lund, Sweden) and anti-CD20 (rituximab) as induction.

HLA typing and antibody analysis

The HLA-A, -B, and -DRβ1 loci of patients and donors were typed by serology or Site-Specific Primer (SSP)-PCR using Olerup SSP® kits (Olerup SSP AB, Salts jöbaden, Sweden) as described by the manufacturer.

The levels of Panel-Reactive HLA class I and II Abs (PRAs) in the pre-transplant sera of patients were determined by Flow Cytometric (FC) analysis using the Flow PRA® test according to the manufacturer's instructions (One Lambda, Inc.). The samples were acquired on a FACScan flow cytometer and analyzed using the Cell Quest Pro software (BD Biosciences, San Jose, CA, USA). In addition, the PRA values were determined by Complement-Dependent Cytotoxicity (CDC) using T- and B-lymphocytes from a panel of 30 donors typed with regard to HLA-A, -B and -DR1.

Crossmatch testing

Before transplantation all patients had a CDC XM performed as previously described [18]. Thirty-two patients also had a T- and B-cell flow cytometric XM performed pre-Tx (Table 1). EPCXM tests were performed on the day of transplantation using Tie-2 (angiopoietin receptor) positive EPC isolated from donor blood with the XM-ONE® kit according to the manufacturer's instructions (AbSorber AB, Stockholm, Sweden) and as previously described [16]. The EPCXM results were not routinely reported to the clinicians responsible for patient care. It could be obtained if asked for, but at our center this did not occur in any case.

Histopathology

Biopsies were performed on clinical indication and processed with standard methods for light microscopy and C4d immunohistochemistry. The 2003 upgraded Banff 97 working classification of renal allograft pathology including AMR was followed because it was the classification used in the MCT [19].

Statistical analysis

Descriptive statistics include mean and standard deviation. A repeated measures ANOVA analysis was performed to evaluate differences between groups over time. Differences in proportions between groups were tested for significance with Fisher's two-sided exact test. A *P- value* of < 0.05 was considered significant. analysis were conducted using the SPSS (IBM Corporation, Somers, NY, USA) and SAS (SAS Institute Inc., Cary, NC, USA) software.

Results

Patient demographics

Demographic data of the 53 patients included in the MCT at our center is given in Table 1 of the recruited patients, all of which had a negative lymphocyte CDC XM, 60% (32/53) were tested in the flow cytometric LXM; 12% (4/32) tested positive and 88% (28/32) tested negative. In the EPCXM positive group, 3 patients also had a positive flow cytometric LXM test. Seven patients (13%) tested positive in the EPCXM test and all had Abs of IgG class. All patients were tested for the presence of HLA-Abs by solid-phase assays and 40/53 (76%) were non-sensitized (PRA < 10%), 8/53 (15%) were sensitized (PRA 10-80%) and 5/53 (9%) were highly sensitized (PRA > 80%). Of the patients with a positive EPCXM test, 3/7 (43%) patients were non-sensitized, 2/7 (29%) sensitized and 2/7 (29%) highly sensitized.

Rejections

The incidence of biopsy-proven acute rejection during the first three months was 71% among patients with a positive EPCXM compared to 11% in those with a negative EPCXM(5/7 vs. 5/46, *P = 0.002*). After the initial three months, three rejections occurred among patients with a negative EPCXM while patients with a positive EPCXM had no late rejections. Chronic Allograft Nephropathy (CAN) was seen in three patients with a negative EPCXM and in no patients with a positive EPCXM. All patients receiving kidneys from deceased donors (n = 25) in the MCT were included at our center. When these patients were analyzed separately the results were the same as for the whole group recruited at our center (data not shown). Types of rejections and mean time to rejection are presented in Table 2.

Kidney graft function

Renal function as assessed by mGFR was significantly higher in the EPCXM negative compared to the EPCXM positive group at 6 months (Figure 1A; 50 vs. 29 ml/min, *P = 0.01*). However, the difference decreased over time and became non-significant one to four years post-Tx. Similarly, SCr levels as a marker for renal function were significantly higher 3 and 6 months post-Tx in the patients with a positive EPCXM when compared to those with a negative EPCXM (Figure 1B; 183 vs. 118 μmol/l, *P = 0.01* and 172 vs. 124 μmol/l, *P = 0.02*). Over time, SCr values in patients with a positive EPCXM decreased slowly, while the SCr increased slightly in those with a negative EPCXM. As a result, the statistical difference between the two groups became non-significant from one to four years post-transplantation (Figure 1B). In conclusion, the statistical difference in renal function between the EPCXM groups found at 3 and 6 months disappeared over time.

Discussion

Like in the whole patient population, EPCXM positive patients recruited at our center had significantly higher SCr levels three and six months after transplantation compared to EPCXM negative patients. The mGFR was lower in the EPCXM positive than in the EPCXM negative group at six months post-Tx. There was no statistically significant difference in SCr or mGFR between EPCXM positive and EPCXM negative patients at one-year post-Tx. In EPCXM negative patients, the SCr seemed to increase slowly over time from 1 to 4 years after transplantation. However, the SCr decreased over the same period in the EPCXM positive group making the difference between the groups statistically non-significant (Figure 1A). Similar observations were made by Jackson and coworkers, who in a study of 60 LD kidney recipients showed that EPCXM positive patients had higher SCr values and incidence of cellular rejection early (mean 50 days) post-Tx compared to EPCXM negative patients – a difference that disappeared late (mean 815 days) post-Tx [20]. A significant finding in that report was that Abs detected in the EPCXM, in contrast to anti-HLA Abs, were enriched for the IgG2 and IgG4 subclasses [20]. Despite the fact that IgG2 and IgG4 are poor complement activators, anti-endothelial cell Abs detected in the EPCXM can cause hyper acute rejection in the absence of complement activation and HLA DSA [21]. It is currently not clear why the presence of Abs against donor EPCs pre-Tx is associated with poor kidney graft function early (< one year), but not late (> one year), post-Tx. One possible explanation could be that the EPCXM positive patients with more early rejections initially received higher doses of CNI than the EPCXM negative patients with fewer early rejections. These higher, possibly nephrotoxic, doses of CNI might have been reduced over time leading to a decrease in SCr.

Of the 53 patients recruited into the MCT at our center and tested with an EPCXM test before kidney transplantation, 7 (13%) tested positive in the EPCXM as compared to 24% in the entire study population [16]. One reason for the lower number of EPCXM positive patients in our cohort may be the absence of patients of Afro-Caribbean origin. In the MCT, 58% of Afro-Caribbeans were EPCXM positive as compared to only 21% of patients of other origins [16]. Immunological and non-immunological factors contribute to the racial disparities observed for renal graft recipients both in terms of time on the waiting list as well as the outcome of the transplantation, with blacks being at a disadvantage compared to whites [22]. Black recipients appear to be stronger immune responders [23] and experience a higher frequency of pre-Tx positive lymphocyte crossmatch tests [24].

As in the MCT, patients recruited at our center with a positive EPCXM test had a higher incidence of acute cellular rejection (ACR) in the first three months. In fact, the incidence of rejection among ECPXM positive patients at our center was 71% compared to 46% in the MCT (P > 0.05) [16]. This may be explained by the

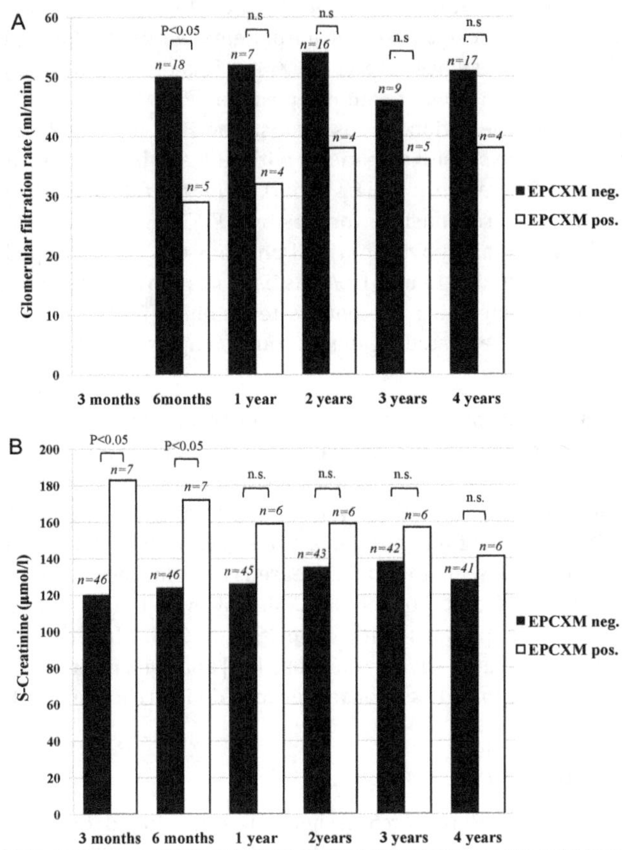

Figure 1: Measured glomerular filtration rate (A) and serum creatinine (B) in patients with positive (empty bars) and negative (filled bars) endothelial precursor cell crossmatch (EPCXM) tests. n.s, not significant

fact that only patients with EPC-reactive antibodies of IgG class were detected in our cohort, while EPC Abs of both IgG and IgM class were found in the entire study population. More sensitized patients were found in our EPCXM positive group compared to the ECPXM positive group of the MCT (57% vs. 29%; *P* > 0.05). No acute rejections were diagnosed more than three months after transplantation in the EPCXM positive patients, while three acute rejections were diagnosed in the EPCXM negative patients. We believe the early rejections, even though not AMR, to be associated with the presence of EPC Abs. Besides a direct effect of EPC antibodies on the ECs of the graft, they may potentiate antigen uptake and presentation, and thereby initiate a T cell-mediated ACR. Chronic Allograft Nephropathy (CAN) was also only seen in the EPCXM negative patient group. This is surprising considering the proposed role of Abs for the development of CAN [4], but a similar observation was made by Jackson and coworkers who showed no statistically significant difference between the IgG EPCXM positive and negative groups with regard to late (> 100 days post-Tx) rejections [20]. Because biopsies were performed on clinical indication and rejections occurred early in the EPCXM positive and late in the EPCXM group, this could explain why we only observed CAN in patients with a negative EPCXM.

The most important limitation of this follow-up study is the relatively low number of patients. However, advantages of the study are that the results of the EPCXM were blinded to the transplant clinicians and that no patients were lost during the follow-up. A weakness of the EPCXM as it is performed today is that also HLA Abs will result in a positive EPCXM test [25]. Thus, unequivocal detection of Abs against non-HLA in the EPCXM is not feasible in sensitized patients with HLA Abs binding to donor EPC [25]. Therefore, it will be important for the future to identify the antigens responsible for positive EPCXM tests such that a solid phase assay with purified antigens can be developed. A number of candidate non-HLA has been described using various approaches including proteomics techniques [26-28]; antigens that should be tested against serum samples positive in the EPCXM test.

In conclusion, a positive pre-transplant EPCXM in kidney transplantation is associated with increased risk of early rejection and decreased renal function three and six months after transplantation. However, the negative effect of a positive EPCXM on renal function seems to disappear after one year and there does not seem to be an increased risk of late acute rejection or CAN based on a positive EPCXM. However, the risk of early rejections in patients with a positive EPCXM should not be neglected, because severe rejections [11] and even graft loss [21] have been reported as a consequence of donor-reactive anti-EPC Abs.

Authorship

MG, JH, MEB: research design, data collection, data analysis, and writing of the manuscript. AAM: data collection and analysis. LR: research design, data collection and analysis. J.H and M.E.B shares joint senior co-authorship.

Funding Sources

This work was supported by the Swedish Research Council (K2011-65X-3031-01-6 to J.H.), the County Council of Västra Götaland (ALF) to JH and MEB, and AbSorber AB. Ayeda AlMahri is fully sponsored by Sheikh Khalifa Medical City (SKMC) in Abu Dhabi, United Arab Emirates.

Conflict of Interest

J.H. is founder of AbSorber AB, the company manufacturing the XM-ONE® test. He is also a shareholder in Allenex AB, the majority owner of AbSorber AB, and his wife receives a royalty from XM-ONE® sales. No other author has any conflict of interest to declare.

References

1. Pei R, Lee JH, Shih NJ, Chen M, Terasaki PI. Single human leukocyte antigen flow cytometry beads for accurate identification of human leukocyte antigen antibody specificities. Transplantation. 2003;75(1):43-49.

2. Zachary AA, Ratner LE, Graziani JA, Lucas DP, Delaney NL, Leffell MS. Characterization of HLA class I specific antibodies by ELISA using solubilized antigen targets: II. Clinical relevance. Hum Immunol. 200;62(3):236-246.

3. Claas FHJ. Clinical relevance of circulating donor-specific HLA antibodies. Curr Opin Organ Transplant. 2010;15(4):462-466. doi: 10.1097/MOT.0b013e32833b9c38.

4. Loupy A, Hill GS, Jordan SC. The impact of donor-specific anti-HLA antibodies on late kidney allograft failure. Nat Rev Nephrol. 2012;8(6):348-357. doi: 10.1038/nrneph.2012.81.

5. Susal C, Opelz G. Impact of HLA matching and HLA antibodies in organ transplantation: a collaborative transplant study view. Methods Mol Biol. 2012;882:267-277. doi: 10.1007/978-1-61779-842-9_15.

6. Dragun D, Müller DN, Bräsen JH, Fritsche L, Nieminen-Kelhä M, Dechend R, et al. Angiotensin II type 1-receptor activating antibodies in renal-allograft rejection. N Engl J Med. 2005;352(6):558-569.

7. Dragun D, Philippe A, Catar R. Role of non-HLA antibodies in organ transplantation. Curr Opin Organ Transplant. 2012;17(4):440-445. doi: 10.1097/MOT.0b013e328355f12b.

8. Sumitran-Holgersson, Suchitra, Holgersson, Jan. Clinical importance of non-HLA antibodies in solid organ transplantation. Current Opinion in Organ Transplantation: August 2006 - Volume 11 - Issue 4 - p 425–432. doi: 10.1097/01.mot.0000236708.13823.d2

9. Zou Y, Stastny P, Susal C, Döhler B, Opelz G. Antibodies against MICA antigens and kidney-transplant rejection. N Engl J Med. 2007;357(13):1293-1300.

10. Kalil J, Guilherme L, Neumann J, Rosales C, Marin M, Saldanha L, et al. Humoral rejection in two HLA identical living related donor kidney transplants. Transplant Proc. 1989;21(1 Pt 1):711-713.

11. Holgersson J, Elsheikh E, Grufman P, Sumitran-Holgersson S, Tydén G. A case of acute vascular rejection caused by endothelial-reactive non-HLA antibodies. Clin Transpl. 2006:535-538.

12. Sumitran-Karuppan S, Tyden G, Reinholt F, Berg U, Moller E. Hyperacute rejections of two consecutive renal allografts and early loss of the third transplant caused by non-HLA antibodies specific for endothelial cells. Transpl Immunol. 1997;5(4):321-327.

13. Opelz G; Collaborative Transplant Study. Non-HLA transplantation immunity revealed by lymphocytotoxic antibodies. Lancet. 2005;365(9470):1570-1576.

14. Glotz D, Lucchiari N, Pegaz-Fiornet B, Suberbielle-Boissel C. Endothelial cells as targets of allograft rejection. Transplantation. 2006;82(1 Suppl):S19-S21.

15. Le Bas-Bernardet S, Hourmant M, Coupel S, Bignon JD, Soulillou JP, Charreau B. Non-HLA-type endothelial cell reactive alloantibodies in pre-transplant sera of kidney recipients trigger apoptosis. Am J Transplant. 2003;3(2):167-177.

16. Breimer ME, Rydberg L, Jackson AM, Lucas DP, Zachary AA, Melancon JK, et al. Multicenter evaluation of a novel endothelial cell crossmatch test in kidney transplantation. Transplantation. 2009;87(4):549-556. doi: 10.1097/TP.0b013e3181949d4e.

17. Vermehren D, Sumitran-Holgersson S. Isolation of precursor endothelial cells from peripheral blood for donor-specific crossmatching before organ transplantation. Transplantation. 2002;74(11):1479-1486.

18. Terasaki PI, McClelland JD. Microdroplet assay of human serum cytotoxins. Nature. 1964;204:998-1000.

19. Racusen LC, Colvin RB, Solez K, Mihatsch MJ, Halloran PF, Campbell PM, et al. Antibody-mediated rejection criteria - an addition to the Banff 97 classification of renal allograft rejection. Am J Transplant. 2003;3(6):708-714.

20. Jackson AM, Lucas DP, Melancon JK, Desai NM. Clinical relevance and IgG subclass determination of non-HLA antibodies identified using endothelial precursors isolated from donor blood. Transplantation. 2011;92(1):54-60. doi: 10.1097/

TP.0b013e31821b60e9.

21. Jackson AM, Kuperman MB, Montgomery RA. Multiple hyperacute rejections in the absence of detectable complement activation in a patient with endothelial cell reactive antibody. Am J Transplant. 2012;12(6):1643-1649. doi: 10.1111/j.1600-6143.2011.03955.x.

22. Malek SK, Keys BJ, Kumar S, Milford E, Tullius SG. Racial and ethnic disparities in kidney transplantation. Transpl Int. 2011;24(5):419-424. doi: 10.1111/j.1432-2277.2010.01205.x.

23. Kerman RH, Kimball PM, Van Buren CT, Lewis RM, Kahan BD. Possible contribution of prestransplant immune responder status to renal allograft survival differences of black versus white recipients. Transplantation. Transplantation. 1991;51(2):338-342.

24. Kerman RH, Kimball PM, Van Buren CT, Lewis RM, Cavazos D, Heydari A, et al. Influence of race on crossmatch outcome and recipient eligibility for transplantation. Transplantation. 1992;53(1):64-67.

25. Alheim M, Johansson SM, Hauzenberger D, Grufman P, Holgersson J. A flow cytometric crossmatch test for simultaneous detection of antibodies against donor lymphocytes and endothelial precursor cells. Tissue Antigens. 2010;75(3):269-277. doi: 10.1111/j.1399-0039.2009.01439.x.

26. Qin Z, Zou Y, Lavingia B, Stastny P. Identification of endothelial cell surface antigens encoded by genes other than HLA. A combined immunoprecipitation and proteomic approach for the identification of antigens recognized by antibodies against endothelial cells in transplant recipients. Hum Immunol. 2013;74(11):1445-1452. doi: 10.1016/j.humimm.2013.05.002.

27. Sigdel TK, Sarwal MM. Moving beyond HLA: a review of nHLA antibodies in organ transplantation. Hum Immunol. 2013;74(11):1486-1490. doi: 10.1016/j.humimm.2013.07.001.

28. Stastny P. Introduction: what we know about antibodies produced by transplant recipients against donor antigens not encoded by HLA genes. Hum Immunol. 2013;74(11):1421-1424. doi: 10.1016/j.humimm.2013.05.004.

6

Changes in Skin Allergy Testing Reactivity observed after a Hurricane: Is the Environment Responsible?

Diego Saporta*

Associates in ENY & Allergy PA, 470 North Ave, Elizabeth NJ 07208, USA

***Corresponding author:** Diego Saporta, Associates in ENY & Allergy PA, 470 North Ave, Elizabeth NJ 07208, USA, Tel: +908-352-6700; E-Mail: saporta.allergy@gmail.com*

Absract

Background: Different patterns of skin reactivity were observed while performing intradermal dilutional allergy tests after the year 2011, the year that a severe storm (Hurricane Irene) affected the author's geographical area.

Methods: In order to study these changes, consecutive patient's allergy charts were arranged in two groups: Group A (Pre-hurricane) contained test results obtained from 2005 to 2010. Group B (Post-hurricane) contained test results obtained in 2012 and 2013.

Results: Statistical analysis showed that there were clear differences in the test results between pre and post hurricane groups.

More tests were positive and at significantly weaker dilutions in Group B. The wheal diameters (and spread) were also larger. Group B had an increased number of patients with lower respiratory symptoms and a significant increase in the number of children who tested positive for allergy.

Conclusions: Allergic sensitivity, as reflected by increased skin reactivity and increased airway symptoms were significantly increased in Post-hurricane patients. The patient population in Group B appears to represent a more sensitized and potentially sicker population, suggesting that the potential role of severe climatic events should be considered as a contributing factor to allergic sensitization.

Introduction

In order to successfully treat allergy, it is important to properly identify to which allergens the patient is reactive to, by using standard allergy tests. The Intradermal Dilutional Test (IDT) [1,2] is routinely used for the diagnosis and management of allergic conditions.

In August 2011, the Northeast of the US was affected by Hurricane Irene. At that time, vast areas of the outdoors environment and a large part of the infrastructure were affected and/or destroyed by wind and water. While this Hurricane was not an isolated climatic event [3,4] it was the most devastating one to affect our area in more than 20 years.

By late 2011, it became evident that patients were presenting with unusual skin reactions during Intradermal Testing (IDT). This consisted of more frequent positive reactions, reactions to weaker allergen concentrations and skin wheals with much larger diameters, indicating a degree of increased sensitivity than previously observed. It was also observed that more patients had Lower Respiratory Symptoms (LRS) and younger patients were seeking allergy evaluation and treatment.

It is recognized that environmental pollution [5,6], including climate changes [7-9], facilitates the development of allergic conditions and asthma. This is one of the few papers to document those changes by comparing pre and post hurricane test results. Environmental exposures have been proposed to lead to a malfunction or sensitization of the immunological system favoring the pro-allergic T- helper type 2 (Th2) allergic pathways [10].

Methods

To evaluate the hypothesis that skin testing reactivity changed after the severe hurricane of August 2011, consecutive allergy charts were gathered and separated, according to testing dates:

a) Tests performed during 2005-2010

b) Tests performed during 2012 and 2013

c) Tests performed during 2011

Since unusual test results began to appear in late 2011 we chose to eliminate results from this year as it could be considered a "transition period" where changes were not yet fully established. Group A served as a control group as all testing in this group was done prior to the hurricane and the reported changes in Group B are compared to the results in Group A.

Data, including date of a test, age, sex, and test results were entered into an Excel spreadsheet. Test results were recorded as diameter in millimeters (mm) of the skin wheals for each dilution. Results were submitted to a statistician for pre and post hurricane test results comparison (See Statistical analysis below). Each chart was assigned a number to assure confidentiality. IRB was obtained from Trinitas Medical Center, 225 Williamson Street, Elizabeth, NJ 07206.

***Corresponding author email:** saporta.allergy@gmail.com*

Intradermal Dilutional Test (IDT)

A brief discussion on the IDT follows to facilitate understanding for those not familiar with the diagnostic allergy techniques taught by the American Academy of Otolaryngic Allergy (AAOA) [1,2]. Successive serial 5-fold dilutions of each allergenic extract are made and conventionally labeled as Dilution 1 (1:5), Dilution 2 (1:25), Dilution 3 (1:125), Dilution 4 (1:625), Dilution 5 (1:3125) and Dilution 6 (1:15,625). IDT consists of injecting 0.01 mL of allergen to produce a 4-mm wheal and measuring the diameter of such wheal 10-15 minutes after injection. Allergen tests are compared to a test control consisting of the same volume of normal saline. The test starts by injecting the weakest dilution (#6) and advancing to more potent dilutions (#2 or#1) until finding the first reactive wheal for the tested allergen. The positive test (reactive wheal) is the first wheal that grows a minimum of 2 mm greater than the control, which is usually 5mm. Therefore, a test growing to 7 mm or more is considered positive. Injection of the next stronger dilution elicits a wheal that will be 2 or more mm larger than the previous reactive wheal, (so it will measure 9 mm or more) therefore showing a continued increased response at the next higher dilution. The first reactive wheal is called the End Point (EP). The wheal resulting from the injection of the next higher dilution is called the Confirmatory Reaction (CR). The EP or first reactive wheal represents the minimal antigen concentration able to elicit a significant skin response. Test injections that yield a response less than 2mm larger than the control are considered to be non-reactive allergens. When the tested allergen is not reactive there will be no response to the injection of even a very large concentration (1:25 dilution or 1:5 dilution of the allergenic extract).

The author personally trains and supervises all nursing personnel and is personally involved in the reading of all test results. Strict testing technique parameters are enforced therefore changes in testing technique over time are unlikely.

Management of the patient with lower respiratory symptoms

It is usually considered safe to begin testing non-asthmatics by injecting the 4th dilution of the allergenic extract. When a patient has symptoms involving the lower airway: cough, dyspnea, chest tightness or wheezing (lower respiratory symptoms or LRS), has used inhalers or nebulizers or has an abnormal spirometry, the test starts at Dilution 6 (1:15,625) otherwise the test starts at Dilution 4 (1:625).Therefore, asthmatics are tested beginning at the 6th dilution. As stated, the objective of the IDT is to determine the lowest dose that will elicit a positive skin reaction, usually without triggering systemic symptoms.

Tested allergens

Allergens were grouped into two panels

Dust and Animal Dander Panel (DD): Dermatophagoides Pteronyssinus (Der p), Dermatophagoides Farinae (Der f), American cockroach, German cockroach, Cat and Dog.

Pollen panel (P): Hickory, Oak, Sycamore, Pigweed, Ragweed, Bahia and Timothy.

Allergens were obtained from the manufacturer as standardized allergens when available (Der p, Der f, Cat, and Timothy), otherwise as weight/volume [11,12].

Patients

Two hundred charts from patients with or without asthma were included in the study. One hundred consecutive charts with tests obtained from patients tested in 2005 to 2010 were identified as Group A (Pre-Hurricane).One hundred consecutive charts with tests from 2012 and 2013 were identified as Group B (Post-Hurricane).

Analyzed parameters

Three parameters were used for this analysis:

1) All positive results (End Points or first reactive wheals) (Table 2).

2) Total Average: Average of the diameter of all reactive wheals (RW) in each panel (RW's included all EP's and CR's, so any wheal that grew to a diameter of 7 mm or larger was included for analysis) (Table 2).

3) Dilution 6: Allergen tests with positive results at Dilution 6 (Table 3)

Analysis consisted of comparing the number of positive reactions, the wheal diameters and diameter variance (See Statistical analysis below).

Statistical analysis

ANOVA test was used to study the difference in the average of wheal diameters between both the groups. The Welch test was used to evaluate the spread or Variance of the wheal diameters in both the groups (In statistics, Variance is defined as the square of the standard deviation: $V = STD^2$). The Chi-Square test was used to determine if the frequency of the results differed between groups. The evaluation was performed for each of the groups (Group A and B) as well as a comparison of one against the other for the parameters above mentioned.

Test examples

Below is an example of a test for DD from Group A on the left and from Group B on the right (figure 1). The allergen dilutions are numbered from 6 (weakest) to 1 (strongest). The numbers in the boxes represent the diameter in millimeters (mm) of each wheal: 5 mm implies a non-reactive wheal, 7 mm or more represents a reactive wheal. By definition when an allergen is reactive (yields a positive test) the diameter of the wheal is expected to increase 2 or more mm between dilutions. It is extremely unusual to have an increase of more than 3-4 mm as the concentration increases. As shown in the example, unusual wheal growth or "hyper-reactive responses" were commonly observed in the Post-Hurricane tests.

The EP's are marked with a circle; the hyper-reactive results are marked in bold. In the above example both dust mites and

Group A (Pre-Hurricane) test example

Dust/Dander	6	5	4	3	2	1
Der p	5		5	7	11	
Der f	5		5	7	11	
Cat	5		5		7	9
Dog	5		5		5	7
A Roach	5		5		5	7
G Roach	5		5		5	7

Group B (Post-Hurricane) test example

Dust/Dander	6	5	4	3	2	1
Der p			5	15	17	
Der f			5	18	25	
Cat			7	9		
Dog			5		5	7
A Roach			5	7	9	
G Roach			5	7	9	

dander panel were reactive and coincidentally the initial level of reactivity (EP) was dilution 3 for both groups. In the Group A example, the wheal growth occurs as expected (5 -7 -11). In the Group B example, the wheal growth is unusual and surpasses the expected pattern of growth: 5 -15-17 and 5-18-25. It was the repeated observation of these hyper reactive results that motivated this analysis.

Results

Both groups, A & B, included one hundred cases (Table 1). There is no statistical difference between the groups for age or gender. There is a statistically significant increase in the number of children and number of patients with asthma or LRS in Group B *(P < 0.01)*.

Although not all the patients were tested for both panels there is no statistical difference in the number of tests run in each group; therefore Groups A & B are similar in this respect (Table 2). The number of positive EPs was analyzed for each panel in each group. Table 2 shows that the percentage of positive tests in Group B was higher for both panels.

The concept of Total Average (TA) was used to compare the degree of reactivity among all patients by considering all reactive wheals. TA is defined as the average diameter of all the reactive wheals (EP and CR) that grew to 7 mm or more. TA was used because it was clinically observed that there was a large disparity in the diameters of the wheals obtained during testing in 2012 and 2013 as compared to the hurricane before. Both the average diameter and the diameter variance (square of the standard deviation) were significantly larger for both the panels in Group B *(P < 0.001)* (Table 2).

When considering tests results only at Dilution 6, it was found that in Group B more tests were required to be done at dilution 6 for both DD and P panels combined because more of the post-hurricane patients presented with LRS. A higher percentage of those tests were reactive *(P < 0.05)* (Table 3). Both findings indicate increased allergic sensitization of the post-hurricane patient population.

Results Summary

Group B had a greater preponderance of children *(P < 0.01)* (Table 1), had more patients with LRS *(P < 0.01)* (Table 1), had more positive test results *(P < 0.001)* (Table 2), both the Total Average and Variance of all RW's was larger *(P < 0.001)* (Table 2), the number of patients required to be tested at Dilution 6 (1:15,625) was larger *(P < 0.05)* (Table 3) and the number of positive reactions at that dilution was significantly larger *(P < 0.05)* (Table 3).

Conclusion

Skin reactivity in patients tested in 2012 and 2013 after the Hurricane was significantly increased as compared to patients tested on or before the Hurricane. This is suggested by the presence of larger (hyperreactive), wider and more frequently positive wheals in the skin test.

The patient population presenting with allergy symptoms in Group B appears to represent a more sensitive and potentially sicker population than the Pre-hurricane cohort. This is suggested by the presence of more positive test results in the overall test, more positive results at the weakest dilution (Dilution 6), and the fact that more children and more patients with LRS were seen in the Post-Hurricane group Group B.

The potential role of severe climatic events should be considered as a factor leading to allergic sensitization.

Discussion

It is not possible to be certain about reasons for the changes here reported, but the potential role of severe climatic events should be considered as a factor leading to allergic sensitization. The increasing prevalence of allergic conditions may be related to the increasing levels of environmental pollution[13]. It is difficult to demonstrate that a specific circumstance in the environment is responsible for a certain change in the level of reactivity when patients are tested. The northeast of the US suffered several storms, two of which are well known for their destructiveness: We first noticed significant changes in our patient population and their hypersensitivity after Hurricane Irene in 2011. Whatever effect this hurricane had, it was amplified by Hurricane Sandy which further brought destruction to our area in 2012.

These storms strongly affected vast areas of the outdoor environment and also affected and/or destroyed a large part of the infrastructure in the eastern part of New Jersey, where the author's office is located. Water damage after hurricanes and floods increases the likelihood of mold contamination in buildings [14] and exposure to fungal contamination can affect the respiratory and other systems [15].

Table 1: Demographic information.

GROUP	Number of patients	Age ± STD (range)	Male	Female	≤ 18	LRS
A (Pre- Hurricane)	100	38 ± 21 (13-89)	46	54	6	28
B (Post-Hurricane)	100	49 ± 18 (5-80)	36	64	30 **	47 **

A: Tests from 2010 and before
B: Tests from 2012 and 2013
Number of patients: 100 in each Group
Age ± STD (range): Mean Age with standard deviation and age range
≤ 18: Child 18 years of age or younger
LRS: Lower respiratory symptoms. Includes patients with diagnosis of asthma, symptoms pertaining the lower airway including exercise induced or abnormal spirometry
**: *P < 0.01*
All values indicate percentages as the sample has 100 individuals.

Table 2: Number of Tests and Patients in each panel, Positive results and Total average.

	Group A DD	Group B DD	Group A Pollen	Group B Pollen
Number of patients	99	100	88	100
Number tests	676	699	609	701#
Positive Result (EP)	448	**540**	310	**419**
% of Total	66.3	**77.3 ****	50.9	**59.9 ****
Total Average				
Number of RW's (Total mm)	800 (6714)	776 (6909)	563 (4839)	601 (5665)
Average (mm)	8.39	**8.90 ****	8.60	**9.43 ****
Variance	3.57	**7.97 ****	5.30	**16.26 ****

Number of patients: Number of patients that were tested in each group
Number tests: Total number of skin tests run for each panel in each group
Group A DD: Dust and Dander panel in Group A (tests from 2010 and before or Pre-Hurricane)
Group B DD: Dust and Dander panel in Group B (tests from 2012 and 2013 or Post-Hurricane)
Group A Pollen: Pollen panel in Group A (tests from 2010 and before or Pre-Hurricane)
Group B Pollen: Pollen panel in Group B (tests from 2012 and 2013 or Post-Hurricane)
Positive Results (EP): Number of skin tests that were positive (reactive)
Total average: Average diameter of all reactive wheals
Number of RW's (Total mm): Total number of wheals that had 7 mm or more (Total cumulative diameters)
Variance: Spread of the diameters (Variance = STD^2)
#: $x^2 = 0.227$, p = N/S (not significant)
**: *P < 0.001*

Table 3: Tests at Dilution 6. Both panels.

DD & Pollen	Tests Dil 6	Pos results (%)
Group A	286	21 (7.3%)
Group B	**590**	**73 (12.4%) ****

DD & Pollen: Dust-Dander and Pollen panels
Group A: Pre-Hurricane (2010 and before)
Group B: Post-Hurricane (2012 & 2013)
Tests Dil 6: All tests performed with the 6th Dilution
Pos results (%): Number of positive results from the total of individual tests (percentage)
**: *P < 0.05*

Environmental pollution [5,6,13], including climate changes [7-9], facilitates the development of allergic conditions and asthma by leading to a malfunction of the immunological system favoring the T- helper type 2 (Th2) allergenic pathway [10]. Hurricane Irene occurred shortly before the time the author started to observe the changes here described. The environmental effects of a hurricane have been well documented. Even though a cause and effect relationship regarding allergy cannot be proven, it is possible that this storm (and Hurricane Sandy which followed 14 months later) have contributed to environmental changes that could explain these findings.

Many other storms have occurred in our area before Irene [3,4]. It is possible that these weather-related events are links in a long chain of environmental events where patients over time become more reactive to allergens to which they are exposed. If environmental pollution is able to affect the immunological system in such a way that people become more susceptible to develop allergic conditions, it is possible that the population of Eastern New Jersey became a more sensitive and reactive population and a sicker population (with more severe allergic disease like asthma or LRS that may suggest bronchial hyper reactivity).

It is very difficult from this type of study to demonstrate whether present day population is sicker than the population evaluated prior to the hurricane exposure. The following findings in our study suggest this could be the case:

When a patient has asthma or LRS, standard intradermal testing begins at the 6th dilution. Our finding that more Group B patients were tested at Dilution 6 (Table 3) attests to the fact that the post-hurricane patient population was clinically diagnosed to have a higher incidence of upper respiratory symptoms and/or asthma. The fact that a higher percentage of IDT were positive in Group B (Tables 2 & 3) also supports the hypothesis that a climatic event altered the sensitivity of the population to a variety of allergens. More children tested in Group B and more patients with LRS in this group (Table 1) suggest that we could be witnessing an earlier onset of a more severe disease in post-hurricane patients.

Furthermore, finding more positive tests with wheals that had larger diameter and wider spread (Table 2) with higher levels of reactivity (Table 3) also suggests that Group B is potentially not only more sensitized but also more reactive than the population from the same patient area as seen before these climatic events.

Published literature supports the possibility that climatic events can affect the environment [7,8,9] and therefore, the health of the population [10,15].

Mold grows in materials that remain wet for more than 48 hours, and flooding, particularly when floodwaters remain for days or weeks, provides an optimal opportunity for mold growth [14]. Increase in asthma incidence has been noted after Hurricane Floyd (that affected North Carolina to New York in 1996) [16]. A strong association between respiratory symptoms and exposure to water-damaged homes has been established after Hurricanes Katrina and Rita (that affected Louisiana in August and September 2005) [17]. Several studies assessed the allergen content of the homes in that area. Some studies were done 1-2 months after the hurricanes [18,19], and some studies were done several months later, while cleanup and remediation efforts had already begun [20-22]. While all these articles agree that wet conditions provided an ideal environment for mold and bacterial growth with subsequent potential health effects, only in the articles that evaluated the affected area 1-2 months after the hurricanes [18,19] was the concentration of indoor mold found to be significantly higher than the mean concentration outdoors.

One study evaluated home allergen concentration as well as skin responses by prick test of the people living in affected homes after Hurricane Katrina [20] but this study had no control data pertaining to the period prior to the hurricane. Those studies, by the nature of their design [18-22] did not compare what was found in the aftermath of those natural disasters to that population's health before the hurricanes.

It is known that dampness and the resultant mold growth that ensures have an adverse health effect on the inhabitants of those buildings. It is logical to assume that the flooding and the subsequent mold growth that followed Hurricanes Irene and Sandy could have been a factor in the findings here reported.

Our data offers a unique opportunity to evaluate changes in a population that live in a stable geographical area, that have attended a single allergy clinic and that has been evaluated by the same practitioner using the same testing techniques, thus increasing the chance that these are significant observations.

Through epidemiological studies it has been shown that the prevalence of asthma has increased from 7.3% in 2001 to 8.4% in 2010 for adults and from 8.7% in 2001 to 9.3% in 2010 for children [23]. While the reasons for this are likely multifactorial, the role of the environment as suggested by this study should be strongly considered. To more definitely answer the question of whether or not a climactic event such as a hurricane can make a population sicker would involve studies involving larger groups of patients belonging to different geographical areas, including not only test results but also symptom-comparison from before and after a major climatic event. Ideally, clinical studies should be conducted in areas that typically are affected by severe storms, not only measuring the amount of allergens in the indoor and outdoor environments but also studying medical records from before and after the storm. While knowing allergen concentration is an important piece of information, not all the individuals will become sensitized in the same way or react in the same way. Studying the symptoms that develop as a consequence of allergic sensitization will concretely give information about this issue.

Other factors that could have contributed to these findings include

a) Antigen lot-to-lot variation: Theoretically a new batch of allergenic extract could be more (or less) reactive than the previous batch. This could theoretically explain a random variation in test results but not a persistent change in test results in one direction only. Present day, allergenic extract manufacturer companies use sophisticated technology for the preparation of allergenic extracts which minimizes the possibility of large batch-to-batch variation.

b) It could be argued that if asthmatic and young children were more reactive, then the presence of this type of patient in Group B would be the reason for the difference in the results between both groups. Given that these patients came from the same geographical patient area, and that the practice has a stable and homogeneous patient base, the sample size is large enough to eliminate the possibility of primarily selecting one type of patient over the other. Rather it is likely that our hypothesis is correct and allergies in the post-hurricane population are developing earlier with such severity to warrant parents and patients to seek help sooner.

The data in this report supports the author's impression that something about the hurricane of 2011 caused a significant change in allergen sensitivity in what had been a stable patient population. Years ago it was rare to see so many allergic children or so many asthmatic patients-child or adult. It is now common at the time of the initial consultation to see pediatric patients who have already been diagnosed with asthma, who are or have used nebulizers or have an exercise- induced asthma and/ or the shortness of breath. This is a distinct change in disease presentation when comparing patients from before and after the hurricanes of 2011 and 2012. It has been shown that after an initial negative answer regarding asthma, if the patient is asked more specifically about the presence of symptoms of lower airway involvement, the presence of nighttime symptoms, exercise triggered symptoms or previous use of inhalers, a much higher percentage of patients will admit to having symptoms suggestive of lower airway hyperreactivity [24].

The information presented here should be considered as an observation, but hopefully this information will provide a stimulus for a large population study with the intention of determining if the changes here described are really happening in other populations exposed to hurricanes. Our findings suggest that in addition to the initial destruction of habitat, hurricanes also somehow (perhaps by stimulating increased mold growth and mycotoxin production) seem to sensitize the affected population for increased reactivity to the many allergens in their environment.

Acknowledgments

Statistical analysis was performed by Yassir M. Samra, Ph.D, Assistant Professor of management, Manhattan College.

References

1. Mabry RL. Skin End Point Titration, AAOA Monograph Series, New York, USA: Thieme Medical Publishers; 1994.
2. Fornadley JA. Skin testing in the diagnosis of inhalant allergy. In: JH Krouse SJ. Chadwick BR. Gordon and MJ. Derebery, Eds. Allergy and Immunology: An Otolaryngic Approach, Philadelphia, Pa, USA, Lippincott Williams & Wilkins, 2002. p. 114-123.
3. List of New Jersey hurricanes. From Wikipedia. In: wikipedia.org/wiki/List_of_New_Jersey_hurricanes. Accessed January 15, 2015.
4. www.weather2000.com/NY_Snowstorms.html. Accessed January 15, 2015.
5. Schildcrout JS, Sheppard L, Lumley T et al. Ambient air pollution and asthma exacerbations in children: an eight-city analysis. Am J Epidemiol. 2006;15;164(6):505-517.
6. Lee SY, Chang YS, Cho SH. Allergic diseases and air pollution. Asia Pacific Allergy 2013;3(3):145-154. doi: 10.5415/apallergy.2013.3.3.145.
7. Pawankar R, Canonica GW, Holgate ST. WAO White Book on Allergy. Milwaukee: World Allergy Organization; 2011. http://www.worldallergy.org/UserFiles/file/WAO-White-Book-on-Allergy_web.pdf.
8. Nasser SM, Pulimood TB. Allergens and thunderstorm asthma. Curr Allergy Asthma Rep. 2009;9(5):384-390.
9. D'Amato G, Baena-Cagnani CE, Cecchi L, Annesi-Maesano I, Nunes C, Ansotegui I, et al. Climate change, air pollution and extreme events leading to increasing prevalence of allergic respiratory diseases. Multidiscip Respir Med. 2013;8(1):12. doi: 10.1186/2049-6958-8-12.
10. Liu J, Ballaney M, Al-alem U, et al. Combined inhaled diesel exhaust particles and allergen exposure alter methylation of T helper genes and IgE production in vivo. Toxicol Sci. 2008;102(1):76-81.
11. Antigen Laboratories, Inc.30 South Main Street. Liberty MO USA (NOTE: Since this data was collected Antigen Laboratories has been acquired by Greer Laboratories, 639 Nuway Circle NE, Lenoir, NC 28645).
12. Alk-Abello, Inc.1700 Royston Lane, Round Rock, TX, USA.
13. Rea WJ. Chemical Sensitivity. CRC Press 1992.
14. Brandt M, Brown C, Burkhart J, Burton N, Cox-Ganser J, Damon S, et al. Mold prevention strategies and possible health effects in the aftermath of hurricanes and major floods. MMWR Recomm Rep. 2006;55(RR-8):1-27.
15. Metts TA. Addressing environmental health implications of mold exposure after major flooding. AAOHN J.2008;56(3):115-120.
16. Morbidity and mortality associated With Hurricane Floyd--North Carolina, September-October 1999. MMWR Morb Mortal Wkly Rep. 2000;49(17):369-372.
17. Cummings KJ, Cox-Ganser J, Riggs MA, Edwards N, Hobbs GR, Kreiss K. Health Effects of exposure to water-damaged New Orleans Homes six months after Hurricanes Katrina and Rita. Am J Public Health. 2008 May;98(5):869-875. doi: 10.2105/AJPH.2007.118398.
18. Solomon GM, Hjelmroos-Koski M, Rotkin-Ellman M, Hammond SK. Airborne Mold and Endotoxin Concentrations in New Orleans, Louisiana, after Flooding, October through November 2005. Environ Health Perspect. 2006; 114(9):1381-1386.
19. Rao CY, Riggs MA, Chew GL, Muilenberg ML, Thorne PS, Van Sickle D, et al. Characterization of Airborne Molds, Endotoxins, and Glucans in Homes in New Orleans after Hurricanes Katrina and Rita. Appl Environ Microbiol. 2007;73(5):1630-1634. doi: 10.1128/AEM.01973-06.
20. Grimsley LF1, Wildfire J, Lichtveld M, Kennedy S, El-Dahr JM, Chulada PC, et al. Few Associations Found between Mold and Other Allergen Concentrations in the Home versus Skin Sensitivity from Children with Asthma after Hurricane Katrina in the Head-Off Environmental Asthma in Louisiana Study. Int J Pediatr. 2012;2012:427358. doi: 10.1155/2012/427358.
21. Mitchell H1, Cohn RD, Wildfire J, Thornton E, Kennedy S, El-Dahr JM, et al. Implementation of Evidence-based Asthma Interventions in Post-Katrina New Orleans: The Head-off Environmental Asthma in Louisiana (HEAL) Study. Environ Health Perspect. 2012;120(11):1607-1612. doi: 10.1289/ehp.1104242.
22. Grimsley LF, Chulada PC, Kennedy S, White L, Wildfire J, Cohn RD, et al. Indoor environmental exposures for children with asthma enrolled in the HEAL study, post-Katrina New Orleans. Environ Health Perspect. 2012;120(11):1600-1606. doi: 10.1289/ehp.1104840.
23. National Surveillance of Asthma: United States, 2001–2010. P3 (Current Asthma Prevalence and Asthma Attack Prevalence).
24. Saporta D. Changes in Peak Flow Value During Immunotherapy Administration. J Environ Public Health. 2012;2012:212867. doi: 10.1155/2012/212867.

7

The PDGFC CUB Domain Enhances Survival in PDGFC Mutant Mice

Brian J. Hayes[1], Kimberly J. Riehle[1,2], Debra G. Gilbertson[3], Wendy R. Curtis[3], Edward J. Kelly[4], Piper M. Treuting[5] and Jean S. Campbell[1*]

[1]*Department of Pathology, University of Washington, Seattle, WA 98195, USA*
[2]*Department of Surgery, University of Washington, Seattle, WA 98195, USA*
[3]*Zymogenetics Inc, a Bristol-Myers Squibb Company, Seattle, WA 98102, USA*
[4]*Department of Pharmaceutics, University of Washington, Seattle, WA 98195, USA*
[5]*Department of Comparative Medicine, University of Washington, Seattle, WA 98195, USA*

***Corresponding author:** Jean S. Campbell, Executive Director of Research and Development, Oncosec Medical, 434 N. 34th St, Suite 100; Seattle, WA 98103, USA, Tel: +(206) 221-5422; E-mail: jcampbell@oncosec.com*

Absract

Platelet-derived growth factor (PDGF) signaling pathways are necessary for normal development. Here we report a homozygous *Pdgf-c* mutant mouse that is viable. In this mouse, alternative splicing gives rise to a truncated transcript containing the entire coding region of the complement components C1r/C1s sea urchin EGF bone morphogenetic protein 1 (CUB) domain of PDGFC but lacking the majority of the Growth Factor Domain (GFD). A mutant protein is translated from the truncated sequence *in vitro*. However, the viability of our homozygous mutant *Pdgf-c* mice depends on the presence of both *Pdgfrα* alleles.

Keywords: PDGFC; Cleavage, Growth factor; Knockout

Introduction

The platelet-derived growth factor (PDGF) signal transduction pathway has important functions in development. Mice lacking PDGF receptor (PDGFR) -α or –β are embryonic lethal [1,2], and deletion of *Pdgf-a, -b, or –c* results in perinatal lethality in homozygous Knockout (KO) animals, though live births are seen in *Pdgf-a* and *Pdgf-c* KO mice [3-5]. *In vitro* studies demonstrate that PDGF-AA, -BB, and -CC induce *Pdgfrα* dimerization, PDGF-BB, and PDGF-DD induce dimerization of *Pdgfrβ*, and -AB, -BB, and -CC induce αβ -receptor dimerization [6]. As predicted by *in vitro* binding studies, *Pdgf-b* KO mice have a phenotype similar to *Pdgf-b* KO mice, with defects in kidney and hematologic development [2,4]. Mice lacking PDGFA have a defect in lung development [3], and interestingly do not phenocopy deletion of *Pdgfrα*, which causes abnormalities in skeletal development and neural crest migration [1,7]. On the other hand, *Pdgf-c* KO animals, referred to as *Pdgfc^{tm1nagy}*, have a defect in palate formation, resulting from abnormal neural crest cell migration and proliferation [5]. Ding, et al. [5], further reported that *Pdgf-a*, *Pdgf-c* double KO mice have a phenotype similar to *Pdgfrα* KO mice, suggesting that PDGF-C is a major activator of *Pdgfrα* signal transduction in the context of development.

PDGF-AA, PDGF-BB, and PDGF-AB are secreted as active dimers [8], while PDGF-CC and PDGF-DD are secreted as inactive homodimers requiring extracellular cleavage of an N-terminal complement components C1r/C1s sea urchin EGF bone morphogenic protein 1 (CUB) domain to allow receptor binding (Figure 1a)[9-12]. All PDGFs share a common growth factor domain (GFD), which is 110 amino acids long and contains 8 conserved cysteines that facilitate intra- and intermolecular disulfide bonds [13]. PDGF signal transduction is mediated through the GFD, and mutations of these cysteines and loss of their disulfide bonds cause a lack of PDGFR signal transduction [14,15]. Furthermore, mutation of sequences in loops I, II, or III of the GFD, the domain that physically interacts with PDGFRs, decreases the GFD's ability to bind PDGFRs [16-18]. Effective signal transduction by *Pdgf-c* thus should require the conserved cysteines and loops I-III of the GFD, as well as cleavage of the CUB domain. We hypothesized that deletion of the majority of the GFD would thus eliminate receptor binding and downstream signal transduction. Contrary to our hypothesis, we describe a homozygous *Pdgf-c* mutant mouse wherein most of the GFD is indeed deleted, but the expression of the CUB domain in these mice is sufficient for viability. Viability, however, depends on genetic background, as has been described for other *Pdgfrα* mutations [19,20].

Methods

Mouse Models

The *Pdgf-c* locus was isolated from a 129S5/SvEvBrd genomic BAC library. The targeting construct replaced a 5 kb genomic region on chromosome 3 with an IRES LacZ MC1 Neo cassette. The homologous arms consisted of a 3 kb 5' fragment and a 2.1 kb 3' fragment. Lex-1 129S5/SvEvBrd ES cells were transfected with the targeting construct and injected into C57BL/6 blastocysts to generate chimeras. Chimeras were bred to albino *C57BL/6J-Tyr^{c-2J}* mice (The Jackson Laboratory) to generate F1 progeny. F1 hybrids were crossed to generate initial Mendelian ratios (Table 1). Mice

***Corresponding author email:** jcampbell@oncosec.com*

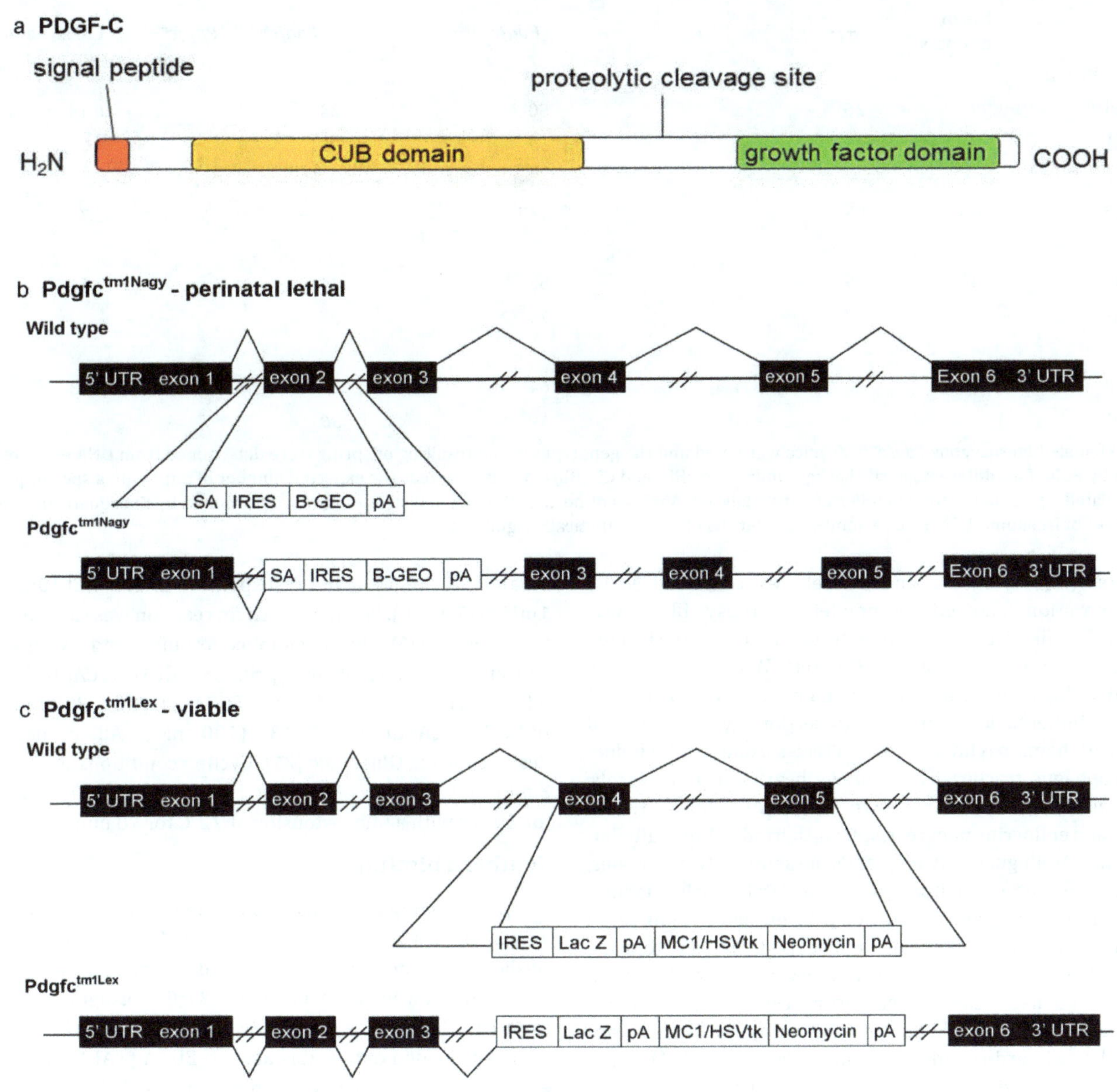

Figure 1: Design of targeting constructs to inactivate PDGF-C. a) A linear diagram of the PDGF-C protein showing the CUB (orange) and GFD (green) above the exons that correspond to these protein domains. b) The targeting strategy for *Pdgfc*^{tm1lex} mice, which removes exons 4 and 5. c) The targeting strategy for the *Pdgf*^{tm1nagy} strain, which removes exon 2 Ding, et al. [5] Abbreviations : splice acceptor (SA), internal ribosomal entry site (IRES), complement components C 1r/C1s sea urchin EGF bone morphogenic protein 1 (CUB), beta galactosidase neomycin fusion protein (β GEO), fusion polyprotein (A) tail (pA), Lac Z gene encoding beta galactosidase (Lac Z), enhancer, promoter, and neomycin phosphotransferase (MC1/ HSVtk and Neomycin).

were further backcrossed to a C57BL/6 background six times to generate the Mendelian ratio for C57BL/6 (Table 1). These mice can be obtained from the Mutant Mouse Regional Resource Centers (http://www.mmrrc.org/index.php) supported by the NIH. Hemizygous *Pdgfrα* mice, *Pdgfrα*$^{tm11(egfp)sor}$, were purchased from The Jackson Laboratory (stock # 007669). Animals were housed at the University of Washington, which is an Association for the Assessment and Accreditation of Laboratory Animal Care approved facility, and all experiments were performed with the University of Washington Institutional Animal Care and Use Committee approval.

Necropsy and Histology

Homozygote *Pdgfc*^{tm1lex} mice (n = 4) and WT C57BL/6

Table 1: Heterozygous *Pdgfc*^{tm1lex} breeding pairs in two backgrounds produce offspring with normal Mendelian distribution.

Cohort	No. of animals	+/+	*Pdgfc*$^{tm1lex/+}$	*Pdgfc*^{tm1lex} / *Pdgfc*^{tm1lex}	Chi sq value
129Sv/EvBrd					
observed	99	26	50	23	0.19
expected		*25*	*50*	*25*	
C57BL/6					
observed	265	55	147	63	3.7
male					
observed	146	29	83	34	3.1
expected		*37*	*73*	*37*	
female					
observed	119	26	64	29	0.83
expected		*30*	*59*	*30*	

Male and female heterozygous *Pdgfc*^{tm1lex}/+ mice were bred and the genotypes of the resulting offspring were determined from DNA extracted from tail snips by PCR. Two different genetic backgrounds, 129SvBrd and C57BL/6, were analyzed. The expected number of pups with a specific genotype was calculated based on the total number of pups analyzed. Analysis of normal Mendelian distribution was determined by Chi square analysis with two degrees of freedom. A Chi square number greater than 5.99 is statistically significant.

littermates (n = 4) aged 5 and 8 weeks were euthanized by CO_2 asphyxiation followed by complete necropsy. Blood was collected for chemistry and complete blood counts performed by Phoenix Central Laboratories (Everett, WA). Tissues were collected for histological analysis with immersion fixation in 10% phosphate-buffered formalin, 4-6 μm sections were made and stained with hematoxylin and eosin. Tissues examined included: lungs, esophagus, trachea, liver, kidneys, heart and great vessels, adrenal glands, gallbladder, brown and white adipose tissue, exocrine and endocrine pancreas, spleen, thyroid, submandibular, parotid and sublingual salivary glands, mesenteric lymph nodes, mesentery, preputial or clitoral glands, skeletal muscle, bladder, male accessory sex glands, testes or ovaries, haired skin, large and small intestine, glandular and non-glandular stomach, uterus and cross section of the head (eyes, middle and internal ears, oral and nasal cavities, cerebrum, cerebellum, olfactory lobes, brain stem, pituitary, tongue and teeth). All tissues were examined by a board-certified Veterinary Pathologist (PMT) and given a morphological diagnosis where applicable. Skeletons were processed by eviscerating, removing the brain and as much muscle tissue and fat as possible, followed by immersion in 95% ethanol for at least 72 hours, placed in acetone overnight, and stained the next night at 37°C with a solution of 70% ethanol, 5% glacial acetic acid, 11.6 μM Alcian blue, and 14.6 μM Alizarin red. The following day the skeletons were washed in 95% ethanol and cleared in 2% KOH for 2 days and transferred to 1% KOH until fully cleared, with the solution refreshed every 2 to 3 days until soft tissues were cleared, at which point the skeletons were transferred to 100% glycerol.

Genotyping

Genomic DNA was extracted from mouse tails by incubation overnight at 55°C in 20 μg/ml Proteinase K (life technologies) in lysis buffer (200 mM NaCl, 1% W/V SDS, 10 mM Tris-HCl, 1 mM EDTA pH 8.0). Phenol chloroform extraction was performed, followed by ethanol precipitation and resuspension in Tris-HCl 10mM EDTA 1mM pH 7.4 (TE). Polymerase chain reaction was carried out with 0.5 U GemTaq (MGQuest), supplied 5x buffer, and dNTPs to a final concentration of 0.2mM with primers 5' CCTGGTCAAGCGCTGTGG 3'(200nM), 5' TCTGGATTCATCGACTGTGG 3'(200nM), 5' ACGGCTAACATGGAGCACG 3' (100 nM). All primers were designed using Oligo Calc [21]. Cycling conditions were 95°C for 3 min, and 35 cycles of 95°C for 20 sec, 60°C for 30 sec, and 72°C for 30 sec with a final extension at 72°C for 10 min.

Southern blotting

The targeting construct was verified using standard methods as described [22]. Briefly, genomic DNA was extracted as described above and purified DNA was restriction digested using EcoR1 HF (New England Biolabs) at 37°C overnight. 10μg of DNA from *Pdgfc*^{tm1lex} homozygous, heterozygous, and wild-Type (WT) mice was loaded onto a Tris-acetate-EDTA (TAE) gel. Following electrophoresis, the gel was denatured in 0.4M NaOH, 1.5 M NaCl for 10min followed by neutralization in Tris-HCl with 1.5M NaCl, pH 7.4. DNA was transferred to Hybond N+ (GE Healthcare) nylon membrane in 10x SSC and cross-linked in a Stratalinker 1800 (Stratagene). Exonic DNA including exon 6 and the 3' UTR was used as a probe, amplified by 5' CCTTTTAGGTCCTTCAGTTGAGACC 3' and 5 ATCTATGCAAACAGGTTGGAGAAATCC 3'. The probe was labeled with 50 μCi [α-^{32}P] dCTP using random decamers (DECAprime II, Ambion) to a specific activity >108 dpm/μg. The blot was prehybridized with Quickhyb (Stratagene) at 68°C for 30 min and then incubated with the denatured probe (106dpm/ml) at 68°C for 2 hr. Labeled membranes were then washed in 2x SSC with 0.1% SDS and 0.2x SSC with 0.1% SDS at 60°C for 30 min each. The sizes of the digested genomic DNA were confirmed by comparison with 32P- dCTP end-labeled λHin III DNA marker (Fermentas).

RT-PCR analysis

Tissues were collected and flash frozen immediately in liquid nitrogen. RNA was extracted with Trizol (Invitrogen) as per the manufacturer's instructions, and cDNA synthesized from 0.5 µgRNA. Primers (F1) 5' CTCACGTGTGCTGCTACGAAGG 3' and (R3) 5' CCCTGCGATTCTCTGCTGCC 3' were used with the following conditions; 95°C for 3 min, and 35 cycles of 95°C for 15 sec, 58°C for 15 sec, and 72°C for 30 sec with a final extension at 72°C for 10min.

Cloning

WT and *Pdgfc*^{tm1lex} cDNA was amplified with primers 5' GCCCTCGCCCCAGTCAGC 3' and 5' CTCACGTGTGCTGCTAC-GAAGG 3' to generate *Pdgf-c* and *Pdgfc*^{tm1lex} sequences. WT cDNA was amplified with primers 5' ATGGTGGTGAATCTAAATCTCCTC 3' and 5' CTCACGTGTGCTGCTACGAAGG 3' to generate the GFD construct. Additionally, a sequence containing a Kozak sequence and the *Pdgf-c* signal peptide for secretion was added to the 5' end of the GFD by Amplification (i.e., 5'GCAGAATTGCCACCAT-GCTCCTCCTCGGCCT-CCTCCTGCTGACATCTGCCCTGGCCGGC-CAAAGAACGGGGACTCGGGCTGAGTCC3').

The *Pdgf-c*, *Pdgfc*^{tm1lex}, and *GFD* sequences were cloned separately into pEF1/*myc*-His B plasmids (Life Technologies), but retained the termination codons which prevent translation of the myc-His tag. The plasmids were linearized with Mlu I (New England Biolabs).

Transfection and production of conditioned media (CM)

Cells [HEK293](ATCC® CRL1573™) were maintained at 37°C 95% humidity with 5% CO_2. Transfections were performed using the calcium phosphate method as described [23], and the cells were subjected to 500 µg/ml G418 (Life Technologies) for two weeks to select for cells expressing the following constructs: empty vector (pEF1empty), *pEF1Pdgfc*^{GFD}, *pEF1Pdgfc*^{mut}, or *pEF1Pdgfc*^{FL} to produce untagged proteins, *Pdgf-c*^{GFD}, *Pdgf-c*^{mut} and *Pdgf-c*^{FL} respectively. Confluent cultures were allowed to condition media for 24 hours, at which point media were collected and frozen at -80°C. Baby Hamster Kidney (BHK) cells, which do not express *Pdgfrα* (BHK 570), and BHK cells transfected with *Pdgfrα* (BHK Rα) were a kind gift from Dr. Daniel Bowen-Pope [24].

Immunoblot analysis

SDS-PAGE and immunoblotting were carried out as described [25]. For PDGF-C detection membranes were incubated with goat anti-mouse PDGF-C antibody (1:1000, AF1447 R&D Systems) followed by incubation with an anti-goat Horseradish Peroxidase (HRP) conjugated secondary antibody and detection performed with ECL (Pierce). For ERK and p-ERK detection, membranes were first incubated with anti-phospho p44/42 MAPK (1:1000, 9101 Cell Signaling Technology) followed by an anti-rabbit conjugated secondary antibody. Primary antibody; secondary antibody complexes were detected with ECL Plus (GE Healthcare) and imaged with a Storm 840 phosphor imager (GE Healthcare). Immunoblots were then stripped with 2% SDS, 62.5 mM Tris, pH 6.8 with 7 µl beta-mercaptoethanol per ml. Blots were then reprobed with anti-ERK1/2 pAb 7884 followed by secondary and ECL as above. Densitometry analyzes were performed using Image J software (NIH) and is presented as the relative value of the phosphorylated protein normalized to the total levels of the same protein after re-probing.

Results

Homologous recombination targeting strategy

We examined a *Pdgf-c* mutant mouse, *Pdgfc*^{tm1lex}, in which exons 4 and 5 of the GFD are deleted using the targeting strategy depicted in Figure 1b. Surprisingly, a viable mouse resulted from this approach; in contrast to the *Pdgf-c* KO mice, which results in perinatal lethality (*Pdgfc*^{tm1nagy}, Figure 1c). Figure 1 shows the protein domain structure of PDGF-C and compares the different targeting approaches used to generate the two mutant strains. *Pdgfc*^{tm1lex} mice were derived from a 129S5/SvEvBrd embryonic stem cell line (Lex1) targeted by homologous recombination to replace exons 4 and 5. This strategy should eliminate transcription of the entire GFD (Figure 1b). *Pdgfc*^{tm1nagy} mice were derived from R1 embryonic stem cells targeted by homologous recombination to replace exon 2 of PDGFC with an SA-IRES-β-geo-pA cassette [5], preventing transcription downstream of β -geo, including the CUB and GFD.

Verification of recombination

To confirm that our selection cassette was targeted correctly, we performed Southern blotting on genomic DNA isolated from homozygous *Pdgfc*^{tm1lex}, heterozygous *Pdgfc*^{tm1lex}, and WT littermates. Digestion of genomic DNA with EcoRI should produce a 12.4 kb band including exons 4, 5, 6, and the 3' untranslated region (UTR) from WT DNA, and a 6.5 kb band including exon 6 and the 3' UTR in correctly targeted *Pdgfc*^{tm1lex} mice (Figure 2a). A ^{32}P labeled DNA probe with a complementary sequence to exon 6 and the 3' UTR of *Pdgf-c* was used to detect these 12.4 and 6.5 kb bands by hybridization and bands of the expected size were detected for homozygous *Pdgfc*^{tm1lex}, heterozygous *Pdgfc*^{tm1lex}, and WT mice (Figure 2b).

Pdgfc^{tm1lex} transcription and translation

3.4 Given our Southern blotting results, we next verified production of the expected *Pdgfc*^{tm1lex} transcript, which should result from transcription of exons 1, 2, and 3 of *Pdgf-c* (Figure 1c). Attempts to amplify a transcript containing exon 3 and the internal ribosomal entry site (IRES) (Figure 3a, primers F1 and R1) or *lacZ* (Figure 3a, F1 and R2) did not produce a product (data not shown), suggesting the targeting construct may have facilitated skipping the KO cassette. RT-PCR of *Pdgfc*^{tm1lex} cDNA using primers F1 and R3 (Figure 3a) generated a 296 bp band, consistent with a transcript in which exons 1, 2, 3 and 6 remain (Figure 3b). Sequencing this amplicon revealed that indeed the region corresponding to WT exons 4 and 5 was removed, but exon 6 was spliced in frame to the 3' end of exon 3 to create a truncated transcript (Figure 4). To detect the full-length coding region of

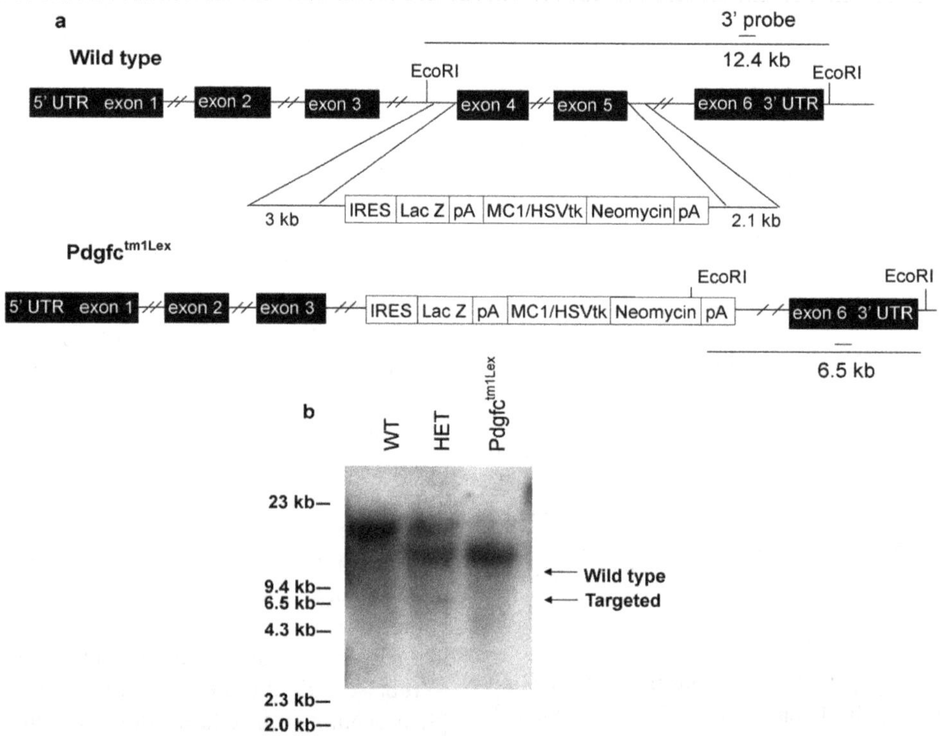

Figure 2: Recombination at the *Pdgf-c* locus in the *Pdgfc*^{tm1lex} mouse. a) The anticipated results from EcoR1 digestion of genomic DNA from a wild type mouse and a *Pdgfc*^{tm1lex} mouse. d) Southern blot of genomic DNA digested with EcoRI from wild type, heterozygous, and homozygous *Pdgfc*^{tm1lex} mice.

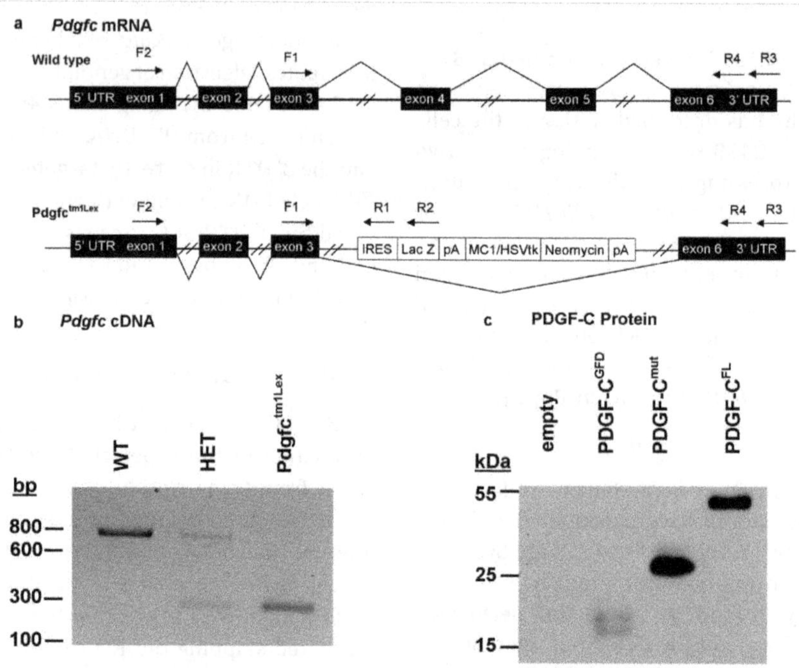

Figure 3: An alternative mRNA is transcribed from the *Pdgf-c* locus in *Pdgfc*^{tm1lex} mice, and protein is expressed from this sequence. a) Diagram of the splicing of wild type and mutant PDGFC with various PCR primers (F1, F2, R1, R2, R3, and R4) used to detect transcripts. b) Amplicons generated by RT-PCR (F1 and R3) of RNA from kidneys of wild type, heterozygous, and homozygous *Pdgfc*^{tm1lex} mice are shown, producing products of 722 bp for *Pdgf-c* and 296 bp for *Pdgfc*^{tm1lex}. c) Immunoblot of conditioned media from transfected HEK293 cells expressing the coding sequences for an empty vector (empty), the *GFD* (*Pdgf-c*^{GFD}), *Pdgfc*^{tm1lex} (*Pdgf-c*^{mut}), or full-length *Pdgf-c* (*Pdgf-c*^{FL}). The size of DNA ladder (b) and protein standards (c) are shown to the left of each image.

PDGFC, we designed primers to anneal to the 5' and 3' UTRs, and found that the *Pdgfc*^{tm1lex} transcript retains the endogenous 5' and 3' UTRs (Figure 3a, F2 and R4) and associated regulatory elements (data not shown).

The *Pdgfc*^{tm1lex} mutant transcript is expected to have an expression pattern similar to WT *Pdgf-c*, given that the selection cassette did not alter the *Pdgf-c* 5' and 3' regulatory elements. The translated protein product of *Pdgfc*^{tm1lex} is expected to contain the C-terminal 38 amino acids of the GFD and the entire CUB domain, resulting in a predicted molecular weight of 22.5 kDa. We found that HEK293 cells transfected with plasmids expressing the coding regions for full length *Pdgf-c* (*Pdgf-c*^{fl}), *Pdgfc*^{tm1lex} (*Pdgf-c*^{mut}), or *Pdgf-c* GFD (*Pdgf-c*^{GFD}) secrete proteins of the predicted molecular weights into culture media, as determined by immunoblot analysis (Figure 3c). Thus *Pdgf-c*^{FL}, *Pdgf-c*^{mut} and *Pdgf-c*^{GFD}, are all stable protein products.

Signal transduction in response to conditioned medium

To demonstrate that the CUB domain of *Pdgf-c* could an activates *Pdgfrα*, we assessed ERK phosphorylation in BHK cells treated with various Conditioned Media (CM). Media were conditioned by HEK293 cells transfected with plasmids expressing *Pdgf-c* CUB, *Pdgf-c* GFD, or an empty vector control. When CM was added to BHK 570 cells, which have no *Pdgfrα*, no phosphorylation of ERK was detected (data not shown). When conditioned medium α was cells added to BHK R α , which have been transfected with *Pdgfrα* , phosphorylation of ERK was detected in a time-dependent manner, with maximal phosphorylation attained after five minutes of exposure (Figure 5a). Media conditioned by HEK293 cells transfected with an empty vector control interestingly did show ERK phosphorylation in BHK R α cells but not in BHK 570 cells, albeit to a lesser extent than the PDGF-C constructs, suggesting that

```
                       exon 3 →                 exon 4 →
Pdgfc           473 TCCACTACAGTATTATCATGCCACAAGTCACAGAAACCAC 512
                    |||||||||||||||||||||||
Pdgfc^tm1Lex    473 TCCACTACAGTATTATCATGCCA----------------- 495
Pdgfc           513 GAGTCCTTCGGTGTTGCCCCCTTCATCTTTGTCATTGGAC 552
Pdgfc^tm1Lex    495 ---------------------------------------- 495
Pdgfc           553 CTGCTCAACAATGCTGTGACTGCCTTCAGTACCTTGGAAG 592
Pdgfc^tm1Lex    495 ---------------------------------------- 495
Pdgfc           593 AGCTGATTCGGTACCTAGAGCCAGATCGATGGCAGGTGGA 632
Pdgfc^tm1Lex    495 ---------------------------------------- 495
Pdgfc           633 CTTGGACAGCCTCTACAAGCCAACATGGCAGCTTTTGGGC 672
Pdgfc^tm1Lex    495 ---------------------------------------- 495
                                                exon 5 →
Pdgfc           673 AAGGCTTTCCTGTATGGAAAAAAAGCAAAGTGGTGAATC 712
Pdgfc^tm1Lex    495 ---------------------------------------- 495
Pdgfc           713 TGAATCTCCTCAAGGAAGAGGTAAAACTCTACAGCTGCAC 752
Pdgfc^tm1Lex    495 ---------------------------------------- 495
Pdgfc           753 ACCCCGGAACTTCTCAGTGTCCATACGGGAAGAGCTAAAG 792
Pdgfc^tm1Lex    495 ---------------------------------------- 495
Pdgfc           793 AGGACAGATACCATATTCTGGCCAGGTTGTCTCCTGGTCA 832
Pdgfc^tm1Lex    495 ---------------------------------------- 495
Pdgfc           833 AGCGCTGTGGAGGAAATTGTGCCTGTTGTCTCCATAATTG 872
Pdgfc^tm1Lex    495 ---------------------------------------- 495
Pdgfc           873 CAATGAATGTCAGTGTGTCCCACGTAAAGTTACAAAAAAG 912
Pdgfc^tm1Lex    495 ---------------------------------------- 495
                             exon 6 →
Pdgfc           913 TACCATGAGGTCCTTCAGTTGAGACCAAAAACTGGAGTCA 952
                             |||||||||||||||||||||||||||||||
Pdgfc^tm1Lex    495 ---------GTCCTTCAGTTGAGACCAAAAACTGGAGTCA 512
```

Figure 4: Sequence of cDNA from the 296 bp *Pdgfc*^{tm1lex} band shown in 3b compared to the consensus mouse sequence for PDGFC, CCDS17425.1. Exon junctions are indicated above the consensus sequence.

transfection of HEK293 cells induces the release of a molecule that can signal through *Pdgfra* When BHK Rα cells are exposed to increasing ratio of CM to fresh medium we see an increase in intensity of ERK phosphorylation just as increasing the amount of recombinant *Pdgf-c* increases ERK phosphorylation (Figure 5b). CM from *Pdgf-c* GFD and *Pdgf-c* CUB domain transfected HEK 293 cells maintains some ability to induce ERK phosphorylation even at low CM ratio to fresh medium.

Anatomic characterization

Originally derived on a 129S5/SvEvBrd background, *Pdgfc*^{tm1lex} mice were backcrossed to a C57BL/6 background. Homozygous F1 C57BL/6 x 129S5/SvEvBrd *Pdgfc*^{tm1lex} progeny had a hunched appearance, and gross skeletal analyzes demonstrated spina bifida (Figure 5a and 5b), which was confirmed by histology (data not shown). All other organs were grossly normal. After six backcrosses, no overt anatomical differences were noted between homozygous *Pdgfc*^{tm1lex} mice and WT C57BL/6 littermates, i.e. the spina bifida phenotype was rescued. Comprehensive histological analyses (see methods) conducted on 5 and 8-week old *Pdgfc*^{tm1lex} homozygous mice were unremarkable or showed incidental lesions, such as dermatitis or mild multifocal extramedullary hematopoiesis in the liver, which are known to be associated with the C57BL/6 background [26]. Brains of *Pdgfc*^{tm1lex} mice were examined histologically and did not differ from those of WT littermates; clinical chemistry and blood counts were also unremarkable (data not shown).

Genotype ratios

We observed normal Mendelian ratios in the offspring of heterozygote *Pdgfc*^{tm1lex} breeding pairs on two genetic backgrounds (Table 1). We did not observe premature death in *Pdgfc*^{tm1lex} mice up to 12 months of age and did not see sex-dependent differences in viability on the C57BL/6 background, as has been reported for *Pdgfc*^{tm1nagy} mice [20]. Similar to Fredriksson L, et al. [6] we observed a statistically significant decrease in body mass in C57BL/6 *Pdgfc*^{tm1lex} mice at 3 and 7 months (a 17 to 24% decrease, data not shown). To determine whether *Pdgfra* necessary for normal development in *Pdgfc*^{tm1lex} mice, we intercrossed hemizygous *Pdgfra* mice with heterozygous *Pdgfc*^{tm1lex} mice. Progeny resulting from this cross did not include mice that were homozygous *Pdgfc*^{tm1lex}; hemizygous *Pdgfra* (Table 2). Neonatal mice homozygous for *Pdgfc*^{tm1lex} and hemizygous for *Pdgfra* were born but failed to thrive and 100% died prior to the age of weaning. This result is surprising, as hemizygous *Pdgfra* mice are phenotypically normal [1,27]; we thus conclude that

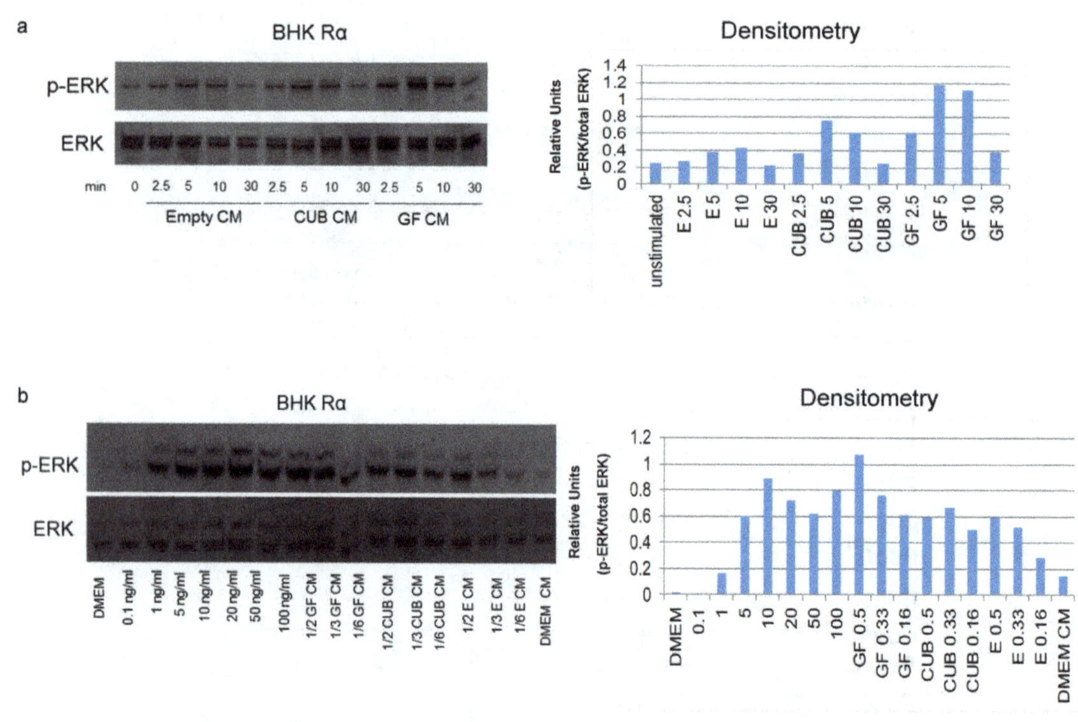

Figure 5: PDGF-C CUB domain stimulates modest ERK phosphorylation in BHK cells expressing PDGFRα. HEK293 cells were transfected with an empty vector (empty), the CUB domain of PDGF-C (CUB), or the growth factor domain of PDGF-C (GF) and the conditioned media (CM) was transferred to BHK cell lines expressing PDGFRα.

a) Time-dependent ERK phosphorylation by PDGF-C CUB, PDGF-C GF, or empty vector (E). Immunoblot phospho-ERK1/2 in BHK Rα cells transfected with *Pdgfra* after addition of CM. corresponding densitometry measurements are shown to the right.

b) Varying amounts of GF CM, CUB CM, and empty vector CM with non-CM stimulated ERK phosphorylation while DMEM (left lane) and DMEM conditioned from non-transfected HEK293 cells weakly phosphorylated ERK. Corresponding densitometry measurements are shown to the right. Note: no phosphorylation of ERK was seen in the parental BHK cells under similar conditions (data not shown).

Table 2: Heterozygous *pdgfc*^{tm1lex} breeding pairs with one copy of *pdgfrα* do not produce *pdgfc*^{tm1lex} homozygous and hemizygous *pdgfrα* offspring.

Cohort	No. of animals	*Pdgfc+/+ Pdgfrα +/-*	*Pdgfc+/- Pdgfrα +/-*	*Pdgfc-/- Pdgfrα +/-*	*Pdgfc+/+ Pdgfrα +/+*	*Pdgfc+/- Pdgfrα +/+*	*Pdgfc-/- Pdgfrα +/+*	Chi sq value
observed	127	14	27	0	20	50	16	30
expected		16	32	16	16	32	16	

Male and female heterozygous *Pdgfr*^{tm1lex} /+ mice, with a single copy of *Pdgfrα*, were bred, and the genotypes of the resulting offspring were determined from DNA extracted from tail snips by PCR. The expected number of pups with a specific genotype was calculated based on the total number of pups analyzed. Analysis of normal Mendelian distribution was determined by Chi squared analysis with five degrees of freedom. A Chi square number greater than 11.07 is statistically significant. *Pdgfc-/-* represents *Pdgfc*^{tm1lex} /^{tm1lex} mice.

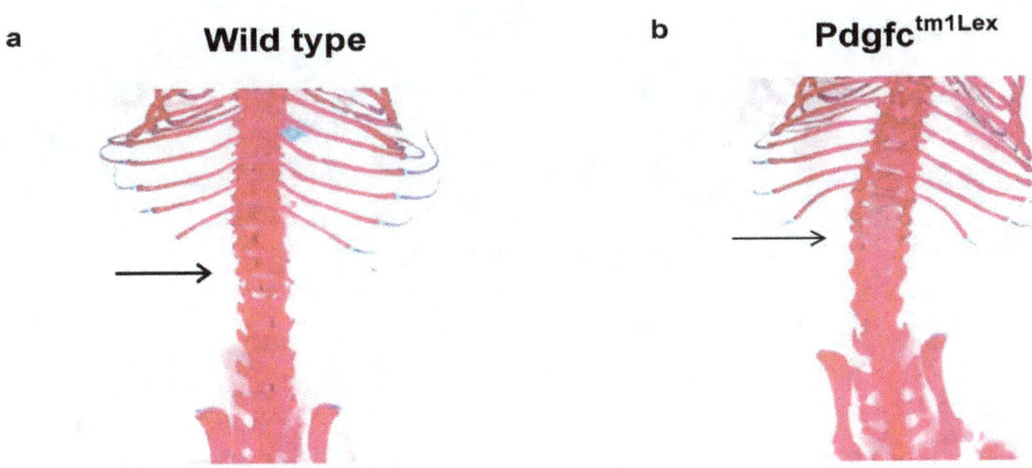

Figure 6: Anatomic characterization of *Pdgfc*^{tm1lex} mice.
a) 129Sv/EvBrd x C57BL/6 F1 wild type mice have normal thoracic and lumbar vertebrae.
b) Spina bifida in the thoracic and lumbar region (arrow) of *Pdgfc*^{tm1lex} homozygous mice.

Pdgfc^{tm1lex} viability requires two alleles of *Pdgfrα*. Histological and gross anatomical analysis of these neonatal animals revealed severe spina bifida encompassing a region from vertebrae T11 to L1 (Figure 6 a-c). Microscopic analysis revealed demyelinated spinal cords with excessive blood infiltration. Like *Pdgfc*^{tm1nagy} mice, homozygous *Pdgfc*^{tm1lex}; hemizygous *Pdgfrα* mice had palatoschisis. These genetic data indicate that PDGF-C^{mut} interacts with *Pdgfrα* in some novel way, as *Pdgfc*^{tm1nagy} mice have reduced viability, while *Pdgfc*^{tm1lex} mice have lethality only in conjunction with hemizygous *Pdgfrα*.

Discussion

PDGF signaling is essential for normal development. In this study we analyzed a new PDGF-C mutant mouse strain and demonstrate that the homologous recombination strategy in this mouse results in an alternatively spliced mutant form of *Pdgf-c* that lacks over 60% of the RNA encoding the GFD, but expresses the entire portion of RNA encoding the CUB domain. This truncated transcript likely results from pre-mRNA splicing, which is a complicated event relying on proper sequence of both the 3' and 5' ends of exons, as well as the proper sequence of the exons themselves [28]. The targeting construct in *Pdgfc*^{tm1lex} used the endogenous 3' splice acceptor sequence of exon 4, but the remainder of exon 4 was replaced with viral and bacterial sequences (IRES and Lac Z, respectively), which are incompatible with eukaryotic splicing machinery. The removal of the 3' end of exon 4 makes splicing at this particular sequence less likely, due to the inability of U1 to enhance splice site recognition [29]. The 14 bases remaining of the endogenous exon 4 sequence are likely insufficient to promote splicing, and thus lead to exon skipping in *Pdgfc*^{tm1lex} mice.

The presumed *Pdgf-c* protein that is translated from *Pdgfc*^{tm1lex} should not have any GFD function, as paralogs of PDGF-C, PDGF-A and PDGF-B, require cysteines for dimerization and intramolecular and intermolecular folding [14,15]. We hypothesized that the *Pdgfc*^{tm1lex} protein product *(Pdgf-c*^{mut}*)*, lacking 6 of the 8 conserved cysteines, would have compromised or absent GFD signal transduction activity. *Pdgf-c*^{mut} is also missing the paralogous loops, I and II of the GFD, which are necessary for receptor binding by the PDGF-C paralog, *Pdgf-b* [16,18]. Loop I has also been shown to be critical for receptor activation by the PDGF paralogs Placental Growth Factor (PlGF) and Vascular Endothelial Growth Factor B (VEGFB), as well as the viral protein VEGFE$_{NZ-7}$ [30,31]. Loop III remains in PDGF-C^{mut}, but is not expected to bind *Pdgfrα* in the absence of loops I, II and the cysteine required to form loop III. Additionally, PDGF-

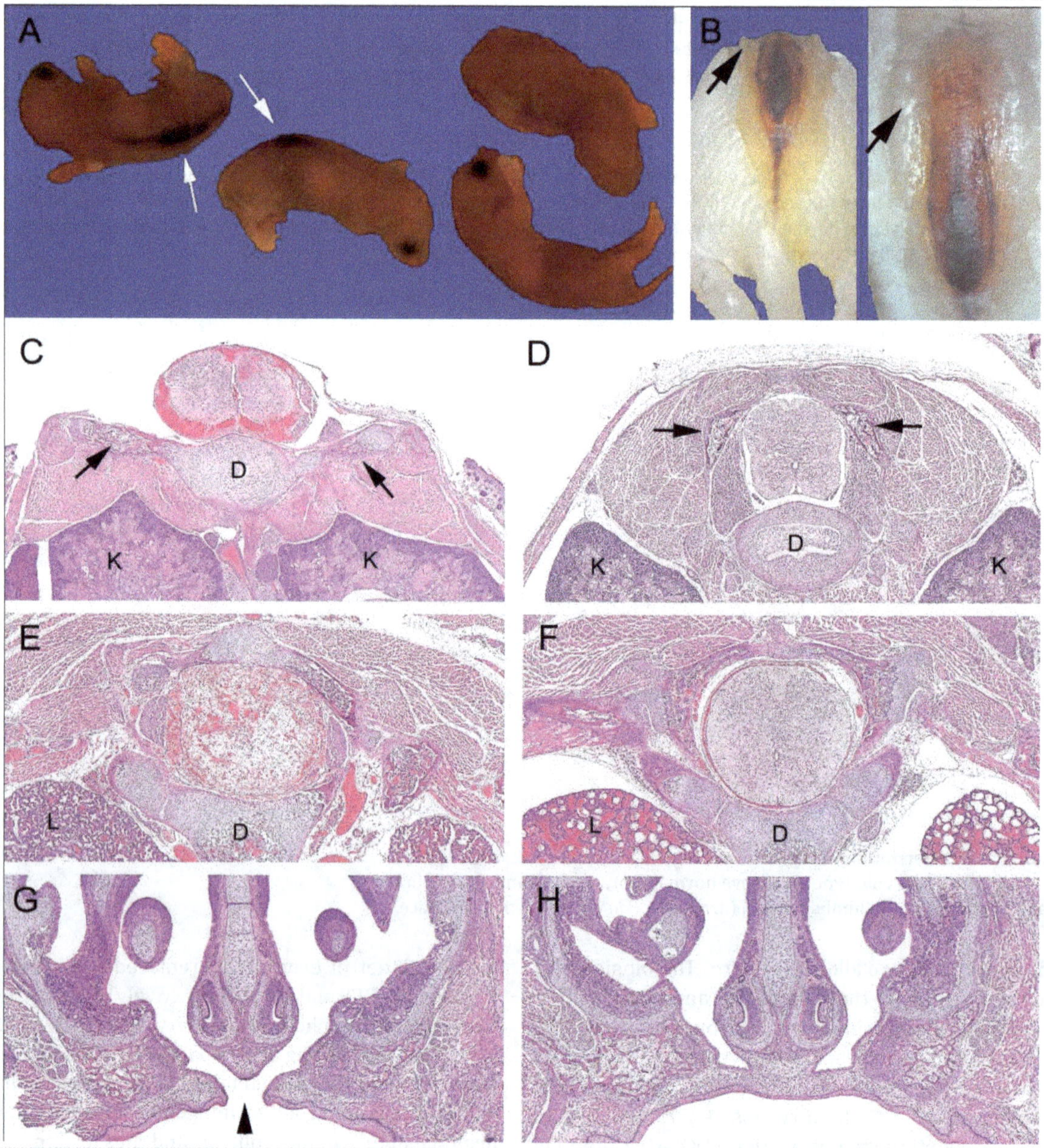

Figure 7: Developmental divergence between wild-type and homozygous *Pdgfc*^{tm1lex}; hemizygous *Pdgfrα* mice.

a) Live P0 pups. Homozygous *Pdgfc*^{tm1lex}, hemizygous *Pdgfrα* mice have dark brown linear protrusions at the level of lumbar spine (white arrows).

b) Formalin-fixed homozygous *Pdgfc*^{tm1lex}; hemizygous *Pdgfrα* P0 pup with dorsal skin dissected away. Note the protruding and incompletely closed vertebral lamina (arrows) and dark meningomyelocele.

c) And d) Histologic sections through the level of the lumbar spine with kidneys (K) and vertebral body or disc (D) indicated for orientation.

c) Homozygous *Pdgfc*^{tm1lex}; hemizygous PDGFRα pup represented in b) the spinal cord and meninges have extensive hemorrhages. The arrow indicates misaligned lamina and there are no dorsal vertebral processes, epaxial skeletal muscles, or abundant subcutaneous tissue covering the spinal cord (contrast to d).

d) Normal thoracic anatomy with proper orientation of vertebral lamina (arrows) and complete enclosure of the spine. Note the skin was removed during dissection.

e) and f) Histological sections of the thoracic spine with lungs (L) indicated for orientation. e) Homozygous *Pdgfc*^{tm1lex}; hemizygous PDGFRα pups have severe ascending hemorrhagic myelomalacia.

f) Corresponding cross section of a wild type mouse. Note that the pulmonary hemorrhage and extravasated red blood cells in the meninges are secondary to euthanasia.

g) and h) Cross section of the sinuses and hard palate at the level of the vomeronasal organ (T1).

g) Palatoschisis (cleft palate) is present in the Homozygous *Pdgfc*^{tm1lex}; hemizygous PDGFRα mouse (arrowhead).

h) Wild-type mouse.

C^{mut} lacks arginine 231, which has specifically been shown to be necessary for cleavage of the CUB domain, and this cleavage has been suggested to be critical for *Pdgfrα* activation. [9,10]. All of these structural differences suggest that PDGF-C^{mut} would not interact with *Pdgfrα* in the same manner as does PDGF-C. Our data suggest CUB-mediated activation of *Pdgfrα* and raise the possibility that the CUB domain interacts with *Pdgfrα* outside of the ligand binding pocket or perhaps interacts with the receptor in conjunction with other proteins to influence cell signaling.

CUB domains of proteins other than PDGF-C are involved in a diverse range of functions, including complement activation, developmental patterning, tissue repair, axon guidance and angiogenesis, fertilization, hemostasis, inflammation, neurotransmission, receptor-mediated endocytosis, and tumor suppression [32-34]. CUB domains have even been reported to be involved in cell signaling [35-37]. The human version of the PDGF-C CUB domain stimulates proliferation of human coronary artery smooth muscle cells [38]. Our results show that the PDGF-C CUB domain simulates modest ERK phosphorylation in BHK cells that express *Pdgfrα*, but not in the parental BHK 570 cells, which do not express the receptor. These results suggest that the CUB domain interacts with *Pdgfrα* to transduce intracellular signaling events. This notion is supported by the observations that pups who are homozygous *Pdgfc*^{tm1lex}; hemizygous *Pdgfrα* (Table 2) were not viable. The precise mechanisms involved with CUB domain-receptor interactions and subsequent signal transduction events are unknown. Relative to the PDGF-C GFD, however, the CUB domain seems to be less potent in activating the receptor. *In vivo,* it is possible that in the absence of the GFD, the CUB domain transduces sufficient *Pdgfrα* activity to prevent perinatal lethality in homozygous *Pdgfc*^{tm1lex} mice.

Genetic background appears to have a profound effect on the PDGF signaling pathway in mice. For example, the Patch mutant mouse (*Ph*), in which a segment of chromosome 5 including *Pdgfrα* is deleted, has a phenotype that varies with background, as C57BL/6 homozygous *Ph* mice die earlier in development than do *Ph* mice on a CBA, or BALB/C background [1]. Likewise, the severity of abnormalities in targeted *Pdgfrα* null mice varies depending on background, with only DBA mice surviving until birth [1,39]. Though our observations may be due in part to strain differences, *Pdgfc*^{tm1lex} and *Pdgfc*^{tm1nagy} mice appear to have different viabilities and phenotypes. *Pdgfcc*^{tm1nagy} mice have developmental abnormalities on a 129S1/Sv*129X/SvJ background, specifically a lethal palate formation defect [5]. These mice are viable on a C57BL/6 background but have brain abnormalities [19,20]. Conversely, *Pdgfcc*^{tm1lex} mice are viable in two backgrounds, C57BL/6 x 129S5/SvEvBrd and C57BL/6 though two alleles of *Pdgfcrα* are necessary for their viability. Mice that are homozygous for *Pdgfcc*^{tm1lex} and hemizygous for *Pdgfcrα* die within days of birth and have severe spina bifida. Our results suggest that the CUB domain of *Pdgfc-c* performs a function in *Pdgfcrα* signaling. Future experiments are needed to determine whether *Pdgfcc*^{tm1nagy} mice on C57BL/6 background require two *Pdgfcrα* alleles. The availability of *Pdgfcc*^{tm1lex} mice will facilitate additional genetic screens, either alone or in conjunction with *Pdgfcc*^{tm1nagy} mice or other mutants, to determine the biological functions of the PDGF-C CUB domain.

Acknowledgments

Grant support: NHLBI, Cardiovascular Pathology training grant (NIH-HL007312) and HHMI program in Molecular Medicine (BJH), Herbert Coe Foundation, American Surgical Association Foundation, and the American College of Surgeons (KJR), Drug Metabolism Transport and Pharmacogenetics Research Fund in the School of Pharmacy, University of Washington (EJK), NCI, mechanisms of PDGF-C induced HCC (NIH-CA127228) (JSC).

Declarations

All animal studies described in this manuscript were performed with University of Washington Institutional Animal Care and Use Committee approval. The University of Washington is an Association for the Assessment and Accreditation of Laboratory Animal Care approved facility.

Sources of Support

This work was supported by the NHLBI, Cardiovascular Pathology training grant (NIH-HL007312) and HHMI program in Molecular Medicine (BJH), Herbert Coe Foundation, American Surgical Association Foundation, Fred Hutchinson CRC New Investigator Award and the American College of Surgeons (KJR), Drug Metabolism Transport and Pharmacogenetics Research Fund in the School of Pharmacy, University of Washington (EJK), NCI, Mechanisms of PDGF-C induced HCC (NIH-CA127228) (JSC).

References

1. Soriano P. The PDGF alpha receptor is required for neural crest cell development and for normal patterning of the somites. Development. 1997;124(14):2691-2700.
2. Soriano P. Abnormal kidney development and hematological disorders in PDGF beta-receptor mutant mice. Genes Dev. 1994;8(16):1888-1896.
3. Boström H, Willetts K, Pekny M, et al. PDGF-A signaling is a critical event in lung alveolar myofibroblast development and alveogenesis. Cell. 1996;85(6):863-873.
4. Levéen P, Pekny M, Gebre-Medhin S, Swolin B, Larsson E, Betsholtz C, et al. Mice deficient for PDGF B show renal, cardiovascular, and hematological abnormalities. Genes Dev. 1994;8(16):1875-1887.
5. Ding H, Wu X, Bostrom H, Kim I, Wong N, Tsoi B, et al. A specific requirement for PDGF-C in palate formation and PDGFR-alpha signaling. Nature genetics. 2004;36(10):1111-1116.
6. Fredriksson L, Li H, Eriksson U. The PDGF family: four gene products form five dimeric isoforms. Cytokine & Growth Factor rev. 2004;15(4):197-204.
7. Tallquist MD, Soriano P. Cell autonomous requirement for PDGFR alpha in populations of cranial and cardiac neural crest cells. Development 2003;130(3):507-518.
8. Betsholtz C, Johnsson A, Heldin CH, Westermark B, Lind P, Urdea MS, et al. cDNA sequence and chromosomal localization of human platelet-derived growth factor A-chain and its expression in tumour cell lines. Nature. 1986;320(6064):695-699.

9. Li X, Pontén A, Aase K, Karlsson L, Abramsson A, Uutela M, et al. PDGF-C is a new protease-activated ligand for the PDGF alpha-receptor. Nat Cell Biol. 2000;2(5):302-9.

10. Gilbertson DG, Duff ME, West JW, Kelly JD, Sheppard PO, Hofstrand PD, et al. Platelet-derived growth factor C (PDGF-C), a novel growth factor that binds to PDGF alpha and beta receptor. J Biol Chem. 2001;276(29):27406-27414.

11. Bergsten E, Uutela M, Li X, Pietras K, Ostman A, Heldin CH, et al. PDGF-D is a specific, protease-activated ligand for the PDGF beta-receptor. Nat Cell Biol. 2001;3(5):512-516.

12. LaRochelle WJ, Jeffers M, McDonald WF, Chillakuru RA, Giese NA, Lokker NA, et al. PDGF-D, a new protease-activated growth factor. Nat Cell Biol. 2001;3(5):517-521.

13. Reigstad LJ, Sande HM, Fluge Ø, Bruland O, Muga A, Varhaug JE et al. Platelet-derived growth factor (PDGF)-C, a PDGF family member with a vascular endothelial growth factor-like structure. J Biol Chem. 2003;278(19):17114-17120.

14. Sauer MK, Donoghue DJ. Identification of nonessential disulfide bonds and altered conformations in the v-sis protein, a homolog of the B chain of platelet-derived growth factor. Mol Cell Biol. 1988;8(3):1011-1018.

15. Kenney WC, Haniu M, Herman AC, Arakawa T, Costigan VJ, Lary J, et al. Formation of mitogenically active PDGF-B dimer does not require interchain disulfide bonds. J Biol Chem. 1994;269(16):12351-9.

16. Kreysing J, Ostman A, van de Poll M, et al. Identification of three amino acid residues in the B-chain of platelet-derived growth factor with different importance for binding to PDGF alpha- and beta-receptors. FEBS Lett. 1996;385(3):181-184.

17. Fenstermaker RA, Poptic E, Bonfield TL, Knauss TC, Corsillo L, Piskurich JF, et al. A cationic region of the platelet-derived growth factor (PDGF) A-chain (Arg159-Lys160-Lys161) is required for receptor binding and mitogenic activity of the PDGF-AA homodimer. J Biol Chem. 1993;268(14):10482-10489.

18. Andersson M, Ostman A, Kreysing J, Bäckström G, Van de Poll M, Heldin CH. Involvement of loop 2 of platelet-derived growth factor-AA and -BB in receptor binding. Growth Factors 1995;12(2):159-64.

19. Fredriksson L, Nilsson I, Su EJ, Andrae J, Ding H, Betsholtz C, et al. Platelet-derived growth factor C deficiency in C57BL/6 mice leads to abnormal cerebral vascularization, loss of neuroependymal integrity, and ventricular abnormalities. Am J Pathol. 2012;180(3):1136-1144. doi: 10.1016/j.ajpath.2011.12.006.

20. Eitner F, Bücher E, van Roeyen C, Kunter U, Rong S, Seikrit C, et al. PDGF-C is a proinflammatory cytokine that mediates renal interstitial fibrosis. J Am Soc Nephrol. 2008;19(2):281-289. doi: 10.1681/ASN.2007030290.

21. Kibbe WA. OligoCalc: an online oligonucleotide properties calculator. Nucleic Acids Res. 2007;35(Web Server issue):W43-6.

22. Crouthamel MH, Kelly EJ, Ho RJ. Development and characterization of transgenic mouse models for conditional gene knockout in the blood-brain and blood-CSF barriers. Transgenic Res. 2012;21(1):113-130. doi: 10.1007/s11248-011-9512-z.

23. Sambrook J, Russell DW. Calcium-phosphate-mediated Transfection of Eukaryotic Cells with Plasmid DNAs. CSH Protoc 2006;2006(1). doi: 10.1101/pdb.prot3871.

24. Ferns GA, Sprugel KH, Seifert RA, Bowen-Pope DF, Kelly JD, Murray M, et al. Relative platelet-derived growth factor receptor subunit expression determines cell migration to different dimeric forms of PDGF. Growth Factors. 1990;3(4):315-3.

25. Campbell JS, Riehle KJ, Brooling JT, Bauer RL, Mitchell C, Fausto N. Proinflammatory cytokine production in liver regeneration is Myd88-dependent, but independent of Cd14, Tlr2, and Tlr4. J Immunol. 2006;176(4):2522-2528.

26. Sundberg JP, Taylor D, Lorch G, Miller J, Silva KA, Sundberg BA et al. Primary follicular dystrophy with scarring dermatitis in C57BL/6 mouse substrains resembles central centrifugal cicatricial alopecia in humans. Vet Pathol. 2011;48(2):513-524. doi: 10.1177/0300985810379431.

27. Hamilton TG, Klinghoffer RA, Corrin PD, Soriano P. Evolutionary divergence of platelet-derived growth factor alpha receptor signaling mechanisms. Mol Cell Biol. 2003;23(11):4013-4025.

28. Reed R, Maniatis T. A role for exon sequences and splice-site proximity in splice-site selection. Cell. 1986;46(5):681-690.

29. Yue BG, Akusjarvi G. A downstream splicing enhancer is essential for in vitro pre-mRNA splicing. FEBS Lett. 1999;451(1):10-14.

30. Anisimov A, Leppänen VM, Tvorogov D, et al. The basis for the distinct biological activities of vascular endothelial growth factor receptor-1 ligands. Sci Signal. 2013;6(282):ra52 doi: 10.1126/scisignal.2003905.

31. Kiba A, Yabana N, Shibuya M. A set of loop-1 and -3 structures in the novel vascular endothelial growth factor (VEGF) family member, VEGF-ENZ-7, is essential for the activation of VEGFR-2 signaling. J Biol Chem 2003;278(15):13453-13461.

32. Thielens NM, Bersch B, Hernandez JF, et al. Structure and functions of the interaction domains of C1r and C1s: keystones of the architecture of the C1 complex. Immunopharmacology. 1999;42(1-3):3-13.

33. Nakamura F, Goshima Y. Structural and functional relation of neuropilins. Adv Exp Med Biol. 2002;515:55-69

34. Mahoney DJ, Mikecz K, Ali T, et al. TSG-6 regulates bone remodeling through inhibition of osteoblastogenesis and osteoclast activation. J Biol Chem. 2008;283(38):25952-25962. doi: M802138200 [pii] 10.1074/jbc.M802138200.

35. Lee HX, Mendes FA, Plouhinec JL, et al. Enzymatic regulation of pattern: BMP4 binds CUB domains of Tolloids and inhibits proteinase activity. Genes Dev. 2009;23(21):2551-2562. doi: 10.1101/gad.1839309.

36. Hollway GE, Maule J, Gautier P, et al. Scube2 mediates Hedgehog signalling in the zebrafish embryo. Dev Biol. 2006;294(1):104-118.

37. Wu YY, Peck K, Chang YL, et al. SCUBE3 is an endogenous TGF-β receptor ligand and regulates the epithelial-mesenchymal transition in lung cancer. Oncogene. 2011;30(34):3682-3693. doi: onc201185 [pii] 10.1038/onc.2011.85.

38. Dijkmans J, Xu J, Masure S, et al. Characterization of platelet-derived growth factor-C (PDGF-C): expression in normal and tumor cells, biological activity and chromosomal localization. The international journal of biochemistry & cell biology. 2002;34(4):414-26.

39. McKinnon RD, Waldron S, Kiel ME et al. PDGF alpha-receptor signal strength controls an RTK rheostat that integrates phosphoinositol 3'-kinase and phospholipase Cgamma pathways during oligodendrocyte maturation. J Neurosci. 2005;25(14):3499-3508.

8

Immunobiology of Allograft Human Leukocyte Antigens in the New Microenvironment

Mepur H. Ravindranath*, Vadim Jucaud, Paul I. Terasaki

Terasaki Foundation Laboratory, 11570 W, Olympic Blvd, Los Angeles, CA 90064, California, USA

***Corresponding author:** *Mepur H. Ravindranath, Terasaki Foundation Laboratory, 11570 W. Olympic Blvd, Los Angeles, CA 90064, USA, E-mail: ravimh@terasakilab.org*

Absract

The antibodies formed by allograft recipients against donors' mismatched allo-human leukocyte antigens (allo-HLA) correlate with graft rejection. This review's aim is to elucidate the immunological events associated with allografts' mismatched HLA from transplantation to acute/chronic rejection. Such clarification may permit more precisely relevant therapeutic strategies that can facilitate better graft survival. The trigger event in the allograft microenvironment is an inflammatory response, promoting overexpression of allo-HLA as dimers or monomers. Overexpression of membrane- bound HLA is accompanied by matrix membrane proteases that dissociate and release monomeric HLA from the cell surface, forming soluble HLA (sHLA), and the recipient's immune components interacting with both membrane-bound and sHLA. Immune recognition of allo-HLA may vary depending on whether the allo-HLA is dimeric or monomeric. An important player in immune rejection of the allograft appears to be sHLA.

Monomeric sHLA that are free of β2-microglobulin (β2m) bind to receptors of allogenic cytotoxic T cells (NKT and CD8+ T cells) to prevent cytolytic destruction of the allograft, and they expose epitopes that β2m made cryptic. The antibodies formed against donor-specific allo-sHLA during acute/chronic rejection cause arteriosclerosis and affect the glomerular filtration rate, leading to allograft rejection. Anti-HLA antibodies can be diverse, some recognizing specific epitopes on the mismatched allo-HLA, others recognizing the cryptic but common shared epitopes β2m previously masked. When monoclonal antibodies to sHLA were generated, it was shown that the functions of monospecific antibodies differ from those of polyreactive antibodies, which recognize epitopes shared with all HLA-I alleles and suppress T-cell proliferation while monospecific antibodies can induce proliferation of CD8+ T cells. Knowing the physicochemical structure of HLA-I/-II is essential for the elucidation of HLA functions. Although the functional potentials of N-linked carbohydrates located at the junction of the antigen-presenting groove remain unexplored, anti-HLA antibodies are monitored by coating recombinant HLA (rHLA) on solid matrices. The coated rHLA are not glycosylated nor is the ratio of monomeric and dimeric rHLA constant for all alleles. This review emphasizes that until these issues are resolved, we cannot comprehend why several allograft recipients do not show any signs of pathogenicity or rejection episodes post-transplantation in spite of the presence of donor specific allo-HLA Abs.

Keywords: Allograft; Alloreactive; Arteriosclerosis; Creatinine; Donor organ; Endothelial cells; Glomeruli; Glycosylresidues; Microenvironment; Monoclonal antibodies (mAbs); Monospecific mAbs; N-linked carbohydrates; Polyreactive mAbs; Proinflammatory cytokines; Recipient; Rejection; Vasculitis; Viremia

Abbreviations

β2m: β2-Microglobulin; β2m-free HC: β2m-free Heavy Chain; CTL: CD8+ Cytotoxic T Lymphocytes; DSA: Donor Specific Antibodies; HC: Heavy Chain Polypeptide; HLA: Human Leukocyte Antigens; HLA-I: Human Leukocyte Antigen Class I; HLA-II: Human Leukocyte Antigen Class II; IFNγ: Interferon-γ; kDa: kilodalton; MHC: Major Histocompatibility Complex; MMP: Matrix Metalloproteinase; rHLA: Recombinant HLA; sHLA: soluble HLA

Introduction

The biological process involved in transplantation is comparable to the ecological concept of "habitat selection." When an organism is exposed to a challenging new environment, survival of the organism depends on a sequence of events: initial shock response, acclimation, adaptation, and, finally, survival: not being rejected or killed. Survival of the transplanted organ in the new host depends on the same sequence of events: initial shock response to inflammation caused by infiltration of inflammatory cells, immune reactions similar to those of hypersensitivity type IV with or without infection, then acclimation (equivalent to tolerance and avoidance of acute rejection), and, finally, adaptation if the allograft is not rejected by the microenvironment of the new host. A comprehensive understanding of the events taking place in the allograft microenvironment is necessary in order to develop strategies to prevent allograft rejection. Several factors are involved in this microenvironment such as augmentation of cell-mediated and antibody-mediated immune responses against allograft-associated antigens, and, even today, our understanding of the sequential events in the immune response is poor or incomplete. That information, however, it is very much needed to render the microenvironment congenial to allograft survival, facilitating allograft acceptance in the new environment; and an understanding of the changes occurring in the allograft microenvironment soon after transplantation is essential. There is also a need to elucidate the all components in the allograft microenvironment, whether natural (like cellular elements, molecules or even pathogens introduced or pre-existing in the microenvironment) or clinically imposed (exogenous components such as immunosuppressive agents

***Corresponding author email:** *ravimh@terasakilab.org*

and other molecules). This review focuses on only one aspect of the initial shock response: the major cellular component (or antigen) known as the Major Histocompatibility Complex (MHC). This complex is expressed by the tissues of the allograft, and is altered under the influence of the new microenvironment. The alteration includes monomerization (e.g., 2m-free HLA-I) and both overexpression and activation of the enzymes (matrix membrane proteinases) that cleave membrane-bound HLA-these being the primary symptoms of shock response to the altered profiles of the new microenvironment. The new host environment then responds to the allograft's HLA by eliciting immune-cell infiltration and augmentation of proinflammatory cytokines. The allograft responds to the new environment by releasing the cell surface allo-HLA into the host microenvironment. The shed allo-HLA interacts with proinflammatory allo-immune cells, resulting in suppression of cytotoxic CD8+ T and NK cells and upregulation of CD4+ T cells and B cells, culminating in the production of antibodies. This interaction between the allograft and the microenvironment is on-going. Monitoring the sequence of events associated with the shock response is critical for developing appropriate treatment strategies and for realizing the possibility of organ-specific therapies pre- and post-transplantation.

The MHC is a group of cell-surface molecules encoded by a large gene family that controls a major part of human immunobiology. MHC molecules interact with several kinds of immune cells; they possess the critical determinants of the compatibility of donors for organ transplant. The MHC gene family is divided into three subgroups: classes I, II, and III. Class I MHC molecules have β2 subunits; hence can be recognized only by CD8 co-receptors. Class II molecules have no β2 subunits so can be recognized by CD4 co-receptors. In humans, the MHC classes I and II are also known as the Human Leukocyte Antigens (HLA). The third region (located between the other two) encodes for MHC class III, the other important proteins involved in the immune system such as CBF, C2, and C4A complement genes, a group of Tumor Necrosis Factors (TNF), and heat shock protein 70 chaperoning genes. The list of the many additional genes includes peptide transporter proteins TAP1 and TAP2, and PSMB8 and PSMB9 genes that code for components of the β-immuno-proteasome.

The HLA system: First thesis

HLA constitute the centerpiece of the immunobiology of transplantation. Jon van Rood first identified what would later be designated "HLA-Bw4 and -Bw6," and he published (with van Leeuwen) "Leukocyte grouping: A method and its application" in 1963 [1]. Two important clinical studies elucidated the relevance of leukocyte grouping in understanding transplant immunology. The first was also by van Rood (with Eernisse and van Leeuwen) [2], involving platelet transfusion. Eernisse gave the first platelet transfusion; it was from an unknown blood donor to a pregnant woman who had developed severe aplastic anemia after chloramphenicol treatment. This patient stopped bleeding after random platelet transfusions, but had developed anti-human leukocyte antibodies, possibly generated against the donor platelets. When platelet recovery dropped to zero, her bleeding recommenced. During this period, nine genetically determined leukocyte groups or HLA were recognized, which prompted verification of whether the patient's eight brothers and sisters had what the authors called a "negative leukocyte agglutination cross match" with the patient's serum. This turned out to be the case; so, every week, one of these siblings donated platelets to their sister, resulting in an excellent recovery.

The second important study was that of Terasaki, et al. [3], presented at the Conference on Histocompatibility testing in Washington, DC, in 1964. After extensive investigations between 1959 and 1961 of antibody responses to homografts [4-10], Paul Terasaki, in collaboration with pioneering renal transplant surgeon, Thomas Starzl, showed that the allo-HLA antibodies pre-existent in a recipient are responsible for the rejection of kidney allografts. This was the first and conclusive proof of the role of allo-HLA antibodies in allograft rejection. Concurrently, Jean Dausset and his team in a Paris blood bank found "humoral antibodies after skin homografting"-as reported at the 6th International Transplantation Conference in New York [11]. After examining the allo-HLA antibodies formed in long-term kidney transplant survivors, Terasaki, et al. [3], highlighted the importance of HLA testing in transplantation, stressing the need for testing HLA groups or types both in donors and organ recipients. This study marked the beginning of the humoral theory of rejection in transplantation.

Elucidation of donor-specific HLA antibodies: Terasaki's humoral theory

The Dausset team had examined anti-HLA antibodies using a "Leuko-agglutination" assay [12]; but soon, Terasaki and his team began to develop [13,14], validate [15-17] and finally establish [18-21] a more sensitive assay to monitor the alloantibodies to HLA that are directly involved in the rejection of organs by allograft recipients. Terasaki postulated the humoral theory of organ rejection [22] and developed strategies for in-depth characterization of antibodies formed against HLA specific to the donor organ-i.e., Donor-Specific Antibodies (DSA) [23]. Elucidating the nature of DSA required chemical characterization of the structure and diversity of HLA. These efforts helped bring about a better understanding of the specificity of auto-HLA antibodies, naturally occurring allo-HLA antibodies and DSA.

A persisting enigma is that even though DSA may exist in HLA-mismatched allograft recipients, not all recipients reject the transplant. What does this mean? Could two different categories of DSA exist-pathogenic and non-pathogenic DSA? If so, how can transplant immunologists or clinicians distinguish them? For this, an in-depth knowledge of the variability of the molecular structure, chemistry and specificity of donor HLA and the nature and diversity of antibodies formed against them is critical.

The current concepts concerning DSA have evolved with the discovery of Donor-Specific (DS) soluble HLA in allograft recipients and the evolution of assays employed in monitoring both HLA and their antibodies. Understanding the structural diversity of HLA class I and II expressed on the membrane, their diversity in different cell populations, and the diversity of those released from the membrane into the internal environment (extracellular domain, circulation or body fluids) is important for elucidation of the immune responses to HLA in general and particularly in transplantation.

Diversity of HLA groups

Ever since van Rood's "Leukocyte grouping" thesis, efforts to clarify the allelic diversity of HLA continued. Recognition of diverse human leukocyte groups led to in-depth analyses of HLA alleles, significantly increasing understanding of the nature and diversity of HLA. The naming of new HLA genes and allele sequences and their quality control has been effected since 1968 by the World Health Organization's Nomenclature Committee for Factors of the HLA System, which meets regularly, and has published several reports documenting HLA, their genes and alleles. The standardization of HLA antigenic specifications is controlled by the exchange of typing reagents and cells at the International Histocompatibility Workshops. The international immunogenetics project (http://www.ebi.ac.uk; or http://www.ebi.ac.uk/ipd/imgt/hla/intro.html) updated HLA genes and alleles in July 2015 [24].

In humans, the HLA genes are located in the short arm of chromosome 6 (6p21.3), which is composed of three regions. The first is the distal region, containing the genome of MHC class I, which includes the classical (Ia) HLA molecules-HLA-A, -B, and -C heavy chains with extensive polymorphis-3,192 alleles at HLA-A; 3,977 at HLA-B; and 2,740 at HLA-C loci (Table 1)-and the non-classical (Ib) HLA- E, -F, and -G. HLA-I is expressed on the surface of nearly all nucleated cells. All HLA-I molecules form heterodimers with β2- microglobulin (β2m) coded by a gene on chromosome 15. In addition, the distal region contains the MHC class I chain-related MICA and MICB. The second region is the proximal (closer to the centromere), containing genes for two heavy chains of MHC class II, consisting of HLA-DR,-DQ, -DP, -DM, and -DO (for known alleles and proteins see Table 1). HLA-II molecules are expressed on Antigen-Presenting Cells (APC) such as macrophages, dendritic cells, Langerhans and Kuepfer cells, as well as B lymphocytes. Almost every cell in the body expresses HLA-II upon inflammation-most notably, thyroid epithelial cells, intestinal epithelial cells and endothelial cells. The third region (located between the other two) encodes for MHC Class III, which includes other important immune system proteins such as CBF, C2, and C4A complement genes, a group of Tumor Necrosis Factors (TNF), and heat shock protein 70 chaperoning genes. The list of the many additional genes includes peptide transporter proteins TAP1 and TAP2, and PSMB8 and PSMB9 genes that code for components of the β-immuno-proteasome.

Each person carries a pair of chromosomes 6 and therefore expresses on the surface of a single cell one or two HLA-I alleles, each, for HLA-A, HLA-B and HLA-C, several HLA-1b, and one to four HLA-II alleles, each, of HLA-DR, -DQ, and -DP, all together comprising hundreds of thousands of different units [24]. Since each person carries a unique set of HLA genes, the array of possible combinations expressed by each individual in a given population is enormous. This HLA diversity-along with the multiplicity of alleles- leads to the extraordinary individuality and potential diversity in antigen presentation.

Given the enormous diversity of HLA types and combinations, it is very difficult to find unrelated individuals who are identical. Information on the specific HLA types of the donor and recipient is required to match the HLA types of the allograft with the recipient's HLA types. The tests available to identify HLA type range from low-resolution, serological typing, to high-resolution, more specific molecular typing. Tissues in the donor organ may be of HLA types totally different from those of recipient's tissues; or the ratio of matched and mismatched HLA types between

Table 1: Numbers of HLA Alleles (as of July 2015) and their proteins.

HLA Class I						
Gene	**A**	**B**	**C**	**E**	**F**	**G**
Alleles	3,192	3,977	2,740	17	22	50
Proteins	2,245	2,938	1,941	6	4	16
HLA Class II						
Gene	**DRA**	**DRB**	**DQA1**	**DQB1**	**DPA1**	**DPB1**
Alleles	7	1,868	54	807	40	550
Proteins	2	1,364	32	539	20	447
HLA Class II - DRB Alleles						
Gene	**DRB1**	**DRB2**	**DRB3**	**DRB4**	**DRB5**	**DRB6**
Alleles	1,764	1	59	16	21	3
Proteins	1,290	0	47	9	18	0
Other non-HLA Genes						
Gene	**MICA**	**MICB**	**TAP1**	**TAP2**		
Alleles	102	41	12	12		
Proteins	80	27	6	5		

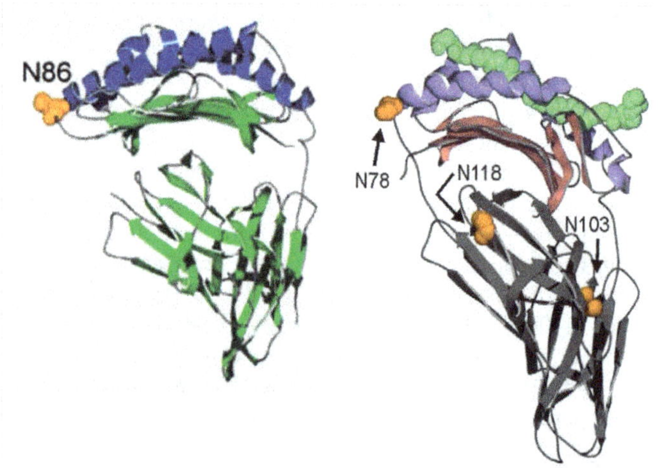

Figure 1: The lateral views of HLA class I (left) and class II (right) molecules. The antigen binding groove (α helices) are blue. Sites of N-glycan sites are orange. Source; Ryan SO and Cobb BA, [25] and Ryan, et al. [30].

donor and recipient may vary in degree. Not all HLA allele types are considered in matching(e.g., HLA-Cw, DP, and HLA-Ib). Our review of the immunobiology of allograft HLA in the new environment addresses the extent to which matching of donor and recipient HLA types is needed for graft survival in that host microenvironment.

Structural diversity of HLA on the cell surface

Understanding cell-surface expression of HLA is key to elucidation of the interaction of HLA with other cells, their receptors, and different kinds of antibodies that may lead to cell survival and proliferation-or to cell death. The expression of each HLA molecule may differ depending on the cell type, physiological status of the cells, and status after shedding from the cell surface into the internal environment. Though it is well known that both HLA class I and II molecules are glycoproteins, the glycosylation of HLA has received the least attention of all factors affecting transplant immunobiology. The true definition of native HLA is a glycosylated α-helical protein. Ryan and Cobb provided an excellent review of the structure and functions of HLA glycosylation in HLA immunobiology [25].

Cell-surface HLA diversity caused by glycosylation

Glycosylation patterns in HLA-I and HLA-II: In general with glycoproteins, two kinds of carbohydrate moieties are attached to amino acids: An asparagine-linked (N-linked) oligosaccharide chain of about 3 kDa and a serine-linked (O-linked) oligosaccharide. In HLA-I molecules, only the N-linked oligosaccharides are common whereas in HLA-II molecules both N- and O-linked (serine/threonine) oligosaccharides may occur. The extracellular region carries a single N-linked oligosaccharide composed of residues of N-acetylglucosamine, mannose, galactose, fucose and sialic acid [26]. One must realize that these complex N-glycans are large and highly flexible structures extending to 30Å, roughly corresponding to the size of an immunoglobulin domain [27,28]. Most important, HLA-I has a single carbohydrate moiety at asparagine (N) position 86 from the NH-terminal end of the α1 domain of the heavy chain, whereas HLA-II has three known glycosylation sites (based on HLA-DR2) on both the α and β chains: N78, N118, and N103 [29,30] (Figure 1). It is clear from these studies that the structural conformation of HLA- crucial for receptor or antibody recognition- may differ depending on the glycosylation of the HLA molecules [31-33].

Glycosylation of a particular HLA molecule may differ with the cell types within an individual: In normal cells, HLA-I molecules are principally fucosylated biantennary structures while HLA-DR class-II molecules have bi-, tri- and tetraantennary structures and high-mannose structures. The sugar residues on a single HLA allele or molecule may differ with cell type within a person [34,35]. For example, the level of sialylation (addition of the number and possibly the nature of sialic acids) of HLA-II molecules may differ between normal B cells and normal monocytes, with the level higher in B cells than in monocytes. The pattern of glycosylation of the same HLA in the B-cell line transformed by an Epstein-Barr virus (EBV) is very different from that of the native B cells: the biantennary structures are decreased concomitant with an increase of the tri- and tetraantennary structure fractions-especially in HLA-II molecules-while both triantennary and high-mannose structures are increased in HLA-I molecules. Moreover, when compared with normal B cells, HLA-I molecules in the EBV-transformed B-cell line are under-sialylated, and HLA-II over-sialylated. It is possible that viremia of the donor or the recipient can induce changes in HLA glycosylation patterns. For example, the inflammation caused by human CMV, a common virus found in both donors and recipients, is capable of stimulating HLA-I transcription and can significantly enhance and alter the expression of HLA-I [36].

It is important to note that the pattern of glycosylation of HLA on the surface of malignant human cancer cells may differ markedly from that of its non-malignant counterpart [37]. Perhaps even more significant, alterations in cellular protein glycosylation have been reported in inflammatory bowel disease and patients with colon cancer [38,39]. It is also well documented that gastric epithelial cells show varied glycosylation of HLA-II compared with B cells [40] at inflamed sites. These observations raise a concern about whether glycosylation of a particular HLA-I or -II allele in the endothelial cells of a donor organ can differ from that of the recipient's cells.

Functions of the glycosylation residues in HLA: The glycosyl residues on glycoproteins serve as a check point for proper protein folding and trafficking to the cell surface [31-33]; they also protect the protein backbone from proteolysis once it is on the cell surface [41] and promote the appropriate geometric spacing of receptors and other molecules on the cell surface for optimized cell-cell attachment and communication with the immune components of the host [42].

Ryan, et al. [30] demonstrated that the presence of the carbohydrate moiety on HLA-II is essential for presentation of a certain class of peptides. In contrast to native glycosylated HLA-DR2, recombinant (bacterially expressed) DR2 failed to associate with peptides containing sugar residues (also known as glycoantigens or GlyAg) despite being properly folded and fully capable of binding to antigenic peptide. Similarly, elimination of native complex N-linked glycans (i.e., unaltered mammalian complex N-glycans) on HLA-II reduced the presentation of some peptides in live APC, nearly eliminated their binding to recombinant HLA-II *in vitro*, and significantly limited *in vitro* and in vivo GlyAg-mediated T-cell recognition and activation despite no detectable defects in peptide and intact protein antigen controls. These findings established the importance of glycosyl residues on HLA-II in the modulation of antigen binding and presentation and this study marked the first time that the N-linked glycans on HLA class II was shown to be integral to antigen binding.

Glycosylation residues can affect antibody recognition of HLA: Monitoring donor-specific HLA antibodies post-transplantation is an essential early step in clinical evaluation, and several tools are available for that monitoring including both an enzyme-linked immunoassay that uses microtiter plates and Luminex single antigen bead assay. HLA-I and HLA-II reactivity of allograft recipients' IgG is analyzed and the data acquired by Single Antigen Bead assay using dual-laser flow cytometry with Luminex xMAP® multiplex technology (LABScan™ 100; One Lambda, Inc., Canoga Park, CA). More than 95% of published studies of HLA antibodies in transplantation used Luminex single antigen beads, and it is important to note not only that the beads are coated with recombinant HLA but also that none of the HLA molecules are glycosylated. So the operative factors in monitoring anti-HLA antibodies are aglycosylated recombinant HLA molecules. Frequently in the literature, these aglycosylated HLA are wrongly thought to be native HLA. Clinicians deduce the existence of DSA based on the reactivity of IgG antibodies to what is usually called the "aglycosylated version" of allo-HLA coated on the beads. Even when present, though, these DSA often do not lead to graft rejection. That is possibly due to the fact that these antibodies may bind to aglycosylated allo-HLA but not to glycosylated HLA in the native state. If the antibody binds to an epitope located near the glycosylation site, then the antibody may bind to beads but not to the native HLA. This possibility impels the need to verify whether the DSA that bind to aglycosylated HLA on beads also bind to glycosylated HLA. As long as the HLA coated on the beads are not glycosylated, true assessment of DSA cannot make.

When the beads (or solid matrix) are coated with aglycosylated recombinant HLA, the coating is a mixture of heavy chains (HC) with β2-microglobulin (β2m). Therefore, each bead may contain both dimeric and monomeric HLA molecules representing a single HLA allele. However, published reports infer that the antibodies bind to native HLA. Certainly, the 2m-associated recombinant HLA-I on the beads do mimic the physical conformation of the native HLA with β2m; however it is unknown what percentage of recombinant HLA-I is with or without β2m on the beads. Recently One Lambda introduced "i-beads" to eliminate β2m-free heavy chains to monitor β2m-associated HLA. In spite of this useful effort, the ratio of β2m-free and β2m-associated heavy chains of HLA-I is not resolved and therefore there is no certainty as to whether the anti-HLA antibodies bind to β2m-free and β2m-associated heavy chains of HLA.

In any case, early investigators were influenced by the observation of Ploegh, et al. [43] that neither the lack of carbohydrate nor the presence of glycosylation affects reactivity with human alloantisera or the mouse monoclonal Antibody (mAb) W6/32 that reacts with all HLA-A and -B specificities. The hybridoma-secreting mAb W6/32 (IgG2a) was established by immunizing a mouse with membrane from human thymocyte preparations [44]. The purified form of mAb was used to confirm that the mAb recognizes HLA-A, -B and -Cw alleles when they are associated with β2m, but not when they are free from β2m [45]. β2m-free HLA are recognized by the mAb HC10 and also by Q1/28 [46-49]. It is of interest that neither the lack of carbohydrate nor the presence of glycosylation affected the binding affinity of these mAbs (W6/32, HC10 and HCA2), which were extensively used to immunolocalize HLA-I. Obviously, these antibodies bind to a sequence of amino acids far from the single glycosylation site (Asn^{86}) in HLA-I. Similar binding was also observed with some of the polyclonal antibodies in human serum [25] although patients' polyclonal anti-HLA sera cannot be restricted only to these antibodies.

Based on these and similar findings, it became dogma that glycosyl residues on the HLA-I molecule do not interfere with antibody recognition of HLA, making it acceptable to define anti-HLA antibodies and DSA by screening the antibodies with solid matrices coated with aglycosylated recombinant HLA molecules. This generalization was also extended to serum anti-HLA antibodies for allograft recipients. From the therapeutic perspective, there exists a possibility that some antibodies in humans, although binding to beads coated with aglycosylated HLA, may not bind to glycosylated native HLA, accounting for the observed fact that, in spite of the presence of such DSA, rejection does not always occur. It is important to visualize the difference in size: the glycoslation residue, which is located near the gate of one end of the peptide groove, is about 30Å, while the diameter of the peptide groove is 12-16Å.

In this regard, the work of Ferrone's group [50] is significant. By employing a diagnostic mAb (Q6/64), they demonstrated that the glycosylation of the HLA-I HC can hinder antibody recognition either directly or by altering the conformation of the α1 helix. There is a need to identify and characterize the mAbs that can distinguish glycosylated from non-glycosylated HC of HLA-I and -II molecules. It should be established whether an antibody that recognizes a native HLA allele on the cell surface can also recognize HLA in their true native state in the presence of glycosylation or aglycosylated HLA. Until that is done, the HLA antibodies claimed to be "HLA-allele-specific" or "donor-HLA-allele-specific" will remain enigmatic. Such antibodies may account for allograft survival in spite of the presence of a high level of DS HLA antibodies [51].

Cell-surface HLA diversity caused by the absence of β2- microglobulin

Another important component of HLA-I immunobiology,

highly relevant to transplantation, is the naturally occurring β2m-free HC of HLA that has immunomodulary potential. To understand the microenvironment of the allograft and how immune responses (both cellular and humoral) are generated against the HLA of the allograft, it is important to recognize and accept the natural existence of β2m-free HC of HLA molecules both on the cell surface and in body fluids. There is often the misconception in the literature that β2m-free HC of HLA is "denatured HLA". In fact, the recombinant (deglycosylated) HLA-I molecules coated onto solid matrix or microbeads-with or without coupling to β2m-are the true denatured molecules of HLA-I since natural β2m-free HLA-I is glycosylated as is β2m-associated HLA-I. Recent better understanding about the proinflammatory cytokine-mediated upregulating of β2m-free HLA-I and potential novel immunoregulatory functions of β2m-free HLA-I elevates those molecules to a pedestal when it comes to understanding the immunobiology of transplantation and the significant role β2m-free HLA-I plays in a variety of immune functions. The following section validates the immunological potential of β2m-free HCs of HLA.

W6/32-negative, anti-H serum-positive HLA-I

The anti-HLA mAb W6/32 is a valuable tool for identifying the HLA-I dimer—the HC, both glycosylated and non-glycosylated, associated with β2m. While studying the biosynthesis of HLA-I molecules in the human B lymphoblastoid cell line T5-1, which is positive for HLA-Al, -A2, -B8, and -B27, Strominger's group noted the failure of W6/32 to recognize β2m-free HLA [52]. They also noticed that anti-H polyclonal serum, which failed to recognize β2m-associated HLA, did recognize β2m-free HLA. Based on their immunostaining and immunoprecipitation with W6/32 and anti-H[2], they confirmed the presence of two distinct populations of HLA HCs. W6/32 recognized the HC population associated with β2m while anti-H specifically precipitated the HLA molecules devoid of β2m. The group reported that "W6/32 reactive material was detected on the cell surface, whereas anti-H-reactive material ... could only sometimes be detected in small quantities on overexposed fluorographs. Thus, at most only 1 or 2% of the heavy chains present on the surface of T5-1 were precipitable by anti-H" [52]. Since this report, cell-surface expression of β2m-free HLA-Ihas received considerable attention; some investigators have even speculated that β2m-free HC could be immunologically inert [53]. In any case, expression of β2m-free HCs is observed across lymphocyte subpopulations (CD3+ T cells, CD19+ B cells, CD56+ NK cells and CD14+ monocytes), and-most important-these HCs are overexpressed particularly on activated cells and on extravillous trophoblast and monocytes [54].

Can β2m-free HC hold and present peptides-as does intact HLA-I?

When HLA-I is assembled in the endoplasmic reticulum and exported to the cell surface, a peptide of appropriate length and sequence is incorporated with β2m -associated HC. Peptide binding to the groove on the α1 and α2 helices of HLA-I results in the formation of a stable heterotrimer from an unstable β2m-associated HC heterodimer. Such a conformational change provides a strong interaction between the three components and also signals the release of stably assembled HLA-I molecules from the endoplasmic reticulum. Very little is known about the peptide-carrying capabilities of β2m-free HC of HLA-I. Rigney, et al. [55] studied the peptide-induced conformational change in HC that lacked β2m: they added synthetic peptides to cell lysates containing 2m-free HC of HLA-I. Based on the pattern of conformational changes, the physiological relevance is implicated. The Rigney group demonstrated that the synthetic peptides, binding the β2m-free platform, is relatively stable in physiological buffers and undergoes a conformational change that is detectable with antibodies. Most important, they showed that the structural features of peptides that induce this conformational change in the platform are the same as those required to observe the conformational change in β2m-associated HC. These observations confirm that the α1 and α2 domains of β2m-free HC-which together form the peptide binding site of class I MHC-are able to act independent of the rest of the molecule. Springer, et al. showed that the α1 and α2 domains of HC devoid of β2m can remain stable and peptide-receptive as long as they are glycosylated [56]. Despite this, the group could not obtain the crystalline structure of the peptide carrying α1 and α2 domains of HC devoid of β2m nor could they successfully perform these experiments on a membrane matrix carrying β2m-free HC [57].

Immunological relevance of β2m-free HC

The first evidence for the immunological relevance of β2m-free HC was provided by Schnable, et al. [58] when they observed that T lymphocytes, activated *in vitro* or in vivo, but not resting, expressed a considerable number of surface β2m-free HLA-I HC molecules. While W6/32 recognized the β2m-associated HC of HLA-I, the mAbs L45 and HC10 bound specifically to β2m-free-but not to β2m-associated-HC of HLA-I. W6/32 did not bind to β2m-free HLA-I. Immunoprecipitation and co-capping experiments showed that LA45 bound to β2m-free HLA-I HC at the cell surface. It is interesting that L45 bound to PHA-activated T cells from a panel of 12 people with different HLA types, suggesting that LA45 may bind to epitopes shared by all HLA-I HCs. The β2m-free HLA-I molecules expressed on the cell surface of activated T cells-or EBV-transformed B cells-are referred to as "peptide-binding empty HLA" [59,60].

Immunologic potential of β2m-free HC as "open conformer"

The expression of β2m-free HLA class I HC was confirmed on activated T cells *in vitro* and in vivo as well as on B cell lines (RAJI, NALM6), EBV-transformed B cells, and the myeloid cell line KG-1A [58,61,62]. The expression of β2m-free HLA observed on the cell surface in normal human T cells upon activation and cell division was found to be proportional to the level of proliferation [63]. Biochemical analysis showed that the β2m-free HCs (called "open conformers") present at the cell surface are fully glycosylated [63]. β2m-free HCs are tyrosine phosphorylated and are associated with kinase activity. The investigators showed that inhibition of tyrosine phosphorylation with the Src-family tyrosine kinase inhibitor PP2 resulted in enhanced release of β2m-free HCs from the cell surface of activated T cells. A review of the literature on open conformers suggests that, as one report put it, the function of the open MHC-I conformers seems to be related to their inherent ability to cis-associate, both with themselves and with other receptors [64]. Data indicate that the open MHC-I conformers are regulators of ligand–receptor interactions and have potential implications for immune activation.

"Self"- and receptor-associating immunopotential of β2m-free HC

In comparison with healthy control tissues, inflamed tissues (such as in spondylo-arthorpathy) show increased levels of β2m-free HCon CD14b+ monocytes compared with other leukocyte subsets. The level of β2m-free HC is also increased on activated dendritic cells on the extravillous trophoblast [54]. The unusual properties of the β2m-free HC of the B27 allele include an ability of this β2m-free HC to maintain the peptide binding groove *in vitro* [65]. β2m-free HC of HLA-B27 may induce arthritis in transgenic mice. It is interesting that β2m-free HC-specific antibodies decreased the disease incidence in this model [66]. β2m-free HCs exist as dimers or in clusters at the cell surface in vivo [65,67-70], which could have profound effects on receptor engagement.

HLA class I molecules are also known as ligands for members of the killer Ig receptor family (KIR) and Ig-like transcript (ILT)/LIR/LILR family (the new LILR nomenclature is at www.genenames.org/genefamilies/LIiLR). Members of these families bound HLA-B27 in both β2m-associated and β2m-free HC. HLA-B27 HC with β2m bound ILT2, ILT4, and LIR6 transfectants but not ILT1, ILT3, or ILT5. β2m-free HC of HLA-B27 bound ILT4 and LIR6. HLA-B27 β2m-free HC bound CD14 cells in peripheral blood lymphocytes from healthy controls, which was consistent with ILT4 expression on monocytes. Alternative recognition of β2m-associated and β2m-free HCs HLA-B27 by KIR or ILT could influence their immunomodulatory function and may imply a role in inflammatory disease [71].

Role of β2m-free HC in antibody-mediated signal transduction

Two important aspects of β2m-free HLA have emerged from the observations. First, on the cell surface, these molecules are capable of reacting with "self" to form homodimers. Matko, et al. [69] made it clear that β2m-free HLA molecules are capable of clustering upon overexpression on the surfaces of activated and transformed human cells. Second, β2m-free HLA can interact with other cell-surface receptors on the activated lymphocytes. On the cell surface of activated lymphocytes, not only are β2m-free HLA upregulated but so are IL-2R [72-74], Fc receptors for IgG (FcγRI/CD64, FcγRII/CD32 and FcγRIII/CD16) [75], IgE (FcεRII)/CD23) [76], insulin receptors, insulin-like growth factors 1R and IL-2R [77], alpha-fetoprotein and transferrin receptors [78], a non-disulphide-linked heterodimer of polypeptide chains 33 kDa and 38 kDa called "Me14/D12" [79], MICA [80], and, finally, HLA class II antigens HLA-DR, -DP and -DQ [81-83].

Investigation at the Terasaki Foundation Laboratory found that different regions of β2m-free HLA can be recognized independently by various epitope-specific mAbs. We have identified two such domains, one comprising peptide sequences of α1 & α2 helices (^{65}RSARDTA^{71} and ^{143}SEQKSNDASE^{152}) and another peptide sequence that remains cryptic in the presence of β2m but is exposed upon loss or absence of β2m and is shared with other HLA class I molecules (^{117}AYDGKDYLT^{125}, ^{126}LNEDLRSWTA^{135}) [84-88,89-91].

While studying the specificity of anti-HLA-E antibodies—both commercial and generated in-house-the Terasaki group observed several anti-HLA-E antibodies that are not specific for HLA-E. Although generated by immunizing recombinant β2m-free HLA-E, they bound to all HLA-I molecules coated onto microbeads, which were monitored on a Luminex platform. Two of these anti-HLA-E mAbs (TFL-006 and TFL-007) bound to HLA-A, -B, -Cw, -F and -G [84,85], and these polyreactive mAbs recognized peptide sequences shared by all HLA classI molecules [86-88].What is most important, both TFL-006 and TFL-007 suppressed secretion of anti-HLA antibodies by activated B cells [92] and also the blastogenesis and proliferation of activated CD4+ T lymphocytes [93]. It was postulated that TFL-006 and TFL-007 were able to bind to the shared epitopes in β2m-free HLA, which were previously hidden in the presence of β2m-associatedHLA. This in turn signaled T-cell deactivation leading to suppression of blastogenesis and proliferation. A signal-transducing function of the cytoplasmic tail is implicated because, in contrast to β2m-associated HC, β2m-free HLA molecules, overexpressed on activated T lymphocytes, have an elongated cytoplasmic tail (the consequence of exon 6 and 7) with phosphorylation sites (tyrosine and serine at, respectively, positions 320 and 335) [94-96].

In contrast to their TFL-006 and TFL-007-which recognize amino acid sequences shared by almost all β2m-free HLA molecules but are masked by β2m in β2m-associated HLA—the Terasaki group [89] also generated mAbs (TFL-033, TFL-034, TFL-073 TFL-074, and TFL-145)that specifically recognize HLA-E, but do not bind HLA-A, -B, -Cw, -F or -G. The group designated these "HLA-E monospecific mAbs," and, using mAb TFL-033, they confirmed the expression of HLA-E on gastric cancer cells [90]. They also noted that these monospecific mAbs are specifically capable of inducing CD8+ cytotoxic T cells with or without co-stimulation. This induction of proliferation of resting CD8+ T cells by the monospecific anti-HLA-E mAbs-which bind to α1 and/or α2 helices expressed by HLA-E-specific amino acid sequences, resulting in a further five-fold increase in the induction of proliferation of PHA-activated CD8+T cells-suggests that the phosphorylation of the tyrosyl and serine residues of the elongated cytoplasmic tail of HLA-E open conformers may be involved in signal transduction. Open conformers are situated on the lipid raft of the bi-layered lipid membrane of non-phosphorylated non-activated CD8+ T cells. Binding of monospecific anti-HLA-E mAbs to the HLA-E open conformers may induce Lymphocyte-Specific Protein Tyrosine Kinase (LCK) to phosphorylate the tyrosyl residue located on the cytoplasmic tail of HLA-E. The enhanced proliferation of PHA-activated CD8+ T cells could be due to clustering of β2m-free HLA molecules on the cell surface as documented by Matko and colleagues [69] onCD8+ T cells activated by PHA or IFN-γ.

Structure and Functions of soluble HLADiversity of soluble HLA

Immune responses to an allograft can start as a consequence of post-surgical injury and inflammation at the allograft site. One can visualize two kinds of immune reaction in the allograft. The first is similar to the hypersensitivity reaction-like Arthus vasculitis-that occurs in experimental settings following the injection of antigens. This is the interaction that may occur between donor antigens in endothelial linings (such as in vessels and glomeruli) and antibodies that may pre-exist in recipients: such antibodies do occur in the circulation of all people during their life span as pre-existing natural antibodies [84,87,97-103], either masked or free [104].The other kind of immune response occurs as a consequence of inflammatory cytokine and/or chemokine being released by allograft-infiltrating immune cells, which may include components of both innate and adaptive immunity. The cytokines may mediate upregulation of β2m-free HLA and HLA-II, which are known to occur on activated lymphocytes, and may mediate shedding of the HLA in the allograft microenviroment and circulation, constituting a pool of soluble (s) HLA. In short, a surgical creation of a "new habitat" for an allograft generates an inflammatory platform. Both injury and inflammation-and in frequent infection-bring about the activation of immune cells, leading to the production of cytokines/chemokines that are capable of promoting overexpression of donor-derived antigens on the cell surface and/or their release into the allograft microenvironment.

Early reports on the immunobiology of sHLA

Van Rood, et al. [105] showed that sera from normal people positive for HLA-A2 could inhibit anti-HLA-A2 antibody while sera from those without HLA-A2 could not-suggesting the presence of sHLA-A2in the sera of the A2-positive. Charlton and Zmijewski [106] proved the same for the A7 sHLA of human leukocytes in the sera of people positive for HLA-A7. To validate the hypothesis proposed by these two independent groups that sHLA (A2 orA7) is responsible for blocking cytotoxicity in A2-and A7-positive patients, the sHLA should be purified from the serum and its immunogenicity documented. Terasaki and his team were the first to isolates HLA-A molecules from serum by ion exchange chromatography [107-109], using them to immunize rabbits to generate antibodies [110,111]. Soluble HLA was also identified in the high-density lipoprotein fractions of human plasma [112,113]. Since platelets contain 73% of the HLA molecules in the entire blood cell population, it was suggested that sHLA is derived primarily from platelets [112]. Soon after sHLA was isolated from both serum and urine, it was characterized for immunogenicity [114,115]. These constitute the early steps involved in isolating sHLA, all of which led to a better understanding of its role in transplantation.

Two forms of sHLA classI: β2m-associated HC and β2m-free HC

HLA molecules, like other cell-surface receptors, are internalized and subsequently re-expressed on the cell surface by "recycling" [116,117]. It was suggested that release of HLA molecules either after several cycles of intracellular degradation or from the cell surface contributes to the serum pool of sHLA-I [118,119].Early investigators thought that shed sHLA might be in association with the membrane moieties [106,120], suggesting that sHLA may occur in association with β2m. However, other investigators distinguished two forms of sHLA-I in human plasma and confirmed them as β2m-associated and β2m-free HLA-I [121-124]. It was postulated that if HLA molecules are shed after antigen presentation, then conformational changes may occur in the α1 and α2 helices in the absence of peptides in their groove, causing dissociation of β2m. Even if they are shed as β2m-associated sHLA, β2m may become dissociated in circulation. A significant correlation between cell-surface expression of β2m-free HC and the amount of soluble HC of HLA was observed, the average amount of β2m-free HC in the sera determined to be 46.9 +/- 38.6 mM/l [119].

Demaria, et al. [125] further examined the nature of the two forms of HLA-I by using HLA-specific membrane-bound matrix metalloprotease (MMP) and determining whether β2m-free HLA could be released as HLA with "peptide-induced conformations". The group showed that dissociation of β2m from the β2m-attached HC on the surface of lymphoblastoid (chimeric immune receptor) cells generated both conformed and non-conformed β2m-free HC as recognized by conformation-dependent antibodies. Conformed HC, having bound the HLA-A2-specific peptide (HTLV-1 tax 11-19), retained proper conformations after dissociation of β2m. Indeed, MMP cleaved both conformed and non-conformed surface β2m-free HC and released HC with preserved conformations. The Demaria team [125] also observed that exogenous β2m binds only to conformed HC, and protects them from further proteolysis. To the best of our knowledge, this is the first report explaining how both β2m-free and β2m-associated sHLA occur in circulation.

Allograft-Derived soluble HLA

Mechanism of release of sHLA from the allograft

The release of β2m-free HC is mediated by a Zn(2+)-dependent MMP that does not cleave HLA-DR, CD4, or CD71 surface receptors and can be activated by phorbol myristate acetate *in vitro* [126]. Specific cleavage by the MMP occurs at a site close to the papain cleavage site in the α3 of HC. Most important, this site is not accessible to the MMP in β2m-associated HC-thereby emphasizing the presence of β2m-free sHLA-I.

The mechanism underlying the genesis of sHLA-I during initiation of allograft rejection was demonstrated and discussed extensively by Burlingham and his co-investigators [127-130]. When DeVito-Haynes, et al. [127], observed donor-derived soluble β2m-associated HLA-I in the sera of allograft recipients who had experienced acute and chronic rejection, they speculated that soluble β2m-associated HLA-I may result from upregulation of the cell-surface expression of β2m-associated HLA-I during allograft rejection. Further observation led to an understanding of activation-induced upregulation of HLA-I *in vitro*, which serendipitously resulted in the expression of β2m-free HLA-I on the cell surface. Most important, these investigators also confirmed that β2m-free HCs, but not β2m-associated HCs, are cleaved by a specific MMP and released into supernatants as soluble 36 kDa proteins. In addition, activated peripheral blood lymphocytes predominantly produced the 36 kDa forms of sHLA proteins that appeared in the culture supernatants as both β2m-free HC and β2m-associated HC. Using an inhibitor of metalloproteases BB-94, the same group inhibited pokeweed mitogen-induced release of not only β2m-free HC, but also β2m-associated HC. Investigations by both the Burlingham team [127-130], Demaria team [126] showed that the possibility of β2m-associated HLA being cleaved directly from the cell surface is highly unlikely since β2m protected the HLA HC from cleavage by the MMP. Ultimately, the presence of β2m-associated HC in culture supernatants or in serum or plasma is attributed to the re-association of shed β2m with β2m-free HC. In support of this contention, it was shown in a rat model that soluble β2m-free HC can rapidly re-associate with β2m [131]. These observations elucidated the role of HLA-I-specific MMP in the healthy and in generating soluble β2m-free and 2m-associated HLA during allograft rejection.

Factors augmenting the release of sHLA from allografts

Cytokines and allograft viremia: Further clarifying the mechanism underlying the formation of soluble β2m-free HLA, the Burlingham group showed that sHLA is actively released from primary Bronchial Epithelial Cells (BEC) by the MMP pathway [129]. The proinflammatory cytokine IFN-γ stimulated the release of β2m-free HLA from the BEC in a time- and concentration-dependent manner while another proinflammatory cytokine, TNF-α, which induces the BEC to release IL-8, had little or no effect on the release of sHLA-I. Based on reports that the inflammation caused by CMV infection augments release of sHLA in patients, the same group hypothesized that CMV infection of Endothelial Cells (EC) may induce host T cells to release IFN-γ, which in turn may activate the MMP-cleavage pathway to generate sHLA-I [130] . They analyzed sHLA-I in the supernatants of cultures of peripheral blood mononuclear cells (PBMC) containing either uninfected EC or CMV-infected EC (EC/CMV). They observed that the responder PBMC became activated and released sHLA-I by the MMP pathway when stimulated by the inflammation caused by allogeneic EC/CMV. In Transwell® cultures, IFN-γ was released by PBMC in response to EC/CMV. In addition, the IFN-γ recovered from the cultures stimulated release of sHLA-I from uninfected allogeneic EC; this release was also shown to be MMP-dependent. These findings implied that the inflammation caused by CMV infection within the transplanted allograft will not only stimulate the release of "self" HLA from responding PBMC, but can also stimulate the release of donor sHLA-I from uninfected bystander EC, both byway of the class I MMP pathway.

Similar to CMV-mediated production of IFN-γ and TNF-α, stimulation of MMP and shedding of soluble HLA-I is observed after infection of EC by Japanese Encephalitis Virus (JEV), which is a single-stranded RNA virus that also leads to the production of IFN-γ and TNF-α, followed by the release/shedding of sHLA-Ib molecules [132]. An MMP inhibitor also blocked this shedding of sHLA-E. While identifying the MMP involved in the release of HLA-I as MMP-9, this team observed JEV-mediated upregulation of MMP. Addition of the UV-inactivated JEV-infected cell culture supernatants stimulated shedding of sHLA-E from uninfected EC, indicating a role for soluble factors/cytokines in the shedding process. Antibody-mediated neutralization of both TNF-α and IFN-γ receptors not only resulted in the inhibition of sHLA-E shedding from uninfected cells but also inhibited HLA-E and MMP-9 gene expression in JEV-infected cells. Shedding of sHLA-E was also observed with purified TNF-α and IFN-β, and adding IFN-β and TNF-α together further potentiated the shedding. These observations provide clear insight into the mechanism of shedding of sHLA mediated by proinflammatory cytokines with or without viral infection.

Donor Specific sHLA in allograft recipients: Observations made of liver allograft recipients elucidate the shedding of sHLA and consequent changes associated with the shed HLA. Davies, et al. [133] observed the presence of comparable levels of graft-derived sHLA in the sera of liver allograft recipients and sHLA in the donor serum. Levels of sHLA in allograft recipients that increased, post-transplantation, from a detectable pre-transplant level, and the persistence of sHLA at a high level as long as the liver allograft functioned confirmed the post-transplantation shedding of sHLA by allografts. Rhynes, et al. [134] observed that the serum sHLA levels were low pre-transplant but increased substantially during the initial 10 days post-transplantation in liver (n = 9) and heart (n = 12) allograft recipients. In renal transplant patients, the increase in sHLA-I was noticed prior to or during 16 of 20 (80%) biopsy-proven rejections and in 9 of 11 (83%) after episodes of infection (bacterial, viral, and fungal).

The most compelling evidence came from the Burlingham group [135]. Using an HLA-A2-specific ELISA, they monitored both pre- and post-transplantation sera from five kidney and eight Simultaneous Pancreas-Kidney (SPK) transplants with HLA-A2-negative recipients and HLA-A2- positive donors. Donor HLA-I proteins have been found in the sera of allograft recipients at high levels (30-300 ng/ml) immediately and continuously after liver transplantation [135]. DeVito-Haynes et al. hypothesized that although donor HLA-I proteins may not be secreted continuously after kidney or pancreas-kidney transplantation, those expressed on the surface of allograft cells might be released under the stress of rejection. The group therefore, meticulously monitored pre- and post-transplantation sera daily from those five kidney and eight SPK transplants of allografts positive for HLA-A2 into HLA-A2-negative recipients. This remarkable study demonstrated that, unlike liver allografts, neither kidney nor SPK allografts continuously secrete donor HLA-I proteins. However, three of four rejection episodes in kidney recipients and all seven rejection episodes in SPK recipients resulted in the augmentation in the serum of soluble donor HLA-A2 (> 5 ng/ml).

While total sHLA levels were also elevated during rejection, the rise in DS sHLA was more dramatic when compared with pre-transplantation levels. The presence of donor sHLA-I in sera is considered "a systemic indication of rejection" of allografts. The authors' concluded that detection of donor sHLA in allograft recipient sera could be a potential noninvasive biomarker of rejection, especially in the pancreas, which is currently difficult to monitor as a single-organ transplant.

McMillan, et al. [136] observed sHLA-I in the sera of 40 liver allograft donor-recipient pairs. They observed sHLA-I in sera obtained from liver donors, as well as from liver transplant recipients with End-Stage Liver Disease (ESLD) before and after transplantation at various intervals up to 3 yrs. The sHLA-I in patients with ESLD had a mean of 909 + 596 ng/ml, greater than that in the general population (643 ng/ml) *(p < 0.05)*; sHLA-I secretion decreased as the severity of liver disease increased. In the recipients, sHLA-I levels stabilized approximately one month after transplant and remained relatively stable thereafter (mean level 950 + 536 ng/ml). These observed levels were also greater than those in the general population *(p < 0.05)*.

It is interestingly that an increase in serum sHLA-I levels was also observed in patients waiting for a renal allograft and undergoing Hemodialysis (HD) [137]. Sera of the HD patients (n =21) were analyzed before and after HD, and HLA-I levels were significantly higher in those undergoing HD than in the normal controls (574.8 +431.1 vs. 415.6 +256.1 ng/ml, *p < 0.05*). Neither HD duration nor pre- and post-HD serum sHLA-I levels were different (574.8 + 431.1 vs. 568.3 + 398.4 ng/ml, *p > 0.05*). After transplantation, the serum sHLA-I levels significantly decreased (574.8 + 431.1 vs. 226.7 + 202.8 ng/ml, *p = 0.0001*) but increased significantly during rejection compared with the pre-rejection phase (642.8 + 296.1 vs. 305.5 + 194.7 ng/ml, *p = 0.0002*). These observations suggest that sHLA-I levels are stable in uremic status and sHLA-I and could serve as a potential biomarker for monitoring acute renal allograft rejection.

Soluble HLA with serum HLA antibodies may form an immune complex: Davies, et al. [133] identified two molecular forms of sHLA-I in the sera of donors, recipients, and normal people. One of the forms of sHLA-I consists of sHLA HC monomers associated with β2m; the other has a molecular weight indicative of HLA antigen/antibody complexes. Rhynes, et al. [134] also noticed the presence of this sHLA-Iin one of the patients who had rejected the allograft. The molecular weight of one of the isolated sHLA forms was higher than the usual molecular weight (near 1,000,000 daltons), which was attributed to the presence of sHLA/antibody complexes in allograft recipients. It is obvious from these studies that sHLA may shed from the allograft and that the shed molecules may remain unattached or may be bound by pre-existing anti-HLA antibodies from the allograft microenvironment or by antibodies generated against sHLA for the purpose of clearing the sHLA from circulation.

Donor-specific sHLA elicits donor-specific anti-HLA antibodies: The presence of HLA antigen/anti-HLA-antibody immune complexes in patients' sera was positively associated with chronic humoral rejection *(p < 0.0001)* [138]. Following depletion of sHLA by magnetic immunoaffinity, Suciu-Foca et al. [138], could identify anti-HLA-antibodies in 57% of the sera obtained from patients undergoing chronic rejection of kidney allografts compared with the 41% identified prior to antigen depletion, thereby confirming the presence of HLA antigen/antibody complexes in patients' sera. The primary role of the HLA antibodies may be to clear the sHLA from the circulation of the host-so anti-HLA antibodies may be generated to perform the function, some of which could be DS anti-HLA antibodies and others non-donor-specific. Zavazava, et al. [139] also noted that during acute rejection episodes-in 20 renal and 30 cardiac allograft recipients-that both serum sHLA-I/-II and the HLA-antigen/antibody complexes were augmented. It was further observed that serum sHLA in 50 renal and 50 cardiac graft recipients was elevated to over 2-5 normal levels up to 10 days before histological evidence of rejection [140]. The investigators confirmed the presence of DS sHLA (A2) in one cardiac recipient. The incidence of cytotoxic anti-HLA antibodies in the sera of cardiac allograft recipients increased from 15% to 42.5% after thermal dissociation (at 56°C) of sHLA from the immune complexes formed in the allograft recipients.

Mohana kumar's group, after purifying and characterizing the sHLA in liver transplant recipients that had originated both in themselves and in the donors, observed that the sHLA consisted of four major polypeptides having the molecular mass of 44, 41, 35-37, and 12 kD complexed with IgM and IgG antibodies [141]. The sHLA-I in the lung Epithelial Lining Fluid (ELF) was also found to be enriched relative to other proteins, and it increased in lung allograft recipients who had developed chronic or acute rejection [142]. The total HLA-I from the ELF was found to contain a mixture of both donor- and recipient-type HLA-A and HLA-B proteins; and the donor-type HLA-A2 was found to be highly enriched in the ELF relative to serum.

Donor-specifics HLA immune complex in allograft rejection by vascular blockage or arteriosclerosis: Several investigators [143-148] demonstrated a strong and consistent association between serum DS HLA antibodies and arteriosclerosis. These investigators examined both the occurrence of DSA in allograft recipients' sera in the context of arteriosclerosis and the strength of the association between the DSA and the severity of arteriosclerosis. Reed, et al. [147] observed that the appearance and persistence of DS serum sHLA (A2) for more than 26 weeks following cardiac allograft put the recipients at significantly increased risk *(p < 0.01)* of developing transplant-related arteriosclerosis, which is attributed to the immunogenicity of the allograft-derived serum sHLA (A2) and production of DS HLA antibodies following transplantation Loupy, et al. [148], documented in 250 patients (33.6% of the total cohort) with severe arteriosclerosis (luminal narrowing >25% due to fibrointimal arterial thickening) a significant association between the circulating DS-anti-HLA antibodies and severe allograft arteriosclerosis (HR = 2.9, *p < 0.0001)*. Allograft endothelial activation, endarteritis, and complement deposition were observed in patients with severe arteriosclerosis and anti-

HLA antibodies (n = 91, 12.2%). High levels of anti-HLA antibodies and their complement binding capacity were associated with increased severity of arteriosclerosis. Most important, the patients with antibody-associated severe arteriosclerosis there was decreased allograft survival and increased mortality *($p < 0.0001$).*

Direct correlation between serum donor-specifics HLA and serum creatinine levels: Glomerular tubules in the kidney are involved in the clearance of creatinine, a product of muscle metabolism. In patients with biopsy-diagnosed humoral rejection (C4d+), serum levels of both DS and Non-DSHLA Antibodies (NDSA) were significantly correlated with patient serum creatinine levels [149]. Therapeutic prevention of rejection successfully reduced both DSA and NDSA and reversed humoral rejection. It is interesting that during the rejection episodes of renal allograft recipients over a 1-3 month period post-transplantation, sHLA increased significantly concomitant with a significant increase in serum creatinine levels. The return to normal serum creatinine levels are reflected in the levels of sHLA-I (0.3-1.5 micrograms/ml), a range seen in healthy persons. Notably, patients without rejection episodes maintained normal levels of sHLA [150].

Donor-specifics HLA level correlates with acute allograft rejection: Examining the sHLA levels in 252 sera of 16 liver allograft recipients, Tilg, et al. [151] observed that the sHLA-I level in six patients with acute rejection reached a peak on day 2 (870+137 nmol/L) from the normal level of 420 + 210 nmol/L. However, in patients without any sign of rejection, the level of sHLA-I decreased to 277+54 nmol/L in the first 10 days post-transplant. Bacterial infection in seven allograft recipients also raised the level of sHLA, though the source of sHLA was not identified.

Similarly, Puppo, et al., [152], while monitoring sHLA-I in 16 liver allograft recipients, noted that six who had no evidence of transplant-related complications showed no change in the level of serum sHLA whereas the serum sHLA-I level were significantly raised in the 10 graft recipients with acute rejection episodes.

Rizzo, et al. [153], showed that the level of sHLA-I was significantly higher in pulmonary allograft recipients with acute rejection than in those with no rejection. Seventy-two percent of the patients with rejection had sHLA levels above the normal range, the levels peaking in the first 2 weeks post-transplantation and decreasing thereafter. Increased levels of sHLA were found in patients with acute rejection but not in those with chronic rejection or those who had infection.

Hagihara, et al. [154], examined sHLA-I changes in 33 pediatric live-donor liver recipients. Three different molecules of sHLA-I (45, 39 and 34-36 kDa) were observed. The investigators noticed a significant elevation of sHLA-I in six of eight patients who suffered episodes of acute rejection. All patients with infectious episodes also had an elevated sHLA-I level. However, increased sHLA-I was observed in 10 of 22 patients (45%) who had no clinical complications. From 1 week to 24 months after grafting, the allograft liver released sHLA-I molecules.

Donor-specifics HLA, but not its immune complex, suppresses cytolytic activity of CD8+ T cells: Cell-mediated immunity is implicated in acute allograft rejection. Survival of transplanted allogeneic organs during acute rejection is strongly dependent on escaping or avoiding T-cell reactivity against foreign MHC products. The particular species of T lymphocyte involved is identified as alloreactive cytotoxic CD8+ T cells. Parham, et al. [155], were the first to document the inhibition of alloreactive Cytotoxic T Lymphocytes (CTL) by peptides from the α2 domain of HLA-A2. Mathew, et al. [141], showed the HLA specificity of the CTL by demonstrating that affinity-purified sHLA-A3, but not sHLA-A2, inhibited the cytolytic activity of an HLA-A3-specific CD8+ T cell line *in vitro*. Smith, et al. [141], showed that sHLA-A3 inhibited CTL activity of an HLA-A3 T-cell line by 53%whereas sHLA-A2 had no effect. Furthermore, sHLA-A3 also increased T-cell death by 77% over the control whereas sHLA-A2 had no significant effect. However, sHLA-A2 induced 21% apoptosis of an anti-HLA-A2 T-cell line whereas sHLA-A3 caused only 3% apoptosis. A significant portion of the sHLA found in the sera of liver transplant patients is complexed with IgG and IgM antibodies. The sHLA from transplant patient sera that are complexed with DS anti-HLA antibodies are less effective in inducing apoptosis than the noncomplexed sHLA. Pre-incubation of T cells with anti-T-cell receptor mAbs protected the T cells.

The molecular mechanism underlying interaction between sHLA-I and CTL is illustrated in the works of Puppo, et al. [156,157], who clearly documented that sHLA-I antigens purified from serum interact through their α3 domain with the α chain of CD8 molecules and that this interaction triggers apoptosis in PHA-activated $CD8^{+}CD95^{+}$ T cells. These observations lend support to the contention that-after dissociation of β_2m-the β2m-free HC of HLA-I do indeed interact with CD8+ CTL through the α3 domain. This finding is confirmed by blocking the inhibitory effect of pre-incubating sHLA-I antigens with W6/32 mAbs that bind between the HLA class I α3 domain and CD8 α chain [158]; it is also confirmed by pre-incubation of PHA-activated $CD8^{+}$ T cells with anti-CD8 α chain mAbs. Notably, anti-α1 domain HLA-I mAbs failed to inhibit apoptosis.

Zavazava and Kronke [159] showed that sHLA-I molecules purified from spleen lymphocytes induce apoptosis in alloreactive CD8+ T cells, upregulated CD95-L and induced apoptosis. Cell death was blocked by a neutralizing anti-CD95L antibody. All these observations clearly pinpoint the role of sHLA-I-notably, the soluble β2m-free HC of HLA-I-in the allograft microenvironment: it is capable of causing apoptosis of alloreactive cytotoxic CD8+ T cells, thereby preventing cell-mediated rejection of allograft. It appears that the level of soluble β2m-free allo-HLA-I is an important factor in the prevention of acute allograft rejection. As mentioned earlier, Hagihara, et al. [154], examining sHLA-I changes in pediatric live-donor liver recipients, noted that increased levels of sHLA-I were observed in 10 of 22 (45%) patients without any clinical complications. From the first week to 24 months after grafting, the allograft liver released sHLA-I molecules. Although augmentation of sHLA has been correlated with occurrence of acute allograft rejection, as noted earlier, the occurrence of sHLA in different molecular sizes shows that there may be several forms of sHLA-I in circulation: β2m-associated sHLA and β2m-free sHLA. Again as noted, Puppo addressed the exposure of the α3 domain, which is cryptic in the presence of β2m and is important for CTL inhibitory function. It therefore appears that while β2m-free sHLAmay be associated with allograft escape from CTL attack [159], the β2m-associated sHLA or sHLA complexed with antibodies may work against survival of the allograft.

Soluble HLA-II in Serum

Ferrone and co-investigators were the first to detect serum sHLA-II molecules using a red blood cell rosette assay [160]. Only one of 19 normal human sera was positive for HLA-DR molecules. In contrast to healthy humans, some patients with acute and chronic lymphocytic leukemia, multiple myeloma, and Waldenstrom's macroglobulinemia had sera that contained higher levels of HLA-DR molecules. Most important, Ferrone's group further documented the immunogenicity of the soluble HLA-DR isolated from the body fluids [161].

Later, Herlyn, et al. [162] developed a detection assay with mAb DDIA to detect HLA-DR in human sera. They used two antibodies binding to different determinants of the HLA-DR molecule as a catcher and a tracer. Of 155 sera from normal people tested, only two had high levels of soluble HLA-DR. Both sera were from siblings of patients with leukemia. About 55-66% of sera from patients with Acute Lymphoblastoid Leukemia (ALL) prior to treatment had high levels of sHLA-DR in their sera. In melanoma patients, detectable levels of sHLA-DR in serum appeared only in the later stage of the disease. The above findings on the increased level of sHLA-DR in sera of patients with ALL were confirmed by Thompson, et al. [163], using a double-determinant ELISA with two mAbs. They further compared sHLA-II in the blood of healthy people and of patients who received either an allogeneic or autologous bone-marrow transplant. Levels were higher in pre-transplant patients than in the healthy, and in the majority of patients, these levels rose even higher at 4 weeks post-transplantation. In addition, the group also observed sHLA-II in the synovial fluid of patients with active rheumatoid arthritis. Later, Westhoff, et al. [164] utilizing two mAbs specific for HLA-DR/-DP (mAb Tii35) and HLA-DQ (mAb Tii22) measured sHLA-II molecules -DR, -DQ, -DP. The population analysis of 209 unrelated, HLA-typed healthy donors showed a mean sHLA-II protein concentration of 1.53 + 2.44 mg/ml in plasma. McDonald, et al. [165] independently developed an ELISA to quantify sHLA-II in 702 sera obtained from normal subjects, patients with ESRD, and recipients of renal, hepatic and cardiac transplants. In transplant recipients, the McDonald group did not observe any differences in the levels of sHLA-II between pre-operative and post-transplant sera but noted that the level of sHLA-II in the sera of liver allograft recipients was significantly higher than in kidney patients, and values for heart patients were lowest of all groups.

Problems associated with identifying and characterizing sHLA

It is of interest that none of these studies addressed the question of whether sHLA-II exists as a monomer or as a dimer (as expressed on the cells).The studies contrasted their findings with those of previous observations of sHLA-I, all of which showed almost the precise reverse. It seems likely that these clear differences in sHLA-II and sHLA-I concentrations relate to different physiologic processes in either production, function, or elimination; or they involve procedural inaccuracy. Non-availability of allele-specific mAbs could be the reason for lack of further progress in sHLA-II studies. On the other hand, the same argument applies to sHLA-I, although observations were rarely made of allele-specific sHLA-I. Most of the investigations used W6/32 as capture antibodies and β2m antibodies as detection antibodies; see Table 2 in [166]. Using W6/32 is ideal for capturing soluble antigens, but it would be necessary to develop monospecific mAbs for every HLA-I allele in order to characterize specifically the nature of the soluble antigens in a solid matrix assay-either ELISA or Luminex microbeads. Generating monospecific mAbs for every one of the HLA alleles would the first step. Recently, the Terasaki Foundation Laboratory has managed to identify and characterize one such HLA-I-specific monospecific mAb [84,85].

Conclusion

Both HLA class I and class II molecules expressed on allograft tissues are implicated in transplant rejection. Antibodies developed against the donor HLA types (DSA) are often associated with allograft rejection. However, not all patients with DSA reject donor organs. It is important to note that the poor allograft survival in patients with de novo DSA is strongly associated with acute rejection episodes; it has been observed that without acute rejection there was no difference in graft survival compared to DSA-negative cases [167]. Furthermore, in a recent study [168] of 503 patients with de novo DSA, only 120 (24%) experienced allograft rejection. Of these, 42 developed acute rejections, 13 had de novo DSA before the acute rejection, and 12 developed DSA long after acute rejection. Comparing the 78 patients with de novo DSA who did not experience acute rejection with the 345 who did not have de novo DSA, no difference was observed in renal function. There are several other such observations in the transplant literature, suggesting that not all DSA are pathogenic. How is that possible since these antibodies, by definition, were developed specifically against the donor antigen? One possibility is that the anti-HLA antibodies may be an indicator of the presence of allograft-directed T cells. Alternatively, the anti-HLA antibodies might not be a result of the immunogencity of mismatched allo-HLA but could be due to HLA-E molecules, which are upregulated upon inflammation [169] and are capable of binding to other HLA-I molecules since they recognize epitopes shared by all HLA [87]. It is in this regard that our earlier report described how-by immunizing a non-classical HLA-Ib antigen, β2m-free recombinant HLA-E-we generated a group of mAbs that reacted not only to HLA-E but also to other HLA-I molecules. The anti-HLA-E mAb TFL-006 reacted with 31 HLA-A alleles,

50 HLA-B alleles and 16 HLA-C alleles in addition to reacting well with HLA-G and HLA-F [84,85]. Such a polyreactive mAb is a result of the immunogenicity of the amino acid sequences in the HLA-E that are common to all other HLA. Using shared epitopes, the polyreactivity of such mAbs was blocked. All of this suggests that a soluble β2m-free heavy chain of HLA can elicit polyreactive antibodies against cryptic, as well as non-cryptic but shared epitopes (amino acid sequence shared by all HLA-I molecules); see Figure 2 in [170]. Such polyreactive antibodies can bind to a solid matrix (such as Luminex microbeads) coated with an HLA molecule, representing that found in allografts or in circulation. If such antibody reactivity shows high mean fluorescent intensity to one of the HLA alleles that belong to the donor HLA type, the antibodies are wrongly conceived to be DSA though they are not. Such "DSA"- even if they occur in the sera of allograft recipients-cannot be pathogenic. Moreover, such anti-HLA antibodies may occur in recipients prior to transplantation (and might be mistaken for "preformed DSA") in normal people with or without infection, inflammation, or different degrees of autoimmune disease. As has been discussed elsewhere [170], only if we can identify an antibody that remains monospecific for donor-derived HLA can it be defined as pathogenic HLA. The first step in screening HLA antibodies in allograft recipients should be to identify indirectly the monospecific DS antibody that occurs only in the absence of any NDSA-specific antibodies. Such DS antibodies could be pathogenic, with the potential of rejecting an allograft.

The earliest event that occurs after transplantation appears to be the presence of sHLA, which might have been derived from the allograft. Again, this review documents the fact that sHLA is highly prevalent prior to acute rejection. So it is important to characterize the nature of the sHLA soon after transplantation, because any-or a combination-of five categories of sHLA may occur following transplantation: β2m-associated sHLA; β2m-free sHLA; membrane lipoprotein-associated β2m-associated sHLA; β2m-associated sHLA/antibody complex; and β2m-free sHLA/antibody complex. It is certain that the α3 and α2 helices of HLA should be free of β2m so they can interact with CD8 receptors to initiate the apoptosis of cytotoxic CD8+ T cells in the allograft microenvironment. Since the presence of alloreactive CD8+ cytotoxic T cells is detrimental to the survival of the allograft, eliminating CD8+ T cells is essential to save the allograft from acute rejection. Release of β2m-free sHLA into the allograft microenvironment by the allograft tissues is therefore the first step in escaping shock response and beginning acclimation. But this strategy for the allograft may not last long because the host immune system counteracts shed β2m-free sHLA as well as β2m-associated sHLA either by neutralizing them with pre-existing HLA antibodies or by generating novel polyreactive and/or monospecific mAbs to form an immune complex with the sHLA.

Anti-HLA antibodies, produced against the donor HLA can complex not only with sHLA but also with the cell-surface HLA of the activated endothelial cells present in the vicinity of the allograft either as arteries or as glomerular tubules. Arteriosclerosis is one of the consequences of the HLA/antibody complex in the vessels. Another complication is that such blockage can affect the glomerular filtration rate leading to increased serum creatinine levels. Whether these events occur soon after transplant or long after transplant, they are bound to cause rejection of the allograft. There are not yet any studies that show whether antibodies are produced against the glycosylation residues of HLA. Often, IgM antibodies are elicited against glycosyl residues and production of such antibodies occurs in a T-cell-independent manner involving CD5+ B1 B cells. Such IgM antibodies can be expected to exert a highly deleterious effect on the allograft if they pass through the endothelial venules.

While the HLA molecule is the centerpiece of the immunobiology of transplantation, the soluble HLA molecule is an important player in allograft survival or rejection. Two considerations make it seem unlikely that the involvement of sHLA in allograft outcome is merely a bystander effect. First, there is the ability of the sHLA to bind to receptors on CD8+ T cells [171,156-159], with such interaction capable of inducing apoptosis of CD8+ cells and arresting of cytolytic capability [141,156,157]. Second, there is the recent report of a monospecific monoclonal anti-HLA-E antibody able to induce proliferation and blastogenesis of CD8+ cells *in vitro* [93]. Both phenomena point to a therapeutic strategy. It is also important to note that it is not just one kind of soluble mismatched HLA that occurs in the blood of allograft recipients; there are, at minimum, two different kinds-β2m-free and β2m-associated heavy chains of soluble allo-HLA. The functional distinctions of these two kinds of sHLA require further study. Furthermore, the role of glycosylation residues of HLA deserves more serious attention in the future to obtain a full picture of organ rejection post-transplantation.

Acknowledgement

This project is supported from the grants from Terasaki Family Foundation and facilities from Terasaki Foundation Laboratory. We profusely thank our research associates at Terasaki Foundation Laboratory: Judy Hopfield, Anh Nguyen, Nubia Banuelos and Curtis Maehara for critically reading the manuscript and offering suggestions for its improvement.

Conflicts of Interest

None of the authors have any conflict of interest or financial support other than that mentioned in the acknowledgements.

References

1. van Rood JJ, van Leeuwen A. Leukocyte grouping. A method and its application. J Clin Invest. 1963;42:1382-1390.

2. Bosch LJ, Eernisse JG, van Leeuwen A, Loeliger EA, van Rood JJ. Treatment of thromobocytopenic patients with repeated platelet transfusions. Rev Belg Pathol Med Exp. 1965;31:139-145.

3. Terasaki PI, Marchioro TH, Starzl TE. Sero-typing of human lymphocyte antigens. In Conference on Histocompatibility Testing. Washington, DC; 1964.

4. Terasaki PI, Cannon JA, Longmire WP Jr. Antibody response to homografts. I. Technic of lymphoagglutination and detection of lymphoagglutining upon spleen injection. Proc Soc Exp Biol Med. 1959;102:280-285.

5. Terasaki PI, Antibody response to homogratfs. II. Preliminary studies of the time of appearance of lymphoagglutinins upon homografting. Am Surg. 1959;25:896-899.

6. Terasaki PI, Cannon JA, Longmire WP Jr. Time-lapse cinematography studies of the in vitro action of serum from homografted animals. Surg Forum. 1960;10:865-868.

7. Terasaki PI, Cannon JA, Longmire WP Jr. Antibody response to homografts. IV. Time of appearance of lymphoagglutinins in partially tolerant animals. Transplant Bull. 1960;7:415-418.

8. Terasaki PI, Cannon JA, Longmire WP Jr. Antibody response to homografts. VI. In vitro cytotoxins produced by skin homografts in rabbits. Proc Soc Exp Biol Med. 1961;106:133-136.

9. Terasaki PI, McClelland JD, Cannon JA, Longmire WP Jr. Antibody response to homografts. VII. The role of complement in cytotoxicity of serum from homografted rabbits. J Immunol. 1961;87:39-45.

10. Terasaki PI, Cannon JA, Longmire WP Jr, Chamberlain, CC. Antibody response to homografts: V. Cytotoxic effects upon lymphocytes as measured by time-lapse cinematography. Ann N Y Acad Sci. 1960;87:258-265.

11. Colombani J, Colombani M, Dausset J. Leukocyte antigens and skin homograft in man. Demonstration of humoral antibodies after homografting by the antiglobulin consumption test. Ann N Y Acad Sci. 1964;120:307-21.

12. Dausset J, Brecy H, Lal VB, Identity of leukocytic antigens detectable from immune iso-leuko-agglutinins in monozygotic twins. C R Seances Soc Biol Fil. 1957;151(2):244-246.

13. Patel R, Mickey MR, Terasaki PI. Serotyping for homotransplantation. XVI. Analysis of kidney transplants from unrelated donors. N Engl J Med. 1968;279(10):501-506.

14. Patel R, Terasaki PI. Significance of the positive crossmatch test in kidney transplantation. N Engl J Med. 1969; 280(14):735-739.

15. Wang GX, Terashita GY, Terasaki PI. Platelet crossmatching for kidney transplants by flow cytometry. Transplantation. 1989;48(6):959-961.

16. Mickey MR, Ayoub G, Terasaki PI. Prediction of negative crossmatch: an aid for cost-effective kidney sharing. Transplant Proc. 1982;14(2):279-281.

17. Ting A, Terasaki PI. Lymphocyte-dependent antibody cross-matching for transplant patients. Lancet. 1975; 1(7902):304-306.

18. Pei R, Lee J, Chen T, Rojo S, Terasaki PI. Flow cytometric detection of HLA antibodies using a spectrum of microbeads. Hum Immunol. 1999;60(12):1293-1302.

19. Pei R, Lee JH, Shih NJ, Chen M, Terasaki PI. Single human leukocyte antigen flow cytometry beads for accurate identification of human leukocyte antigen antibody specificities. Transplantation. 2003;75(1):43-49.

20. Qiu J, Cai J, Terasaki PI, El-Awar N, Lee JH. Detection of antibodies to HLA-DP in renal transplant recipients using single antigen beads. Transplantation. 2005;80(10):1511-1513.

21. El-Awar N, Lee J, Terasaki PI. HLA antibody identification with single antigen beads compared to conventional methods. Hum Immunol. 2005;66(9):989-997.

22. Terasaki PI. Humoral theory of transplantation. Am J Transplant. 2003;3(6):665-673.

23. Terasaki PI. A personal perspective: 100-year history of the humoral theory of transplantation. Transplantation. 2012;93(8):51-6. doi: 10.1097/TP.0b013e3182483713.

24. Robinson J, Halliwell JA, Hayhurst JD, Flicek P, Parham P, Marsh SG. The IPD and IMGT/HLA database: allele variant databases. Nucleic Acids Res. 2015;43:D423-D431. doi: 10.1093/nar/gku1161.

25. Ryan SO, Cobb BA. Roles for major histocompatibility complex glycosylation in immune function. Semin Immunopathol. 2012;34(3):425-441.doi: 10.1007/s00281-012-0309-9.

26. Parham P, Alpert BN, Orr HT, Strominger JL. Carbohydrate moiety of HLA antigens. Antigenic properties and amino acid sequences around the site of glycosylation. J Biol Chem. 1977;252(21):7555-7567.

27. Rudd PM, Wormald MR, Stanfield RL, Huang M, Mattsson N, Speir JA, et al. Roles for glycosylation of cell surface receptors involved in cellular immune recognition. J Mol Biol. 1999;293(2):351-366.

28. Rudd PM, Elliott T, Cresswell P, Wilson IA, Dwek RA. Glycosylation and the immune system. Science. 2001 ;291(5512):2370-2376.

29. Gauthier L, Smith KJ, Pyrdol J, Kalandadze A, Strominger JL, Wiley DC, et al. Expression and crystallization of the complex of HLA-DR2 (DRA, DRB1*1501) and an immunodominant peptide of human myelin basic protein. Proc Natl Acad Sci USA. 1998;95(20):11828-11833.

30. Ryan SO, Bonomo JA, Zhao F, Cobb BA. MHCII glycosylation modulates Bacteroides fragilis carbohydrate antigen presentation. J Exp Med. 2011;208(5):1041-1053. doi: 10.1084/jem.20100508.

31. Bergeron JJ, Zapun A, Ou WJ, Hemming R, Parlati F, Cameron PH, et al. The role of the lectin calnexin in conformation independent binding to N-linked glycoproteins and quality control. Adv Exp Med Biol. 1998;435:105-116.

32. Trombetta ES, Helenius A. Lectins as chaperones in glycoprotein folding. Curr Opin Struct Biol. 1998;8(5):587-592.

33. Saito Y, Ihara Y, Leach MR, Cohen-Doyle MF, Williams DB. Calreticulin functions in vitro as a molecular chaperone for both glycosylated and non-glycosylated proteins. EMBO J. 1999;18(23):6718-6729.

34. Néel D, Merlu B, Turpin E, Rabourdin-Combe C, Mach B, Goussault Y, et al. Characterization of N-linked oligosaccharides of an HLA-DR molecule expressed in different cell lines. Biochem J. 1987;244(2):433-442

35. Merlu B, Neel D, Goussault Y, Charron DJ. Glycosylation of human leukocyte locus A molecules is dependent on the cell type. Eur J Biochem. 1989;181(3):755-760.

36. Wilkinson GWG, Aicheler RJ, Wang ECY. In: Reddehasse M (ed). Cytomegaloviruses: From molecular pathogenesis to intervention. Wymondham, Norfolk: Caister Academic Press; 2013. p.172-190.

37. Alexander S, Hubbard SC, Strominger JL. HLA-DR antigens of autologous melanoma and B lymphoblastoid cell lines: differences in glycosylation but not protein structure. J Immunol. 1984;133(1):315-320.

38. Campbell BJ, Yu LG, Rhodes JM. Altered glycosylation in inflammatory bowel disease: a possible role in cancer development. Glycoconj J 2001;18:851–858.

39. Bodger K, Halfvarson J, Dodson AR, Campbell F, Wilson S, Lee R, et al. Altered colonic glycoprotein expression in unaffected monozygotic twins of inflammatory bowel disease patients. Gut. 2006;55:973–977.

40. Barrera, C, Espejo R, Reyes VE. Differential glycosylation of MHC class II molecules on gastric epithelial cells: implications in local immune responses. Hum Immunol 2002;63(5):384-393.

41. Wormald MR , Dwek RA. Glycoproteins: glycan presentation and

protein-fold stability. Structure. 1999;7(7): R155-60.

42. Dustin ML, Golan DE, Zhu DM, Miller JM, Meier W, Davies EA, et al. Low affinity interaction of human or rat T cell adhesion molecule CD2 with its ligand aligns adhering membranes to achieve high physiological affinity. J Biol Chem. 1997;272(49):30889-30898.

43. Ploegh HL, Orr HT, Stominger JL. Biosynthesis and cell surface localization of nonglycosylated human histocompatibility antigens. J Immunol. 1981;126(1):270-5.

44. Barnstable CJ, Bodmer WF, Brown G, Galfre G, Milstein C, Williams AF, et al. Production of monoclonal antibodies to group A erythrocytes, HLA and other human cell surface antigens-new tools for genetic analysis. Cell. 1978;14(1): 9-20.

45. Parham P, Barnstable CJ, Bodmer WF. Use of a monoclonal antibody (W6/32) in structural studies of HLA-A, B, C, antigens. J Immunol. 1979;123(1):342-9.

46. Quaranta V, Walker LE, Ruberto G, Pellegrino MA, Ferrone S. The free and the beta 2microglobulin-associated heavy chains of HLA-A, B alloantigens share the antigenic determinant recognized by the monoclonal antibody Q1/28. Immunogenetics. 1981;13(4):285-95.

47. Benjamin RJ, Abrams JR, Parnes JR, Madrigal JA, Parham P. Polymorphic specificity of Q1/28, a monoclonal antibody that preferentially reacts with free class I heavy chains. Immunogenetics. 1992;37(1):73-76.

48. Stam NJ, Spits H, Ploegh HL. Monoclonal antibodies raised against denatured HLA-B locus heavy chains permit biochemical characterization of certain HLA-C locus products. J Immunol. 1986;137(7):2299-2306.

49. Brodsky FM, Parham P, Barnstable CJ, Crumpton MJ, Bodmer WF. Monoclonal antibodies for analysis of the HLA system. Immunol Rev. 1979;47:3-61.

50. Wilson BS, Glassy MC, Quaranta V, Ng AK, Ferrone S. Effect of tunicamycin on the assembly and antigenicity of HLA antigens: analysis with monoclonal antibodies. Scand J Immunol. 1981;14(2):201-205.

51. Filippone EJ, Farber JL. Humoral Immune Response and Allograft Function in Kidney Transplantation. Am J Kidney Dis. 2015;66(2):337-347. doi: 10.1053/j.ajkd.2015. 03. 033l.

52. Krangel MS, Orr HT, Strominger JL. Assembly and maturation of HLA-A and HLA-B antigens in vivo. Cell. 1979; 18(4):979-991.

53. Rock KL, Gamble S, Rothstein L, Gramm C, Benacerraf B. Dissociation of beta 2microglobulin leads to the accumulation of a substantial pool of inactive class I MHC heavy chains on the cell surface. Cell. 1991;65(4):611-620.

54. Raine T, Brown D, Bowness P, Hill Gaston JS, Moffett A, Trowsdale J, et al. Consistent patterns of expression of HLA class I free heavy chains in healthy individuals and raised expression in spondyloarthropathy patients point to physiological and pathological roles. Rheumatology (Oxford). 2006;45(11):1338-1344.

55. Rigney E, Kojima M, Glithero A, Elliott T. A soluble major histocompatibility complex class I peptide-binding platform undergoes a conformational change in response to peptide epitopes. J Biol Chem. 1998;273(23):14200-14204.

56. Springer S, Döring K, Skipper JC, Townsend AR, Cerundolo V. Fast association rates suggest a conformational change in the MHC class I molecule H-2Db upon peptide binding. Biochemistry, 1998;37:3001–3012.

57. Yaneva R, Schneeweiss C, Zacharias M, Springer S. Peptide binding to MHC class I and II proteins: new avenues from new methods. Mol Immunol. 2010;47(4):649-657.

58. Schnabl E, Stockinger H, Majdic O, Gaugitsch H, Lindley IJ, Maurer D, et al. Activated human T lymphocytes express MHC class I heavy chains not associated with beta 2microglobulin. J Exp Med. 1990;171(5):1431-1442.

59. Schumacher TN, Heemels MT, Neefjes JJ, Kast WM, Melief CJ, and Ploegh HL. Direct binding of peptide to empty MHC class I molecules on intact cells and in vitro. Cell. 1990;62:563 -567.

60. Benjamin RJ, Madrigal JA, and Parham P. Peptide binding to empty HLA-B27 molecules of viable human cells. Nature. 1991;351(6321):74-77.

61. Majdic O, Schnabl E, Stockinger H, Gadd S, Maurer D, Radaszkiewics T, et al. LA45, an activation-induced human lymphocyte antigen with strong homology to MHC class I molecules. In Leukocyte Typing IV. Oxford University Press, 1989; p. 511.

62. Madrigal JA, Belich MP, Benjamin RJ et al. Molecular definition of a polymorphic antigen (LA45) of free HLA-A and -B heavy chains found on the surfaces of activated B and T cells. J Exp Med. 1991;174(5):1085–1095.

63. Santos SG, Powis SJ, Arosa FA. Misfolding of major histocompatibility complex class I molecules in activated T cells allows cis-interactions with receptors and signaling molecules and is associated with tyrosine phosphorylation. J Biol Chem. 2004;279(51):53062-53070.

64. Arosa FA, Santos SG, Powis SJ. Open conformers: the hidden face of MHC-I molecules. Trends Immunol 2007; 28(3):115–23.

65. Allen RL, O'Callaghan CA, McMichael AJ, Bowness P. Cutting edge: HLA-B27 can form a novel 2microglobulin-free H chain homodimer structure. J. Immunol. 1999;162:5045-5048.

66. Khare SD, Hansen J, Luthra HS, David CS. HLA-B27 heavy chains contribute to spontaneous inflammatory disease in B27/human beta2microglobulin (beta2m) double transgenic mice with disrupted mouse beta2m. J Clin Invest. 1996; 98(12):2746-2755.

67. Capps GG, Robinson BE, Lewis KD, Zúñiga MC. In vivo dimeric association of class I MHC heavy chains. Possible relationship to class I MHC heavy chain-beta 2microglobulin dissociation. J Immunol. 1993;151(1):159-169.

68. Chakrabarti A, Matko J, Rahman NA, Barisas BG, Edidin M. Self-association of class I major histocompatibility complex molecules in liposome and cell surface membranes. Biochemistry. 1992;31(31):7182-9.

69. Matko J, Bushkin Y, Wei T, Edidin M. Clustering of class I HLA molecules on the surfaces of activated and transformed human cells. J Immunol. 1994;152(7):3353-3360.

70. Triantafilou K, Triantafilou M, Wilson KM, Fernandez N. Human major histocompatibility molecules have the intrinsic ability to form homotypic associations. Hum Immunol. 2000;61(6):585-98.

71. Allen RL, Raine T, Haude A, Trowsdale J, Wilson MJ. Leukocyte receptor complex-encoded immunomodulatory receptors show differing specificity for alternative HLA-B27 structures. J Immunol. 2001;167(10):5543-5547.

72. Rodriguez MA, De Sanctis JB, Blasini AM, Leon-Ponte M, Abadi I. Human IFN-gamma up-regulates IL-2 receptors in mitogenactivated T lymphocytes. Immunology. 1990;69:554–557.

73. Modai D, Berman S, Sheleg Y, Cohn M, Weissgarten J, Averbukh Z.

Interleukin-2 receptor is similarly expressed by activated lymphocytes from patients on chronic hemodialysis and healthy subjects. Clin Immunol Immunopathol. 1990;55(2):237–241.

74. Stentz FB, Kitabchi AE. Transcriptome and proteome expression in activated human CD4 and CD8 T-lymphocytes. Biochem Biophys Res Commun. 2004;324(2):692–696.

75. Engelhardt W, Matzke J, Schmidt RE. Activation-dependent expression of low affinity IgG receptors Fc gamma RII(CD32) and Fc gamma RIII(CD16) in subpopulations of human T lymphocytes. Immunobiology . 1995;192(5):297–320.

76. Kawabe T,Maekawa N,Maeda Y, Hosoda M, Yodoi J. Induction of Fc epsilon RII/CD23 on phytohemagglutinin-activated human peripheral blood T lymphocytes. I. Enhancement by IL-2 and IL-4. J Immunol. 1991;147:548–553.

77. Stentz FB, Kitabchi AE. Hyperglycemia-induced activation of human T-lymphocytes with de novo emergence of insulin receptors and generation of reactive oxygen species. Biochem Biophys Res Commun. 2005;335(2):491–495.

78. Torres JM, Laborda J, Naval J, Darracq N, Calvo M, Mishal Z, et al. Expression of alpha-fetoprotein receptors by human T-lymphocytes during blastic transformation. Mol Immunol. 1989;26(9):851–857.

79. Carrel S, Salvi S, Isler P, Rapin C, Hayoz D, Gallay P, et al. gp33-38, an early human T cell activation antigen. J Immunol. 1990;144:2053–2062.

80. Molinero LL, Fuertes MB, Rabinovich GA, Fainboim L, Zwirner NW. Activation-induced expression of MICA on T lymphocytes involves engagement of CD3 and CD28. J Leukoc Biol. 2002;71:791–797.

81. Tate G, Katagiri M. Analysis of HLA-class II antigen mRNAs on T lymphoblasts. Hokkaido Igaku Zasshi. 1986; 61(2):249–256.

82. Moriya N, Sanjoh K, Yokoyama S, Hayashi T. Mechanisms of HLA-DR antigen expression in phytohemagglutinin-activated T cells in man. Requirement of T cell recognition of self HLA-DR antigen expressed on the surface of monocytes. J Immunol. 1987;139:3281–3286.

83. Puppo F, Gurreri G, Pattarini R, Ranise A, Indiveri F. Proliferation in autologous mixed lymphocyte reactions, expression of HLA class II antigen and serum immunomodulatory activity in patients with renal insufficiency on chronic dialysis. Allergol Immunopathol (Madr). 1991;19(4):167–173.

84. Ravindranath MH, Terasaki PI, Pham T, Jucaud V, Kawakita S. Therapeutic preparations of IVIg contain naturally occurring anti-HLA-E antibodies that react with HLA-Ia (HLA-A/-B/-Cw) alleles. Blood. 2013;121(11):2013-2028. doi: 10.1182/blood-2012-08-447771.

85. Ravindranath MH, Zhu D, Pham T, Jucaud V, Hopfield J, Kawakita S, et al. Anti-HLA-E monoclonal antibodies reacting with HLA-la and lb alleles like IVIg as potential IVIg-immunomimetics: an evolving therapeutic concept. Clin Transpl. 2013;293-305.

86. Ravindranath MH, Taniguchi M, Chen CW, Ozawa M, Kaneku H, El-Awar N, Cai J, Terasaki PI. HLA-E monoclonal antibodies recognize shared peptide sequences on classical HLA class Ia: relevance to human natural HLA antibodies. Mol Immunol. 2010;47(5):1121-1131. doi: 10.1016/j.molimm.2009.

87. Ravindranath MH, Kaneku H, El-Awar N, Morales-Buenrostro LE, Terasaki PI. Antibodies to HLA-E in nonalloimmunized males: pattern of HLA-Ia reactivity of anti-HLA-E-positive sera. J Immunol. 2010;185(3):1935-1948. doi: 10.4049/ jimmunol.1000424.

88. Ravindranath MH, Pham T, El-Awar N, Kaneku H, Terasaki PI. Anti-HLA-E mAb 3D12 mimics MEM-E/02 in binding to HLA-B and HLA-C alleles: Web-tools validate the immunogenic epitopes of HLA-E recognized by the antibodies. Mol Immunol. 2011;48(4):423-430. doi: 10.1016/j.molimm.2010.09.011.

89. Allen RL, Trowsdale J. Recognition of classical and heavy chain forms of HLA-B27 by leukocyte receptors. Curr Mol Med. 2004;4(1):59–65.

90. Santos SG, Antoniou AN, Sampaio P, Powis SJ, Arosa FA. Lack of tyrosine 320 impairs spontaneous endocytosis and enhances release of HLA-B27 molecules. J Immunol. 2006;176:2942–2949.

91. Ravindranath MH, Terasaki PI, Pham T, Jucaud V. The Monospecificity of Novel Anti-HLA-E Monoclonal Antibodies Enables Reliable Immunodiagnosis, Immunomodulation of HLA-E, and Upregulation of CD8+ T Lymphocytes. Monoclon Antib Immunodiagn Immunother. 2015;34(3):135-53. doi: 10.1089/mab.2014.0096.

92. Sasaki T, Ravindranath MH, Terasaki PI, Freitas MC, Kawakita S, Jucaud V. Gastric cancer progression may involve a shift in HLA-E profile from an intact heterodimer to β2microglobulin-free monomer. Int J Cancer. 2014;134(7):1558-1570. doi: 10.1002/ijc.28484.

93. Ravindranath MH, Pham T, Ozawa M, Terasaki PI. Antibodies to HLA-E may account for the non-donor-specific anti-HLA class-Ia antibodies in renal and liver transplant recipients. Int Immunol. 2012;24(1):43-57.

94. Zhu D, Ravindranath MH, Terasaki PI, Miyazaki T, Pham T, Jucaud V. Suppression of allo-human leucocyte antigen (HLA) antibodies secreted by B memory cells in vitro: intravenous immunoglobulin (IVIg) versus a monoclonal anti-HLA-E IgG that mimics HLA-I reactivities of IVIg. Clin Exp Immunol. 2014;177(2):464-477. doi: 10.1111/cei.12307

95. Ravindranath MH, Terasaki PI, Pham T, Jucaud V, Kawakita S. Suppression of blastogenesis and proliferation of activated CD4(+) T cells: intravenous immunoglobulin (IVIg) versus novel anti-human leucocyte antigen (HLA)-E monoclonal antibodies mimicking HLA-I reactivity of IVIg. Clin Exp Immunol. 2014;178(1):154-177. doi: 10.1111/cei.12391.

96. Guild BC, Strominger JL. Human and murine class I MHC antigens share conserved serine 335, the site of HLA phosphorylation in vivo. J Biol Chem. 1984;259:9235–9240.

97. Billing R, Terasaki PI. Human leukemia antigen. I. Production and characterization of antisera. J Natl Cancer Inst. 1974;53(6):1635-1638.

98. Aster RH, Miskovich BH, Rodey GE. Histocompatibility antigens of human plasma. Localization to the HLD-3 lipoprotein fraction. Transplantation. 1973;16(3):205-210.

99. Rittner C, Täuberecht I. Demonstration of HL-A inhibitors in human serum. I. Does an association exist between HL-A and serum lipoproteins? Z Immunitatsforsch Exp Klin Immunol. 1973;146(2):123-135.

100. Oh SK, Pellegrino MA, Ferrone S, Sevier ED, Reisfeld RA. Soluble HL-A antigens in serum. I. Isolation and purification. Eur J Immunol. 1975;5(3):161-166.

101. Reisfeld RA, Allison JP, Ferrone S, Pellegrino MA, Poulik MD. HL-A antigens in serum and urine: isolation, characterization, and immunogenic properties. Transplant Proc. 1976;8(2):173-178.

102. Machy P, Truneh A, Gennaro D, Hoffstein S. Major histocompatibility complex class I molecules internalized via coated pits in T lymphocytes. Nature. 1987;328(6132):724-726.

103. Machy P, Truneh A, Gennaro D, Hoffstein S. Endocytosis and de novo expression of major histocompatibility complex encoded class I molecules: kinetic and ultrastructural studies. Eur J Cell Biol. 1987;45(1):126-136.

104. Reid PA, Watts C. Cycling of cell-surface MHC glycoproteins through primaquine-sensitive intracellular compartments. Nature. 1990;346(6285):655-657.

105. Pickl WF, Majdic O, Faé I, Reuschel R, Holter W, Knapp W. The soluble pool of beta 2microglobulin free HLA class I alpha-chains. Qualitative and quantitative characterization. J Immunol. 1993;151(5):2613-2622.

106. Allison JP, Pellegrino MA, Ferrone S, Callahan GN, Reisfeld RA. Biologic and chemical characterization of HLA antigens in human serum. J.Immunol. 1977;118(3):1004-1009.

107. Krangel MS. Two forms of HLA class I molecules in human plasma. Hum Immunol. 1987;20(2):155-165.

108. Pollard SG, Davies HF, Calne RY: Preoperative appearance of serum class I antigen during liver transplantation. Transplantation. 1990;49(3):659-660.

109. Demaria S, Schwab R, Bushkin Y.The origin and fate of beta 2m-free MHC class I molecules induced on activated T cells. Cell Immunol. 1992;142(1):103-113.

110. Demaria S, Bushkin Y. CD8 and beta 2microglobulin-free MHC class I molecules in T cell immunoregulation. Int J Clin Lab Res. 1993;23(2):61-69.

111. Demaria S, DeVito-Haynes LD, Salter RD, Burlingham WJ, Bushkin Y. Peptide-conformed beta2m-free class I heavy chains are intermediates in generation of soluble HLA by the membrane-bound metalloproteinase. Hum Immunol. 1999;60(12):1216-1226.

112. Demaria S, Schwab R, Gottesman SR, Bushkin Y. Soluble beta 2microglobulin-free class I heavy chains are released from the surface of activated and leukemia cells by a metalloprotease. J Biol Chem. 1994;269(9):6689-6694.

113. DeVito-Haynes LD, Demaria S, Bushkin Y, Burlingham WJ. The metalloproteinase-mediated pathway is essential for generation of soluble HLA class I proteins by activated cells in vitro: proposed mechanism for soluble HLA release in transplant rejection.Hum Immunol. 1998;59(7):426-434.

114. DeVito-Haynes LD, Jankowska-Gan E, Meyer KC, Cornwell RD, Zeevi A, Griffith B, et al. Soluble donor HLA class I and beta 2m-free heavy chain in serum of lung transplant recipients: steady-state levels and increases in patients with recurrent CMV infection, acute rejection episodes, and poor outcome. Hum Immunol. 2000;61(12):1370-1382.

115. Haynes LD, Bushkin Y, Love RB, Burlingham WJ. Interferon-gamma drives the metalloproteinase-dependent cleavage of HLA class I soluble forms from primary human bronchial epithelial cells. Hum Immunol. 2002;63(10):893-901.

116. Haynes LD, Waldman WJ, Bushkin Y, Love RB, Burlingham WJ. CMV-infected allogeneic endothelial cells initiate responder and bystander donor HLA class I release via the metalloproteinase cleavage pathway. Hum Immunol. 2005; 66(3):211-221.

117. Zhai Y, Knechtle S. Two distinct forms of soluble MHC class I molecules synthesized by different mechanisms in normal rat cells in vitro. Hum Immunol. 1998;59(7):404-414.

118. Shwetank, Date OS, Kim KS, Manjunath R. Infection of human endothelial cells by Japanese encephalitis virus: increased expression and release of soluble HLA-E. PLoS One. 2013;8(11):e79197. doi: 10.1371/journal.pone.0079197.

119. Davies HS, Pollard SG, Calne RY. Soluble HLA antigens in the circulation of liver graft recipients.Transplantation. 1989;47(3):524-527.

120. Rhynes VK, McDonald JC, Gelder FB, Aultman DF, Hayes JM, McMillan RW, et al. Soluble HLA class I in the serum of transplant recipients. Ann Surg. 1993;217(5):485-489.

121. DeVito-Haynes LD, Jankowska-Gan E, Sollinger HW, Knechtle SJ, Burlingham WJ. Monitoring of kidney and simultaneous pancreas-kidney transplantation rejection by release of donor-specific, soluble HLA class I. Hum Immunol. 1994;40(3):191-201.

122. McMillan RW, Gelder FB, Zibari GB, Aultman DF, Adamashvili I, McDonald JC. Soluble fraction of class I human histocompatibility leukocyte antigens in the serum of liver transplant recipients. Clin Transplant. 1997;11(2):98-103.

123. Yang CW, Kim TG, Kim YS, Han H, Chang YS, Yoon YS, et al. Serum soluble HLA class I antigen levels in Hemodialysis patients and following renal transplantation. Am J Nephrol. 1995;15(4):290-294.

124. Suciu-Foca N, Reed E, D'Agati VD, Ho E, Cohen DJ, Benvenisty AI, et al. Soluble HLA antigens, anti-HLA antibodies, and antiidiotypic antibodies in the circulation of renal transplant recipients. Transplantation. 1991;51(3):593-601.

125. Zavazava N, Hausmann R, Kraatz E, Müller-Ruchholtz W. Clinical relevance of soluble HLA and interaction of papain derived class I molecules with alloreactive CTL. Transpl Int. 1992;1:S606-8.

126. Zavazava N, Bottcher H, Ruchholtz WM. Soluble MHC class I antigens sHLA and anti-HLA antibodies in heart and kidney allograft recipients. Tissue Antigens. 1993;42:20-26.

127. Mathew JM, Shenoy S, Phelan D, Lowell J, Howard T, Mohanakumar T. Biochemical and immunological evaluation of donor-specific soluble HLA in the circulation of liver transplant recipients. Transplantation. 1996;62(2):217-223.

128. DeVito-Haynes LD, Jankowska-Gan E, Heisey DM, Cornwell RD, Meyer KC, Love RB, et al. Soluble HLA class I in epithelial lining fluid of lung transplants: associations with graft outcome. Hum Immunol. 1997;52(2):95-108.

129. Colvin RB. Antibody-mediated renal allograft rejection: Diagnosis and pathogenesis. J Am Soc Nephrol. 2007;18(4):1046–1056.

130. Cornell LD, Smith RN, Colvin RB. Kidney transplantation: Mechanisms of rejection and acceptance. Annu Rev Pathol. 2008;3:189–220.

131. Shimizu T, Ishida H, Shirakawa H, Omoto K, Tsunoyama K, Tokumoto T, et al. Clinicopathological analysis of acute vascular rejection cases after renal transplantation. Clin Transplant. 2010;22:22-26. doi: 10.1111/j.1399-0012.2010.01277.x.

132. Lefaucheur C, Loupy A, Vernerey D, Duong-Van-Huyen JP, Suberbielle C, Anglicheau D, et al. Antibody-mediated vascular rejection of kidney allografts: a population-based study. Lancet. 2013;381(9863):313-319. doi: 10.1016/S0140-6736(12)61265-3.

133. Reed EF, Hong B, Ho E, Harris PE, Weinberger J, Suciu-Foca N. Monitoring of soluble HLA alloantigens and anti-HLA antibodies identifies heart allograft recipients at risk of transplant-associated coronary artery disease. Transplantation. 1996;61(4):566-572.

134. Loupy A, Vernerey D, Viglietti D, Aubert O, Van Huyen JP, Empana JP, et al. Determinants and Outcomes of Accelerated Arteriosclerosis:

Major Impact of Circulating Antibodies. Circ Res. 2015;117(5):470-82. Doi: 10.1161/CIRCRESAHA.117.306340.

135. Cai J, Terasaki PI, Bloom DD, Torrealba JR, Friedl A, Sollinger HW, et al. Correlation between human leukocyte antigen antibody production and serum creatinine in patients receiving sirolimus monotherapy after Campath-1H induction. Transplantation. 2004;78(6): 919-924.

136. Zavazava N, Leimenstoll G, Müller-Ruchholtz W. Measurement of soluble MHC class I molecules in renal graft patients: a noninvasive allograft monitor. J Clin Lab Anal. 1990;4(6):426-429.

137. Tilg H, Westhoff U, Vogel W, Aulitzky WE, Herold M, Margreiter R, et al. Soluble HLA class I serum concentrations increase with transplant-related complications after liver transplantation. J Hepatol. 1992;14(2-3):417-419.

138. Puppo F, Pellicci R, Brenci S, Nocera A, Morelli N, Dardano G, et al. HLA class-I-soluble antigen serum levels in liver transplantation. A predictor marker of acute rejection. Hum Immunol. 1994;40(3):166-170.

139. Rizzo M, Sundaresan S, Lynch J, Trulock EP, Cooper J, Patterson GA, et al. Increased concentration of soluble human leukocyte antigen class I levels in the bronchoalveolar lavage of human pulmonary allografts. J Heart Lung Transplant. 1997;16(11):1135-1140.

140. Hagihara M, Munkhbat B, Kanai N, Shimura T, Kiuchi T, Inomata Y, et al. Serum soluble human leucocyte antigen class I in paediatric liver transplantation with live, related donors. Transpl Immunol. 1997;5(3):219-224.

141. Parham P. Clayberger C, Zorn SL, Ludwig DS, Schoolnik GK, Krensky AM. Inhibition of alloreactive cytotoxic T lymphocytes by peptides from the α-2 domain of HLA-A2. Nature. 1987;325:625-628.

142. Smith MA, Naziruddin B, Poindexter NJ, Haynes AE, Howard T, Mohanakumar T. Liver transplant recipient sera derived soluble HLA mediates allele specific CTL apoptosis. Transplantation. 2000;69(1):157-162.

143. Puppo F, Contini P, Ghio M, Brenci S, Scudeletti M, Filaci G, et al. Soluble human MHC class I molecules induce soluble Fas ligand secretion and trigger apoptosis in activated CD8(+) Fas (CD95)(+) T lymphocytes. Int Immunol. 2000;12(2):195-203.

144. Puppo F, Contini P, Ghio M, Indiveri F. Soluble HLA class I molecules/ CD8 ligation trigger apoptosis of CD8+ cells by Fas/Fas-ligand interaction. Scientific World Journal. 2002;2:421-423.

145. Norment AM, Salter RD, Parham P, Engelhard VH, Littman DR. Cell-cell adhesion mediated by CD8 and MHC class I molecules. Nature. 1988;336(6194):79-81.

146. Zavazava N, Krönke M. Soluble HLA class I molecules induce apoptosis in alloreactive cytotoxic T lymphocytes. Nature Med. 1996;2(9):1005–1010.

147. Wilson BS, Indiveri F, Pellegrino MA, Ferrone S. Production and characterization of DR xenoantisera: use for detection of serum DR antigens. J Immunol. 1979;122(5):1967-71.

148. Indiveri F, Wilson BS, Natali PG, Pellegrino MA, Ferrone S. Immunogeneity of DR antigens isolated from body fluids. Transplant Proc. 1980;12(1):121-4.

149. Coupel S, Moreau A, Hamidou M, Horejsi V, Soulillou JP, Charreau B. Expression and release of soluble HLA-E is an immunoregulatory feature of endothelial cell activation. Blood. 2007;109(7):2806-2814.

150. Herlyn M, Lange B, Bennicelle J. Increased levels of circulating HLA-DR antigens in sera of patients with acute lymphoblastoid leukemia. Leuk Res. 1984;8(3):323-334.

151. Thompson S, Wareham M, Pearson ADJ, Sviland L, Turner GA. A preliminary study of serum class II levels in healthy individuals and bone marrow transplant patients. Clin Chim Acta. 1989;185(1):45-52.

152. Westhoff U, Thinnes FP, Gotz H, Grosee-Wilde H. Quantitation of soluble HLA class II molecules by an enzyme-linked immunosorbent assay. Vox Sang. 1991;61(2):106-110.

153. McDonald JC, Adamashvili I, Hayes JM, Aultman DF, Rhynes VK, Gelder FB. Soluble HLA class II concentrations in normal individuals and transplant recipients. Comparison with soluble HLA class I concentrations. Transplantation. 1994;58(11):1268-1272.

154. Pouletty P, Ferrone S, Amesland F, Cohen N, Westhoff U, Charron D, et al. Summary report from the first international workshop on soluble HLA antigens. Paris, August 1992. Tissue Antigens. 1993;42(1):45-54.

155. Cooper JE, Gralla J, Cagle L, Goldberg R, Chan L, Wiseman AC. Inferior kidney allograft outcomes in patients with de novo donor-specific antibodies are due to acute rejection episodes. Transplantation. 2011;91(10):1103-1109.

156. DeVos JM, Gaber AO, Teeter LD, Graviss EA, Patel SJ, Land GA, et al. Intermediate-term graft loss after renal transplantation is associated with both donor-specific antibody and acute rejection. Transplantation. 2014;97(5):534-540.

157. Vadim J, Ravindranath MH, Terasaki PI. Immunobiology of HLA Class-Ib molecules in transplantation. SOJ Immunology, 2015;3(4).

158. Tongio MM, Falkenrodt A, Mitsuishi Y, et al. Natural HLA Abs. Tissue Antigens, 1985;26(5):271-285. doi: 10.1111/j.1399-0039.1985.tb02225.x.

159. Ameglio F, Saba F, Bitt A, Aceti, N. Tanigaki, R. Sorrentino, A, et al. Antibody reactivity to HLA Classes I and II in sera from patients with hydatidosis. J Infect Dis. 1987;156(4):673-676.

160. Májský A. Natural HLA-A, B and DR Abs in the serum of non-immunized men. Vnitr Lek. 1989;35(5):467-471.

161. Luscher MA, Choy G, Embree JE, Nagelkerke NJ, Bwayo JJ, Njenga S, et al. Anti-HLA alloantibody is found in children but does not correlate with a lack of HIV type 1 transmission from infected mothers. AIDS Res Hum. Retroviruses. 1988;14(2):99-107.

162. Zhou B, Saito S, Nakazawa Y, Kobayashi N, Matsuda M, Matsumoto Y, et al. Existence of an immunoglobulin G component of naturally occurring HLA class I antibodies that are not directed against self-antigens in human serum. Tissue Antigens. 2008;72(2):98-104. doi: 10.1111/j.1399-0039.2008.01074.x.

163. Lee J, Romero R, Xu Y, Miranda J, Yoo W, Chaemsaithong P. Detection of anti-HLA antibodies in maternal blood in the second trimester to identify patients at risk of antibody-mediated maternal anti-fetal rejection and spontaneous preterm delivery. Am J Reprod Immunol. 2013;70(2):162-75. doi: 10.1111/aji.12141.

164. Ravindranath MH, Terasaki PI, Maehara CY, Jucaud V, Kawakita S, Pham T, et al. Immunoglobulin (Ig)G purified from human sera mirrors intravenous Ig human leucocyte antigen (HLA) reactivity and recognizes one's own HLA types, but may be masked by Fab complementarity-determining region peptide in the native sera. Clin Exp Immunol. 2015;179(2):309-328. doi: 10.1111/cei.12450.

165. van Rood JJ, van Leeuwen A, van Santen MC. Anti HL-A2 inhibitor in

normal human serum. Nature. 1970;226(5243):366-367.

166. Charlton RK, Zmijewski CM. Soluble HL-A7 antigen: localization in the beta-lipoprotein fraction of human serum. Science. 1970;170(3958):636-637.

167. Billing RJ, Mittal KK, Terasaki PI. Isolation of soluble HL-A antigens from normal human sera by ion exchange chromatography. Tissue Antigens. 1973;3(4):251-256.

168. Billing RJ, Terasaki PI. Purification of HL-A antigens from normal serum. J Immunol. 1974;112(3):1124-30.

169. Billing R, Terasaki PI. Human leukemia antigen. II. Purification. J Natl Cancer Inst. 1974;53(6):1639-1643.

170. Billing RJ, Terasaki PI. Rabbit antisera to HL-A9 isolated from normal serum. Transplantation. 1974;17(2):231-234.

171. Jucaud V, Ravindranath MH, Terasaki PI. Immunobiology of HLA Class-Ib Molecules in Transplantation. SOJ Immunol. 2015;3(4):1-15.DOI: dx.doi.org/10.15226/2372-0948/3/4/00137.

9

Immunobiology of HLA Class-Ib Molecules in Transplantation

Vadim Jucaud*, Mepur H. Ravindranath, Paul I. Terasaki

Terasaki Foundation Laboratory, 11570 W. Olympic Blvd, Los Angeles, CA 90064, California, USA

***Corresponding author:** Vadim Jucaud, Terasaki Foundation Laboratory, 11570 W. Olympic Blvd, Los Angeles, CA 90064, USA, E-mail: vjucaud@terasakilab.org*

Absract

The objective of this review is to elucidate the role of HLA-Ib molecules in transplantation after elucidating their immunobiological potential. Structurally, the non-classical HLA-Ib molecules (HLA-E, HLA-F and HLA-G) are less polymorphic than HLA-Ia molecules. In transplantation, HLA-Ib molecules are emerging as immune regulators, functioning as ligands for immunomodulatory cell surface receptors expressed by the subsets of NK and CD8+ T cells-the major players in allograft rejection.

HLA-E is the most pleiotropic molecule in an allograft setting; it can interact with both inhibitory (CD94/NKG2A) and activating (CD94/NKG2C) receptors expressed by NK and CD8+ T cells. This interaction is dependent on the nature and source of peptides presented by HLA-E. When HLA-Ia-derived peptides are presented, HLA-E interacts with CD94/NKG2A, inhibiting the cytotoxic cell functions that promote graft survival. When the HLA-G leader sequence is presented, it interacts mainly with CD94/NKG2C to activate the cytotoxic cells, leading to graft rejection. In addition, HLA-E can present viral and bacterial peptides that can bind to both CD94/NKG2 receptors, and also can interact with the other receptors of CD8+ cells, enhancing the risk of allograft rejection.

HLA-G and HLA-F can promote graft acceptance by binding to another family of receptors: the Ig-like transcripts (ILT2/ILT4) expressed by NK cells. Consequently, higher levels of HLA-G on the cell surface and in circulation promote graft acceptance, survival, and immunosuppression-free status for graft recipients. In contradistinction, overexpression of HLA-Ib results in higher risk of developing Graft-versus-Host Disease (GvHD) in cell transplantation.

Soluble HLA-Ib (sHLA-Ib) molecules are augmented in circulation during injury, inflammation and transplantation. The sHLA-Ib can be free or associated with β2-microglobulin. In addition, sHLA-Ib molecules occur in a variety of conformations (isoform); as a result, HLA-Ib exposes epitopes to different immune components, contributing to antibody production. These antibodies can be monospecific or polyreactive. Binding of the monospecific antibodies can block the interaction of HLA-Ib with inhibitory receptors on cytotoxic cells, affecting graft survival. Polyreactive antibodies that bind to both cryptic and non-cryptic domains can affect routine HLA-Ia antibody screening and organ allocation. The antibodies binding to a cryptic domain are capable of suppressing blastogenesis and proliferation of CD4+ T cells as well as secretion of HLA antibodies by B cells; both could prolong graft survival.

Keywords: HLA-Ib; HLA-E; HLA-F; HLA-G; Transplantation outcome; HLA antibodies

Abbreviations

APC: Antigen-Presenting cells; β2m: Beta2-Microglobulin; BMT: Bone Marrow Transplantation; DSA: Donor-Specific Antibody; ER: Endoplasmic Reticulum; GvHD: Graft-versus-Host Disease; HLA: Human Leukocyte Antigen; HSCT: Hematopoietic Stem Cell Transplantation; IFN: Interferon; ILT2: Ig-like Transcript 2; ILT4: Ig-like Transcript 4; MHC: Major Histocompatibility Complex; MLR: Mixed Lymphocyte Reaction; mAbs: Monoclonal Antibodies; NFκB: nuclear factor κB; NK: Natural Killer; sHLA: Soluble HLA; TCR: T Cell Receptors; URR: Upstream Regulatory Region; UTR: Untranslated Region.

Introduction

In humans, the Major Histocompatibility Complex (MHC) is called the "Human Leukocyte Antigen" (HLA). It consists of more than 200 genes located on chromosome 6 (6p21.31). There are three groups in this large gene family: class I, class II, and class III. The MHC class I genes are divided into classical HLA-Ia (HLA-A, -B and -Cw) and non-classical HLA-Ib (HLA-E, -F and -G). The MHC class II genes include HLA-DR, -DQ and -DP. The MHC class III genes include components of the complement system involved in inflammation and other immune system activities. Every individual inherits a pair of each MHC gene, one maternal and the other paternal.

The classical HLA-Ia molecules are expressed on the surface of all nucleated cells except those of specialized tissues such as the brain [1]. HLA-II molecules are expressed on the surface of most endothelial cells and other specialized tissues and cells, such as B cells [2]; indeed, HLA-II molecules are primarily expressed by Antigen-Presenting Cells (APC) [3] and are critically involved in promoting adaptive immunity [4,5]. Most importantly, the endothelium, which acts as a functional barrier between the vessel wall and the blood stream, expresses both HLA-I and -II on its surface. In transplant immunology, cell surface HLA-Ia and HLA-II influence the outcome of the allograft survival [6,7] due to their high polymorphism [8,9]. Without immunosuppression, the fate of the allograft in the host depends on the degree of genetic compatibility between the donor and recipient, and is dependent on the effect of the sum of matches or mismatches [10].

HLA genotypes of donor and recipient [11] and the antibodies

***Corresponding author email:** vjucaud@terasakilab.org*

generated against donor HLA in the recipient are critical for understanding the post-transplantation course of an allograft [12]. When a donor and a recipient do not share one or more HLA alleles of their genotype, it is referred to as a mismatch. A minimal mismatch between donor and recipient HLA genotypes favors better allograft survival since mismatched HLAs become "non-self" antigenic targets recognized by the recipient immune system, resulting in a cell- and antibody-mediated immune response, consequently damaging allograft integrity and function and leading to allograft rejection.

Antibody response against a donor-derived HLA allele-known as Donor-Specific Antibody (DSA)-can be detrimental to allograft survival, so the evidence that, DSA is present prior to transplantation, is critical for an appropriate organ allocation since post-transplant DSA places the allograft at the risk of rejection [12]. Therefore, HLA-Ia and -II, and the antibodies generated by them in the circulation of recipients, have been the major focus in the immunobiology of clinical transplants. It is in this context that the recent discovery of non-classical HLA-Ib molecules brings a new dimension to the immunobiology of transplantation and a reappraisal of graft tolerance and/or acceptance.

Unlike classical HLA-Ia, non-classical HLA-Ib genes and molecules are oligomorphic, with restricted and selective tissue distribution [13,14]. Our understanding of the structural dynamics and functionality of HLA-Ib is progressively increasing, though HLA-Ib has yet to be fully characterized. Recent reports have attributed immunomodulatory functions to these molecules, and antigen presentation for HLA-E only; they may play an important role in immune responses and regulation, both in innate and adaptive immunity-i.e., through interactions with, respectively, Natural Killer (NK) cells, dendritic cells, and naive/ effector T and B lymphocytes, antibodies [15]. The HLA-Ib molecules are capable of interacting with cell-surface receptors present on specific immune-cell subsets, inducing activation or inhibition of signaling cascades within such specific immune cells as NK cells [15,16]. These functions are apart of the innate immunity [17]; e.g., HLA-Ib is expressed during pregnancy, playing a major role in tolerance of the fetus-the only true physiological analog of a semi-allograft [18]. Therefore, the innate immunity of HLA-Ib is often attributed to those molecules' tolerance-inducing functions, by interacting with cell surface receptor located on cells that are part of the innate immune system, such as NK cells and macrophages.

As part of adaptive immunity [18], HLA-E is capable of binding to peptides for antigen presentation in order to monitor cellular stress, integrity and function [19]. HLA-G is capable of modulating the adaptive immunity [20], as well as exerting antigen-presenting functions, although the latter is not its primary function [21,22]. HLA-Ib molecules also can generate a pool of antibodies *in vivo*, which can be either monospecific or polyreactive (cross-reactive with other HLA-I molecules) [23,24], thus pointing out the immunogenic potential of HLA-Ib molecules *in vivo* [25-27]. The specificity of such antibodies can impact the outcome of transplantation, as well as the allocation of organs. HLA-Ib molecules are thus capable of playing a role in the adaptive immunity linked to organ transplantation.

The low polymorphism and relative invariance of non-classical HLA-Ib molecules seems to define their functions in the allograft microenvironment [28]. Their precise role, whether protective or pathogenic, remains to be clarified. As observed, HLA-Ia and -Ib and the antibodies elicited by them may play a role in organ transplantation. This review sheds light on this aspect of innate and adaptive immunity in understanding the immunobiology of HLA-Ib in transplantation.

HLA-Ib mediated innate immunity

HLA-Ib is naturally occurring molecules; they are primarily characterized during pregnancy and are considered to play a role in the tolerance of the fetus-the physiological equivalent of a semi-allograft [29,30]. The engineering of HLA-Ib tetrameric complexes was used to study their function and potential ligands [31]. Their interaction with different immunomodulatory (activating and/ or inhibiting) cell-surface receptors present on NK cells and macrophages signify their role in the innate immunity; these receptors include CD94/NKG2, Ig-Like Transcript 2 (ILT2), Ig-like transcript 4 (ILT4), KIR2DL4, and CD160.

HLA-E

HLA-E has 17 alleles (1 null allele), encoding 6 proteins [9], and is expressed in most tissues at low levels. Like HLA-Ia, HLA-E occurs on the surface of cells as a trimolecular complex comprised of HLA-E heavy chain, β2-microglobulin (β2m) and a peptide. In addition, HLA-E can be released into the microenvironment of the allograft and in circulation as soluble HLA (sHLA)-E [32,33]. In contrast to HLA-Ia, HLA-E binds to a restricted set of peptides, commonly derived from the leader sequence of HLA-Ia and HLA-G molecules [34-36]; it can also bind to viral, bacterial and stress protein-derived peptides [15,19,37]. Two variants of HLA-E are well studied: the difference between them lies in one amino acid at position 107, located on a loop between β-strands in the α2 domain of the heavy chain, one with arginine at position 107- HLA-E*01:01 (HLA-E^{R107})-and the other with glycine at the same position, HLA-E*01:03 (HLA-E^{G107}) [38]. The difference in the amino acid residue is significant in terms of charge, size and conformation [28], affecting the relative peptide binding affinity of each allele, and consequently the cell-surface expression of HLA-E.

Several factors can upregulate HLA-E expression on normal and malignant cells. Proinflammatory cytokines-e.g., interferon (IFN)-γ [28,39]- and the availability of several peptides [19] upregulate HLA-E expression. HLA-E acts as both an inhibitory and activating ligand for CD94/ NKG2 receptors on NK cells and T lymphocytes [31]. The interaction between HLA-E and CD94/ NKG2 receptors is influenced by the sequence of peptide presented by HLA-E [37,40,41]. When HLA-E is loaded with a peptide derived from the HLA-Ia leader sequence, it interacts with the inhibitory CD94/NKG2A receptor [42,43] whereas, when loaded with HLA-G leader sequence peptide, HLA-E interacts mainly with the activating receptor CD94/NKG2C [36]. HLA-E loaded with viral peptides interacts with CD94/NKG2A *in vivo*, showing a higher affinity, while, *in vitro*, increasing the population of NK cells that express CD94/NKG2C [41]. These observations illuminate the immunomodulatory potential of HLA-E. Furthermore, the HLA-E/ peptide complex plays a crucial role in immunosurveillance mediated by NK cells and T lymphocytes, and is thus involved in both innate and adaptive immunity [37].

HLA-G

HLA-G has 50 alleles (2 null alleles), encoding 16 proteins [9] with 7 protein isoforms [44]. Four of these proteins are membrane-bound (HLA-G1, -G2, -G3 and -G4), while 3 are soluble (HLA-G5, -G6 and -G7) [45]. HLA-G1 is the full-length heavy chain, having α1, α2 and α3 helices, a transmembrane domain, and a cytoplasmic tail associated with β2m. HLA-G2 lacks the α2 domain, HLA-G3 has only the α1 domain, and HLA-G4 lacks the α3 domain [44]. The three sHLA isoforms, HLA-G5, -G6 and -G7, are the counterparts, respectively, of HLA-G1, G-2 and -G3. HLA-G5 and -G6 retain intron 4, and HLA-G7 retains intron 2, both including a stop codon that prevents the translation of the transmembrane domain and cytoplasmic tail [44]. The high diversity of HLA-G molecular structures is due to the alternative splicing of the primary transcript [46], suggesting the versatility of HLA-G molecules *in vivo*.

HLA-G immunoregulatory properties are attributed mainly to the inhibition of different immune-cell populations. The interaction of HLA-G with immune effector cells is mediated by ILT2 on T cells, B cells, NK cells and APC, by ILT4 on APC (myeloid cells), by KIR2DL4 on T cells and NK cells, and by CD160 on T cells, NK cells and endothelial cells [45]. However, earlier studies using engineered HLA-G tetrameric complexes showed positive binding of HLA-G on CD14+ monocytes, especially a CD16+ subset; in contrast, this HLA-G tetramer did not significantly bind to CD56+ (NK cells), CD3+ (T cells) or CD19+ (B cells) [31]. HLA-G tetramers may fail to bind to certain ligands if the affinity is insufficient or if cell-surface expression is low, which explains the negative staining of immune cells having HLA-G ligand on their surface, and also implies that the factors augmenting both the ligand and receptors are prerequisites for initiating immunomodulation.

HLA-F

HLA-F is the least investigated and understood of non-classical HLA molecules, but it has a structure similar to that of HLA-Ia and –Ib, and has 22 alleles, encoding 4 proteins [9]. Unlike HLA-Ia, it has an intracytoplasmic domain [47]. Although HLA-F associates with β2m [48], no information is available on peptide presentation. It is thought to be biologically similar to other HLA-Ib molecules, such as in pregnancy [29,30] and cancer [32,49]; HLA-F also occurs in circulation as sHLA-F [32] like other sHLA-I molecules. However, the exact function of HLA-F remains to be elucidated. As with other HLA-I molecules, HLA-F expression is upregulated by IFN-γ. The IFN-stimulated response element motif of HLA-F displays higher homology to HLA-Ia loci than to HLA-Ib loci [50]. In addition, HLA-F expression is inducible by nuclear factor κB (NFκB) through the κB1 site of enhancer A, located in the proximal promoter region [51].

No cell-surface expression of HLA-F is observed on any of the resting T-cell, B-cell, or NK-cell subsets despite the presence of intracellular HLA-F [52]; however, it is upregulated on activated lymphocytes [52] although there is no difference in the total level of HLA-F proteins before or after activation. The intracellular expression of HLA-F is independent of lymphocyte activation, but surface expression is promoted upon activation. What are the most important, regulatory T cells do not express HLA-F upon activation, in contrast to memory T cells. Moreover, HLA-F was found to be entirely dependent on its cytoplasmic tail for export from the Endoplasmic Reticulum (ER) [53]- in contrast to HLA-Ia molecules-suggesting an alternative function for HLA-F that is independent of loading with peptides in the ER.

The HLA-F molecule is known to bind to the surface of monocytes, including cells in both CD14+high and CD14+mid populations [54] and to a subpopulation of CD19+ B cells, but not to CD56+ NK cells or CD3+ T cells. HLA-F was shown to bind to ILT2 and ILT4 receptors [31,54]; however, the interaction with ILT2 and ILT4 was only partially inhibited by monoclonal antibodies (mAbs) specific for ILT2 and ILT4. So HLA-F may also interact with other receptors expressed on B cells and monocytes [31]. Interestingly enough, HLA-F may dimerize or combine with β2m-free HLA-Ia to present exogenous peptide and/or to interact with NK cells [55-57]. This is indeed a significant finding, and deserves attention, because of its significance in organ transplantation.

In conclusion, HLA-Ib molecules are essential for the normal development of human fetuses, and are able to induce the acceptance of the fetus by the mother even though they probably share only one HLA haplotype. It certainly would seem that the cell-surface expression of HLA-Ib molecules on the allograft-as well as soluble HLA-Ib in circulation-may play a role in allograft survival, particularly considering the major actors in allograft rejection (NK cells, CD4+ and CD8+ T cells) [58]. Figure 1 summarizes how HLA-Ib molecules expressed on the surface of an allograft can affect transplantation outcome. Only the innate immunity mediated by HLA-Ib molecules has been described in a transplant setting. HLA-E expressed on the allograft interacts with CD94/NKG2A inhibitory receptors or CD94/NKG2C activating receptors, and this interaction depends on the peptide loaded onto HLA-E [36]. When HLA-E is loaded with a peptide derived from the leader sequence of HLA-Ia molecules, the HLA-E/HLA-Ia-peptide complex interacts with CD94/NKG2A expressed on NK and T cells [36,37], which leads to the inhibition of cytotoxicity [40] and cell-mediated damage of the allograft tissue and, in turn, leads towards graft acceptance. In contrast, HLA-E loaded with a peptide derived from the leader sequence of HLA-G molecules will preferentially bind to CD94/NKG2C activating receptors [41], leading to the activation of cytotoxicity and secretion of cytokines, in turn promoting antibody production and eventually graft loss. This is important from the perspective of viremia (CMV, BK virus) associated with the allograft as part of adaptive immunity. HLA-G expressed on the allograft interacts with ILT2, ILT4 and KIR2DL4 receptors on NK and T cells, leading to the inhibition of cytotoxicity and possibly towards graft acceptance [29,45]. Similarly, HLA-F will interact with ILT2 and ILT4

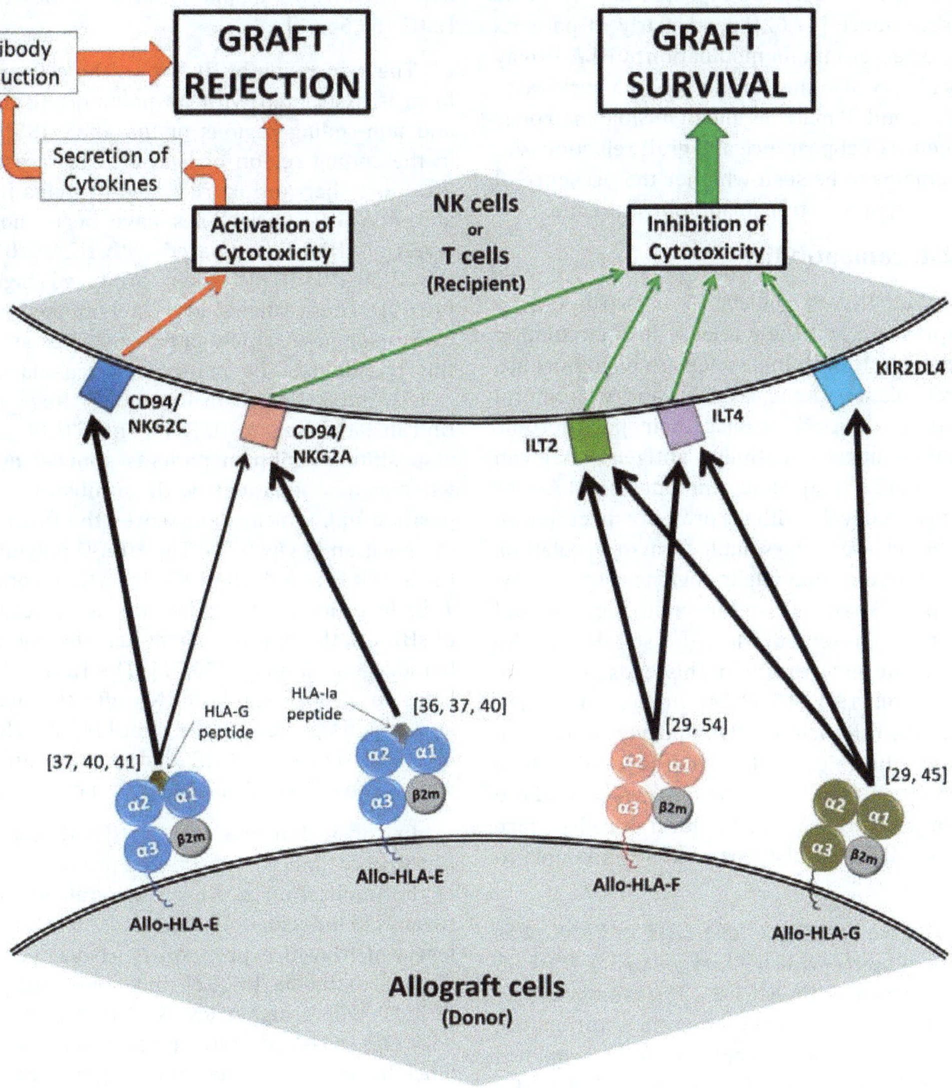

Figure 1: The innate interaction between HLA-Ib molecules and their respective immune receptors can affect transplantation outcomes. HLA-E molecules expressed on the surface of the allograft, can present peptides derived from the leader sequence of HLA-Ia and HLA-G. When HLA-E is presenting a non-allogeneic peptide, allograft cells inhibit the cytotoxic activity of NK and T cells by interacting with CD94/NKG2A receptors, inducing graft acceptance. However, HLA-E loaded with a peptide derived from HLA-G leader sequence interacts mainly with CD94/NKG2C receptors, leading to the activation of cytotoxicity, and thus rejection. HLA-G molecules expressed on the surface of the allograft can interact with ILT2, ILT4, and KIR2DL4 receptors expressed on the surface of NK and T cells, thus inhibiting the cytotoxic activity of NK and T cells, which can induce graft acceptance. Similarly, HLA-F molecules expressed on the surface of the allograft can interact with ILT2 and ILT4 receptors expressed on the surface of NK and T cells, possibly inducing graft acceptance; however, cell surface expression of HLA-F was not demonstrated on allograft.
Abbreviations: ILT2: Ig-like transcript 2; ILT4: Ig-like transcript 4; KIR2DL4: Killer cell immunoglobulin-like receptor 2DL4; β2m: beta 2-microglobulin.

receptors [29,54] leading to the inhibition of cytotoxicity and graft acceptance; however, it remains to be elucidated whether allograft-associated vascular and endothelial cells, infiltrating leukocytes, can express HLA-F on their surface. In the case of Bone Marrow Transplantation (BMT) or Hematopoietic Stem Cell Transplantation (HSCT), HLA-F can be expressed on donor immune cells only if they become activated [52]. A remaining question arises with these observations: what is the influence of a specific HLA-Ib genotype on the outcome of transplantation? Should HLA-Ib typing be considered in evaluating the degree of risk in organ transplantation?

Non-Classical HLA-Ib and Transplantation Outcome

In spite of the low polymorphism of HLA-Ib molecules, their allelic variants produce protein variants with divergent cellular expression, peptide presentation and soluble forms. While HLA-Ia and HLA-II typing of both recipients and donors is routinely performed prior to organ transplant, the importance of typing HLA-Ib has not yet become widely acknowledged. It is interesting that HLA-E mismatches were observed in patients with matched HLA-Ia [59-61]. Paradoxically, the immunomodulatory mechanisms mediated by HLA-Ib can be expected to play a role in promoting graft acceptance [47,62], particularly in patients without viremia. This aspect of immunomodulation by HLA-Ib may clarify the variability observed in the duration of graft survival in transplants with HLA-Ia and -II matches and in mismatched ones when the time between DSA appearance and graft rejection was substantial [12]. It remains to be seen whether the presence of HLA-Ib specific alleles can predict transplantation outcome.

HLA-Ib: Role in histocompatibility

Histocompatibility of HLA-Ib molecules is attributed to their cell-surface expression and their release into circulation as sHLA [38,62]. Although, HLA-Ib molecules are oligomorphic, they can induce in transplant patients a cellular and/or humoral immunological response directed against their polymorphic residues; and, like minor histocompatibility antigens, they can affect allograft survival [12,47]. Specific genotypes of HLA-E or HLA-G are associated, respectively, with the protein expression on the cell surface and the release of the soluble form in circulation. Indeed, the coding and non-coding allelic regions-respectively, exons and introns (plus 5' and 3' untranslated regions)-and the nature of the peptide presented affect the stability of the messenger (m) RNA and the mature protein. This leads to variable levels of protein expression [19,43,45,62-64]. On the other hand, HLA-F is expressed on the cell surface of lymphocytes only upon activation [52] and can be released into the circulation in a soluble form [32]; however, there is not yet any evidence of the effect of specific genotypes of HLA-F or the level of cell-surface expression. The role of HLA-F in histocompatibility remains to be evaluated.

The level of cell-surface expression of HLA-E in PBMC was significantly higher in subjects with homozygous HLA-E*01:03 than in those with homozygous HLA-E*01:01. The same was true for peptide-induced HLA-E surface expression. In addition, the level of cell-surface expression was dependent on the sequence of the peptide presented by HLA-E [19]. Indeed, the CMV, HLA-B7 and HLA-B15 peptides induced higher expression of cell-surface HLA-E than those derived from Hsp60, and the induction of surface HLA-E is always significantly higher in patients with homozygous HLA-E*01:03 regardless of the peptide presented. The expression of cell-surface HLA-E is regulated at the post-transcriptional level since there is no difference in mRNA expression levels between HLA-E genotypes or between peptide-induced cell-surface HLA-E [19]. These observations indicate that the cell-surface expression of HLA-E is dependent on HLA-E genotypes and the nature of peptides presented by the HLA-E molecule. There are functional implications for the allelic differences in the surface expression of HLA-E. Higher cell surface expression of HLA-E*01:03 may result in a greater capability of inhibiting immune cells through CD94/NKG2A receptors [65] provided that the peptide presented by HLA-E is derived from the leader sequence of HLA-Ia. Therefore, HLA-E*01:03 homozygous recipients may be better graft "acceptors" than HLA-E*01:01 patients provided that the allograft is devoid of viremia. Withal, a HLA-E genotype match should be beneficial in transplantation because the HLA-E homozygous state has been shown to be associated with transplantation outcomes, especially in BMT and HSCT [59,66,67].

The expression of HLA-G on the cell surface and its soluble form is associated with its polymorphism, both in the coding and non-coding regions of the allele [35]. The polymorphism in the coding region of HLA-G genes creates protein variants, including aberrant or truncated proteins [68]. HLA-G*01:01:03 and HLA-G*01:05N alleles have been shown to produce low levels of sHLA-G compared with HLA-G*01:01:01; in contrast, the HLA-G*01:01:04 allele produced high levels of sHLA-G [46,69]. These alleles are likely linked to their cell-surface expression level, although more studies are needed to elucidate this possibility. The non-coding regions of HLA-G alleles are called "the 5' Upstream Regulatory Region (URR)" and the "3' Untranslated Region (UTR)." The 5'URR polymorphism (C→G) at position -725bp in patients showed no detectable sHLA-G whereas that of patients with no substitution (C→G) at the same position did, which underscores the functional implications of the position *in vivo* [70]. The HLA-G polymorphism observed in the 3'UTR is called "the 14bp insertion" (or deletion). The 3'UTR 14bp insertion homozygous state is associated with lower levels of sHLA-G, the opposite being true for the 3'UTR 14bp deletion homozygous state [44,70,71]. The HLA-G 3'UTR 14bp insertion leads to a more stable mRNA affecting its rate of translation, and thus its protein expression [44,72]. The non-coding region of HLA-G alleles-5' URR and 3' UTR affects the cell-surface expression of HLA-G and the presence of sHLA-G in circulation.

In conclusion, the role of HLA-Ib genes and molecules in histocompatibility has attracted interest in recent years because of the implication of HLA-Ib in graft acceptance, especially in regards to inducing tolerance. Most often, it is agreed that higher levels of HLA-Ib expression is associated with better allograft survival outcome [47,62], but some exceptions exist, such as in BMT, where higher levels of HLA-G expression is associated with higher risk of acute graft-versus-host disease (GvHD) [71]. Since heart, liver, kidney and lung allografts express HLA-Ib on their cell surface [47,62,73], even if the expression of HLA-Ib is low compared with that of HLA-Ia, both cell-surface expression and release in circulation of HLA-Ib are capable of affecting transplantation outcome. Selected patients with a high level of HLA-Ib expression, based on HLA-Ib genotypes, could benefit from lower immunosuppressive protocols while others could benefit from alternate therapeutic strategies.

HLA-E, viremia and transplantation

As part of adaptive immunity [15], HLA-E can present viral and bacterial-derived peptide from CMV, HCV, HIV, EBV, influenza, salmonella enterica and mycobacterium; and the HLA-E/peptide complex can interact with T-Cell Receptors (TCR) [40]. In transplantation, the potent immunosuppressive treatment regimen employed currently improves graft survival, but at the cost of increasing the recipient's susceptibility to infections [74] as well as to pre-existing viremia. Indeed, the viremia status of both donor and recipient plays a role in allograft failure. Poor survival of allografts in recipients with CMV or BK viremia is well known [75,76], although the mechanism underlying the association between viremia and allograft rejection has not been clarified. However, it has been shown that allograft-infiltrating CD8+ T cells can recognize and kill allogeneic (CMV-infected) endothelial cells by the interaction of their TCR with HLA-E that has been loaded with a viral peptide [77]. The direct interaction between that HLA-E loaded with a viral peptide and activating receptors (CD94/NKG2C) on cytotoxic NK and CD8+ T cells, serves to activate antiviral activity capable of controlling infection and also has the potential of destroying the allografts that harbor viruses in the recipients. Figure 2 illustrates this process, which contributes to allograft rejection. Terasaki and his co-investigators [78] developed a strategy to control the cytotoxic activity of activated CD8+ T cells to prevent allograft rejection. They reported that heart transplant recipients treated with pravastatin showed a decrease in the incidence of clinically severe acute rejection episodes, in the incidence and progression of transplant coronary vasculopathy, and in rejection mediated by NK cell cytotoxicity. It is possible that allograft recipients with CMV or BK viremia may be much benefited by statin treatment, which indirectly counteracts HLA-E-mediated augmentation of NK and CD8+ T cell cytotoxicity.

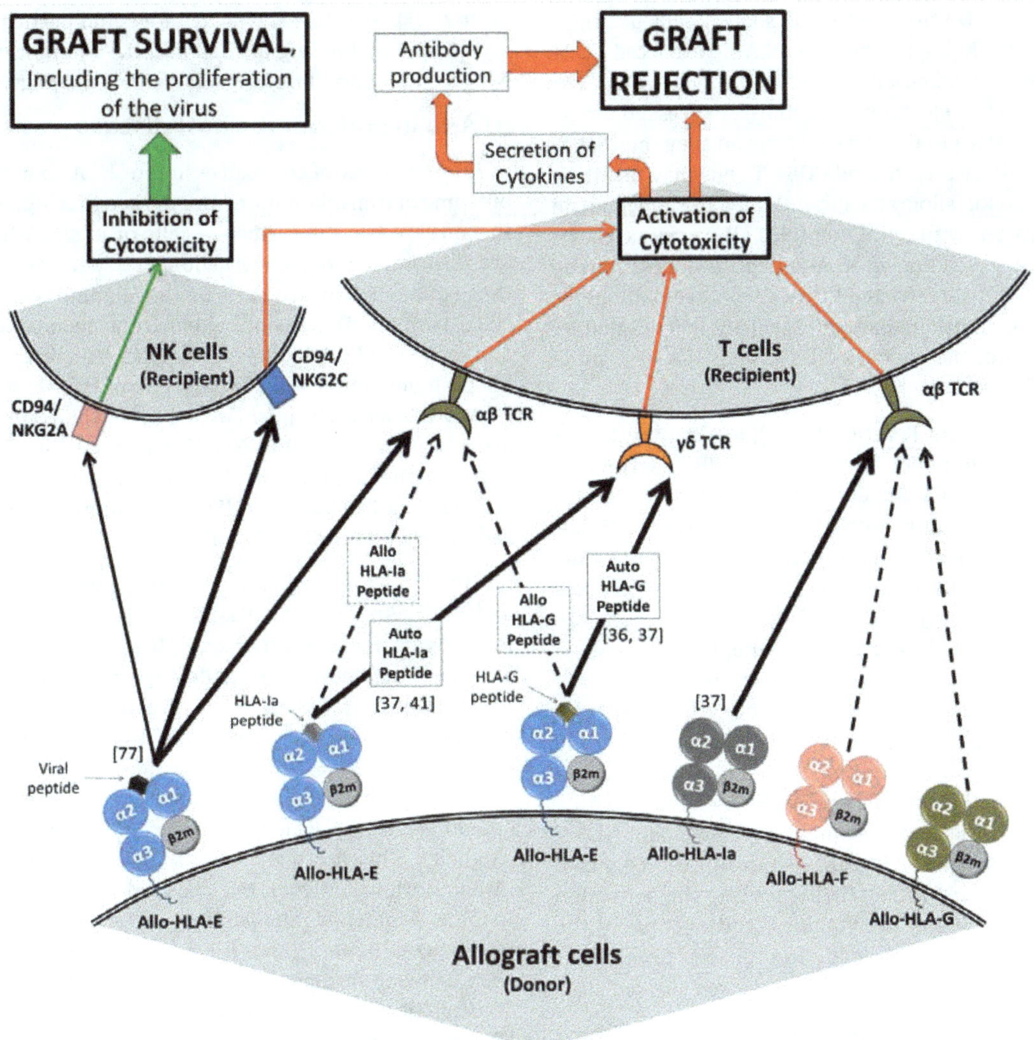

Figure 2: The adaptive interaction between HLA-Ib molecules and their respective immune receptors can affect transplantation outcomes. Allogeneic HLA-Ia molecules, expressed on the surface of the allograft, are recognized by T cells as "non-self," which will eventually lead to graft rejection. The same principle can be applied to HLA-G and HLA-F if they are recognized by T cells as "non-self"-although there is no evidence of this interaction. HLA-E molecules expressed on the surface of the allograft, can present peptides derived from the leader sequence of HLA-Ia and HLA-G. Only when HLA-E is loaded with auto-peptide (from HLA-Ia and/or HLA-G), donors T cells use their γδTCR to recognize HLA-E on the allograft, leading to the production of proinflammatory cytokines, and thus to rejection. When HLA-E is presenting allogeneic peptides (from HLA-Ia and/or HLA-G), donors T cells use their αβTCR to recognize HLA-E on the allograft, thus inducing the cytotoxic activity of CD8+ T cells, promoting graft rejection. Finally, HLA-E presents viral peptides when the allograft is infected. Viral peptides, presented by HLA-E, interact with αβTCR on donor's T cells or CD94/NKG2C on NK cells, leading to the activation of cytotoxicity and eventually graft loss. Depending on the nature of the viral peptides presented by HLA-E, the complex HLA-E/viral-peptide can interact with CD94/NKG2A on NK cells leading to the inhibition of cytotoxicity and thus graft acceptance; however, this mechanism leads to the proliferation of the viruses, which is detrimental to recipient survival. Dashed lines indicate possible interactions. Abbreviations: TCR: T cell receptors; β2m: beta 2-microglobulin.

HLA-Ib, endothelium and transplantation

The vascular endothelium of allografts is the main target for cellular- and/or antibody- mediated vasculopathy and rejection. HLA-E expressed on endothelial cells is overexpressed and then released into circulation upon activation of the cells by IFN-γ [33], thus highlighting the relevance of HLA-E in transplantation. On one hand, a recipient who is HLA-E*01:03- homozygous and receives an HLA-E-matched transplant can be expected to accept the graft better because higher cell-surface expression of HLA-E*01:03 than of HLA-E*01:01 may result in a higher capability to inhibit NK cells and CD8+ T cells through CD94/NKG2A receptors [65]. On the other hand, some T cells use their TCR to recognize HLA-E [31,41]. *In vitro*, HLA-E-restricted CD8+ T cells recognized and killed allogeneic endothelial cells [77], which can mediate graft rejection. If the HLA-E, expressed on the allograft is loaded with a peptide derived from an allogeneic HLA-Ia leader sequence, HLA-E-restricted CD8+ T cells may use their TCR to recognize and kill allograft cells. Although this mechanism is well demonstrated with viral peptides, it remains to be elucidated with other well known HLA-Ia peptides presented by HLA-E. Therefore, HLA-E-restricted CD8+ T cells are pathogenic during organ transplantation, that pathogenicity is mediated by the TCR recognition of the complex HLA-E/allo-HLA-Ia-peptide [37] as shown in Figure 2.

In a Mixed Lymphocyte Reaction (MLR), loading the leader sequence of HLA-G1 onto HLA-E of a donor's PBMC produced different patterns of NK-CTLs-mediated cytotoxicity [37]. NK-CTLs recognize and kill donor's cells with far greater efficacy. The HLA-E/HLA-G1 leader-sequence peptide complex prevents the cytotoxicity of NK-CTLs by binding to the inhibitory receptor CD94/NKG2A. However, NK-CTLs cytotoxicity is promoted when the HLA-E/HLA-G1 leader sequence-peptide complex is recognized as "non-self" as a result of the interaction with CD94/NKG2C or TCR. The MLR shows high levels of cytotoxicity with only one donor, possibly because the recipient's NK-CTLs and those of the donor do not share their HLA-G alleles; but more studies are needed to elucidate this point. HLA-E loaded with a peptide derived from the leader sequence of allogeneic HLA-G molecules (expressed in the allograft cells) can be recognized by CD8+ T cells by means of their TCR, inducing the activation of cytotoxicity, the secretion of cytokines, and eventually the production of antibodies, which leads to graft rejection-all, as shown in Figure 2.

HLA-G was shown to be expressed by endothelial cells and its expression and secretion upregulated in a dose-dependent manner by progesterone [79]. The induction of HLA-G expression in endothelial cells of the allograft can clearly benefit the patient by inhibiting effector cells responsible for rejection and vasculopathy. In addition, it was shown that hypoxia can modulate HLA-G expression depending on cell type and culture conditions. During pregnancy, hypoxia can upregulate HLA-G [80], but in melanoma the opposite happens [81]. In transplantation, hypoxia is a common feature during organ procurement and re-implantation [73]; this impels the conclusion that the expression of HLA-G induced by hypoxia should be investigated in a transplant setting. Even before that happens, though, the induction of HLA-G expression should not be used in patients mismatched for HLA-G because the allograft will express "non-self" antigens that could induce allograft rejection during solid organ transplantation or GvHD in BMT or HSCT.

Under certain conditions, HLA-F can be expressed on the surface of endothelial cells. Indeed, human brain microvascular endothelial cells infected by the Japanese encephalitis virus activate NFκB leading to functional HLA-F gene induction and cell-surface expression [82]. HLA-F expression may be linked to viral infection. It remains to be elucidated whether allografts with viremia express HLA-F. If so, HLA-F could participate in allograft acceptance by binding to ILT2 and ILT4 inhibitory receptors on NK cells to induce immune evasion [56] of allograft cells.

HLA-Ib in graft-versus-host disease

Most studies of the expression of HLA-Ib molecules and the outcome of transplantation have agreed that higher levels of HLA-Ib expressed on the surface of cells-or of sHLA-Ib in circulation-are associated with an overall longer allograft survival. However, those patients who are at risk of developing GvHD (e.g, BMT and HSCT recipients) may not benefit from the increased expression and release of HLA-Ib molecules. These can be expressed on both allograft and donor cells, and despite of their low polymorphism, HLA-Ib alleles can be mismatched even when donor and recipient are HLA-Ia identical [59-61]. The mismatches in HLA-Ib alleles increase the number of allogeneic targets recognized by the recipient, and, in BMT and HSCT, the number of allogeneic targets recognized by the donor's cells, that lead to GvHD.

The HLA-E genotype is associated with GvHD in both BMT and HSCT. An HLA-E*01:01-homozygous state was a risk factor for early bacterial infections and transplant-related mortality in unrelated-donor BMT [59]. Allele frequencies within the studied cohort were similar to those previously reported in various population groups (donor group HLA-E*01:01 was 60% and HLA-E*01:03 40%; recipient group HLA-E*01:01 was 56.5% and HLA-E*01:03 43.5%). Among the 77 pairs in that study, 39% had different HLA-E alleles despite being identical for HLA-C and -A alleles. This further confirms the existence of weak linkage disequilibrium between the classical HLA and HLA-E loci [59]. In HLA-E-matched allogeneic HSCT, the HLA-E*01:03/01:03 homozygous genotype has a protective role since it is associated with a lower frequency of acute and chronic GvHD, and it is associated with improved overall survival compared with other HLA-E genotypes [66,67]. Therefore, lower HLA-E expression on allograft-associated with the HLA-E*01:01 homozygous state [19] may lead to complications such as GvHD because the recipient may be less capable of inhibiting immune cells through CD94/NKG2A receptors. It is clear that HLA-E should be matched in BMT or HSCT because it leads to a transplant that is less at risk of developing GvHD. More studies are needed to evaluate how much HLA-E must be expressed on allograft cells to sustain acceptance before the antigenic load of allogeneic HLA-E triggers an allogeneic immune response leading to rejection. Of course, other factors (e.g., epigenetic factors regulating HLA-I expression) [83] may influence the outcome of BMT and HSCT, with their risk of developing GvHD, or of transplants between monozygotic twins because the level of DNA methylation-an epigenetic feature that changes in response to the environment-is able to differentiate between monozygotic (hence HLA-identical) twins [84].

The HLA-G polymorphism observed in the 3'UTR-which, as noted is called "14bp insertion or deletion"-is associated with a high risk of acute GvHD [44]. It was shown that recipients, homozygous for HLA-G 3'UTR 14bp deletion, had a high risk of acute GvHD [71], and that patients, homozygous for HLA-G 3'UTR 14bp insertion, had a moderate association with a lower risk of acute GvHD compared to HLA-G 3'UTR 14bp heterozygous and HLA-G 3'UTR 14bp deletion homozygous patients [85]. They produced high levels of sHLA-G [70,86]; in solid organ transplantation they experienced a low rate of rejection [44], although in BMT it was the opposite [71]. Higher levels of sHLA-G in patients homozygous for HLA-G 3'UTR 14bp deletion indicate that HLA-G expression is higher than with other 3'UTR genotypes, thus increasing the number of antigenic target available, promoting GvHD.

There is no information about HLA-F alleles and their role in GvHD, but HLA-F is a marker of activated lymphocytes [52], and upon immunological activation in patients experiencing GvHD, the donor's and recipient's immune cells will express HLA-F on their surface, so an HLA-F genotype mismatch may increase the severity of GvHD. More studies are needed to confirm this hypothesis.

These observations suggest that in BMT and HSCT, HLA-E and other HLA-Ib alleles should be matched in order to reduce the number of possible antigenic targets recognized by the donor cells. If HLA-Ib polymorphism is not matched in BMT and HSCT, then donor cells have more potential recipient targets to attack, which will result in severe GvHD in those HLA-Ib mismatched patients.

To sum up about HLA-Ib molecules, they clearly play a role in histocompatibility and they impact transplantation outcomes. The presence of high levels of HLA-Ib molecules (on the cell surface or in circulation) is usually associated with better graft acceptance, longer graft survival, and immunosuppressive-free patients; however, in the case of cell transplantation (BMT and HSCT), HLA-Ib alleles mismatches may increase the risk of developing GvHD by increasing the foreign antigenic load recognized by the donor cells. Therefore, it is still debatable whether HLA-Ib alleles should be matched for all transplants, and if so, whether the immunosuppressive function of HLA-Ib molecules is mediated by the donor's molecules or those of the recipient. Some studies emphasize that the allograft (donor cells) should be expressing HLA-Ib molecules-and at high levels-in order to evade the cytotoxic activity of NK cells and CD8+ T cells that lead to graft rejection [18,87-89] because cell-to-cell contact is mandatory for the interaction of HLA-Ib molecules and their inhibitory receptors to prevent the lysis of allograft cells by NK cells and CD8+ T cells [90]. Promoting the expression of HLA-Ib-rather than sHLA-Ib-by the donor's cells therefore provides the best and most specific strategy for helping allograft cells evade cell-mediated damage (or rejection). Soluble HLA-Ib will most likely participate in the regulation, differentiation, and proliferation of immune effector cells, thus participating in graft acceptance, but not directly mediating it.

Immunobiology of Soluble HLA-Ib in Transplantation

The importance of sHLA-Ib molecules is characterized by their ability to interact with different lymphocyte subsets and, what is most important, with NK and CD8$^+$ T cells. Soluble HLA-Ib can bind to both inhibiting and activating receptors-respectively, (CD94/NKG2A, ILT2, ILT4) and (CD94/NKG2C, ILT2, ILT4). The inhibition of the cytotoxic activity of NK and CD8$^+$ T cells, mediated by sHLA-Ib, may favor allograft survival while the activation of their cytotoxic capabilities would promote allograft rejection.

Soluble HLA-Ib can be released in the circulation by three different mechanisms: exosomal shedding, alternative splicing, and proteolytic processing [91,92]. Each mechanism will produce different isoforms of sHLA-Ib, contributing to a pool of sHLA-Ib in the allograft microenvironment and circulation. Soluble HLA-I levels are increased in several physiological and pathological conditions such as pregnancy, acute rejection episodes following allografts, and acute GvHD following BMT [93]. Indeed, the increased level of sHLA-I may be a result of immune activation, as shown by a rapid decrease in the level of sHLA-I following immunosuppressive therapy in patients with acute rejection episodes [94].

Previous studies have shown that sHLA-G is associated with an increased graft survival and a decrease in rejection episodes in such organ transplantation as heart [95], liver [96], kidney [96-98], lung [89] and HSCT [62]. HLA-G expressed on the allograft, and sHLA-G in circulation, may also play a role in T-cell anergy as well as inducing immunosuppressive T cells [18], suggesting that the donor's molecular typing of HLA-G should be considered in allograft transplantation because HLA-G is regulated by genetic and cytokines depending on the viral or non-viral microenvironment [62]. Compared with all other sHLA-Ib molecules, HLA-G has a high diversity of molecular structures due to alternative splicing of the primary transcript [44,45]. As noted earlier, the expression of HLA-G on the cell surface and its soluble form is associated with its polymorphism in both the coding and non-coding regions of the HLA-G allele. There is a need to elucidate the kind of isoforms of HLA-G that are more potent in inducing immunosuppressive T cells. There may be a good correlation between higher cell-surface expression of some isoforms with higher levels of sHLA-G release from the allograft, which facilitates allograft tolerance in the recipient due to those isoforms' specific affinity for inhibitory receptors (ILT2, ILT4 and KIR2DL4) [45].

HLA-G can form disulfide-linked dimeric complexes on the cell surface and in circulation because of high preferential binding to immune inhibitory receptors; sHLA-G dimers were also found to be significantly higher in renal allograft recipients who had experienced no rejection episodes than in patients with chronic rejection [99]. In this study, the investigators evaluated the levels of sHLA-G1 and HLA-G5 dimers and monomer isoforms. In confirmation of their finding, the increased cell-surface expression of HLA-G on monocytes of patients with no rejection episodes was found to be associated with high levels of plasma sHLA-G dimers [100]. These observations underscore the potential of HLA-G in graft acceptance; but it remains to be determined whether this potential is due to cell-surface expression of HLA-G1 or whether its soluble form, HLA-G5, induces graft acceptance in the course of its specific mechanism of action. Possibly HLA-G5 molecules may be involved in the inhibition of immune functions by their direct interaction with ILT2 and ILT4 receptors present on the main effector cells responsible for graft rejection [18,62,64]. HLA-G1, expressed on the surface of allograft cells, inhibits the cytotoxic function of NK and CD8⁺ T cells through cell-to-cell contact, preventing the cell-mediated allograft damage that leads to cellular rejection. In contrast, HLA-G5 molecules interact with immune effector cells in circulation, and probably have a more potent effect in allograft rejection on CD4⁺ T cells and APC than do NK and CD8⁺ T cells by inhibiting alloreactivity and proliferation and by up-regulation of inhibitory receptors on their surface [18,62].

The presence of sHLA during rejection episodes [101] creates a pool of soluble antigens that can lead to the production of HLA antibodies [102,103], which supports the immunogenic potential of sHLA molecules *in vivo*. The greatest challenge in characterizing sHLA (classical Ia, or non-classical Ib) is the degree of homology that these molecules have. Indeed, only mAbs-particularly those that are monospecific (directed against an epitope that is unique for a specific HLA allele)-should be used to identify specific sHLA isoforms or alleles present in the sera. For example, it was shown that commercially available HLA-E mAbs recognize shared peptide sequence on HLA-Ia molecules [104]; therefore, such mAbs bind to more than one HLA-Ia or -Ib allele, decreasing the reliability of such mAbs in binding to a specific HLA-I molecule, affecting the interpretation of the results. Rarely have HLA-Ia or -Ib mAbs been characterized for their unique specificity either by screening their reactivity with a panel of HLA molecules (such as single antigen HLA beads) or by using a peptide sequence unique for the allele to inhibit the mAb's specific binding [23,104-106].

Humoral immune response to HLA-Ib

The humoral theory of transplantation proposed by Terasaki [12] envisages a primary role for HLA antibodies in post-transplantation allograft rejection. The increasing sensitivity of HLA antibody screening has dramatically improved transplantation outcomes. Routinely, only HLA-Ia and -II antibodies are tested in transplant recipients, and HLA-Ib antibodies are never considered. However, various investigations [23,24,39,104-107] of HLA-Ib antibodies showed that one molecule of HLA-Ib (with or without β2m) can elicit different kinds of mAbs, some recognizing strictly the HLA immunogen used in immunization. Alternatively, these antibodies (e.g., HLA-E mAb TFL-033) [23,39] can be considered monospecific since they react to only one HLA-Ib molecule. The other antibodies generated by an HLA-Ib molecule (e.g., HLA-E mAbs TFL-006 and TFL-007) [24,106,107] can react to other HLA-I molecules at various levels. Such polyreactive binding (reacting with more than one HLA-Ib and/or -Ia molecule) can be a consequence of the antibodies recognizing the amino acid sequences common to the HLA molecules [23-25,104]. So the presence of HLA-Ib antibodies in the serum can affect the screening of HLA-Ia antibodies. Figure 3 illustrates how HLA-Ib antibodies can affect routine HLA antibody screening and organ allocation, and it shows the possible transplantation outcomes of having HLA-Ib antibodies in circulation. There remains a need to elucidate the immunogenicity and antigenicity of HLA-Ib molecules and the complex pool of antibodies that can be produced *in vivo*.

In addition to its other abilities, one HLA-G gene can produce several different protein isoforms that can expose cryptic epitopes usually hidden in the full-length protein associated with β2m [68]. Also, HLA-Ib molecules have been shown to homodimerize and heterodimerize on the cell surface or in circulation [55-57], which can expose cryptic epitopes recognizable by the immune system.

Immunogenicity of HLA-Ib molecules

A native HLA-Ib molecule (a trimolecular complex comprised of HLA heavy chain, β2m and a peptide) may become immunogenic once the heavy chain of HLA-Ib detaches from β2m or after it has undergone proteolytic alteration. Both loss of β2m and proteolytic alteration will expose amino acid sequences or epitopes that have previously remained cryptic on HLA. Exposure of cryptic antigens makes the molecule immunogenic, eliciting production of antibodies. The presence of polyreactive HLA-E antibodies was observed in renal and liver transplant recipients as well as in melanoma patients [26,27]. In addition, antibodies reacting against HLA-Ib molecules (HLA-E, -F and -G) and HLA-Ia (HLA-A, -B and -Cw) were present in different IVIg preparations and in the IgG-purified serum fraction from non-alloimmunized males [108]. Together, these findings provide evidence of the immunogenicity of HLA-Ib *in vivo*.

The immunogenicity of HLA-Ib molecules is driven primarily by polymorphic amino acids residues located on the α1 and α2 helices, leading to an immunological response directed at a unique HLA-Ib molecule as shown in Figure 3. Meanwhile, non-native forms of HLA-Ib molecules (produced by proteolytic degradation, alternatively spliced transcript, or shedding) can expose epitopes that are shared by HLA-Ib and/or HLA-Ia molecules. Such shared or common epitopes may trigger an immunological response when the immunogen is encountered, and may also trigger one when these epitopes are present and exposed on HLA-Ia or -Ib molecules other than the immunogen. Such shared epitopes may be exposed on the molecular surface (e.g., those recognized by W6/32) or may be cryptic, like those recognized by HC10, MEM-02, MEM-06, MEM-07, 3D12 or by TFL mAbs TFL-006 and TFL-007 [24,39,104-106,108]. The fact that HLA-Ib molecules carry epitopes shared by HLA-Ia molecules points up the possibility that shared epitopes cannot be immunogenic because if they were they should be recognized as "self"; however, the presence of natural HLA antibodies [109], HLA-E antibodies [25] in non-alloimunized males, and HLA-Ia/-Ib antibodies in IVIg [108] shows that HLA-I self-shared epitopes can become immunogenic *in vivo*.

HLA-Ia or -Ib cryptic epitopes, usually hidden in a non-inflammatory state, can elicit an immune response by exposing hidden amino acid residues upon inflammation; indeed, the dissociation of β2m from the HLA heavy chain creates potential, newly exposed epitopes for antibody production, possibly to clear sHLA heavy chain from circulation, thus preventing possible immunomodulatory function of sHLA-Ia or -Ib.

Antigenicity of HLA-Ib molecules

Antibodies generated after the immunization of HLA-Ib molecules can produce a variety of antibodies with a variable panel of reactivity towards HLA-Ia and -Ib molecules. Indeed, the production of HLA-E mAbs, generated after the immunization of HLA-E heavy chain, produced eight types of HLA-E mAbs. These were classified based on their reactivity to a panel of HLA-Ia and -Ib molecules [24]. Monospecific antibodies-called "type 1 mAbs" in that report-bound only to the HLA-Ib molecule used as the immunogen. Low polyreactive antibodies (called "types 2, 3 and 4 mAbs") bound to more than one HLA-Ib molecule but not to HLA-Ia. Moderate polyreactive antibodies ("types 5, 6 and 7 mAbs") bound to one or more HLA-Ib molecules and to most HLA-Ia molecules. Finally, high polyreactive antibodies ("type 8 mAbs") bound to all HLA-Ia and -Ib molecules [24].

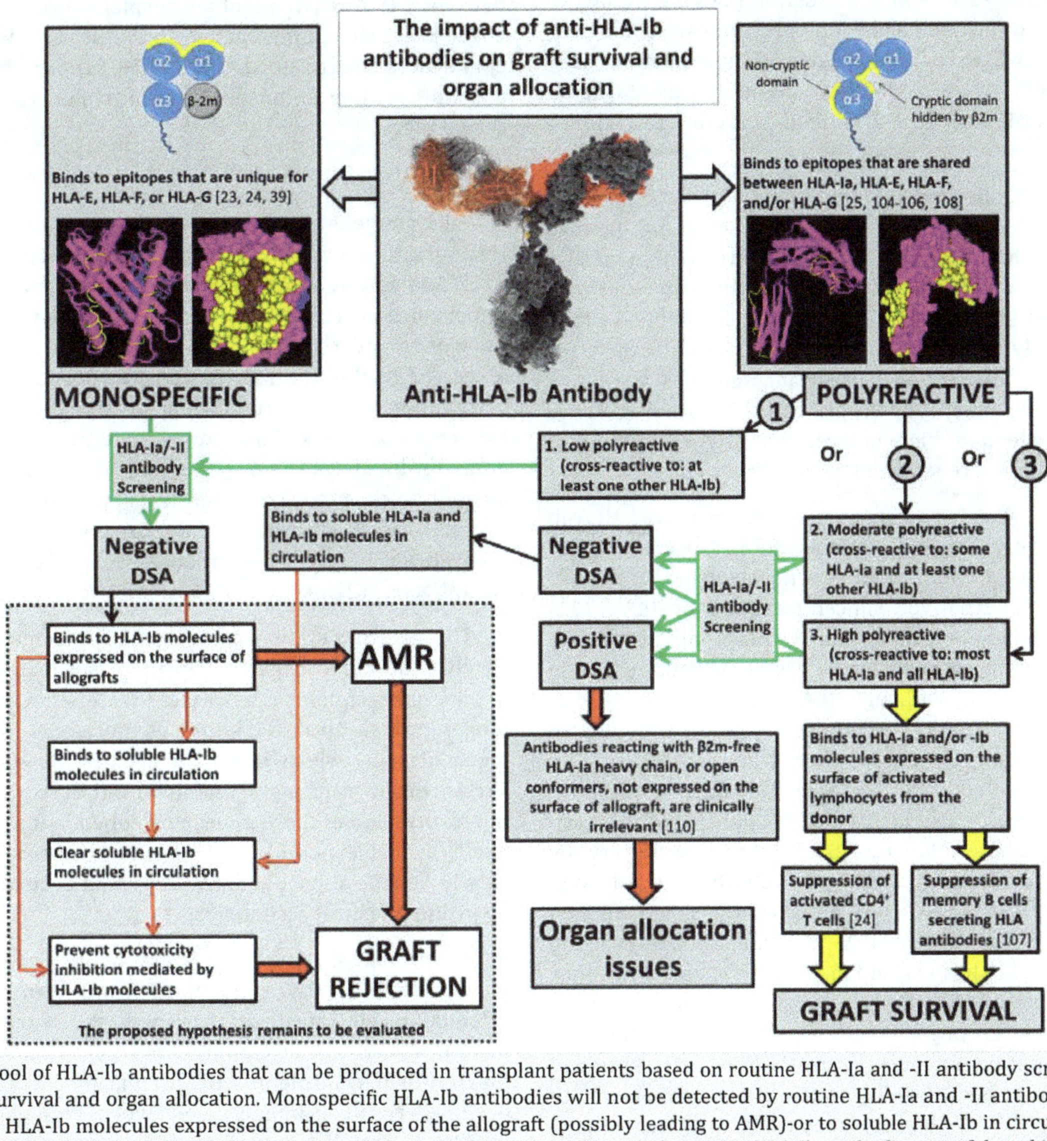

Figure 3: The pool of HLA-Ib antibodies that can be produced in transplant patients based on routine HLA-Ia and -II antibody screening, and their effect on graft survival and organ allocation. Monospecific HLA-Ib antibodies will not be detected by routine HLA-Ia and -II antibody screening, but they can bind to HLA-Ib molecules expressed on the surface of the allograft (possibly leading to AMR)-or to soluble HLA-Ib in circulation-to prevent the inhibition of cytotoxicity mediated by HLA-Ib molecules, thus leading to graft loss. Polyreactive HLA-Ib antibodies are of three kinds: (1) low polyreactive HLA-Ib antibodies that may contribute to graft lost in the same way as monospecific antibodies; (2) moderate polyreactive HLA-Ib antibodies that can affect routine HLA-Ia and -II antibody screening and thus organ allocation, and contribute to graft lost in the same way as monospecific antibodies; and (3) high polyreactive HLA-Ib antibodies that can affect routine HLA-Ia and -II antibody screening and thus organ allocation. But high polyreactive antibodies can bind to open conformers expressed on the surface of activated lymphocytes, promoting graft acceptance by suppressing activated CD4+ T cells and suppressing memory B cells' HLA antibodies secretion.
Abbreviations: AMR: Antibody-Mediated Rejection; DSA: Donor-Specific Antibodies; β2m: beta 2-microglobulin.

As shown in Figure 3, HLA-Ib molecules can produce a complex pool of antibodies that can affect routine HLA antibody screening, organ allocation, and transplantation outcomes. Monospecific (type 1) and low polyreactive (types 2, 3, and 4) HLA-Ib antibodies will not affect HLA antibody screening because HLA-Ib molecules are not represented in the routine HLA-I and -II antibody screening panel. Therefore, the presence of those types 1-4 in the serum will not affect the organ. Nonetheless, such monospecific and low polyreactive antibodies could have a negative impact on organ transplantation outcome. Both can prevent the inhibition of cytotoxicity mediated by HLA-Ib molecules by binding to sHLA-Ib molecules and clearing them from circulation. Only monospecific antibodies can bind to HLA-Ib expressed on the surface of the allograft, possibly leading to antibody-mediated rejection and eventual graft loss. In addition, monospecific HLA-E mAbs are capable of upregulating CD8+ T lymphocytes [23] *in vitro*, which would be detrimental to an allograft recipient if such HLA-E antibodies are in the patient's serum.

On the other hand, moderate and high polyreactive HLA-Ib antibodies will affect routine HLA antibody screening because they can bind to a shared epitope exposed in the recombinant HLA protein coated on a solid matrix. Such antibodies can react with many HLA-Ia and -Ib molecules, increasing the likelihood that a patient with a polyreactive HLA-Ib antibody will be positive for DSA. That means the allocation of organ and immunosuppressive treatment is greatly affected by the pre-transplantation presence of those moderate and high polyreactive HLA-Ib antibodies because the antibodies react with DSA *in vitro*.

In addition, HLA-Ia screening of polyreactive HLA-Ib antibodies will show high levels of non-DSA because antibodies produced after an immunization event involving immature HLA-E [59]-or an alternatively spliced transcript of HLA-G [71] or an HLA-F/HLA-Ia heterodimer [26,43]- react with non-native HLA molecules, which are known to be present in routine HLA-Ia antibody screening [110]. Although HLA antibodies reacting with β2m-free HLA heavy chain or open conformers are thought to be clinically irrelevant to the outcome of transplantation [110], it has been shown that such β2m-free HLA heavy chains are expressed on the surface of activated cells [111], and they can homodimerize and heterodimerize [55-57] thus exposing cryptic epitopes available for the recognition of β2m-free, HLA heavy-chain antibodies [24,57,107,108] detected by single antigen beads assay [24]. This suggests that HLA-Ib antibodies reacting against β2m-free HLA heavy chain may have a clinical relevance because polyreactive antibodies may react with donor/recipient HLA, and, although reacting with non-native molecules, they may impact transplantation outcome (Figure 3). The aforementioned moderate polyreactive HLA-Ib antibodies-cross-reactive to most but not all HLA-Ia/-Ib alleles-may exert a negative impact on transplantation by binding to sHLA-Ib, thus preventing the tolerance-inducing function of sHLA-Ib associated with graft acceptance. High polyreactive (type 8) HLA-Ib antibodies can be protective. Indeed, those antibodies, raised against HLA-E molecules and produced *in vivo*, can bind to open conformers expressed on the surface of activated lymphocytes, and were able to suppress both activated CD4+ T cells [24] and memory B-cell HLA antibody secretion [107] *in vitro*. Clearly, then, high polyreactive HLA antibodies, raised against HLA-Ib molecules, play a role in graft acceptance by suppressing alloreactive T and B cells. In addition, it was shown that alloantigen-activated CD4+ T cells express HLA-E on their surface, resulting in an increased resistance to NK cells [90], thus suggesting a role in the regulation and differentiation of CD4+ T cells *in vivo*; those T cells also use HLA-E as a potential immunoregulatory therapeutic target for transplant patients. More studies are needed to evaluate the effect of polyreactive and monospecific HLA-Ib antibodies in transplantation, especially in terms of the presence of non-DSA in the sera of transplanted patients. We can hypothesize that more non-DSA reactivity of a transplant serum probably relates to the presence of polyreactive antibodies that will bind to open conformers on activated lymphocytes, suppressing their activity, and thus leading to longer graft survivalor, rather, better graft acceptance.

Conclusion

HLA antigens are the major influence on the immunological response to allografts in recipients. Matching or mismatching of HLA-Ia and -II alleles may lead to variable graft survival. The importance of HLA-Ib can be best seen in an allograft in which the donor and recipient HLA-Ia /-II types are identical. Such a case highlights the role played by HLA-Ib genes in allograft acceptance or rejection, an insight strongly reinforced by the existence of weak linkage disequilibrium between HLA-Ia and HLA-Ib alleles [59-61]. Soluble HLA-Ib and the antibodies elicited by them are critical in transplantation because of such actions as characterization of donor-specific HLA-Ia antibodies, the role played by HLA-Ib antibodies in clearing the sHLA-Ia and -Ib antigens [12].

HLA-Ib genes may be involved in the extreme variations in the survival rates of transplants. The increased expression of HLA-Ib in allograft recipients is associated with better graft acceptance, longer graft survival, and immunosuppressive-free patients. The mechanism by which HLA-Ib molecules are expressed on the surface of allografts is still unknown, but the consequences of the overexpression of these molecules-dependent as they are on the allele, available peptide, and cytokines microenvironment-are clearly beneficial for patients, protecting against early rejection by promoting graft acceptance.

Nevertheless, there are exceptions such as in cell transplantation (BMT and HSCT) and transplantation with viremia. In cell transplantation, increased expression of HLA-Ib increases the risk of developing GvHD only in patients mismatched for HLA-Ib. In transplants with viremia, the increased expression of HLA-Ib results in an increased level of viral peptide presented by HLA-E. The HLA-E/viral-peptide complex can be recognized by the TCR of CD8+T cells or by the activating receptor CD94/NKG2C on NK cells, both of which induce cytotoxicity in allograft cells, leading to accelerated graft loss.

HLA-Ib antibodies, generated after the immunization of HLA-Ib molecules, can be either monospecific or polyreactive.

Monospecific antibodies can affect graft acceptance by neutralizing HLA-Ib molecules and preventing their interaction with inhibitory receptors on NK and CD8⁺ T cells. In contrast, polyreactive antibodies affect routine HLA antibody screening, and thus organ allocation; however, they can also inhibit the proliferation of CD4+ T cells and the memory B-cell secretion of HLA antibodies, which could promote graft acceptance.

Acknowledgments

The entire research is supported by grants from a non-profit organization, the Terasaki Family Foundation, mediated through the Terasaki Foundation Laboratory.

Conflict of Interest

None of the authors have any conflicts of interest or financial interests.

References

1. Natali PG, Bigotti A, Nicotra MR, Viora M, Manfredi D, Ferrone S. Distribution of human Class I (HLA-A,B,C) histocompatibility antigens in normal and malignant tissues of nonlymphoid origin. Cancer Res. 1984; 44(10):4679-4687.

2. Daar AS, Fuggle SV, Fabre JW, Ting A, Morris PJ. The detailed distribution of MHC Class II antigens in normal human organs. Transplantation. 1984;38(3):293-298.

3. Benacerraf B. A hypothesis to relate the specificity of T lymphocytes and the activity of I region-specific Ir genes in macrophages and B lymphocytes. J Immunol. 1978;120(6):1809-1812.

4. Unanue ER. Antigen-presenting function of the macrophage. Annu Rev Immunol. 1984;2:395-428. doi: 10.1146/annurev.iy.02.040184.002143.

5. Cresswell P. Assembly, transport, and function of MHC class II molecules. Annu Rev Immunol. 1994;12: 259-293. doi: 10.1146/annurev.iy.12.040194.001355.

6. Petersdorf EW. The major histocompatibility complex: a model for understanding graft-versus-host disease. Blood. 2013;122(11):1863-1872. doi: 10.1182/blood-2013-05-355982.

7. Trowsdale J. HLA genomics in the third millennium. Curr Opin Immunol. 2005;17(5):498-504. doi: 10.1016/j.coi.2005.07.015.

8. Horton R, Wilming L, Rand V, Lovering RC, Bruford EA, Khodiyar VK, et al. Gene map of the extended human MHC. Nature Reviews Genetics. 2004;5(12):889–899. doi: 10.1038/nrg1489.

9. Robinson J, Halliwell JA, Hayhurst JH, Flicek P, Parham P, Marsh SGE. The IPD and IMGT/HLA database: allele variant databases. Nucleic Acids Research. 2015. 43:D423-431.

10. Dausset J. Leucocytes, platelets and human homografts. Vox Sang. 1962;7:257-266.

11. Mayor NP, Robinson J, McWhinnie AJ, Ranade S, Eng K, Midwinter W, et al. HLA Typing for the Next Generation. PLoS ONE. 2015;10(5):e0127153. doi: 10.1371/journal.pone.0127153.

12. Terasaki PI. Humoral theory of transplantation. Am J Transplant. 2003;3(6):665–673.

13. Geraghty DE, Koller BH, Orr HT. A human major histocompatibility complex class I gene that encodes a protein with a shortened cytoplasmic segment. Proc Natl Acad Sci USA. 1987;84(24):9145-9149.

14. Koller BH, Geraghty DE, Shimizu Y, DeMars R, Orr HT. HLA-E. A novel HLA class I gene expressed in resting T lymphocytes. J Immunol. 1988;141(3):897–904.

15. Rodgers JR, Cook RG. MHC class Ib molecules bridge innate and acquired immunity. Nat Rev Immunol. 2005;5(6):459–471. doi: 10.1038/nri1635.

16. Kochan G, Escors D, Breckpot K, Guerrero-Setas D. Role of non-classical MHC class I molecules in cancer immunosuppression. Oncoimmunology. 2013;2(11):e26491. doi: 10.4161/onci.26491.

17. Sullivan LC, Hoare HL, McCluskey J, Rossjohn J, Brooks AG. A structural perspective on MHC class Ib molecules in adaptive immunity. Trends Immunol. 2006;27(9):413–20. doi: 10.1016/j.it.2006.07.006.

18. Rouas-Freiss N, Naji A, Durrbach A, Carosella E. Tolerogenic Functions of Human Leukocyte Antigen G: From Pregnancy to Organ and Cell Transplantation. Transplantation. 2007;84(1 Suppl):S21-5. doi:10.1097/01.tp.0000269117.32179.1c.

19. Lauterbach N, Wieten L, Popeijus HE, Vanderlocht J, van Zon PM, Voorter CE, et al. Peptide-induced HLA-E expression in human PBMCs is dependent on peptide sequence and the HLA-E genotype. Tissue Antigens. 2015;85(4):242–51. doi: 10.1111/tan.12525.

20. Morandi F, Pistoia V. Soluble HLA-G modulates miRNA-210 and miRNA-451 expression in activated CD4+ T lymphocytes. Int Immunol. 2013;25(5):279–285. doi: 10.1093/intimm/dxs108.

21. Le Bouteiller P, Lenfant F. Antigen-presenting function(s) of the non-classical HLA-E, -F and -G class I molecules: the beginning of a story. Res Immunol. 1996;147(5):301–313.

22. Le Bouteiller P, Solier C. Is antigen presentation the primary function of HLA-G? Microbes Infect. 2001;3(4): 323–332.

23. Ravindranath MH, Terasaki PI, Pham T, Jucaud V. The Monospecificity of Novel Anti-HLA-E Monoclonal Antibodies Enables Reliable Immunodiagnosis, Immunomodulation of HLA-E, and Upregulation of CD8+ T Lymphocytes. Monoclon Antib Immunodiagn Immunother. 2015;34(3):135–153. doi: 10.1089/mab.2014.0096.

24. Ravindranath MH, Terasaki PI, Pham T, Jucaud V, Kawakita S. Suppression of blastogenesis and proliferation of activated CD4(+) T cells: intravenous immunoglobulin (IVIg) versus novel anti-human leucocyte antigen (HLA)-E monoclonal antibodies mimicking HLA-I reactivity of IVIg. Clin Exp Immunol. 2014;178(1):154–177. doi: 10.1111/cei.12391.

25. Ravindranath MH, Kaneku H, El-Awar N, Morales-Buenrostro LE, Terasaki PI. Antibodies to HLA-E in nonalloimmunized males: pattern of HLA-Ia reactivity of anti-HLA-E-positive sera. J Immunol. 2010; 185(3):1935-1948. doi: 10.4049/jimmunol.1000424.

26. Ravindranath MH, Pham T, Ozawa M, Terasaki PI. Antibodies to HLA-E may account for the non-donor-specific anti-HLA class-Ia antibodies in renal and liver transplant recipients. Int Immunol. 2012;24(1):43-57.doi: 10.1093/intimm/dxr094.

27. Ravindranath MH, Selvan SR, Terasaki PI. Augmentation of anti-HLA-E antibodies with concomitant HLA-Ia reactivity in IFNγ-treated autologous melanoma cell vaccine recipients. J Immunotoxicol. 2012;9(3):282–291. doi: 10.3109/1547691X.2011.645582.

28. Moscoso J, Serrano-Vela JI, Pacheco R, Arnaiz-Villena A. HLA-G, -E and -F: allelism, function and evolution. Transpl Immunol. 2006;17(1):61–64. doi: 10.1016/j.trim.2006.09.010.

29. Ishitani A, Sageshima N, Hatake K. The involvement of HLA-E and -F in

pregnancy. Journal of Reproductive Immunology. 2006;69(2):101–13. doi: 10.1016/j.jri.2005.10.004.

30. Shobu T, Sageshima N, Tokui H, Omura M, Saito K, Nagatsuka Y , et al. The surface expression of HLA-F on decidual trophoblasts increases from mid to term gestation. J Reprod Immunol. 2006;72(1-2):18–32. doi: 10.1016/j.jri.2006.02.001.

31. Allan DS, Lepin EJ, Braud VM, O'Callaghan CA, McMichael AJ. Tetrameric complexes of HLA-E, HLA-F, and HLA-G. J Immunol Methods. 2002;268(1):43-50.

32. Morandi F, Cangemi G, Barco S, Loredana Amoroso, Maria Giuliano, Anna Rita Gigliotti, et al. Plasma Levels of Soluble HLA-E and HLA-F at Diagnosis May Predict Overall Survival of Neuroblastoma Patients. Biomed Res Int. 2013; 956878. doi: 10.1155/2013/956878.

33. Coupel S, Moreau A, Hamidou M, Horejsi V, Soulillou JP, Charreau B. Expression and release of soluble HLA-E is an immunoregulatory feature of endothelial cell activation. Blood. 2007;109(7):2806-2814. doi: 10.1182/blood-2006-06-030213.

34. Braud VM, Allan DS, Wilson D, McMichael AJ. TAP- and tapasin-dependent HLA-E surface expression correlates with the binding of an MHC class I leader peptide. Curr Biol. 1998;8(1):1–10.

35. Lee N, Goodlett DR, Ishitani A, Marquardt H, Geraghty DE. HLA-E surface expression depends on binding of TAP-dependent peptides derived from certain HLA class I signal sequences. J Immunol. 1998;160(10): 4951–4960.

36. Llano M, Lee N, Navarro F, García P, Albar JP, Geraghty DE, et al. HLA-E-bound peptides influence recognition by inhibitory and triggering CD94/NKG2 receptors: preferential response to an HLA-G-derived nonamer. Eur J Immunol. 1998;28(9):2854-2863. doi: 10.1002/(SICI)1521-4141.

37. Moretta L, Romagnani C, Pietra G, Moretta A, Mingari MC. NK-CTLs, a novel HLA-E-restricted T-cell subset. Trends Immunol. 2003;24(3):136-143.

38. Kraemer T, Blasczyk R, Bade-Doeding C. HLA-E: a novel player for histocompatibility. J Immunol Res. 2014; 2014: 352160. doi: 10.1155/2014/352160.

39. Sasaki T, Ravindranath MH, Terasaki PI, Freitas MC, Kawakita S, Jucaud V. Gastric cancer progression may involve a shift in HLA-E profile from an intact heterodimer to β2-microglobulin-free monomer. Int J Cancer. 2014;134(7):1558-15570. doi: 10.1002/ijc.28484.

40. Iwaszko M, Bogunia-Kubik K. Clinical significance of the HLA-E and CD94/NKG2 interaction. Arch Immunol Ther Exp (Warsz). 2011;59(5):353–67. doi: 10.1007/s00005-011-0137-y.

41. Pietra G, Romagnani C, Moretta L, Mingari MC. HLA-E and HLA-E-bound peptides: recognition by subsets of NK and T cells. Curr Pharm Des. 2009;15(28): 3336-3344.

42. Petrie EJ, Clements CS, Lin J, Sullivan LC, Johnson D, Huyton T, et al. CD94-NKG2A recognition of human leukocyte antigen (HLA)-E bound to an HLA class I leader sequence. J Exp Med. 2008;205(3):725-735. doi: 10.1084/jem.20072525.

43. Brooks AG, Borrego F, Posch PE, Patamawenu A, Scorzelli CJ, Ulbrecht M, et al. Specific recognition of HLA-E, but not classical, HLA class I molecules by soluble CD94/NKG2A and NK cells. J Immunol. 1999;162(1):305-313.

44. Donadi EA, Castelli EC, Arnaiz-Villena A, Roger M, Rey D, Moreau P. Implications of the polymorphism of HLA-G on its function, regulation, evolution and disease association. Cell Mol Life Sci. 2011;68(3):369-395. doi: 10.1007/s00018-010-0580-7.

45. Pistoia V, Morandi F, Wang X, Ferrone S. Soluble HLA-G: Are they clinically relevant? Semin Cancer Biol. 2007;17(6):469-479. doi: 10.1016/j.semcancer.2007.07.004.

46. Di Cristofaro J, Reynaud M, Carlini F, , Roubertoux P, Loundou A, Basire A, et al. HLA-G*01:04~UTR3 Recipient Correlates With Lower Survival and Higher Frequency of Chronic Rejection After Lung Transplantation. Am J Transplant. 2015;15(9):2413-2420. doi:10.1111/ajt.13305.

47. Pabón MA, Navarro CE, Osorio JC, , Gómez N, Moreno JP, Donado AF, et al. Impact of human leukocyte antigen molecules E, F, and G on the outcome of transplantation. Transplant Proc. 2014;46(9):2957–2965. doi: 10.1016/j.transproceed.2014.07.010.

48. Wainwright S, Biro P, Holmes C. HLA-F is a predominantly empty, intracellular, TAP-associated MHC class Ib protein with a restricted expression pattern. J Immunol. 2000;164(1):319–328. doi:10.4049/jimmunol.164.1.319

49. Xu Y, Han H, Zhang F, Lv S, Li Z, Fang Z. Lesion human leukocyte antigen-F expression is associated with a poor prognosis in patients with hepatocellular carcinoma. Oncol Lett. 2015;9(1):300–304. doi: 10.3892/ol.2014.2686.

50. Gobin Sj, van Zutphen M, Woltman AM, van den Elsen PJ. Transactivation of classical and nonclassical HLA class I genes through the IFN-stimulated response element. Journal of immunology. 1999;163(3):1428-1434.

51. Gobin SJ, van den Elsen PJ. Transcriptional regulation of the MHC class Ib genes HLA-E, HLA-F, and HLA-G. Hum Immunol. 2000;61(11):1102–1107. doi: 10.1016/S0198-8859(00)00198-1.

52. Lee N, Ishitani A, Geraghty DE. HLA-F is a surface marker on activated lymphocytes. Eur J Immunol. 2010; 40(8):2308-2318. doi: 10.1002/eji.201040348.

53. Boyle L, Gillingham A, Munro S, Trowsdale J. Selective export of HLA-F by its cytoplasmic tail. J Immunol. 2006;176(11):6464–6472.

54. Lepin EJ, Bastin JM, Allan DS, Roncador G, Braud VM, Mason DY, et al. Functional characterization of HLA-F and binding of HLA-F tetramers to ILT2 and ILT4 receptors. Eur J Immunol. 2000;30(12):3552–61. doi: 10.1002/1521-4141.

55. Goodridge JP, Burian A, Lee N, Geraghty DE. HLA-F and MHC class I open conformers are ligands for NK cell Ig-like receptors. J Immunol. 2013;191(7):3553-62.doi: 10.4049/jimmunol.1300081.

56. Goodridge JP, Lee N, Burian A, Tykodi SS, Warren EH, Yee C, et al. HLA-F and MHC-I open conformers cooperate in a MHC-I antigen cross-presentation pathway. J Immunol. 2013;191(4):1567-1577. doi: 10.4049/jimmunol.1300080.

57. Goodridge JP, Burian A, Lee N, Geraghty DE. HLA-F complex without peptide binds to MHC class I protein in the open conformer form. J Immunol. 2010; 184(11): 6199–6208. doi: 10.4049/jimmunol.1000078.

58. Rouas-Freiss N, LeMaoult J, Moreau P, Dausset J, Carosella ED. HLA-G in transplantation: a relevant molecule for inhibition of graft rejection? Am J Transplant. 2003;3(1):11–16.

59. Tamouza R, Rocha V, Busson M, Fortier C, El Sherbini SM, Esperou H, et al. Association of HLA-E polymorphism with severe bacterial infection and early transplant-related mortality in matched unrelated bone marrow transplantation. Transplantation. 2005;80(1):140-144. CX, et al.Influence of HLA-E polymorphism on cytomegalovirus

infection after HLA-matched hematopoietic stem cell transplantation. Zhongguo Shi Yan Xue Ye Xue Za Zhi. 2013;21(4):990–994. doi: 10.7534/j.issn.1009-2137.2013.04.034.

61. Romero V, Larsen CE, Duke-Cohan JS, Fox EA, Romero T, Clavijo OP, et al. Genetic fixity in the human major histocompatibility complex and block size diversity in the class I region including HLA-E. BMC Genet. 2007; 12(8):14. doi: 10.1186/1471-2156-8-14.

62. Rebmann V, da Silva Nardi F, Wagner B, Horn PA. HLA-G as a tolerogenic molecule in transplantation and pregnancy. J Immunol Res. 2014;2014:297073. doi: 10.1155/2014/297073.

63. Felício LP, Porto IO, Mendes-Junior CT, Veiga-Castelli LC, Santos KE, Vianello-Brondani RP, et al. Worldwide HLA-E nucleotide and haplotype variability reveals a conserved gene for coding and 3′ untranslated regions. Tissue Antigens. 2014;83(2):82-93. doi:10.1111/tan.12283.

64. Carosella ED, Favier B, Rouas-Freiss N, Moreau P, Lemaoult J. Beyond the increasing complexity of the immunomodulatory HLA-G molecule. Blood. 2008;111(10): 4862–4870. doi: 10.1182/blood-2007-12-127662.

65. Nguyen S, Dhedin N, Vernant J-PP, Kuentz M, Al Jijakli A, Rouas-Freiss N et al. NK-cell reconstitution after haploidentical hematopoietic stem-cell transplantations: immaturity of NK cells and inhibitory effect of NKG2A override GvL effect. Blood. 2005; 105(10):4135–4142. doi: 10.1182/blood-2004-10-4113.

66. Hosseini E, Schwarer AP, Ghasemzadeh M. The impact of HLA-E polymorphisms in graft-versus-host disease following HLA-E matched allogeneic hematopoietic stem cell transplantation. Iran J Allergy Asthma Immunol. 2012;11(1):15–21.

67. Tamouza R, Busson M, Rocha V, Fortier C, Haddad Y, Brun M, et al. Homozygous statusfor HLA-E*0103 confers protection from acute graft-versus-host disease and transplant-related mortality in HLA-matched sibling hematopoietic stem cell transplantation. Transplantation. 2006;82(11):1436-1440.doi: 10.1097/01.tp.0000244598.92049.dd.

68. Hunt JS. Stranger in a strange land. Immunol Rev. 2006;213(1)36–47. doi: 10.1111/j.1600-065X.2006.00436.x.

69. Rebmann V, van der Ven K, Pässler M, Pfeiffer K, Krebs D, Grosse-Wilde H. Association of soluble HLA-G plasma levels with HLA-G alleles. Tissue Antigens. 2001; 57(1):15–21.

70. Hviid TV, Rizzo R, Christiansen OB, melchiorri L, Lindhard A, Baricordi OR. HLA-G and IL-10 in serum in relation to HLA-G genotype and polymorphisms. Immunogenetics. 2004;56(3):135-41. doi: 10.1007/s00251-004-0673-2.

71. La Nasa G, Littera R, Locatelli F, Lai S, Alba F, Caocci G, et al. The human leucocyte antigen-G 14-basepair polymorphism correlates with graft-versus-host disease in unrelated bone marrow transplantation for thalassaemia. Br J Haematol. 2007;139(2):284–8. doi: 10.1111/j.1365-2141.2007.06779.x.

72. Rousseau P, Discorde M Le, Mouillot G, Marcou C, Carosella ED, Moreau P. The 14 bp deletion-insertion polymorphism in the 3′ UT region of the HLA-G gene influences HLA-G mRNA stability. Hum Immunol. 2003;64(11):1005–1010.

73. Deschaseaux F, Delgado D, Pistoia V, Giuliani M, Morandi F, Durrbach A. HLA-G in organ transplantation: towards clinical applications. Cell Mol Life Sci. 2010;68(3):397–404. doi: 10.1007/s00018-010-0581-6.

74. Dharnidharka VR, Caillard S, Agodoa LY, Abbott KC. Infection frequency and profile in different age groups of kidney transplant recipients. Transplantation. 2006;81(12):1662–1667. doi: 10.1097/01.tp.0000226068.66819.37.

75. Ramanan P, Razonable RR. Cytomegalovirus infections in solid organ transplantation: a review. Infect Chemother. 2013;45(3):260–271. doi: 10.3947/ic.2013.45.3.260.

76. Egli A, Binggeli S, Bodaghi S, Dumoulin A, Funk GA, Khanna N, et al. Cytomegalovirus and polyomavirus BK posttransplant. Nephrol Dial Transplant. 2007;22(8):viii72–viii82. doi: 10.1093/ndt/gfm648.

77. Allard M, Tonnerre P, Nedellec S, Oger R, Morice A, Guilloux Y, et al. HLA-E-restricted cross-recognition of allogeneic endothelial cells by CMV-associated CD8 T cells: a potential risk factor following transplantation. PLoS ONE. 2012;7(11):e50951. doi: 10.1371/journal.pone.0050951.

78. Katznelson S, Wang XM, Chia D, Chia D, Ozawa M, Zhong HP, et al. The inhibitory effects of pravastatin on natural killer cell activity in vivo and on cytotoxic T lymphocyte activity in vitro. J Heart Lung Transplant. 1998;17(4):335–340.

79. Sheshgiri R, Rao V, Tumiati LC, Xiao R, Prodger JL, Badiwala M, et al. Progesterone induces human leukocyte antigen-g expression in vascular endothelial and smooth muscle cells. Circulation. 2008;118(14):S58–64. doi: 10.1161/CIRCULATIONAHA.107.757781.

80. Genbacev O, Zhou Y, Ludlow JW, Fisher SJ. Regulation of human placental development by oxygen tension. Science. 1997;277(5332): 1669–1672.

81. Mouillot G, Marcou C, Zidi I, Guillard C, Sangrouber D, Carosella ED, Moreau P, et al. Hypoxia modulates HLA-G gene expression in tumor cells. Hum Immunol. 2007;68(4):277–85. doi: 10.1016/j.humimm.2006.10.016.

82. Kumar G, Date OS, Kim KS, Manjunath R. Infection of human amniotic and endothelial cells by Japanese encephalitis virus: Increased expression of HLA-F. Virology. 2014;471-473:29–37. doi: 10.1016/j.virol.2014.09.022.

83. Suárez-Alvarez B, Rodriguez RMM, Calvanese V, Miguel ABG, Blanco G, Steve TS, et al. Epigenetic mechanisms regulate MHC and antigen processing molecules in human embryonic and induced pluripotent stem cells. PLoS ONE. 2010;5(4): e10192. doi: 10.1371/journal.pone.0010192.

84. Stewart L, Evans N, Bexon KJ, van der Meer DJ, Williams GA. Differentiating between monozygotic twins through DNA methylation-specific high-resolution melt curve analysis. Anal Biochem. 2015;476:36–39. doi: 10.1016/j.ab.2015.02.001.

85. Sizzano F, Testi M, Zito L, , Crocchiolo R, Troiano M, Mazzi B, et al. Genotypes and haplotypes in the 3′ untranslated region of the HLA-G gene and their association with clinical outcome of hematopoietic stem cell transplantation for beta-thalassemia. Tissue Antigens. 2012;79(5):326–32. doi: 10.1111/j.1399-0039.2012.01862.x.

86. Torres MI, Luque J, Lorite P, Isla-Tejera B, Palomeque T, Aumente MD, et al. 14-Base pair polymorphism of human leukocyte antigen-G as genetic determinant in heart transplantation and cyclosporine therapy monitoring. Hum Immunol. 2009;70(10):830–5. doi: 10.1016/j.humimm.2009.07.012.

87. Castellaneta A, Mazariegos G, Nayyar N, Zeevi A, Thomson AW. HLA-G Level on Monocytoid Dendritic Cells Correlates With Regulatory T-Cell Foxp3 Expression in liver transplant tolerance. Transplantation. 2011; 91(10):1132-1140. doi: 10.1097/TP.0b013e31821414c9.

88. Chen H-F, Yu C-Y, Chen M-J, Chou SH, Chiang MS, Chou WH, et al. Characteristic expression of major histocompatibility complex

and immune privilege genes in human pluripotent stem cells and their derivatives. Cell Transplant. 2015;24(5):845-864. doi: 10.3727/096368913X674639.

89. Brugière O, Thabut O, Pretolani M, Krawice-Radanne I, Dill C, Herbreteau A, et al. Immunohistochemical Study of HLA-G Expression in Lung Transplant Recipients. Am J Transplant. 2009;9(6):1427-1438. doi:10.1111/j.1600-6143.2009.02650.x.

90. Takao S, Ishikawa T, Yamashita K, Uchiyama T. The rapid induction of HLA-E is essential for the survival of antigen-activated naive CD4 T cells from attack by NK cells. J Immunol. 2010;185(10):6031-40. doi: 10.4049/jimmunol.1000176.

91. Bresciani A, Pirozzi G, Spera M, Lombardi ML, Ambrosone L, Migliaresi S, et al. Increased level of serum HLA class I antigens in patients with systemic lupus erythematosus. Correlation with disease activity. Tissue Antigens. 1998;52(1):44-50. doi: 10.1111/j.1399-0039.1998.tb03022.x.

92. Adamashvili I, Wolf R, Aultman D, Milford EL, Jaffe S, Hall V, et al. Soluble HLA-I (s-HLA-I) synthesis in systemic lupus erythematosus. Rheumatol Int. 2003;23(6):294-300. doi: 10.1007/s00296-003-0306-3.

93. Puppo F, Contini P, Ghio M, Indiveri F. Soluble HLA class I molecules/CD8 ligation trigger apoptosis of CD8+ cells by Fas/Fas-ligand interaction. ScientificWorldJournal. 2002;12(2):421-3. doi: 10.1100/tsw.2002.122.

94. Puppo F, Scudeletti M, Indiveri F, Ferrone S. Serum HLA class I antigens: markers and modulators of an immune response? Immunol Today. 1995;16(3):124-7.

95. Luque J, Torres MI, Aumente MD, Marín J, García-Jurado G, González R, et al. Soluble HLA-G in heart transplantation: their relationship to rejection episodes and immunosuppressive therapy. Hum Immunol. 2006;67(4-5):257-63. doi: 10.1016/j.humimm.2006.02.034.

96. Créput C, Durrbach A, Menier C, Guettier C, Samuel D, Dausset J, et al. Human leukocyte antigen-G (HLA-G) expression in biliary epithelial cells is associated with allograft acceptance in liver-kidney transplantation. J Hepatol. 2003 ;39(4):587-94.

97. Xiao L, Zhou WQ, Shi BY, Feng K, He XY, Wei YX, et al. HLA-G expression in the peripheral blood of live kidney transplant recipients. Chin Med J. 2013;126(14):2652-5.

98. Qiu J, Terasaki PI, Miller J, Mizutani K, Cai J, Carosella ED. Soluble HLA-G expression and renal graft acceptance. Am J Transplant. 2006;6(9):2152-6. doi: 10.1111/j.1600-6143.2006.01417.x.

99. Boyson J, Erskine R, Whitman M, Chiu M, Lau JM, Koopman LA ,et al. Disulfide bond-mediated dimerization of HLA-G on the cell surface. Proc Natl Acad Sci U S A. 2002;99(25):16180-5. doi:10.1073/pnas.212643199.

100. Ezeakile M, Portik-Dobos V, Wu J, Horuzsko DD, Kapoor R, Jagadeesan M, et al. HLA-G dimers in the prolongation of kidney allograft survival. J Immunol Res. 2014;2014:153981. doi: 10.1155/2014/153981.

101. Rhynes VK, McDonald JC, Gelder FB, et al. Soluble HLA class I in the serum of transplant recipients. Ann Surg. 1993;217(5):485-9; discussion 489-91.

102. Reed EF, Hong B, Ho E, Harris PE, Weinberger J, Suciu-Foca N. Monitoring of soluble HLA alloantigens and anti-HLA antibodies identifies heart allograft recipients at risk of transplant-associated coronary artery disease. Transplantation. 1996;61(4):566-72.

103. Zavazava N, Böttcher H, Ruchholtz WM. Soluble MHC class I antigens (sHLA) and anti-HLA antibodies in heart and kidney allograft recipients. Tissue Antigens. 1993;42(1):20-6.

104. Ravindranath MH, Taniguchi M, Chen CW, Ozawa M, Kaneku H, El-Awar N, et al. HLA-E monoclonal antibodies recognize shared peptide sequences on classical HLA class Ia: relevance to human natural HLA antibodies. Mol Immunol. 2010;47(5):1121–31. doi: 10.1016/j.molimm.2009.10.024.

105. Ravindranath MH, Pham T, El-Awar N, Kaneku H, Terasaki PI. Anti-HLA-E mAb 3D12 mimics MEM-E/02 in binding to HLA-B and HLA-C alleles: Web-tools validate the immunogenic epitopes of HLA-E recognized by the antibodies. Mol Immunol. 2011;48(4):423-30. doi: 10.1016/j.molimm.2010.09.011.

106. Ravindranath MH, Zhu D, Pham T, Jucaud V, Hopfield J, Kawakita S, et al. Anti-HLA-E monoclonal antibodies reacting with HLA-la and lb alleles like IVIg as potential IVIg-immunomimetics: an evolving therapeutic concept. Clin Transpl. 2013:293-305.

107. Zhu D, Ravindranath MH, Terasaki PI, Miyazaki T, Pham T, Jucaud V. Suppression of allo-human leucocyte antigen (HLA) antibodies secreted by B memory cells in vitro: intravenous immunoglobulin (IVIg) versus a monoclonal anti-HLA-E IgG that mimics HLA-I reactivities of IVIg. Clin Exp Immunol. 2014;177(2):464-77. doi: 10.1111/cei.12307.

108. Ravindranath MH, Terasaki PI, Pham T, Jucaud V, Kawakita S. Therapeutic preparations of IVIg contain naturally occurring anti-HLA-E antibodies that react with HLA-Ia (HLA-A/-B/-Cw) alleles. Blood. 2013; 121(11):2013-28. doi: 10.1182/blood-2012-08-447771.

109. Morales-Buenrostro LE, Terasaki PI, Marino-Vázquez LA, Lee JH, El-Awar N, Alberú J. "Natural" human leukocyte antigen antibodies found in nonalloimmunized healthy males. Transplantation. 2008;86(8): 1111-5. doi: 10.1097/TP.0b013e318186d87b.

110. Visentin J, Guidicelli G, Bachelet T, Jacquelinet C, Audry B, Nong T, et al. Denatured class I human leukocyte antigen antibodies in sensitized kidney recipients: prevalence, relevance, and impact on organ allocation. Transplantation. 2014;98(7):738-44. doi: 10.1097/TP.0000000000000229.

111. Demaria S, Schwab R, Bushkin Y. The origin and fate of beta 2m-free MHC class I molecules induced on activated T cells. Cell Immunol. 1992;142(1):103-13.

10

Progressive Insight into Tumor Promotion with Pro-inflammatory Cytokines

Hirota Fujiki[1]*, Masami Suganuma[2], Tatsuro Watanabe[1], Eisaburo Sueoka[1]

[1]*Department of Clinical Laboratory Medicine, Faculty of Medicine, Saga University, Saga, Japan*
[2]*Graduate School of Science and Engineering, Saitama University, Saitama, Japan*

***Corresponding author:** Hirota Fujiki, Department of Clinical Laboratory Medicine, Faculty of Medicine, Saga University, 5-1-1 Nabeshima, Saga 849-8501, Japan, E-mail: uv4h-fjk@asahi-net.or.jp*

Absract

In 2013, we published a review article entitled "Tumor promoters: from chemicals to inflammatory proteins", describing that pro-inflammatory cytokines and chemokines in the cancer microenvironment induce tumor-promoting inflammation in human cancer development. We present here some significant topics based on our experiments: (i) The okadaic acid class of compounds, which are potent inhibitors of Protein Phosphatases 1 (PP1) and 2A (PP2A), showed a common mechanism of tumor promotion on mouse skin, in rat glandular stomach and in rat liver, (ii) Both 12-*O*-tetradecanoylphorbol-13-acetate (TPA) and okadaic acid commonly induced gene expressions of TNF-α, IL-1α and IL-1β on $TNF^{+/+}$ mouse skin, and these pro-inflammatory cytokines induced clonal growth of *v-Ha-ras*-transfected BALB/3T3 cells (Bhas 42), (iii) *H. pylori* genome contained a new *TNF-α-inducing protein* (*Tipα*) gene, and Tipα protein induced the Epithelial-Mesenchymal Transition (EMT) in human gastric cancer cells and (iv) Nucleolin on cell surface was identified as a receptor for *Tipα*, emphasizing gastric tumor promotion and progression of *H. pylori*. Finally we discussed the broad roles of pro-inflammatory cytokines in tumor progression in relation to inflammasomes reported by other investigators.

Keywords: Cytokine; Okadaic acid; Inflammasome; Nucleolin; Tipα

Introduction

Tumor promotion was intensively studied in the 1980s, and tumor promoters were found in various organs of rodents in two-stage carcinogenesis experiments [1]. However, tumor promotion is not accepted in clinical cancer development, because tumor promotion is thought to be mechanistically different from progression. In order to close up the relationship between tumor promotion and progression, a common mechanism of tumor promoters should be generalized in various organs. Based on our evidence that okadaic acid, an inhibitor of protein phosphatases 1 (PP1) and 2A (PP2A), is a potent tumor promoter on mouse skin initiated with 7,12-dimethylbenz [a] anthraxene (DMBA) as does 12-*O*-tetradecanoylphorbol-13-acetate (TPA), we intensively studied the mechanisms of okadaic acid in relation to TPA. The significant results of our experiments are as follows: Okadaic acid and TPA commonly induced gene expressions of TNF-α, IL-1α and IL-1β on mouse skin [2,3]; The study of tumor promotion by both okadaic acid and TPA revealed that TNF-α is an essential tumor promoter in TNF-α-deficient mouse skin initiated with DMBA [3]; Inhibitors of PP1 and PP2A induced tumor promotion on mouse skin, in rat glandular stomach and rat liver initiated with carcinogens [4]; TNF-α, IL-1α and IL-1β induced clonal growth of *v-Ha-ras*-transfected BALB/3T3 cells (Bhas 42) [3]; Treatment with tumor promoters commonly released TNF-α from target organs [5]. In light of our evidence showing that TNF-α, IL-1α and IL-1β act as endogenous tumor promoters and/or cancer mediators [6,7], we specifically found a *TNF-α–inducing protein (Tipα)* gene in *H. pylori* genome [8,9]. Tipα dimer bind to nucleolin on cell surface of human gastric cancer cells as a receptor [10], and are secreted in large quantities from *H. pylori* of gastric cancer patients [11]. The Epithelial-Mesenchymal Transition (EMT) was induced in human gastric cancer cells by Tipα [12]. The Tipα and nucleolin complex shows a strong link between tumor promotion and clinical cancer development. Since TNF-α and IL-1 are tumor-promoting cytokines, we discussed the role of inflammasome signaling for tumor promotion using apoptosis-associated speck-like protein containing a caspase recruitment domain (ASC)-KO and caspase (CASP)-1-KO mice reported by Campanelli's group [13].

Methods

The main experimental results are cited from our publications, and the relevant literatures reviewed in this article are screened from journals and PubMed.

Results

Common mechanism of tumor promotion by inhibitors of PP1 and PP2A

Tumor promoters of the okadaic acid class, which are all potent inhibitors of PP1 and PP2A, include okadaic acid, dinophysistoxin-1 (35-methylokadaic acid), calyculin A, microcystin-LR and nodularin [4]. Okadaic acid induced clonal growth of DMBA-initiated cells on CD-1 mouse skin as potently as did TPA. Okadaic acid also induced tumor promotion in male SD rat glandular stomach initiated with *N*-methyl-*N'*-nitro-*N*

***Corresponding author email:** uv4h-fjk@asahi-net.or.jp*

nitrosoguanidine. Microcystin-LR and nodularin were identified as new tumor promoters in rat liver initiated with diethyl nitrosamine. All of these results showed for the first time that inhibition of PP1 and PP2A is a common mechanism of tumor promotion in various organs [4,7].

Human fibroblasts treated with both okadaic acid and TNF-α biochemically showed a similar phosphorylation pattern of over 140 proteins in two-dimensional gel electrophoresis, reported by Guy, *et al.* [14]. Next we found that treatment with okadaic acid, which acts as tumor promoter on mouse skin and rat glandular stomach, induced gene expressions of TNF-α, IL-1α and IL-1β on mouse skin, and released TNF-α from KATO III cells of human stomach cancer cell line [3,15]. The i.p. administration of microcystin-LR and nodularin induced *TNF-α* gene expression in rat liver [5], and treatments with microcystin-LR and nodularin induced TNF-α release from rat hepatocytes into the medium, while TPA did not apparently release TNF-α, nor showed tumor-promoting activity in rat liver [5]. To demonstrate the significance of TNF-α in tumor promotion, we conducted experiments with TNF-α deficient mice initiated with DMBA by topical applications of okadaic acid and TPA: Okadaic acid did not induce any tumors on skin of TNF-$\alpha^{-/-}$ mice up to 19 weeks, but TNF-$\alpha^{+/+}$ mice showed 100% of tumor incidence. In addition,TPA-treated TNF-$\alpha^{-/-}$ mice showed 2.8 in average number of tumors, compared with 11.8 for TNF-$\alpha^{+/+}$ CD-1 mice [3]. Considering the residual tumor-promoting activity in TNF-$\alpha^{-/-}$ mice, we found significant expression of *IL-1α* and *IL-1β* genes as additional inflammatory cytokines for tumor promotion in TNF-$\alpha^{-/-}$ mouse skin [3]. On the other hand, okadaic acid and TPA showed a similar potency of tumor promotion in IL-$6^{-/-}$ and IL-$6^{+/+}$ mice initiated with DMBA [15], suggesting that TNF-α and IL-1 are essential tumor promoters, whereas IL-6 supplements functions of other cytokines.

Carcinogenic role of TNF-α in cell transformation and tumorigenicity

Human TNF-α at a concentration of 10 ng/ ml (0.6 nM) induced 1.83 average No. of foci/dish, whereas okadaic acid at a concentration of 20 ng/ ml (0.02 μM) showed 1.58 and TPA at a concentration of 300 ng/ml (0.5 μM) 1.92 foci/dish in BALB/3T3 cells initiated with 3-methylcholanthrene (MCA), respectively [6]. This is important to note that human TNF-α secreted from various cells has a more potent transforming activity than chemical tumor promoters. TNF-α also induced clonal growth in *v-Ha-ras*-transfected BALB/3T3 cells (Bhas 42 cells), whereas it did not induce any growth of BALB/3T3 cells without *v-Ha-ras* gene, indicating that TNF-α is a tumor promoter or cancer mediator in tumor development [3].

In addition, treatment of BALB/3T3 cells with human TNF-α at a concentration of 10 ng/ml (0.6 nM) significantly induced 0.33 foci/dish, even without initiation of MCA, whereas MCA at concentration of 0.1 μg/ml (0.4 μM) alone showed 0.08, and DMSO in medium was 0 [6]. Moreover, the clones from TNF-α-transformed foci of BALB/3T3 cells showed strong tumorigenicity in sites of injected mice, and also induced expression of *IL-1α, IL-6,* and *TGF-β* genes [15]. The results indicate that human TNF-α has both potent initiating and promoting activity.

TNF-α-inducing protein (Tipα) of *H. pylori* for gastric cancer progression

If tumor promotion can be understood as up-regulated expression of *TNF-α, IL-1* and *chemokine* genes that are associated with activation of NF-κB, tumor promotion significantly can be closed up to the stage of progression in human cancer development. In order to prove this concept, we studied a link between TNF-α and *H. pylori* infection with *Helicobacter pylori* membrane protein 1 (HP-MP1) isolated from *H. pylori* strain SR 7791 [16]. HP-MP1 induced release of various inflammatory cytokines, such as TNF-α, IL-1α and IL-8, and also macrophage inflammatory protein 1α from human monocytes, although HP-MP1 is structurally unrelated to virulence factors such as cagA, vacA, and urease [16]. Next, *HP-MP1* gene was transfected into *v-Ha-ras*-transfected BALB/3T3 cells (Bhas 42 cells, named Bhas/mpl) and *v-Ha-ras*-non-transformed BALB/ 3T3 cells (BALB/mp1), and clones of *HP-MP1*-transfected Bhas/ mpl cells induced *TNF-α* gene expression more strongly that did BALB/ mpl clones, and also induced cell transformation and tumors in nude mice [8].

In order to more broadly understand the function of *HP-MP1* gene, we isolated *HP0596* gene from *H. pylori* strain 26695, which was 94.5% homologous to *HP-MP1* gene, and designated HP0596 protein as the TNF-α-inducing protein (Tipα) [11]. Tipα showed the same activity as did HP-MP1, and induced extensive expression of the *chemokine* genes, such as *Ccl2, Ccl7, Cxcl1, Cxcl2, Cxcl5* and *Cxcl10* in mouse gastric cancer cell line MGT-40 [17], while the inactive form of Tipα (del-Tipα has deletion of six amino acids containing two cysteine residues) showed only marginal expression. Since Tipα is produced in various *H. pylori* strains, the amount of Tipα in culture broth of *H. pylori* was determined depending on gastric ailment. *H. pylori* cultured from gastric cancer patients secreted Tipα at 1.4-13.4 relative units, and those from gastritis patients secreted at 0.8-6.7 relative units [11].

Like TNF-α, IL-1β and IL-6 are critical mediators in tumorigenesis, and the stomach-specific expression of human IL-1β in transgene mice induced spontaneous gastric inflammation and cancer [18]. Since HP-MP1 (Tipα) induced release of various inflammatory cytokines, such as TNF-α and IL-1α from human monocytes [16], we conclude that pathologic elevation of a single pro-inflammatory cytokine is sufficient to induce cancer.

Nucleolin on cell surface as a receptor for Tipα

Tipα binding protein was found in cell lysates of mouse gastric cancer cell line MGT-40 using FLAG-pulldown assay and was identified as cell surface nucleolin by LC-MS analysis and flow cytometry using anti-nucleolin antibody. Nucleolin was then co-precipitated with Tipα–FLAG, but not with del-Tipα-FLAG (an inactive mutant) [10]. The direct binding of Tipα to recombinant His-tagged nucleolin fragment (284-710) was also confirmed. Pretreatment with anti-nucleolin antibody enhanced Tipα-incorporation into the cells through nucleolin internalization [10], indicating a new mechanistic insight into gastric cancer development with Tipα.

The rdel-Tipα protein is an inactive monomer, with biological activity 10-50 fold weaker than that of rTipα. The crystal structure of rdel-Tipα was determined using Multiple Isomorphous Replacements with Anomalous Scattering (MIRAS) [19], since recombinant Tipα (rTip) did not grow to the crystal. The results suggest that rTipα is secreted from *H. pylori* as a soluble homodimer through a secretion mechanism different from the Type IV secretion system, because neither HP-MP1 nor rTipα has a hydrophobic surface that interacts with the lipid layer of the membrane [19].

Epithelial-Mesenchymal Transition (EMT) in human gastric cancer cells by Tipα

Epithelial-Mesenchymal Transition (EMT) is theoretically understood as acquisition by epithelial cells of the phenotype of mesenchymal cells, such as fibroblasts. For example, cancer cells phenotypically are more likely to become mesenchymal cells associated with metastatic states, such as changes in cell shape and gene expression [20]. Tipα binding to surface nucleolin resulted in induction of the early metastatic states of cancer cells, such as migration, elongation and formation of filopodia associated with invasive changes in human gastric cancer cell line MKN-1, which were inhibited by the nucleolin-targeted siRNAs [12]. Tipα enhanced phosphorylation of 11 cancer-related proteins and activation of MEK-ERK signal cascade. It is of interest to note that Tipα reduced cell stiffness and increased cell motility, determined by atomic force microscopy. We think Tipα is a new inducer of EMT associated with tumor progression in human gastric carcinogenesis mediated through TNF-α [12].

Discussion

This review provided evidence that TNF-α and IL-1 are directly involved in tumor-promoting inflammation and progression. The secretion of pro-inflammatory cytokines, such as IL-1β, and IL-18 is mediated by cysteine protease caspase-1, IL-1β-converting enzyme, in activation of inflammasomes [21]. Inflammasomes are large protein complexes (>700 kDa) typically consisting of caspase-1, a Nucleotide Binding-Oligomerization Domain (Nod)-like Receptor (NLR), an adaptor protein apoptosis-associated speck-like protein containing a Caspase Activation and Recruitment Domain (Card) domain (ASC) [22], and play divergent roles in different types of cancer reflecting the complexity of inflammation [23]. As previously reported, both IL-1α and IL-1β contribute to tumor angiogenesis and invasiveness, but the role of IL-1β is more evident in these processes [24]. In inflammatory cells, caspase-1 activation can fuel a cycle that leads to sterile inflammation and carcinogenesis, whereas in antigen-presenting cells, inflammasomes can stimulate anticancer immune responses [25].

The results of Campanelli's group are worthwhile to note that the two-stage carcinogenesis experiments with DMBA and TPA induced the first tumor at the 5th week on caspase-1(CASP)-1-KO mice, at the 7th week on ASC-KO mice, and at the 10th week on wild type mice, and that the levels of pro-inflammatory cytokinses, such as IL-1β, IL-18 and TNF-α in the tumors from CASP-1-KO and ASC-KO mice decreased when compared with wild type mice, indicating that loss of inflammasome function enhanced the process of tumor development [13]. The results indicate that inflammasomes play dual roles in the control of cancer development [13].

Acknowledgment

This work was supported by Japan Society for the Promotion of Science, the Smoking Research Foundation, Tokushima Bunri University, the Urakami Foundation, and Grants-in Aid for Cancer Research: Special Cancer Research, from the Ministry of Education, Science, Sports and Culture, Japan, the Research on HTLV-1 Associated diseases from Saga Prefecture, Japan, and the Princess Takamatsu Cancer Research Fund.

References

1. Fujiki H, Hecker E, Moore RE, Sugimura T, Weinstein IB. Cellular Interaction by Enviromental Tumor Promoters. Tokyo/ Utrecht: Jap Sci Soc Press/ VNU Sci Press; 1983.

2. Fujiki H, Suganuma M. Tumor necrosis factor-α, a new tumor promoter, engendered by biochemical studies of okadaic acid. J Biochem. 1994;115:1-5.

3. Suganuma M, Okabe S, Marino MW, Sakai A, Sueoka E, Fujiki H. Essential role of tumor necrosis factor α (TNF-α) in tumor promotion as revealed by TNF-α-deficient mice. Cancer Res. 1999;59(18):4516-4518.

4. Fujiki H, Suganuma M. Tumor promotion by inhibitors of protein phosphatases 1 and 2A: the okadaic acid class of compounds. Adv Cancer Res. 1993;61:143-194.

5. Sueoka E, Sueoka N, Okabe S, Kozu T, Komori A, Ohta T, et al. Expression of the tumor necrosis factor α gene and early response genes by nodularin, a liver tumor promoter, in primary cultured rat hepatocytes. J Cancer Res Clin Oncol. 1997;123:413-419.

6. Komori A, Yatsunami J, Suganuma M, Okabe S, Abe S, Sakai A, et al. Tumor necrosis factor acts as a tumor promoter in BALB/3T3 cell transformation. Cancer Res. 1993;53:1982-1985.

7. Fujiki H, Sueoka E, Suganuma M. Tumor promoters: from chemicals to inflammatory proteins. J Cancer Res Clin Oncol. 2013;139(10):1603-14. doi: 10.1007/s00432-013-1455-8.

8. Suganuma M, Kurusu M, Okabe S, Sueoka N, Yoshida M, Wakatsuki Y, et al. *Helicobacter pylori* membrane protein 1: a new carcinogenic factor of *Helicobacter pylori*. Cancer Res. 2001;61(17):6356-6359.

9. Suganuma M, Kurusu M, Suzuki K, Nishizono A, Murakami K, Fujioka T, et al. New tumor necrosis factor-alpha-inducing protein released from *Helicobacter pylori* for gastric cancer progression. J Cancer Res Clin Oncol. 2005;131(5):305-313.

10. Watanabe T, Tsuge H, Imagawa T, Kise D, Hirano K, Beppu M, et al. Nucleolin as cell surface receptor for tumor necrosis factor-α inducing protein: a carcinogeic factor of of *Helicobacter pylori*. J Cancer Res Clin Oncol. 2010 Jun;136(6):911-21. doi: 10.1007/s00432-009-0733-y.

11. Suganuma M, Yamaguchi K, Ono Y, Matsumoto H, Hayashi T, Ogawa T, et al. TNF-α-inducing protein, a carcinogenic factor secreted from *H. pylori*, enters gastric cancer cells. Int J Cancer. 2008;123:117-122.

12. Watanabe T, Takahashi A, Suzuki K, Kurusu-Kanno M, Yamaguchi K, Fujiki H, et al. Epithelial-mesenchymal transition in human gastric

cancer cell lines induced by TNF-α-inducing protein of *Helicobacter pylori*. Int J Cancer. 2014;134(10):2373-82. doi: 10.1002/ijc.28582.

13. Gasparoto TH, de Oliveira CE, de Fraitas LT, Pinheiro CR, Hori JI, Garlet GP, et al. Inflammasome activation is critical to the protective immune response during chemically induced squamouse cell carcinoma. PLoS One. 2014;9(9):e107170. doi: 10.1371/journal.pone.0107170.

14. Guy GR, Cao X, Chua SP, Tan YH. Okadaic acid mimics multiple changes in early protein phosphorylation and gene expression induced by tumor necrosis factor or interleukin-1. J Biol Chem. 1992;267(3):1846-52.

15. Suganuma M, Okabe S, Kurusu M, Iida N, Ohshima S, Saeki Y, et al. Discrete roles of cytokines, TNF-α, IL-1, IL-6 in tumor promotion and cell transformation. Int J Oncol. 2002;20(1):131-6.

16. Yoshida M, Wakatsuki Y, Kobayashi Y, Itoh T, Murakami K, Mizoguchi A, et al. Cloning and characterization of a novel membrane-associated antigenic protein of *Helicobacter pylori*. Infect Immun. 1999;67:286-293.

17. Kuzuhara T, Suganuma M, Kurusu M, Fujiki H. *Helicobacter pylori*-secreting protein Tipα is a potent inducer of chemokine gene expressions in stomach cancer cells. J Cancer Res Clin Oncol. 2007;133:287-296.

18. Tu S, Bhagat G, Cui G, Takaishi S, Kurt-Jones EA, Rickman B, et al. Overexpression of interleukin-1β induces gastric inflammation and cancer and mobilizes myeloid-drived suppressor cells in mice. Cancer Cell. 2008;14(5):408-19. doi: 10.1016/j.ccr.2008.10.011.

19. Tsuge H, Tsurumura T, Utsunomiya H, Kise D, Kuzuhara T, Watanabe T, et al. Structural basis for the *Helicobacter pylori*-carcinogenic TNF-α-inducing protein. Biochem Biophys Res Commun. 2009;388(2):193-8. doi: 10.1016/j.bbrc.2009.07.121.

20. Polyak K, Weinberg RA. Transitions between epithelial and mesenchymal states: acquisition of malignant and stem cell traits. Nat Rev Cancer. 2009;9(4):265-73. doi: 10.1038/nrc2620.

21. Shurin MR. Dual role of immunomodulation by anticancer chemotherapy. Nat Med. 2013;19(1):20-22.

22. Martinon F, Burns K, Tschopp J. The inflammasome: a molecular platform triggering activation of inflammatory caspases and processing of proIL-β. Mol Cell. 2002; 10(2): 417-26.

23. Janowski AM, Kolb R, Zhang W, Sutterwala FS. Beneficial and detrimental roles of NLRs in carcinogenesis. Front Immunol. 2013 Nov 12;4:370. doi: 10.3389/fimmu.2013.00370.

24. Voronov E, Shouval DS, Krelin Y, Cagnano E, Benharroch D, Iwakura Y, et al. IL-1 is required for tumor invasiveness and angiogenesis. Proc Natl Acad Sci USA. 2003;100(5):2645-2650.

25. Zitvogel L, Kepp O, Galluzzi L, Kroemer G. Inflammasomes in carcinogenesis and anticancer immune responses. Nat Immunol. 2012;13(5):343-51. doi: 10.1038/ni.2224

11

The Inflammatory Cytokine IL-21 is Expressed by Splenic Neutrophils in Response to Transplantation of Allogeneic Cells

John S. Thompson[1,2]*, Debra L. Hardin[2], Judy F. Glass[1], Joshua Dziba[1], Jeffrey Campion[3] and Stephen A. Brown[1,2]

[1]VA Medical Center, Lexington Kentucky 40502

[2]Department of Internal Medicine, College of Medicine, Lexington, Kentucky 40536

[3]College of Pharmacy Lexington Kentucky 40536

***Corresponding author:** *John S. Thompson, VA Medical Center 1101 Veterans Drive, Lexington KY 40502 and Department of Internal Medicine, University of Kentucky 800 Rose Street Lexington KY 40536, E-mail: jsthom1@uky.edu*

Absract

We have previously reported that GR-1 neutrophil/monocytes rose dramatically in the spleen, peaked by day 7 and declined through day 14. This period corresponded to the peak of acute Graft-Versus-Host Disease (aGVHD) in BALB/c mice transplanted with allogeneic donor cells. We now asked: what cytokines did these splenic neutrophil/monocytes express on day 7 and 14 post transplant?

BALB/c mice were transplanted with allogeneic B6 or syngeneic BALB/c donor cells. Long term survival was recorded through day 31. Other groups were sacrificed on days 3, 5, 7, 14, 21 and 31 days post transplant to record the total number of cells in the spleens and their phenotypes. Neutrophils were isolated from the spleens of mice transplanted with B6 and BALB/c cells on days 7 and 14.

Daily body weight demonstrated a transient drop in the syngeneic transplants on day 2 but a much greater drop with its nadir at day 7 and never fully recovering through 31 days. CD8/CD4 T lymphocytes peaked in the spleen on day 5 and were followed on day 7 by GR-I cells in all of the allogeneic transplants. In syngeneic transplants this early rise in lymphocytes did not occur and GR-1 cells peaked on day 14. Highly purified neutrophils were isolated in two separate experiments from the spleens on days 7 and 14 post transplant. In both experiments day 7 allogeneic neutrophils expressed significantly elevated levels of Interleukin -21 (IL-21) mRNA whereas the day 7 and 14 syngeneic cells expressed lower but significant levels of TNFα. Intracellular IL-21 was demonstrated in the allogeneic neutrophils on day 7 before and after in vitro stimulation.

In conclusion Purified neutrophils isolated from the spleen on day 7, the early peak of allogeneic transplantation a GVHD, express high levels of IL-21 message and intracellular IL-21.

Keywords: Allogeneic neutrophils; IL-21; Graft-versus-host disease; Transplantation

Abbreviations

Graft-versus-Host-Disease (GVHD); acute Graft-versus-Host Disease (aGVHD)

Introduction

We have previously demonstrated in a murine stem cell transplant model, the rise of splenic IL-17 mRNA on day 3, its peak on day 5 and fall by day 7 corresponding to the rise and fall of CD3/CD8 T cells [1]. Day 5 marked the beginning of a larger and more sustained increase in GR-1 myeloid cells. This sequence suggested that the earlier rise in CD3/CD8 T cells secreted signals that stimulated GR-1 related myelopoiesis. Since the neutrophil rise occurred at the approximate peak of acute Graft-Versus-Host Disease (aGVHD), we asked what inflammatory factors could these neutrophils be secreting that might enhance the pathology of GVHD?

Herein we show that purified neutrophils isolated from the spleen on day 7, the early peak of aGVHD following allogeneic bone marrow transplantation in this donor/host pair, express high levels of IL-21 mRNA and intracellular IL-21.

Materials and Methods

Animals

Male BALB/c H2^d and C57BL/6 H2^b (B6) mice were purchased from Jackson Laboratories (Bar Harbor, ME). After receipt, they were held for 1 or 2 weeks until they were 9-10 weeks of age. B6 mice were chosen for their MHC incompatibility with BALB/c mice. BALB/c and B6 healthy 9-10 week old males were used as transplant donors. Prior to and after radiation the animals were housed 4 to a cage in micro-isolator cages under the supervision of licensed veterinarians. Sentinel mice to detect viral and bacterial infections in the colonies were always present in the rooms with the mice. This study was carried out in strict accordance with the recommendations in the Guide for the Care and Use of Laboratory Animals of the National Institutes of Health. The protocol was approved by the Institutional Animal Care and Use Committee 596, Lexington, Kentucky. Protocol (THO-12011-AF).

***Corresponding author email:** jsthom1@uky.edu*

Bone Marrow, T cell isolation and quantification of the total number of T cells transplanted

EasySep Mouse T Cell Enrichment Kit (StemCell Technologies (Vancouver, Canada, #19751) was used to negatively select for T Cells in freshly prepared spleen cell suspensions. The reagents for purification were added according to the kit instructions. The viability was 99% (trypan blue).

Flow cytometry was performed on both the isolated T Cells and the bone marrow cells in order to calculate and adjust the number of total T Cells injected into the recipients. Approximately 1 x 10^6 cells were simultaneously stained with PE-conjugated CD3 (Clone145-2C11, BD Biosciences, San Jose, CA #553064) and FITC-conjugated CD45 (Clone 30-F11, Invitrogen #4501) Mabs. Flow cytometric analysis showed that the bone marrow cells and the enriched cells were 7% and 98%, respectively, positive for CD3/CD45.

Transplantation

Allogeneic $H2^b$ B6 or syngeneic BALB/c donor cells were transplanted into 9-10 week old male $H2^d$ BALB/c hosts that had received 8.5 Gy TBI delivered by a Mark IV 137Cesium irradiator (J.L Shepard, Glendale CA) at a dose rate of 1.65 Gy/min. The final suspension of donor cells injected into each recipient consisted of 10 x 10^6 bone marrow cells plus splenic T cells for a total of 1.5 x 10^6 allogeneic T-cells or 1.5 x 10^6 syngeneic T cells. The cells were injected via a tail vein and contained in a volume of 0.25 ml of PBS. Following AAALAC and our local guidelines, animals that met prescribed criteria were euthanized and counted as an experimental death. Each cage contained BALB/c mice transplanted with two each of the allogeneic transplants or syngeneic cells to eliminate a possible confounding "cage effect".

Neutrophil isolation

After CO_2 anesthesia, the mice were then killed by cervical dislocation. The spleens were removed by sterile dissection.

Preparation of spleen cells: After rinsing 3 times in sterile PBS, the spleens were perfused with 1-2 ml of RPMI+10% FCS, placed in a sterile plastic bag containing 5 ml of media and dissociated with a Stomacher (Model 80, Seward Limited, Norfolk, U.K.) for 30 seconds on low setting. Cells were filtered through a 70 μm cell strainer, centrifuged at 550 x g for 7 minutes and adjusted to 1x10^6 cells/ml of RPMI 1640 (Invitrogen, Grand Island, NY) + 10% FBS (Hyclone, Logan, UT) + 2mM L-Glutamine + 100units/ml Penicillin and 100ug/ml Streptomycin (Sigma, St. Louis, MO). All operations were performed at room temperature.

Preparation of purified neutrophils: The Stem Cell Technology's EasySep Mouse Neutrophil Enrichment Kit was used to negatively select for neutrophils in freshly prepared spleens. 2 x 10^8 nucleated cells contained in 2 ml of Phosphate Buffered Saline (PBS) were added according to the kit instructions. The viability was checked with trypan blue and purity was verified by 2-color flow cytometry with anti-Ly6G FITC (Clone 1A8, BD Biosciences, San Jose, CA, #551460) and anti-CD11β PE (Clone M1/70, BD Biosciences, #559911) and hamster anti mouse CD3 (Life Technologies, Grand Island NY). Appropriate isotype controls were also run.

RNA isolation, cDNA synthesis and real-time PCR

Total RNA isolation: An aliquot of the isolated neutrophils was finely minced and homogenized (Tissue Tearor, Biospec Products, Inc. Bartlesville, OK) at RT in RLT buffer (RNeasy isolation kit Qiagen Inc, Valencia, CA) + B-mercaptoethanol (Sigma-Aldrich, St. Louis, MO). Total RNA was eluted in 54-60 ʋl of diethyl procarbonate (DEPC) H_2O. Quantitation (μg/ml) and purity (260/280 ratio) was assessed by use of a spectrophotometer and RNA quality was analyzed on an Agilent Bioanalyzer.

cDNA synthesis: 200 μg of total RNA was used for each 20μl cDNA synthesis reaction. Reagents were purchased from Invitrogen Corporation (Carlesbad, CA); Temperatures were used for synthesis reactions according to protocol. Each 20μl cDNA reaction was then diluted 1:5 for use as template for Real-Time PCR reactions.

Real-Time PCR Amplification: 4 μl of cDNA was used as template for each 20 μl PCR reaction. GAPDH primers were used as an endogenous control. All reagents and equipment were purchased from Applied Biosystems: TaqMan Fast Universal PCR Master Mix (2X) and TaqMan Gene Expression Assays for IL-21, GM-CSF, IFNγ, TNFα, IL-12 (p40) G-CSF, IL-6, and IL-22 were utilized in the 7500 Fast Real-Time PCR System.

Real-Time PCR Analysis: ΔΔCt analysis was performed as follows:

1) Duplicate/triplicate Ct values were obtained. After subtracting the GAPDH average endogenous gene expression, the fold difference between the gene expressions in the experimental sample was compared to the control sample (a pool of completely normal BALB/c mice).

Intracellular cytokine

For the intracellular cytokine analysis, an aliquot of the purified neutrophils was washed once in cold D-PBS. The pellet was resuspended in RPMI+10% FCS (supplemented with 2mM Glutamine and 100 U/ml Penicillin + 100ug/ml Streptomycin), aliquoted into 5 ml polypropylene tubes @2 x 10^6cells/ml. Cells were then treated either with Cell Stimulation Cocktail (ionomycin plus phorbol) plus protein transport inhibitor eBioscience, San Diego, CA) or with just the protein transport inhibitor alone and incubated for 6 hours inside a 5% CO_2 incubator at 37°C. After the incubation, the cells were centrifuged, washed 2X in cold D-PBS+1% FCS (Staining Buffer), blocked with CD16/32, and then stained on ice with GR-1 PE, F4/80 FITC, or CD45 FITC cell surface Moabs (Life Technologies, Grand Island, NY). After washing 2X with Staining Buffer, cells were then intracellularly stained with Anti-Mouse IL-21 APC Moab or its Isotype Control using an Intracellular Fixation and Permeabilization Kit according to the kit instructions (eBioscience, San Diego, CA). All amounts used for staining either surface marker or intracellular IL-21, inhibition of protein, or cell activation were those recommended by the companies. Cell pellets were then suspended in 0.2ml Staining Buffer and analyzed on a FACSCalibur Flow Cytometer (BD Biosciences, San Jose, CA) using CellQuest Pro software.

Statistical analysis

Mean mRNA levels were compared by using an analysis of variance for a 3 x 10 factorial design with factors at three levels. Post hoc comparison of means emphasized a comparison of the donor groups on the day of sacrifice provided the interaction between day of sacrifice and donor was significant at the $p < 0.05$.

Results

BALB/c mice transplanted with allogeneic WT B6 BM plus added splenic T lymphocytes develop acute GVHD

Transplantation of allogeneic bone marrow plus added splenic T cells results in acute GVHD (Figure 1). As the weight curve indicates, this is evident by day 5 and weight loss reaches a nadir by day 7. In other studies some of these animals will die during the day 5-10 periods with severe bowel injury but others will survive and gradually regain some of their lost weight. However, by day 14 weight loss begins to recur, signs of chronic GVHD in the skin appear and diarrhea may also begin to become severe by day 21. Many of these animals will die. To the contrary, the mice transplanted with syngeneic cells only lose a little weight by day 3 secondary to total body radiation. They regain much of their weight and, by day 10, maintain nearly normal weights thereafter and all survive. This figure is shown to serve as the correlative structure for the time points chosen to obtain samples for analysis.

There is a consistent rise in CD3/CD8/CD4 T cells at day 5 followed by a consistent rise in GR-1 cells by day 7 in the spleens of BALB/c mice transplanted with B6 cells

On days 3, 5, 7, 14, 21 and 31 groups of 5-6 mice in each group were euthanized for analysis of the phenotypes of the cells in their spleen. Figure 2 presents the flow data from BALB/c spleens of the mice transplanted with cells from the allogeneic donors as compared to that from the syngeneic donors. Whereas the total number of nucleated cells in the spleens of the mice transplanted with allogeneic cells rose above the day 3 nadir, it never returned to the pre-transplant baseline (top left panel). To the contrary, the total number of nucleated cells rose in the syngeneic transplants above the normal control baseline on day 14 post transplant before gradually declining by day 31. The top right panel illustrates that CD19 B cells dropped in all of the B6 transplants to very few cells and never recovered during the 31days. Total CD19 B cells rose dramatically in the syngeneic transplants between day 7 and 14 to levels even greater than that of the normal controls. The middle 2 panels illustrate a significant sharp rise in CD8 and to a lesser extent CD4 T lymphocytes peaking on day 5. However, note that by day 7 the number of CD8 cells had fallen back almost 4 fold toward baseline and the CD4 cells 2 fold in the spleens from the B6 donors. This sharp transient increase did not occur in the syngeneic transplants. The bottom frame illustrates the pattern of GR-1 reconstitution in the spleen. In contrast, a sharp peak in GR-1 cells arose on day 7 in the spleen cells from the allogeneic transplants at the time that the number of CD8 and CD4 T cells had significantly decreased. The peak of GR-1 cells in the syngeneic transplants occurred on day 14 corresponding to the peak in total cells in the syngeneic transplanted spleens.

In the allogeneic transplants, the spleen was essentially repopulated with H-2^b + donor cells by day 7: day 3, 32.5 +/- 24.02%; day 5, 98.71 +/- 0.15%; day 7, 99.78 +/- 0.11%.

On day 7 post transplant, B6 allogeneic highly purified GR-1 splenic neutrophils express high levels of IL-21 and significant levels of GM-CSF whereas syngeneic neutrophils express significant levels of TNFα

As illustrated in Figure 2, there were two distinct peaks in the GR-1 cells corresponding to the allogeneic peak on day 7 and the syngeneic peak on day 14. We asked were their differences in the cytokine expression on these two days? Therefore, neutrophils

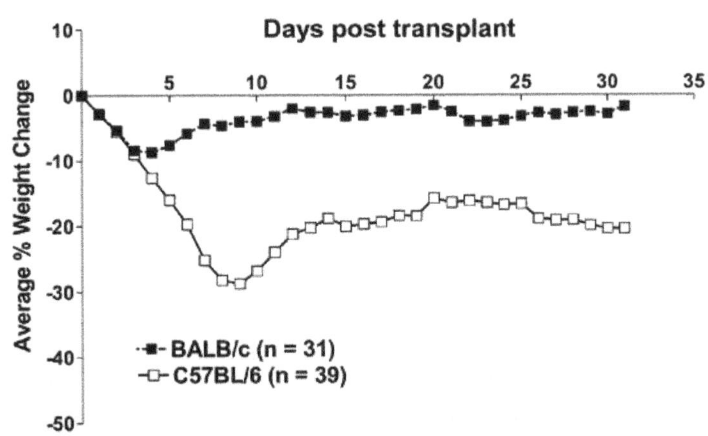

Figure 1: Body weights from 3 consolidated experiments.
H-2^d BALB/c mice transplanted with cells from H-2^b B6 allogeneic or syngeneic BALB/c donors. Each point represents the mean of all of the live animals in their transplant groups weighed on that day.

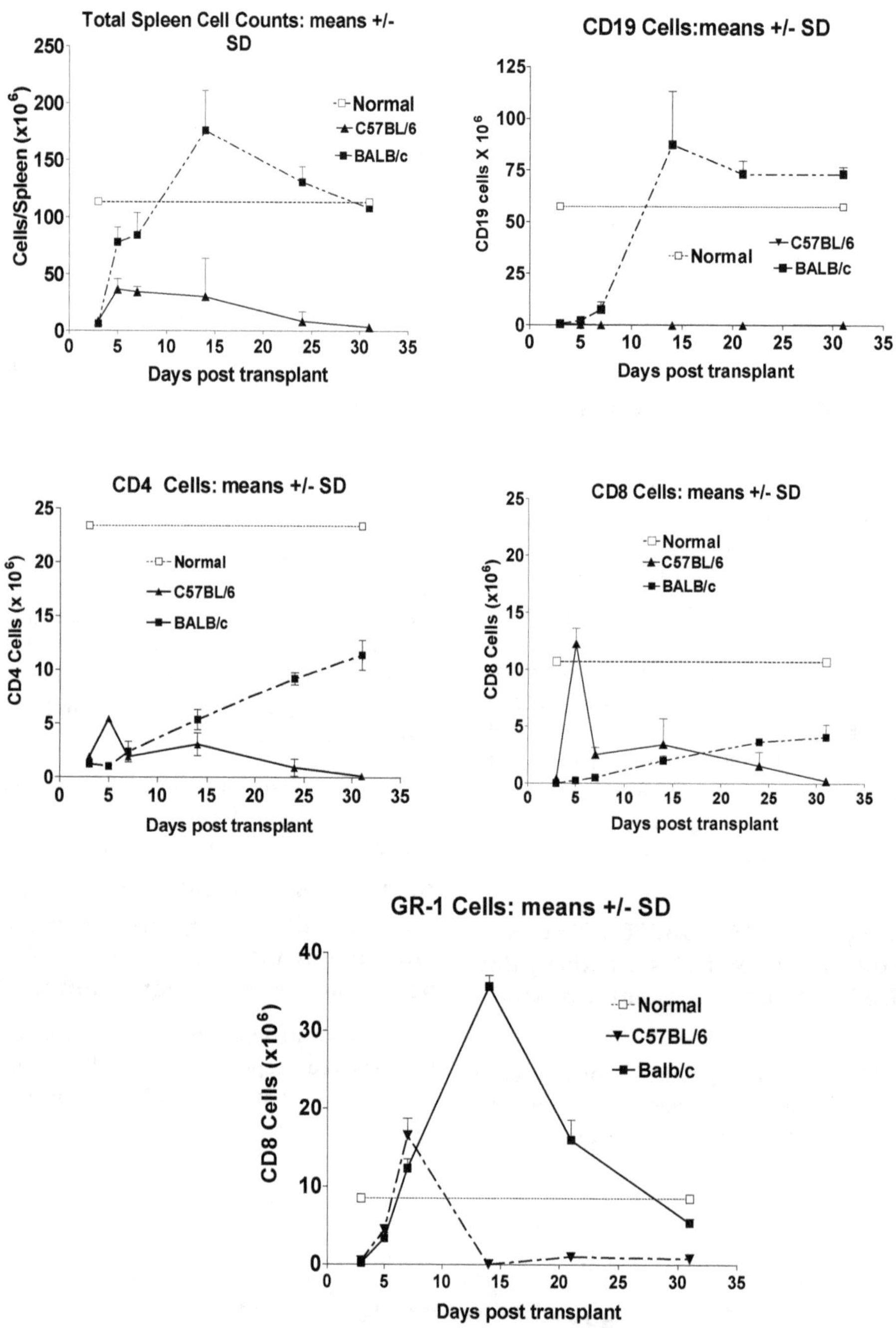

Figure 2: Flow cytometry demonstration of the time and height of phenotypic cells in the spleen of BALB/c mice transplanted with allogeneic and syngeneic cells.
Time and the absolute number of total, CD4, CD8, CD19, and GR-1 cells obtained from the spleens of BALB/c mice transplanted with B6 or syngeneic BALB/c cells on days 3, 5, 7, 14, 21 and 31 days post transplant: Total spleen cells (upper left panel), CD19 B lymphocytes (upper right panel), CD4 and CD8 T lymphocytes (left and right middle panels) and GR-1 cells (lower single panel). The dashed horizontal line in each panel represents the mean of the absolute number of the cells of that phenotype obtained from 6 completely normal BALB/c animals. Means +/- SD (n = 4 - 6).

were isolated from the spleens of BALB/c mice transplanted with allogeneic and syngeneic donor cells on these two days. Purification with LY6G antibody isolates highly purified GR-1 neutrophils and does not include GR-1 macrophages. Figure 3 presents the step wise purification of two completely independent experiments. The resulting number of cells at each step is compared to that from completely normal BALB/c animals. It illustrates two points. First, although the total number of cells in the spleens of the B6 transplanted animals was much less than that from the syngeneic transplants, and second, ,the number of GR-1 cells was greater in the allogeneic on day 7 from that of the 7 day syngeneic and control spleens.

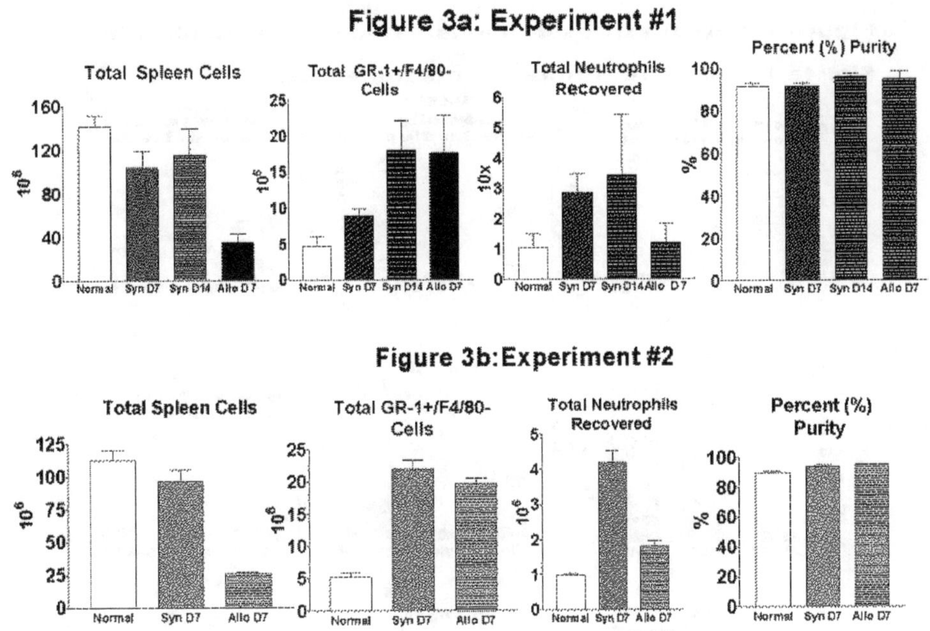

Figure 3: Step-wise isolation of purified neutrophils.
The isolation steps from the spleens of allogeneic (Allo) and syngeneic (Syn) transplants are presented beginning with the total number of cells isolated from the spleen in the upper panel to the percentage of recovery of neutrophils and the purity of each sample on the bottom frames. For comparison the step-wise isolation from the spleens of completely normal BALB/c mice is included but not compared in the bar graphs illustrated in each panel. Means +/- SD. Experiment #1 (allogeneic day 7 n=5), syngeneic day 7 n = 4, syngeneic day 14, n = 5). Experiment #2, both allogeneic and syngeneic day 7 (n = 6).

The purification of GR-1 neutrophils from the spleens of mice transplanted with allogeneic B6 cells was quite adequate for analysis on day 7. However, there was a major decrease in the number of allogeneic cells available for purification on day 14 (see Figure 3a) and it was not possible to obtain a sufficient number of highly purified neutrophils for analysis on that day. Therefore, a day 14 analysis was not attempted in the second experiment (Figure 3b). In this experiment we tested specifically for residual CD3 T cells. The final total number of CD3 + T cells in the allogeneic neutrophil isolate was 0.0044 +/- 0.0011 x 10^6 and in the syngeneic isolate was 0.0008 +/- 0.0007 x 10^6. Based on this data, we felt confident that the mRNA expressions and intracellular cytokines observed were not coming from activated T cells.

mRNA expression in purified neutrophils

In the first experiment (Figure 3a) highly purified (95.0 +/- 3.24 %) GR-1 positive cells isolated 7 days after allogeneic transplantation expressed high levels of IL-21. The very small number of CD3 + T cells in the both the allogeneic and syngeneic samples strongly supports our conclusion that they were not the source of the high IL-21 expression.

GM-CSF was significantly expressed in the allogeneic but not the syngeneic transplants on day 7 in the first experiment but the level was much lower in the second. Similarly IFNγ was significantly increased in both experiments as compared to that in the syngeneic neutrophils on day 7 but the expression level were just above that in the untreated control cells.

This pattern of allogeneic greater than syngeneic expression was reversed with respect to the expression of TNFα which modestly but significantly increased above that in the syngeneic transplants and in the untreated controls. Whereas the levels of IL-12p40 were significantly greater in the syngeneic than the allogeneic neutrophils, the level was less than that in the untreated controls. The expressions of IL-6, G-CSF and IL-22 on day 7 in the allogeneic were not statistically different from that in the syngeneic transplants on day 7.

In the second experiment (see Figure 4b) the purified neutrophils (94.8 +/- 1.53) again expressed high levels of IL-21 (Figure 4b). As noted above, GM-CSF and IFNγ expressions were significantly greater in the allogeneic as compared to the syngeneic transplants but the expressions were just at or below that in the untreated BALB/c controls. Taken together the experiments demonstrate that not only do the syngeneic neutrophils peak 7 days later than that of the allogeneic neutrophils but their expression of cytokines is strikingly different than that in the allogeneic transplants.

Intracellular IL-21

The next question was did these highly purified neutrophils produce IL-21 protein? Unfortunately, IL-21 has only occasionally been detected in the serum of patients and experimental animals [2], even when there was clear evidence of the critical role this cytokine was playing in the pathogenesis of the clinical or the experimental models. Therefore, we elected

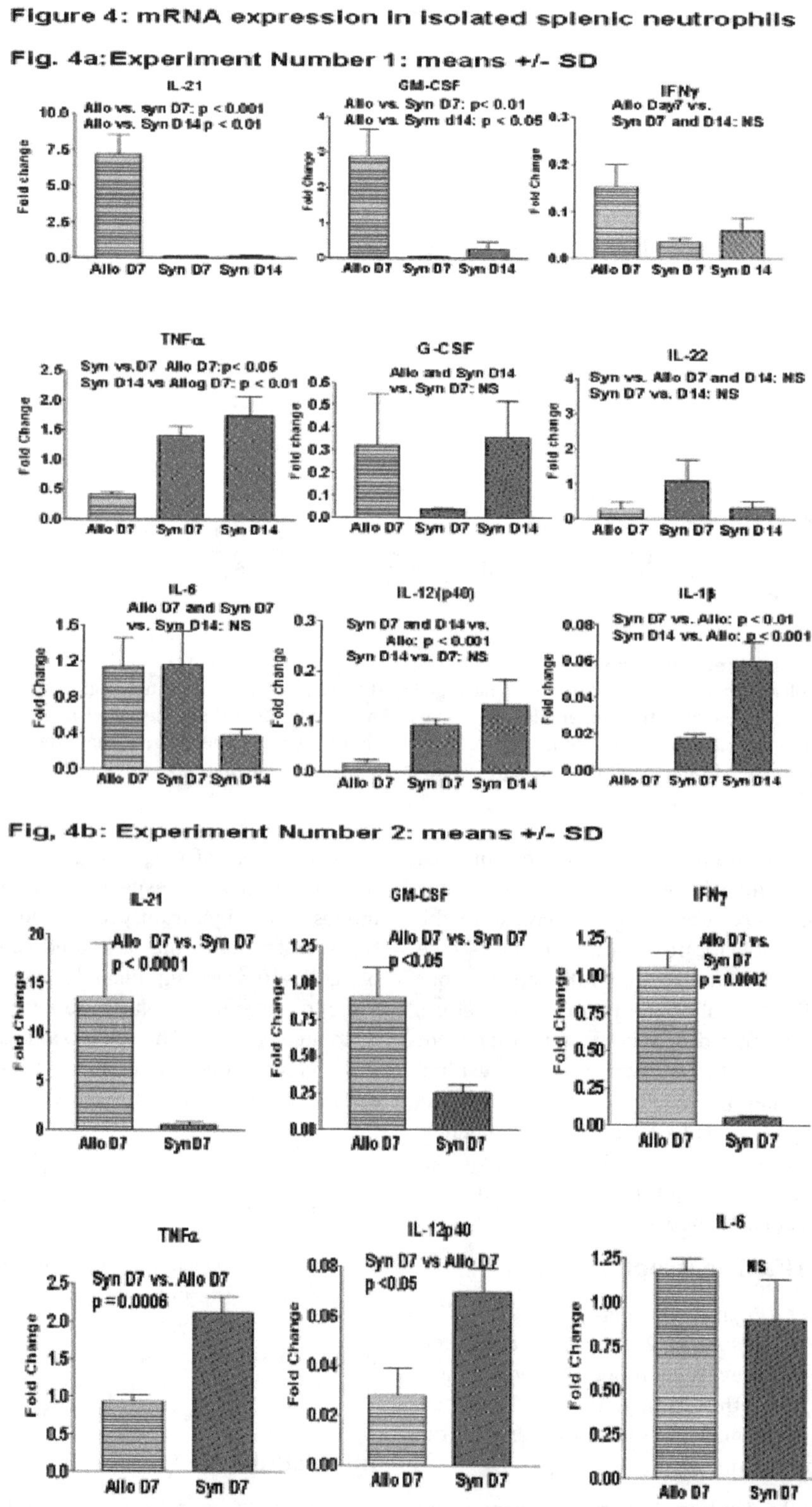

Figure 4: The cytokine expressions of IL-21, GM-CSF, IFNγ, TNFα, IL-12p40 (β), IL-22, G-CSF and IL-6 in neutrophils isolated from BALB/c mice transplanted with syngeneic (Syn) BALB/c cells (tangential hatched bars to the left, day 7; tangential hatched bars to the right day14) or Allogeneic (Allo) B6 cells (horizontal hatched bars day 7). Note the fold change scale differs with each cytokine: Means +/- SD. Experiment #1 (allogeneic day 7, n = 5); syngeneic day 7 n = 4; syngeneic day 14, n = 5).
Experiment # 2. Both allogeneic (Allo) and syngeneic (Syn) cells on day 7 (n=6).
Tangentially hatched, syngeneic day 7; horizontally hatched, allogeneic day 7. NS = not statistically significant.

to directly determine its intracellular presence in the highly purified splenic neutrophils obtained 7days after allogeneic transplantation. Figure 5 illustrates the presence of intracellular IL-21 in these isolated neutrophils before and 6 hours after additional stimulation. Although the difference is relatively small, it is statistically significant. Furthermore, the extremely small number of CD3 positive cells in this preparation (0.0044 +/- 0.0011 x 10^6) could not account for even this change.

Discussion

Neutrophils are the earliest responders in innate immunity [3]. Although they have been traditionally considered as being critical for the elimination of toxins, bacteria, other organisms and foreign bodies [4-6], they also have important inflammatory roles in thermal injury [7], ischemia reperfusion [8], a number of immune related disorders [9-10] and hyperacute solid organ transplant rejection [11]. Depending on the experimental or clinical situation, neutrophils have been reported to secrete IL-8 [12], IL-18 [13], IL-17 [2,6-8] and IFNγ [14]. Although, per cell, neutrophils produce much less cytokine protein than monocytes or lymphocytes [2,15], they are more numerous and may be a major cell identified in inflammatory reactions. For example, neutrophils are prominent in sclerosing cholangitis and biliary duct inflammation of the liver [16,17]. With respect to GVHD, extensive neutrophil infiltration has been observed in pathology of the colon [18,19] and in the lung [20]. Furthermore, Schwab, et al. [21] demonstrated that physical or genetic depletion of neutrophils improved survival in a mouse model of GVHD.

In our model of BALB/c hosts transplanted with fully allogeneic B6 cells, there is an initial marked loss in total cellularity associated with nearly complete loss of lymphoid tissue in the spleen. This is followed by a new increase in spleen weight and cellularity beginning on day 5 post transplant. It is during this period that neutrophils comprise the major splenic cell population. The striking increase in neutrophils recorded in the allogeneic transplants on day 7 corresponded to the maximum weight loss during acute GVHD. These neutrophils expressed IL-21 and produced IL-21 cytokine. Although the question of the biologic importance of the splenic neutrophil production of IL-21 cannot be resolved with these experiments, it is during this period that neutrophils comprise the major splenic cell population and in total could account for an important source of this cytokine.

IL-21 is an inflammatory cytokine produced by activated Th17 cells [22,23], follicular helper T cells [24,25] and by activated NK cells [26]. It is also expressed in Hodgkin's lymphoma [27]. Although it has been identified in neutrophils in association with B lymphocytes in the marginal zone of normal spleens [28], by day 7 post allogeneic transplant, lymphoid follicles in the spleen have been destroyed and B cells are essentially absent.

It may have a major role in the pathogenesis of psoriasis [29-31] and inflammatory bowel disease [32,33]. Distler, et al. [34] observed a 4.7 fold increased expression of IL-21 receptor in the epidermis of the skin of patients with scleroderma as compared to that in normal skin. IL-21 may also be a factor in the pathogenesis of GVHD. Experimentally, survival has been significantly improved with transplantation of IL-21-/- donor cells as compared to WT cells [35] and by blockade of IL-21 [36]. Zhao, et al. [37] concluded that IL-17 and IL-21 may be critically important for the pathogenesis of chronic GVHD.

Although GM-CSF was significantly increased in day 7 allogeneic neutrophils, the expression levels were much less than that of IL-21. To our knowledge GM-CSF has not been implicated in the GVHD reaction but it is important to note that its expression was elevated in the allogeneic but not in the syngeneic neutrophils. It is a cytokine secreted by macrophages, T cells, mast cells, NK cells, endothelial cells and fibroblasts. In general it functions as a white blood cell growth factor. In addition, it has been shown to have roles in cell signaling [38] and in promoting inflammatory dendritic cells [39] and macrophages [40]. Furthermore, it has been implicated in the pathology of lung injury [41] and arthritic pain [42].

Although this study does not establish that neutrophil secretion of IL-21 played a critical role in the pathogenesis of aGVHD, it does document that allogeneic but not syngeneic neutrophils in the spleen express and produce this cytokine. The evidence that strikingly increased IL-21 expression occurred

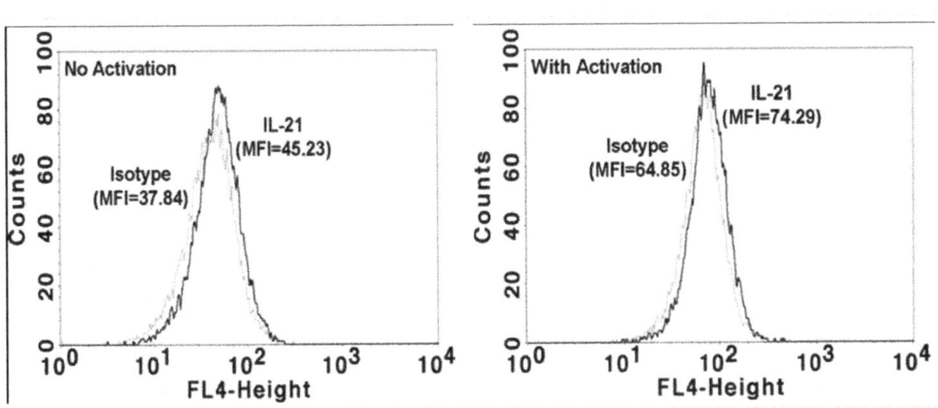

Figure 5: IL-21 intracellular cytokine analysis.
Unstimulated on the left and stimulated with on the right. MFI = Mean Fluorescent Intensity. The solid line is IL-21 and the more faint dotted line is the isotype control.

only in allogeneic neutrophils suggests that there may be other target organs where neutrophils [18-21] could be producing these cytokines and contributing to their organ pathology.

Conflict of Interest

The authors declare no conflict of interest.

Support

Supported by a VA Merit Review Grant

References

1. Thompson JS, Chu Y, Glass JF, Brown SA. Absence of IL-23/p19 in donor allogeneic cells reduces mortality from graft-versus-host disease. Bone Marrow Transplant. 2010;45(4):712-22. doi: 10.1038/bmt.2009.215.

2. Tecchio C, Micheletti A, Cassatella MA. Neutrophil-derived cytokines: facts beyond expression. Front Immunol. 2014;21(5):508. doi: 10.3389/fimmu.2014.00508.

3. Jaillon S, Galdiero MR, Del-Prete DD, Cassatella MA, Garlanda C, Mantovani A. Neutrophils in inate and adaptive immunity. Semin Immunopathol. 2013;35(4):377-94. doi: 10.1007/s00281-013-0374-8.

4. Shibata K, Yamada H, Hara H, Kishihara K, Yoshikai Y. Resident V delta1+ gamma delta T cells control early infiltration of neutophils after Escherichia coli infection via IL-17 production. J Immunol.2007;178(7):4466-4472.

5. Miyamoto M, Prause O, Sjöstrand M, Laan M, Lötvall J, Lindén A.. Endogenous IL-17 as a mediator of neutrophil recruitment caused by endotoxin exposure in mouse airways J Immunol. 2003;170(9):4665-72.

6. Werner JL, Gessner MA, Lilly LM, Nelson MP, Metz AE, Horn D, et al. Neutrophils produce IL17 (IL-17A) in a dectin-1-and IL-23-dependent manner during invasive fungal infection. Infect Immun. 2011;79(10):3966-77. doi: 10.1128/IAI.05493-11.

7. Chen LW, Huang HL, Lee IT, Hsu CM, Lu PJ. Thermal injury-induced priming effect of neutrophil is TNF-α and p38 dependent. Shock. 2006;26:69-76.

8. Li L, Huang L, Vergis AL, Ye H, Bajwa A, Narayan V, et al. IL-17 produced by neutrophils regulates IFN-y-mediated neutrophil migration in mouse kidney ischemia-reperfusion injury. J Clin Invest. 2010;120(1):331-42. doi: 10.1172/JCI38702.

9. Nakagome K, Matsushita S, Nagata M. Neutrophilic inflammation in severe asthma. Int Arch Allergy Immunol. 2012;158 Suppl 1: 96-102. doi: 10.1159/000337801.

10. Lukacs NW, Strieter RM, Chensue SW, Widmer M, Kunkel SL. TNF-α mediates recruitment of neutrophils and eosinophils during airway inflammation. J Immunol. 1995;154(10):5411-7.

11. Miura M, El-Sawy T, Fairchild RI. Neutrophils mediate parenchymal tissue necrosis and accelerate the rejection of complete major histocompatibility complex-disparate cardiac allografts in the absence of interferon-gamma. Am J Pathol 2003;162:509-519.

12. Montemurro P, Mariggiò MA, Barbuti G, Cassano A, Vincenti A, Serio G, et al. Increase in Interleukin-8 production from circulating neutrophils upon antibiotic therapy in cystic fibrosis. J Cyst Fibrosis. 2012;11:518-24.

13. Sporriz R, Joller N, Hibi H, Oxenus A. A novel role for neutrophils as critical activators of NK cells. J Immunol. 2008;181(10):7121-30.

14. Ellis TN, Bearman BL. Muirine polymorhonuclear neutrophils produce interferon-g in response to pulmonary infections with Nocardia asteroids. J Leukoc Biol. 2002;72(2):373-81.

15. Scapini P, Calzetti F, Cassatella MA. On the detection of neutrophil-derived vascular endothelial growth factor (VEGF). J Immunol Methods. 1999;232(1-2):121-9.

16. C Barbatis, P Grases, H A Shepherd, R W Chapman, J Trowell, D P Jewell, et al. Histological features of sclerosing cholangitis in patients with chronic ulcerative colitis. J Clin Path. 1985;38:778-784.

17. Licata LA, Nguyen CT, Burga RA, Falanga V, Espat NJ, Ayala A, et al. Biliary obstruction results in PD-1-dependent liver T cell dysfunction and acute inflammation mediated by Th17 cells and neutrophils. J Leukoc Biol. 2013;94(4):813-23. doi: 10.1189/jlb.0313137.

18. Eigenbrodt ML, Eigenbrodt EH, Thiele D. Histologic similarity of murine colonic graft-versus-host disease (GVHD) to human colonic GVHD and inflammatory bowel disease. Am J Pathol. 1990;137(5):1065-76.

19. Socié G, Mary JY, Lemann M, Daneshpouy M, Guardiola P, Meignin V, et al. Prognostic value of apoptotic cells and infiltrating neutrophils in graft-versus-host disease of the gastrointestinal tract in humans. Blood. 2004;103(1):50-57.

20. Gowdy KM, Nugent JL, Martinu T, Potts E, Snyder LD, Foster WM, et al. Protective role of T-bet and Th1 cytokines in pulmonary graft-versus-host disease and peribronchiolar fibrosis. Am J Respir Cell Mol Biol. 2012;46(2):249-56. doi: 10.1165/rcmb.2011-0131OC.

21. Schwab L, Goroncy L, Palaniyandi S, Gautam S, Triantafyllopoulou A, Mocsai A, et al. Neutrophil granulocytes recruited upon translocation of intestinal bacteria enhance graft-versus-host disease via tissue damage. Nat Med. 2014;20(6):648-54. doi: 10.1038/nm.3517.

22. Liu SM, King C. IL-21-producing Th cells in immunity and autoimmunity. J Immunol. 2013;191(7):3501-6. doi: 10.4049/jimmunol.1301454.

23. Wei L, Laurence A, Elias KM, O'Shea JJ, et al. IL-21 is produced by Th17 cells and drives IL-17 production in a STAT3-dependent manner. J Biol Chem. 2007;282(48):34605-10.

24. Vogelzang A, McGuire HM, Yu D, Sprent J, Mackay CR, King C. A fundamental role for interleukin-21 in the generation of T follicular helper cells. Immunity. 2008;29(1):127-37. doi: 10.1016/j.immuni.2008.06.001.

25. Lüthje K, Kallies A, Shimohakamada Y, Belz GT, Light A, Tarlinton DM, et al. The development and fate of follicular helper T cells defined by an IL-21 reporter mouse. Nat Immunol. 2012;13(5):491-8. doi: 10.1038/ni.2261.

26. Coquet JM, Kyparissoudis K, Pellicci DG, Besra G, Berzins SP, Smyth MJ, et al. IL-21 is produced by NKT cells and modulates NKT cell activations and cytokine production. J Immunol 2007;178:2827-2834.

27. Scheeren FA, Diehl SA, Smit LA, Beaumont T, Naspetti M, Bende RJ, et al. IL-21 is expressed in Hodgkin lymphoma and activates STAT5: evidence that activated STAT5 is required for Hodgkin lymphomagenesis. Blood. 2008;111:4706-4715.

28. Puga I, Cols M, Barra CM, He B, Cassis L, Gentile M, et al. B cell-helper neutrophils stimulate the diversification and production of immunoglobulin in the marginal zone of the spleen. Nat Immun. 2012;13:170-180.

29. Caruso R, Costanzo A, Monteleone G. Pathogenic role of interleukin-21

in psoriasis. Cell Cycle. 2009;8:3629-3630.

30. He Z, Jin L, Liu ZF, Hu L, Dang EL, Feng ZZ, et al. Elevated serum levels of interleukin 21 are associated with disease severity in patients with psoriasis. Br J Dermatol 2012;167:191-193. doi: 10.1111/j.1365-2133.2012.10899.x.

31. Botti E, Spallone G, Caruso R, Monteleone G, Chimenti S, Costanzo A. Psoriasis, from pathogenesis to therapeutic strategies: IL-21 as a novel potential therapeutic target. Curr Pharm Biotechnol. 2012;13:1861-1867.

32. Pallone F, Fina D, Caruso R, Monteleone G. Role of IL-21 in inflammatory bowel disease. Expert Rev Clin Immunol. 2010;6(4):537-41. doi: 10.1586/eci.10.44.

33. MacDonald TT, Bell I, Monteleone G. The opposing roles of IL-21 and TGFβ1 in chronic inflammatory bowel disease. Biochem Soc Trans. 2011;39:1061-1066. doi: 10.1042/BST0391061.

34. Distler JH, Jüngel A, Kowal-Bielecka O, Michel BA, Gay RE, Sprott H, et al. Expression of Interleukin-21 receptor in epidermis from patients with systemic sclerosis. Arth Rheum. 2005;52:856-864.

35. Meguro A, Ozaki K, Oh I, Hatanaka K, Matsu H, Tatara R, et al. IL-21 is critical for GVHD in a mouse model. Bone Marrow Transpl. 2010;45:723-729.

36. Bucher C, Koch L, Vogtenhuber C, Goren E, Munger M, Panoskaltsis-Mortari A, et al. IL-21 blockade reduces graft-versus-host disease mortality by supporting inducible T regulatory cells generation. Blood. 2009;114:5375-5384.

37. Zhao XY, Lv M, Xu LL, Qian X, Huang XJ, et al. Donor IL-17 cells and IL-21 may contribute to the development of chronic graft-versus-host disease after allogeneic transplantation. Eur J Immunol. 2013;43(3):838-50. doi: 10.1002/eji.201242816.

38. Parajuli B, Sonobe Y, Kawanokuchi J, Doi Y, Noda M, Takeuchi H, et al. GM-CSF increases LPS-induced production of proinflammatory mediators via upregulation of TLR4 and CD-14 in murine microglia. J Neuroinflammation. 2012;9:268. doi: 10.1186/1742-2094-9-268.

39. Blyszczuk P, Behnke S, Lüscher TF, Eriksson U, Kania G. GM-CSF promotes inflammatory dendritic cell formation but does not contribute to disease progression in experimental autoimmune myocarditis. Biochim Biophys Acta. 2013;1833(4):934-44. doi: 10.1016/j.bbamcr.2012.10.008.

40. Sorgi CA, Rose S, Court N, Carlos D, Paula-Silva FW, Assis PA, et al. GM-CSF priming drives bone marrow-derived macrophages to a pro-inflammatory pattern and down modulates PGE2 in response to TLR2 ligands. PLoS One. 2012;7(7):e40523. doi: 10.1371/journal.pone.0040523.

41. Choi JC, Jung JW, Kwak HW, Song JH, Jeon EJ, Shin JW, et al. Granulocyte macrophage-colony stimulating factor (GM-CSF) augments acute lung injury via its neutrophil's priming effects. J Korean Med Sci. 2008;23(2):288-95. doi: 10.3346/jkms.2008.23.2.288.

42. Cook AD, Pobjoy J, Sarros S, Steidl S, Dürr M, Lacey DC, et al. Granulocyte-macrophage colony-stimulating factor is a key mediator in inflammatory and arthritic pain. Ann Rheum Dis. 2013;72(2):265-70. doi: 10.1136/annrheumdis-2012-201703.

12

Histopathology of Intestinal Transplant Rejections

Tatsuaki Tsuruyama[1,2]*

[1]*Department of Pathology, Graduate School of Medicine, Kyoto University, Yoshida-konoe-cho, 606-8501 Kyoto, Japan*

[2]*Center for Anatomical, Pathological, and Forensic Medical Research, Graduate School of Medicine, Kyoto University, Yoshida-konoe-cho, Kyoto 606-8501, Japan*

***Corresponding author:** Tatsuaki Tsuruyama, Department of Pathology, Graduate School of Medicine, Kyoto University, Yoshida-konoe-cho, 606-8501 Kyoto, Japan*

Absract

Small bowel transplantation is one of the standard therapies for short-bowel syndrome. Nevertheless, histological rejection is still a main cause for failure of intestinal transplants. The present review aims to elucidate the histologic findings for diagnosis of acute cellular rejection (ACR). We review immunohistologic findings along with assessment of patients' clinical courses. In addition to crypt apoptosis, which is considered as a sensitive histologic finding in ACR, T-lymphocyte apoptosis and phagocytosis of apoptotic bodies in the lamina propria of villi were common findings of ACR. Recent research in variable T cell populations may contribute to the immunological understanding of ACR. Therefore, in the future, earlier diagnosis of ACR may be achievable.

Introduction

Small Bowel Transplantation (SBT) is one of the standard therapies for patients with complications of irreversible requirement of total parenteral nutrition associated with Short-Bowel Syndrome, which is a serious condition with considerable morbidity, because of the deficiencies and metabolic imbalances created when deprived from a regular diet [1-3]. Moreover, SBT is a suitable treatment option for patients with Hirschsprung's disease and related diseases such as chronic idiopathic intestinal pseudo-obstruction syndrome [1], megacystis microcolon intestinal hypoperistalsis syndrome [4], desmoid tumors associated with familial adenomatous polyposis [5], and Crohn's disease [6]. Owing to highly effective immunosuppressive medication and improvements in post-operative care, outcomes of SBT have considerably improved [7,8]. For prolonged post-transplantation control, novel immunologic suppressants, such as mycophenolate mofetil, tacrolimus, and steroids have been used.

Despite improvements in the outcomes, acute cellular rejection (ACR) remains the major cause of intestinal graft failures following SBT [9-12]. For most patients that experience severe ACR, adequate recovery of mucosal function is difficult; bacterial and viral opportunistic infections are inevitable because of defensive mucosal barriers. Therefore, diagnosing ACR in the early phase is essential during postoperative care [13].

Microorganisms and ACR

The intestine is host to bacterial and microorganism flora, and its mucosa includes its native lymphoid tissue. These complicated factors contribute to the high frequency and severity of ACR. Recent studies have indicated novel techniques for identification of the flora, using profiling as a diagnostic marker of rejection [14]. Cytomegalovirus (CMV) and Epstein - Barr virus (EBV) [15] infection are also causes of implications following an increased dose of steroid pulse, tacrolimus, and other immunosuppressive reagents. In particular, CMV enteritis is persistent and erosion in an immunocompromised status leads to the destruction of the mucosal architecture after severe inflammation. Immunohistochemical assessment is one of the available methods for identification of CMV. Recently, a DNA extraction technique has been developed and PCR tests for the identification of CMV have become available using formalin-fixed, paraffin-embedded (FFPE) tissue [16]. The mucosal damage in CMV intestinal enteritis is frequently severe and it is difficult to make a differential diagnosis from ACR. In an immunocompromised state, patients are at increased risk of post-transplantation lymphoproliferative disorders (PTLDs) due to EBV infection. Monomorphic PTLDs have potential to develop into B cell lymphomas, and reduction of immunosuppression results in normalization and loss of EBV-associated expression. The expression of EBV can be analyzed by real-time PCR, and immunohistochemistry of LMP-1 or in situ hybridization for EBER (EBV encoded small RNA) [17]. Ramos, et al. [18] reported that the frequency of graft removal due to EBV infection is higher than 40%, and the subsequent patient survival rate is less than 70%.

Candidate biomarkers of ACR

The candidate biomarkers of ACR have been investigated in peripheral blood of intestinal transplant patients, in which ribosomal proteins such as RPL13A [3], markers IL1R2, ICAM1, GZMB, CCL3 [19,20], and citrulline levels [21] have been reported.

The production of IL-5 increases significantly relative to other cytokines in the allograft tissue during ACR [22]. In parallel with this, eosinophil infiltrates have frequently been observed [22], as well as mixed cellular inflammation [11]. C - reactive protein (CRP) is another indicator of ACR (Figure 1A); this protein is known to rise in inflammation following IL-6 secretion by macrophages. A CRP test has been shown to measure 1.0-3.0 mg/10^{-1}L in patients without administration of an immunosuppressive reagent;

***Corresponding author email:** tsuruyam@kuhp.kyoto-u.ac.jp* however, this value elevates to over 3.0 mg/10^{-1}L at the onset of ACR. Following immunosuppressive therapy, this value promptly decreased to 1.0 mg/10^{-1}L [22,23]. Nonetheless, low copy numbers of CMV infection in the transplants did not significantly elevate the CRP value. Therefore, CRP value is a differential diagnostic maker of ACR from CMV enteritis.

Histologic assessment of apoptotic bodies in crypts and lamina propria of villi

Various laboratory tests have been developed for assessment of ACR; however, the significance of histologic tests remains important. Histologic criteria are shown in Table. According to these criteria, lymphocytic infiltrates (Figure 2A) and crypt apoptosis are commonly observed in ACR (Figure 2B). In practice, the diagnosis of intestinal ACR in the early phase is frequently difficult, particularly due to the complicated interactions between lymphocytes and other immunological cells resulting from the transient coexistence of donor-derived and recipient-derived cells in the graft.

Among histologic criteria, crypt apoptosis in the mucosa is one of the most reliable observations; severe ulceration follows this and sufficient recovery of the mucosa becomes difficult when graft damage reaches the submucosal area [13]. Repeat occurrences of this severe damage leads to chronic graft rejection in which fibrosis proceeds leading to irreversible inefficiency of absorption in the intestine. Apoptosis in the crypt is detectable using the TUNEL method or by the caspase-cleaved keratin fragment marker.

In addition, Tsuruyama, et al. [10] reported that apoptotic bodies cluster in the Lamina Propria (LP) of villi at the onset of ACR (Figure 2C). Apoptotic crypts in grafts undergoing ACR are shown as intensely stained. In addition, macrophages phagocytosing apoptotic T cells are frequently observed with clustering in the LP (Figure 2D). These apoptotic cell clusters are significantly decreased following steroid pulse administration [10]. Therefore, this apoptotic response in LP is one of the immunological reactions associated with ACR. Scoring of the degree of apoptosis in the LP is available for evaluation (score 0, no signals; score 1, scant and isolated signals; score 2, a few signal aggregations; score 3, signal aggregates surrounding

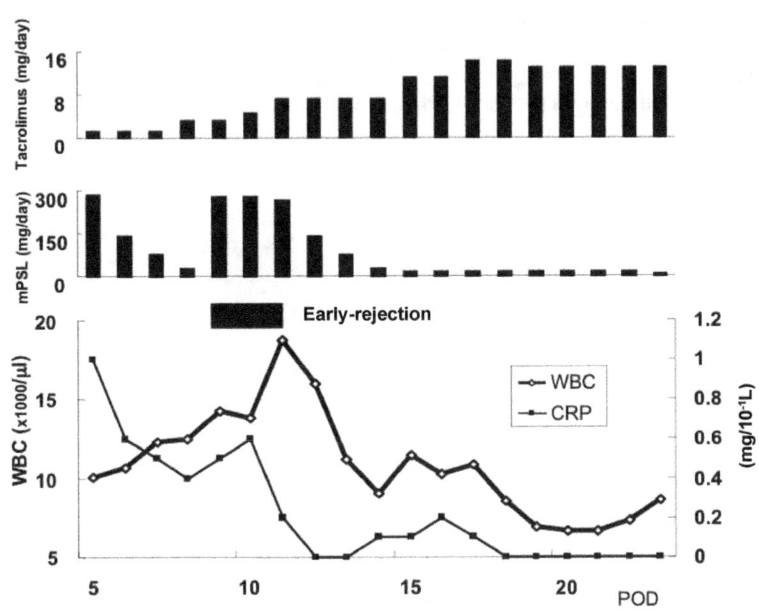

Figure 1: Clinical course of post-transplantation status. Administrative doses of tacrolimus and methylpredinisolone (mPSL). Bottom plots represent white blood cell (WBC) number and concentration of C-reactive protein (CRP) in the peripheral blood.

Table 1: Histologic schema of acute cellular rejection (ACR) of intestinal allograft.

Histologic grade		
	Crypt apoptosis and related findings	Lymphocytic apoptosis in LP
Indeterminate	up to 6 apoptotic bodies per 10 crypts	None
Mild	>6 apoptotic bodies per 10 crypts	Isolated apoptotic bodies
Moderate	Confluent apoptosis Increased Inflammation, Epithelial injury	A few apoptotic body cluster
Severe/ Exfoliative	Mucosal ulceration	Apoptotic bodies aggregate

LP: Lamina Propria

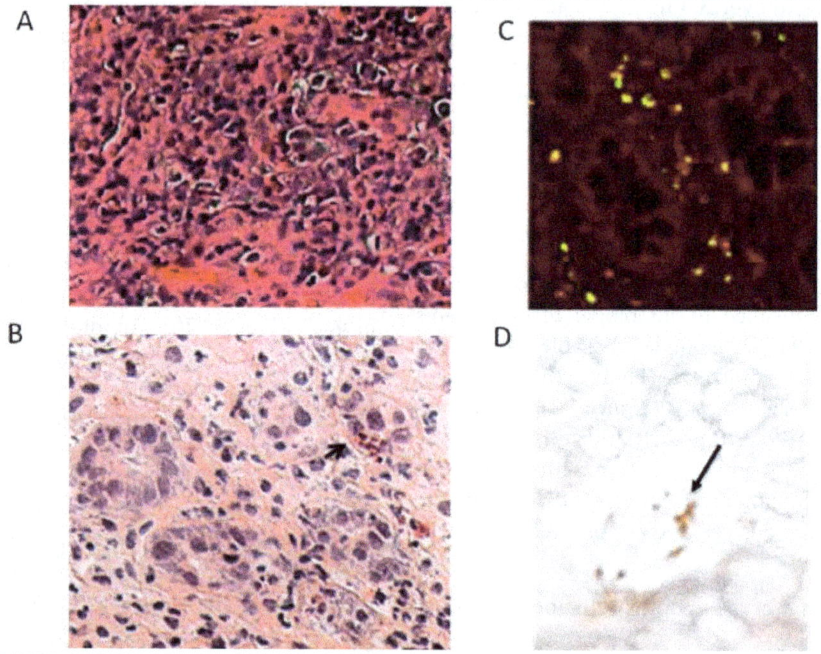

Figure 2: Histology of ACR in the intestine.
A) Lymphocyte infiltrates including eosinophils and neutrophils (× 200).
B) Crypt apoptosis (×200). An Arrow indicates the eosinophilic apoptotic bodies in the crypt.
C) Double staining of NKT cells with TCRValpha24 (red, PE) and TUNEL (green, FITC) (× 200).
D) Apoptotic clusters stained with TUNEL. DAB was used for visualization. An arrow represents the cluster.

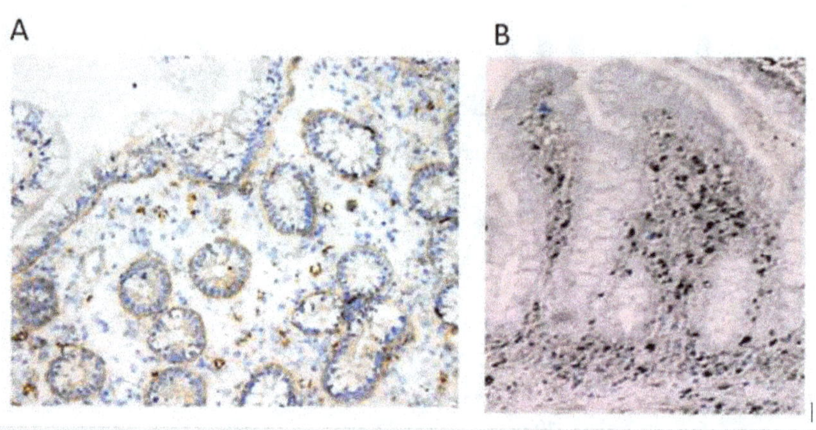

Figure 3: Immunohistochemistry of FasL. A) Intact allograft (× 200), and B) allograft with ACR (× 200). DAB was used for visualization. FasL-positive cells were observed in the lamina propria.

the crypt) [11]. The T cell apoptosis score closely correlates with the crypt apoptosis count. In addition, the receptor Fas is also available for Immunostaining for the identification of apoptosis [10]. It is probable that the FasL-Fas interaction contributes to the apoptosis reaction in LP. Fas Ligand (FasL) staining is useful for the identification of apoptotic bodies in allograft (Figure 3A, B).

Asaoka, et al. [20] reported the activation of cytotoxic T cells (CTLs) in granzyme B/ perforin-mediated graft injury. Unlike apoptosis in graft versus host disease that may be associated with elevation in TNF-alpha production [24], Fas ligand (FasL) is frequently stained in cases of ACR [10]. Therefore, the ACR of intestinal graft includes various activations of apoptosis-related molecules, by expression and release. After T cells undergo apoptosis [10], apoptotic bodies are then phagocytosed by macrophages. Similar findings in a liver allograft have been reported [25]. Thus, these phagocytic findings may be common to multiple allografts in ACR.

Time course of apoptosis in ACR and recovery of the mucosa

To our experience, crypt apoptosis and mixed cellular infiltrates appear at the late stage of ACR of small intestine and are irreversible changes. Once crypt apoptosis develops, even after steroid pulse, apoptosis increases in number and ulceration is inevitable. In contrast, T cell apoptosis in LP appears in the early phase of ACR and immunosuppressive therapy is sufficiently effective, and in fact, apoptotic T cells disappear within a day following the therapy [10,22]. For this reason, T cell apoptosis is more sensitive to treatment procedures. Thus, CRP test and histology likely provides effective prospects for follow-up. After treatment of rejection ACR erosive mucosa is edematous; however, Paneth cells at the crypt regenerate in association with mucosal recovery. Cell cycle markers, such as Ki67 and PCNA, show recovery of stainability in foveolar and crypt epithelial cells. Therefore, this staining method provides a marker of better prognosis after ACR. Nonetheless, for diagnosis of cases in which humoral rejection is suspected, complement staining such as C4d using frozen section may provide more reliable information [12].

Histologic finding in Peyer's patch (PP) and Isolated Lymphoid Follicle (ILF)

After engraftment, host-derived T cells traffic into the intestinal allograft across the high endothelial veins (HEVs) located in the inter-PP follicular region. By endoscopy examination, elevation of mucosa is frequently observed in the intestinal transplant at the onset of ACR. The elevated site includes PPs that consist of B cells, T cells, and dendritic Cells.

The entry of cytotoxic T cells CTLs into PPs via HEVs is observed within 7–10 days after transplantation and ACR occurs in cases. When ACR persists, PPs disintegrate in severe ACR and mucosal recovery is not sufficient when fibrosis develops. This is one of the poor prognoses of intestinal transplants, because once the mucosal defensive mechanism is lost in the erosive site, the graft becomes susceptible to bacterial and viral infection. Therefore, the immunosuppressive therapy before PP disintegration is essential for the control of post-transplantation success rate [23].

Follicular B cells in PPs are stimulated by antigens in the lumen and differentiate from Immunoglobulin M (IgM+) to IgA+ B cells by class switching, which is mediated by Activation-Induced Cytidine Deaminase (AID). IgA+ B cells in PPs circulate throughout the body via the thoracic duct and differentiate into IgA+ B cells by the effect of IL-6 produced by intestinal epithelial cells. The primary antibodies secreted into the intestinal tract mucus are of the IgA class, and are transported to the gut luminal side by binding to multimeric antibody receptors that are retained on intestinal epithelial cells. Notably, host lymphocytes rapidly repopulate allograft PPs/ILFs within two years in the absence of ACR [26]. Allograft ILFs revealed a higher maturation state than control samples, and IgA+ plasma cells were increased in a number in allograft mucosa [26]. AID gene expression in allograft PPs/ILFs that the immunological burden may promote the maturation of B cells [26]. Histological examination showed hyperplastic changes of PPs with an increase in expression of CD20, a mature B cell marker at the onset of ACR [23].

T cells infiltrate allografts at the onset of ACR

It appears to be controversial which types of lymphocytes induce ACR of intestinal allografts. In general, CTLs are considered to induce ACR of intestinal allografts. However, several experimental studies have not supported this. In fact, apoptosis and ACR of intestinal allograft was observed in the absence of CTLs in a rat model [27]. In fact, our immunohistochemistry did not consistently reveal CTL infiltrates in allografts except in the first episode of rejection [10,22,23,28]. CTL has the potential to express FasL. In autoimmune diseases such as inflammatory bowel disease (IBD) such as ulcerative colitis (UC), increases in FasL+ T cells are observed, but do not correlate with increases in Fas+ T cells, indicating that the increased expression of FasL in IBD colonic LP is not paralleled by Fas expression by T cells, and that Fas/FasL-mediated apoptosis is not the main factor. In contrast, expression of perforin, which is another apoptosis-inducing molecule, is correlated with tissue damage and may represent the enhancement of a distinct cytotoxic pathway in UC [29]. Unlike UC, Fas and FasL expression correlates well with ACR in intestinal grafts. Therefore, the immunological status is probably not similar to autoimmune colitis. Then, which types of T cells expressing FasL, or other cell populations, are effectors in ACR of intestinal allografts? The helper T cell (Th)1/ Th2 paradigm in mucosal immunology has been shift-

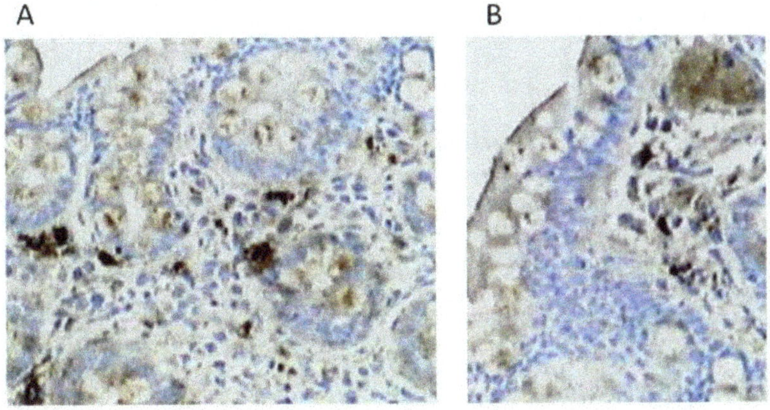

Figure 4: Immunohistochemistry of the restricted T cell receptor (TCR) repertoire. TCR beta11 subunit was stained for detection of NKT cells in the lamina propria of villi (A, × 200; and B, × 200). NKT cells were observed in the lamina propria.

ed [30], and recent discovery of novel subsets of natural killer T (NKT) cells, regulatory T (Treg) cells [31,32] and effecter T helper cells that produce interleukin (IL)-17 (Th17) have been reported [31]. Of note, Th17 cells are potent inducers of inflammation and autoimmune diseases.

At the onset of ACR, natural killer (NK) cells and natural killer T (NKT) cells transiently increase in number, as well as iNKT cells (Figure 4A, B), and both rapidly decrease following steroid pulse therapy [23]. The iNKT cells have the potential to produce IL-4, which contributes to the development of Th2 cells and antagonizes Th1 and CTL responses. Higher levels of IL-4 prior to and shortly after kidney transplantation have been reported, and IL-4 may have a protective effect on renal graft survival [33]. Indeed, NKT cells have been implicated in allograft tolerance in experimental mouse models [34], in induction of chimerism in allogeneic cardiac transplant models [35], and in acceptance of rat-islet xenografts in mice [36]. Interferon-γ production remains low relative to normal donor intestine and does not change during the course of ACR development [37]. Therefore, infiltration of NKT cells may be involved in the protection of allografts in the response to ACR. Therefore, release of Th2-related cytokines by NKT cells may antagonize the proceeding of ACR. It is likely that humoral factors, such as candidates IL-10 and transforming growth factor (TGF)-β [23], recruit NKT cells to the graft mucosa in order to suppress allograft ACR. Using TUNEL and restricted-T cell receptor (TCR) alpha 24 staining, NKT cells are found to increase and undergo apoptosis (Figure 2C). Therefore, the apoptotic reaction involves the NKT cell population. Decrease of NKT cells may lead to the deterioration of ACR. However, because the released IL-4 and IL-5 may damage the allograft via eosinophilic enteritis, NKT cells are a double-edged effector [38].

FoxP3+ Treg are another immunological modulator of intestinal allograft, and the graft-protective mechanism has been extensively investigated [39]. Treg are recruited to the liver allograft at the onset of ACR and are maintained in tolerated liver allograft [40]. Introduction of bone marrow mesenchymal cells into the intestinal allograft increased Treg in parallel with IL-10 and TGF-β [41]. In clinical cases, the roles of Treg are not understood with respect to tolerance or induction of ACR of intestinal allografts. In experimental studies linked with human biopsy samples, IL-17 plays a critical role in ACR of intestinal transplantation [37,42], and may be a target for inhibiting ACR.

Conclusion

ACR remains to be the main cause of intestinal graft failure. Research in molecular immunological responses by T cell populations in allograft has developed, and a greater understanding of the pathogenesis of ACR is expected in the future. With respect to these achievements, histologic diagnosis of rejection at an earlier phase should be possible.

References

1. Bond GJ, Reyes JD. Intestinal transplantation for total/near-total aganglionosis and intestinal pseudo-obstruction. Semin Pediatr Surg. Semin Pediatr Surg. 2004;13(4):286-92.
2. Reyes J. Intestinal transplantation for children with short bowel syndrome. Semin Pediatr Surg. 2001;10(2):99-104.
3. Pirenne J, Hoffman I, Miserez M, Coosemans W, Aerts R, Monbaliu D, et al. Selection criteria and outcome of patients referred to intestinal transplantation: an European center experience. Transplant Proc. 2006;38(6):1671-2.
4. Masetti M, Rodriguez MM, Thompson JF, Pinna AD, Kato T, Romaguera RL, et al. Multivisceral transplantation for megacystis microcolon intestinal hypoperistalsis syndrome. Transplantation. 1999;68(2):228-32.
5. Chatzipetrou MA, Tzakis AG, Pinna AD, Kato T, Misiakos EP, Tsaroucha AK, et al. Intestinal transplantation for the treatment of desmoid tumors associated with familial adenomatous polyposis. Surgery. 2001;129(3):277-81.
6. Gerlach UA, Vrakas G, Reddy S, Baumgart DC, Neuhaus P, Friend PJ, et al. Chronic intestinal failure after Crohn disease: when to perform transplantation. JAMA Surg. 2014;149(10):1060-6.
7. Misiakos EP, Weppler D, Bakonyi A, Nery JR, Pinna AD, Kato T, et al. Clinical outcome of intestinal transplantation at the University of Miami. Transplant Proc. 1999;31(1-2):569-71.
8. Kato T, Nishida S, Mittal N, Levi D, Nery J, Madariaga J, et al. Intestinal transplantation at the University of Miami. Transplant Proc. 2002;34(3):868.
9. Ruiz P, Bagni A, Brown R, Cortina G, Harpaz N, Magid MS, et al. Histological criteria for the identification of acute cellular rejection in human small bowel allografts: results of the pathology workshop at the VIII International Small Bowel Transplant Symposium. Transplant Proc. 2004;36(2):335-7.
10. Tsuruyama T, Okamoto S, Fujimoto Y, Yoshizawa A, Yoshitoshi E, Egawa H, et al. Histology of intestinal allografts: lymphocyte apoptosis and phagocytosis of lymphocytic apoptotic bodies are diagnostic findings of acute rejection in addition to crypt apoptosis. Am J Surg Pathol. 2013;37(2):178-84.
11. Wu T, Abu-Elmagd K, Bond G, Nalesnik MA, Randhawa P, Demetris AJ, et al. A schema for histologic grading of small intestine allograft acute rejection. Transplantation. 2003;75(8):1241-8.
12. Yoshitoshi EY, Yoshizawa A, Ogawa E, Kaneshiro M, Takada N, Okamoto S, et al. The challenge of acute rejection in intestinal transplantation. Pediatr Surg Int. 2012;28(8):855-9. doi: 10.1007/s00383-012-3110-x.
13. Selvaggi G, Gaynor JJ, Moon J, Kato T, Thompson J, Nishida S, et al. Analysis of acute cellular rejection episodes in recipients of primary intestinal transplantation: a single center, 11-year experience. Am J Transplant. 2007;7(5):1249-57.
14. Oh PL, Martinez I, Sun Y, Walter J, Peterson DA, Mercer DF, et al. Characterization of the ileal microbiota in rejecting and nonrejecting recipients of small bowel transplants. Am

J Transplant. 2012 Mar;12(3):753-62. doi: 10.1111/j.1600-6143.2011.03860.x.

15. Lauro A, Pinna AD, Pellegrini S, Bagni A, Zanfi C, Dazzi A, et al. Long-term endoscopic follow-up in small bowel transplant recipients: single-center series. Transplant Proc. 2014;46(7):2325-8.

16. Mills AM, Guo FP, Copland AP, Pai RK, Pinsky BA, et al. A comparison of CMV detection in gastrointestinal mucosal biopsies using immunohistochemistry and PCR performed on formalin-fixed, paraffin-embedded tissue. Am J Surg Pathol. 2013;37(7):995-1000.

17. Sarkar S, Selvaggi G, Mittal N, Cenk Acar B, Weppler D, Kato T, et al. Gastrointestinal tract ulcers in pediatric intestinal transplantation patients: etiology and management. Pediatr Transplant. 2006;10(2):162-7.

18. Ramos E, Hernandez F, Andres A, Martinez-Ojinaga E, Molina M, Sarria J, et al. Post-transplant lymphoproliferative disorders and other malignancies after pediatric intestinal transplantation: incidence, clinical features and outcome. Pediatr Transplant. 2013;17(5):472-8. doi: 10.1111/petr.12103.

19. Andreev VP, Tryphonopoulos P, Blomberg BB, Tsinoremas N, Weppler D, Neuman DR, et al. Peripheral blood gene expression analysis in intestinal transplantation: a feasibility study for detecting novel candidate biomarkers of graft rejection. Transplantation. 2011;92(12):1385-91. doi: 10.1097/TP.0b013e3182370db1.

20. Asaoka T, Island ER, Tryphonopoulos P, Selvaggi G, Moon J, Tekin A, et al. Characteristic immune, apoptosis and inflammatory gene profiles associated with intestinal acute cellular rejection in formalin-fixed paraffin-embedded mucosal biopsies. Transpl Int. 2011;24(7):697-707. doi: 10.1111/j.1432-2277.2011.01259.x.

21. David AI, Selvaggi G, Ruiz P, Gaynor JJ, Tryphonopoulos P, Kleiner GI, et al. Blood citrulline level is an exclusionary marker for significant acute rejection after intestinal transplantation. Transplantation. 2007;84(9):1077-81.

22. Tsuruyama T, Fujimoto Y, Yonekawa Y, Miyao M, Onodera H, Uemoto S, et al. Invariant natural killer T cells infiltrate intestinal allografts undergoing acute cellular rejection. Transpl Int. 2012;25(5):537-44.

23. Tsuruyama T, Aini W. The roles of invariant NKT cells in bowel immunity: suppression of tumor progression and rejection of intestinal transplant. "Immune Response Activation". edited by Guy Huynh Thien Duc, chapter 8, 2014: p. 227-45.

24. Stuber E, Buschenfeld A, von Freier A, Arendt T, Folsch UR. Intestinal crypt cell apoptosis in murine acute graft versus host disease is mediated by tumour necrosis factor alpha and not by the FasL-Fas interaction: effect of pentoxifylline on the development of mucosal atrophy. Gut. 1999;45(2):229-35.

25. Miyagawa-Hayashino A, Tsuruyama T, Egawa H, Haga H, Sakashita H, Okuno T, et al. FasL expression in hepatic antigen-presenting cells and phagocytosis of apoptotic T cells by FasL+ Kupffer cells are indicators of rejection activity in human liver allografts. Am J Pathol. 2007;171(5):1499-508.

26. Meier D, Docena GH, Ramisch D, Toscanini U, Berardi G, Gondolesi GE, et al. Immunological status of isolated lymphoid follicles after intestinal transplantation. Am J Transplant. 2014;14(9):2148-58.

27. Ogura Y, Martinez OM, Villanueva JC, Tait JF, Strauss HW, Higgins JP, et al. Apoptosis and allograft rejection in the absence of CD8+ T cells. Transplantation. 2001;71(12):1827-34.

28. Fujimoto Y, Uemoto S, Inomata Y, Egawa H, Fujita S, Kawanami T, et al. Small bowel transplantation using grafts from living-related donors. Two case reports. Transpl Int. 2000;13 Suppl 1:S179-84.

29. Souza HS, Tortori CJ, Castelo-Branco MT, Carvalho AT, Margallo VS, Delgado CF, et al. Apoptosis in the intestinal mucosa of patients with inflammatory bowel disease: evidence of altered expression of FasL and perforin cytotoxic pathways. Int J Colorectal Dis. 2005;20(3):277-86.

30. Neurath MF, Finotto S, Glimcher LH. The role of Th1/Th2 polarization in mucosal immunity. Nat Med. 2002;8(6):567-73.

31. Bettelli E, Carrier Y, Gao W, Korn T, Strom TB, Oukka M, et al. Reciprocal developmental pathways for the generation of pathogenic effector TH17 and regulatory T cells. Nature. 2006;441(7090):235-8.

32. La Cava A, Van Kaer L, Fu Dong S. CD4+CD25+ Tregs and NKT cells: regulators regulating regulators. Trends Immunol. 2006;27(7):322-7.

33. Karczewski J, Karczewski M, Glyda M, Wiktorowicz K. Role of TH1/TH2 cytokines in kidney allograft rejection. Transplant Proc. 2008;40(10):3390-2.

34. Jiang X, Kojo S, Harada M, Ohkohchi N, Taniguchi M, Seino KI. Mechanism of NKT cell-mediated transplant tolerance. Am J Transplant. 2007;7(6):1482-90.

35. Seino KI, Fukao K, Muramoto K, Yanagisawa K, Takada Y, Kakuta S, et al. Requirement for natural killer T (NKT) cells in the induction of allograft tolerance. Proc Natl Acad Sci U S A. 2001;98(5):2577-81.

36. Liu Y, Luan X, Li J, He Y, Li M. The role of invariant NKT cells in liver transplant tolerance in rats. Transplant Proc. 2012;44(4):1041-4.

37. Yang JJ, Feng F, Hong L, Sun L, Li MB, Zhuang R, et al. Interleukin-17 plays a critical role in the acute rejection of intestinal transplantation. World J Gastroenterol. 2013;19(5):682-91.

38. Lisbonne M, Leite-de-Moraes MC. Invariant Valpha14 NKT lymphocytes: a double-edged immuno-regulatory T cell population. Eur Cytokine Netw. 2003;14(1):4-14.

39. Ceulemans LJ, Braza F, Monbaliu D, Jochmans I, De Hertogh G, Du Plessis J, et al. The Leuven immunomodulatory protocol promotes T-regulatory cells and substantially prolongs survival after first intestinal transplantation. Am J Transplant. 2016.

40. Li Y, Zhao X, Cheng D, Haga H, Tsuruyama T, Wood K, et al. The presence of Foxp3 expressing T cells within grafts of tolerant human liver transplant recipients. Transplantation. 2008;86(12):1837-43.

41. Yang Y, Song HL, Zhang W, Wu BJ, Fu NN, Zheng WP, et al. Reduction of acute rejection by bone marrow mesenchymal stem cells during rat small bowel transplantation. PLoS One. 2014;9(12):e114528.

42. Acosta-Rodriguez EV, Napolitani G, Lanzavecchia A, Sallusto F. Interleukins 1beta and 6 but not transforming growth factor-beta are essential for the differentiation of interleukin 17-producing human T helper cells. Nat Immunol. 2007;8(9):942-9.

Functional and Phenotypic Analysis of Two-Day Monocyte-Derived Dendritic Cells Suitable for Immunotherapy Purposes

David Andrés Bernal-Estévez[1,2,#], Diana Rocio Tovar Murillo[1,2,#], and Carlos Alberto Parra-López[2*]

[1]Immunology and Clinical Oncology Research Group (GIIOC). Fundación Salud de los Andes, Bogotá-Colombia South-America.
[2]Immunology and Translational Medicine Research Group, Universidad Nacional de Colombia, Graduated School in Biomedical Sciences. Department of Microbiology. Carrera 30 Calle 45, Bogotá-Colombia South-America.
[#]Both authors contribute equally to this work.

***Corresponding author:** Carlos Alberto Parra-López, Universidad Nacional de Colombia, Facultad de Medicina, Departamento de Microbiología oficina, Carrera 30 Calle 45-03, Bogotá, Colombia, E-mail: caparral@unal.edu.co*

Abstract

Background: Optimize the generation of clinical-grade monocyte-derived dendritic cells (DC) suitable for cancer immunotherapy affordable to cancer patients worldwide remains a formidable challenge. Although vaccination with DC pulsed with tumor-associated antigens (TAA) has been thoroughly evaluated in Caucasian individuals, the immunogenicity of DCs in Colombian individuals is yet to be explored.

Methodology: Mononuclear cells obtained from buffy coats of healthy Colombian volunteers were used to compare the phenotype and proficiency as Antigen Presenting Cells (APC) of matured DCs obtained in two vs. seven days from monocytes purified either by adherence to plastic or by negative selection used fresh or after cryopreservation. Furthermore, the functional capacity of two types of 2d-DCs maturated with two different cytokine-cocktails: Standard (2d-stDCs) or Type I Alpha (2d-aDCs), to prime or boost TAA specific CD8+ T cells and to stimulate Th1 stem cells and follicular T helper CD4+ T-cells (TFH), was examined.

Results: Both 2d-stDCs and 2d-aDCs exhibited a maturation profile similar to standard DCs obtained in seven days. Despite both types of DCs were efficient in stimulating antigen specific CD4+ and CD8+ T cells, only 2d-aDCs secreted high amounts of IL-12p70 and fostered more efficiently than 2d-stDCs the expansion of stem memory Th1, TFH cells and effector TAA specific CD8+ T cells.

Conclusions: Two-day derived DCs are phenotypically and functionally equivalent to seven day derived DCs and are suitable for cancer immunotherapy in Colombian individuals.

Key Words: Dendritic cells; T Lymphocytes; Flow Cytometry; Immunotherapy; HLA-A*0201-restricted Tumor-Associated Antigen.

Abbreviations

DCs: Dendritic Cells; 2d-DCs: two-day monocyte-derived DCs; 7d-stDCs: seven-day monocyte-derived DCs; iDCs: immature DCs; PBMCs: Peripheral Blood Mononuclear Cells; FC: Flow Cytometry; DC-SIGN: Dendritic cell-specific ICAM-3 grabbingnon-integrin (CD209); CFSE: Carboxy Fluorescein Succinimidyl ester; TAAs-A2: Tumor associated antigens.

***Corresponding author email:** caparral@unal.edu.co*

Introduction

Since the initial description of dendritic cells (DCs) more than 40 years ago [1], their role in T cell mediated immunity in particular in anti-tumor immunity have been deeply studied [2]. The clinical use of autologous DCs for therapy in cancer patients have shown that DCs are a secure therapeutic strategy that elicit variable clinical responses with low adverse effects in treated patients[3]. This strategy aims the recovery of the immune system unresponsiveness promoted by the tumor microenvironment that ends up with the reestablishing of the anti-tumor capacity of cytotoxic T cells (CTLs) [4-10]. Characterizing the phenotype of monocyte-derived DCs after maturation continue being a valuable tool to examine in vitro the potential benefits of these cells in generating protective immunity against tumors in cancer patients.

In humans, there are two groups of primary DCs, the myeloid DCs characterized by the expression of CD11c and CD33, and lymphoid derived or plasmacytoid DCs which express CD123 and CD45RA. While the number of myeloid DCs in peripheral blood or infiltrating DCs in tumors is decreased, the number of plasmacytoid DCs is not altered. Robust evidence suggests that myeloid DCs in cancer patients have an immature phenotype [11, 12] that induce immune tolerance within tumor microenvironment[13, 14]. That DCs isolated from patients recover responsiveness to maturation stimuli *in vitro* or that DCs from normal individuals express an immature phenotype when exposed to tumor cells supernatant, suggests that in cancer patients a wide variety of stimuli secreted by tumors hampers functionality of myeloid DCs[15]. Therefore, vaccines based on DCs has been actively explored for the last several years to surmount both the poor presentation of tumor antigens by tumor cells likewise the deficient performance as APCs of primary DCs in cancer patients [15].

The ability of DCs to activate T cells capable of recognizing and destroying tumor cells in an antigen-specific manner has been demonstrated in clinical protocols carried out in cancer patients in which DCs pulsed with tumor antigens has been used as vaccines [16]. It has been observed that this type of vaccine is safe and more tolerable than other treatments such as chemotherapy or radiation therapy[3], with variable clinical and immunological responses [5, 17-19]. The clinical evidence suggests that immune stimulation of cancer patients with autologous DCs pulsed in vitro with tumor antigens is possible and that the use of this class of therapeutic vaccines in a variety of invasive tumors opens the possibility of prolonging the life expectancy of treated patients while improving their quality of life.

Because DCs are in a very low amount in peripheral blood, during the last decade it has been described several methods for their generation in vitro from peripheral blood monocytes cultured with IL-4 and GM-CSF and maturated with various combinations of pro-inflammatory cytokines and/or Toll-Like Receptors (TLRs) agonists [20]. Most of clinical trials conducted to date have used monocyte-derived DCs generated in seven days (7d-DCs): five days in the presence of GM-CSF and IL-4 followed by the addition of the maturation stimuli for two days. Additionally, there are some reports proposing the generation of DCs in a shorter period of time: DC induction from monocytes treated with GM-CSF and IL-4 for 24 hours is followed by 24 hours incubation with a pro-inflammatory stimulus (2d-DCs), a timing that allows to achieve competent DCs that elicits antigen-specific Th1 immune responses[21, 22]. Translation of this modification in timing into the clinic would make DC production less laborious and will drop the costs of mass production of DCs, two features of DC production that must be considered to make DC-based immunotherapy more accessible to cancer patients in undeveloped countries. Finally, a newly described combination of cytokines for maturation of DCs that includes type I interferons and TLR ligands [23] that induces mature DCs denominated Type I alpha DCs or alpha-DCs, that are characterized by high production of IL-12 and the efficient activation of CD8+ T cells against tumors.

Given the importance of implementing new alternatives for the management of cancer in Colombia, in this work, we evaluated yield of production, phenotypic and functional characteristics of matured 2d-DCs from healthy Colombian individuals obtained through two different combinations of maturation cytokines. Our results demonstrate that DCs derived in two days, have a clear maturation phenotype and the capacity of activating T cells (both cytotoxic and follicular helper - TFH cells), suggesting that 2d-DCs are suitable to be used in Colombian individuals for cancer immunotherapy.

Materials and Methods

Patients and monocyte enrichment: This study was approved through the Ethics Committee of the Medical School – Universidad Nacional de Colombia (CE-14, 9 August 2012, Act. 107). All participants provided informed consent before the blood samples were collected. Heparinized blood samples (60 mL) were obtained from healthy volunteers or latent *Mycobacterium tuberculosis* (*M. tb.*) infected (LTBI) donors (QuantiFERON TB Gold test in tube positive and Tuberculin Skin Test positive) and a blood sample (20 mL) was obtained from a breast cancer patient diagnosed with ductal breast cancer (HLA-A*02:01) before chemotherapy. PBMCs were purified using density gradient ficoll and cryo-preserved in freezing medium (50% RPMI + 40% FCS + 10% DMSO). Monocytes were enriched after the adherence to plastic for 2 hours or by negative selection (RosetteSep - STEMCELL Technologies).

Dendritic cell generation: Immature DCs (iDCs) were differentiated from purified monocytes or total PBMCs using 1000 UI/mL of GM-CSF and 750 UI/mL of IL-4 (Cellgenix - Germany) for 24 hours (for 2d-aDCs or 2d-stDCs generation) based on the methodology described by Dauer et al.[21] or 5 days (for 7d-stDCs generation) as described by Jonuleit et al. [24]. iDCs were matured using two different cytokine cocktails: the standard cocktail, containing IL-1β, IL-6, TNF-α (all from Cellgenix) and PGE2 (Sigma-Aldrich) for 24 (2d-stDCs) or 48 hours (7d-stDCs); or the Type I alpha cocktail, containing IFN-α (Intron-A; Schering Plough Corp., Kenilworth, NJ), IFN-γ (R&D Systems), Poly I:C (Sigma-Aldrich), TNF-α, IL-6, and IL-1β (Cellgenix - Germany) for 24 hours (2d-aDCs) as previously described by Mailliard et al.[23]. All cells (fresh or after cryopreservation in liquid nitrogen) were cultured in AIM-V medium (Life Technologies).

Mixed Leukocyte Reaction (MLR) and proliferation assays: Highly purified CD4+ T cells by magnetic beads system (Miltenyi Biotec – Germany) were obtained from PBMCs of healthy donors and stained with 2μM. CFSE (Life technologies); labeled cells were stimulated with autologous 2d-stDCs or 7d-stDCs, pulsed with tetanus toxoid or stimulated with 10μM phytohemagglutinin (PHA)-M (Sigma-Aldrich). After 6 days of cell culture, CD4+ T cells were labeled with anti-CD69 (BD). For MLR, purified CD4+ T cells from three different donors were labeled with CFSE and co cultured with allogeneic 2d-stDCs for 6 days, after harvesting the cell T cells were labeled with antiCD69 and analyzed by flow cytometry in a FACS Aria II (BD).

T cells stimulation and cytotoxicity assays: Two different T cell stimulation methods were used [Figure S1]. Total PBMCs were enriched with 2d-stDCs or 2d-aDCs based on the methodology of Martinuzzi et al.[25]. Briefly, PBMCs were cultured with IL-4 and GM-CSF (as described above) for 24 hours in the presence of 5 μM of Tumor-Associated Antigens (TAAs) Her2/neu, Melan-A (MART-1), NY-ESO-1, Telomerase, *M. tb.* peptides as described by Goletti et al.[26], purified ESAT-6 protein or *M. tb.* H37Rv sonicate, and *L. major* peptide as control. After 24 hours the iDCs were subsequently maturated using either 2d-stDCs or 2d-aDCs maturation cocktails, with or without the addition of 5 μM of the corresponding peptide(s) for 6 days [Figure S1A]. The second method is based on the methodology of Moser et al.[27]; briefly, frozen PBMCs were sorted (FACS Aria II - BD) into three different populations, CD14+ (monocytes), CD4+ and CD8+ T cells, using the BD FACS Aria II System (BD Biosciences). The monocytes were differentiated into 2d-stDCs or 2d-aDCs, pulsed or unpulsed with corresponding peptides or proteins (5 μM each) and subsequently cultured with purified CD4+ or CD8+ T cells for 14 days at a ratio of 50:1 (T cell: DCs) in AIM-V culture media (Life Technologies), for CD8+ T cell culture 30 UI/mL IL-2 was added. After priming, CD4+ or CD8+ T cells were stimulated with corresponding peptide-pulsed 2d-stDCs or 2d-aDCs and cultured for 6 additional days [Figure S1B] For cytotoxicity assays, CD8+ T cells from healthy donors were stimulated as described above, and labeled with CD107a or CD107b for 12 hours before flow cytometry analysis.

Peptide synthesis: All peptides were generated through solid phase peptide synthesis (21st Century Biochemicals, CPC Scientific and in the Fundación Instituto de Inmunología de Colombia - FIDIC), with high purity >85%. The lyophilized peptides were dissolved in DMSO and diluted in PBS to a working concentration of 2 mM each.

Flow cytometry and cytokine quantification: The DCs were labeled with CD80 PE, CD83 PE-Cy5, CCR7 FITC, HLA-DR APC, CD14 Alexa Fluor 700, CD209 FITC, CD123 PE or CD11c APC (all from BD). Purified T cells or total PBMCs were stained with CD3 Pacific Blue, CD4 ECD (Beckman coulter) or CD8 PE-Texas Red (eBiosciences), CD45RO FITC, CD45RA PE-Cy5, CD62-L PE (BD), CD154 APC (eBiosciences), CD95 APC, or CCR7 FITC (R&D Systems). The intracellular staining of IFN-γ PE, TNF-α APC, and IL-2 PE (BD biosciences) was performed using cytofix/cytoperm and perm/wash reagents from BD Biosciences. Biotinylated HLA-A*02:01 tetramers were synthesized at the Lawrence Stern Laboratory, University of Massachusetts Medical school by CP, Melan-A tetramer was kindly gifted by Professor Pedro Romero, Ludwig Cancer Research Institute – Switzerland, and the tetramers were labeled with streptavidin-PE (Invitrogen) at a 4:1 molar ratio before use. Cytokine secretion (TNF-α, IFN-γ, IL-6 or IL-12p70) was measured in the culture supernatants using human Th1/Th2 and Inflammatory CBA kits (BD Biosciences). The samples were acquired using the FACS Aria II System at the Universidad Nacional de Colombia - Medical School. The flow cytometry data were exported in FCS format v3 and analyzed using Flow Jo software (Treestar Inc.). The graphics were generated using Prism v5 software (Graph Pad).

Microscopy: The images of iDCs, 2d-aDCs and 2d-stDCs were obtained using an inverted microscope (Nikon Eclipse TS100) coupled to a digital camera (Canon Powershot S120) at 20x magnification. The pictures were cropped in Photoshop CS5 (Adobe).

Statistical analysis: The data was analyzed by Two-way ANOVA with Bonferroni posttest to compare the mean between groups, and Mann-Whitney test for non-parametric analysis between immature and mature cells. p value was significant if below 0.05 (*), 0.01 (**), 0.001 (***) or 0.0001 (****). Data was analyzed in Prism v5 software (Graph Pad).

Results

Two-day monocyte-derived dendritic cells (2d-DCs) have similar characteristics than Dendritic cells derived in seven days (7d-DCs)

The optimization process of generating DCs involves the reduction of cultivation time, usually DCs are generated within 7 days, but reducing the time to two-day culture, reduces costs, increases cell viability and the total number of DCs that can be recovered to be used in immunotherapy. Initially, we compared side by side the immune phenotype and functional characteristics of immature DC (iDC) and after treatment for 24 or 48 hours in the presence of the standard cytokine cocktail (IL-1β, IL-6, TNF-α, and PGE2) based on the methodology described by Dauer et al.[21, 28]. Monocytes purified from healthy donors were cultured for 24 hours or 5 days in the presence of IL-4 (1000 IU/mL) and GM-CSF (750 IU/mL). Subsequently, to induce maturation of 2d-stDCs and 7d-stDCs, the iDCs were cultured in the presence of the maturation cocktail containing IL-1β, IL-6, TNF-α and PGE2 [10]. As shown in Figure 1A, changes in cell morphology were evident by direct observation under microscope, initially noted that 7d-DCs have a larger size compared with the size of 2d-DCs despite the maturation state, however, after 24 hours of maturation, DCs change from a round shape to present elongated dendrites in 2d-stDCs, and present multiple dendrites in 7d-stDCs. This difference of size and complexity was confirmed by Flow Cytometry (FC) using SSC-A vs. FSC-A analysis [Figure 1B], where 7d-DCs have the greatest distribution of light dispersion in contrast to 2d-stDC or monocytes. Subsequently, multi-parametric FC was used to compare the expression of CD83, CD209, CD14, CCR7, CD80, HLA-DR, and CD86 in iDCs, and mature DCs 2d-stDCs and 7d-stDCs. The standard cocktail of cytokines in 7d-stDCs clearly induces a maturation phenotype as expected [Figure 1C], with increased expression of CD80, CD83, CCR7, CD86 and HLA-DR and decreased expression of CD14 and CD209. This phenotype was consistent with previous studies [29, 30]. In particular, treating iDCs with maturation stimuli for 24 hours generates 2d-stDCs with a phenotype very similar to that of 7d-stDCs maturation in terms of increase, CD80, CD86, CCR7, CD83, and HLA-DR and decreasing of CD209 and CD14 expression [Figure 1C]. These results suggest that the time reduction necessary to generate mature DCs in vitro does not affect the phenotype of mature DCs.

2d-stDCs have a similar mature phenotype in comparison to 7d-stDCs regardless of monocyte purification system

Despite several procedures available to obtain monocytes from peripheral blood such as (i) positive or negative selection of CD14 cells (either by flow cytometry or magnetic beads) and (ii) cell adhesion to plastic, plastic adherence of monocytes has been the procedure most widely used to produce DCs from monocytes *in vitro*. During the optimization process of DC production, we compared two procedures to obtain monocytes from peripheral blood, first, by negative selection and second, by adherence to plastic that is a low-cost procedure and a rather "standard method" for monocyte enrichment. Additionally, we evaluated if the cryopreservation of purified monocytes compared to freshly isolated monocytes (isolated by negative selection or adherence to plastic) did affect maturation of DCs.

We evaluated the effect of pro-inflammatory cytokines standard cocktail to induce mature DCs (Figure 2). Initially, we evaluate the phenotype of DCs derived from monocytes purified by negative selection, as expected the pro-inflammatory cytokines induce a significant increase in the expression of CCR7,

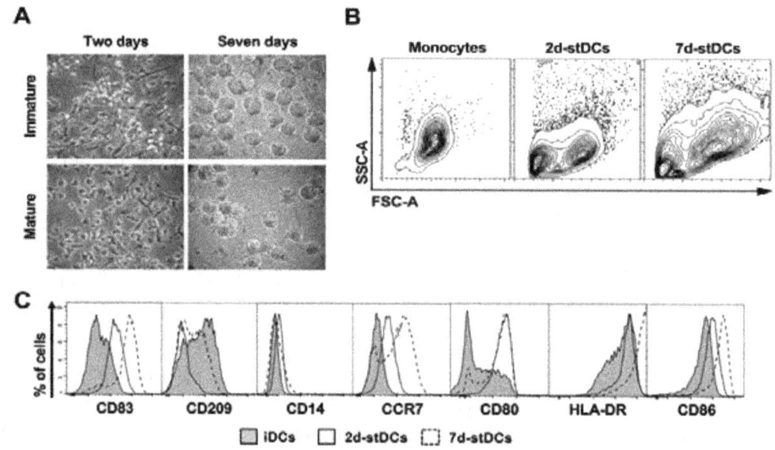

Figure 1: Morphological and phenotypic differences between dendritic cells matured in seven days (7d) and two days (2d). A. Microscopic view (40X) of the DC-derived in vitro from purified monocytes, mature or immature 2d, and 7d (2d-iDCs, 7d-iDC). B. Contour plots of FSC-A vs. SSC-A of purified monocytes, dendritic cells matured in 2d and 7d-DCS (from left to right respectively). C. Histograms representing the expression of maturation markers, CD83, CD209, CD14, CCR7, CD80, HLA-DR, and CD86 on immature dendritic cells (gray histogram), two days mature DCs (empty histogram) and seven days mature DCs (Histogram with dashed line).

CD86, CD83, and CD80 with the expected reduction of CD209 in 2d-stDCs and 7d-stDCs [Figure 2, panels A and E, respectively]. When we evaluate the phenotype of DCs derived in two and seven days from adherent cells suitable for clinical use, we found that the purification method does not affect the phenotype in mature DCs. Interestingly, 2d-stDCs have a more pronounced difference between immature and mature cells than 7d-stDCs [Figure 2, panels B and F, respectively]. Finally, it is important to compare the effect of monocyte cryopreservation in the phenotype of mature DCs, to evaluate this, we analyze the same phenotype as described before in freshly isolated and cryopreserved monocytes in either two days and seven days DCs (Figure 2), panels C, D, G, and H. The phenotype found in these cells evidence that the cryopreservation of monocytes does not affect the capacity of monocytes to respond to pro-inflammatory cytokines, even more, 2d-stDCs have a more significant difference between immature and mature DCs compared to 7d-stDCs. These results also show that the quantification of HLA-DR and CD14 are not suitable for maturation markers of 2d-stDCs or 7d-stDCs. Finally, the mechanism for obtaining and preservation of monocytes does not affect their ability to respond adequately to the cocktail of cytokines.

Based in FDA approved Sipuleucel-T (Provenge®) that consist of a fusion of GM-CSF with the Prostatic Acid Phosphatase (PAP) specific antigen of prostate cancer, used in total cells to induce antigen presentation by Antigen Presenting Cells (APC) thanks to the function of GM-CSF, we wanted to evaluate the advantages of inducing the differentiation and maturation of DCs from total PBMC without prior purification process of monocytes using a similar approach to that described by Martinuzzi et al., who evaluated the ability of IL-4 and GM-CSF followed by a maturation stimulus to amplify memory T cells in total PBMC [25]. Using two-days DCs, we compared the expression of CCR7, CD80, CD83, and CD14, in CD11c+ cells in adherent and non-adherent fraction, exposed or not to the cocktail of cytokines [Figure S2], after collecting the cells we found that in the non-adherent fraction there are cells that have a phenotype of mature DCs with similar phenotype to cells that remain attached to plastic [Figure S2].

Finally, when we analyze the phenotype of 2d-stDCs in comparison to immature DCs, our results confirm that 2d-stDCs [Figure 2D] have a similar phenotype compared to 7d-stDCs [Figure 2H]. This phenotype is characterized by a marked reduction of CD209 and increased expression of CCR7, CD80, and CD83 in 2d-stDCs. Despite differences in the morphology of DC [Figure 1], we found that DCs derived in two days, have a more pronounced difference between immature and mature DCs compared to DCs derived in seven days.

Multiparametric analysis of DCs maturation markers confirms a similar phenotype between 2d- and 7d-stDCs

The study of DCs phenotype includes the expression of different profiles that includes up regulation of co-stimulatory and MHC class II molecules, chemokine receptors, among others, that make the definition of a mature DC difficult to compare between studies. In order to determine the value of the analysis of one or more maturation markers in the phenotype of 2d- and 7d-stDCs we performed a multivariable analysis of two or more parameters to define the mature phenotype of DCs. Initially, we compared the phenotype of immature and 7d-stDCs to establish the combination that describes the best difference between immature and mature DC. Notably, combinations of three parameters that includes CD80, CD209 negative, and HLA-DR or two parameters: CCR7 and CD83 or CD83 and CD86 expression, exhibited the most significant difference [Figure 3A]. The same analysis process was also done with 2d-stDCs; In this case, most combination of markers (including two, or four parameters simultaneously), showed significant differences between mature and immature DCs [Figure 3B]. Finally, when comparing mature 2d-stDCs with 7d-stDCs [Figure 3C], we evidenced that most of the combinations of markers (except for CD80 and CD209 negative cells) were not statistically different between these two types of DCs suggesting that monocyte derived DCs obtained in either two and or seven days have a very similar maturation phenotype.

Mature 2d-stDCs are capable of inducing specific T cell activation and proliferation

Based on the DCs phenotype obtained in 2d-stDCs we assess the capacity of these cells to induce T cell responses. To achieve this, mature 2d-stDCs from healthy individuals vaccinated against tetanus were pulsed with Tetanus Toxoid (TT) recombinant protein (5μM) and PHA-M (5μM) was added to a control culture

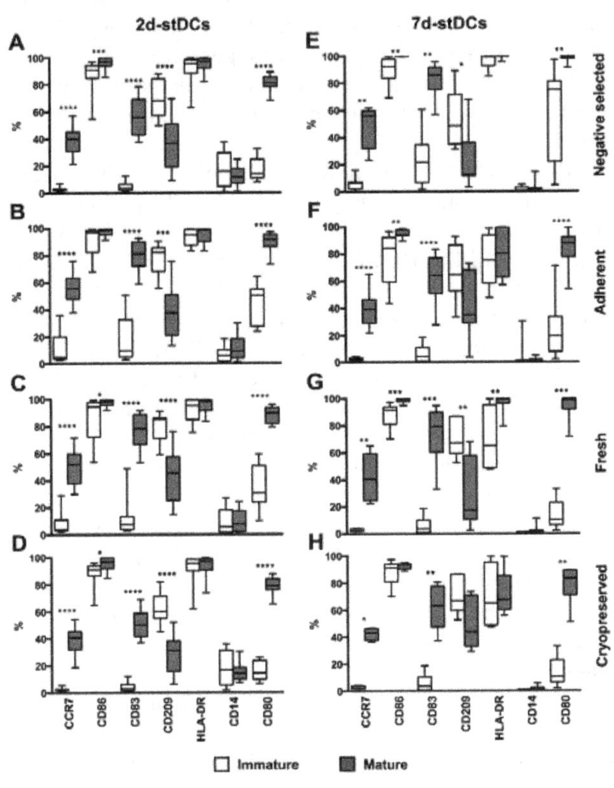

Figure 2: Monocyte collection system and preservation system do not affect the generation of dendritic cells. Comparative expression percentage by flow cytometry analysis of CCR7, CD86, CD83, CD209, HLA-DR, CD14 and CD80 in immature (white box) and mature DCs (gray box) derived in two or seven days (left and right columns respectively). Panel A (n=18 vs. 18) and E (n=11 vs 11), corresponds to DCs derived from negative selected monocytes; B (n=9 vs. 14) and F (n=7 vs. 7), DCs derived from adherent monocytes; C (n= 15 vs. 20) and G (n=6 vs.12), DCs derived from fresh monocytes; D (n=12 vs.12) and H (n=6 vs. 6), DCs derived from cryopreserved monocytes. Data showed in box and whiskers (top and bottom whiskers 90% and 10%), the comparison between immature and mature DCs done with Mann-Whitney test *p<0.05, **p<0.01, ***p<0.001, **** p<0.0001.

as positive control, the pulsed 2d-stDCs were co-cultured with autologous purified CD4+ T cells (positive selection - Miltenyi Biotec) labeled with CFSE (2μM), after 6 days of co-cultivation, proliferation and expression of CD69 by T cells were assessed by FC and IFN-γ was measured by ELISA in culture supernatants. 2d-stDCs showed to be efficient in inducing proliferation of CD4+ T cells [Figure S3] This figure shows that 50% of T cells proliferate in response to the stimulus compared to the unpulsed 2d-stDCs (19.6%). Besides, 2d-stDCs were able to activate CD4+ cells in response to a nonspecific stimulus such as PHA-M (92.9% in cell proliferation). A slight difference in CD69 expression when CD4+ T cells were stimulated with 2d-stDCs pulsed with TT in comparison to negative control (13.4 vs. 8.29% respectively) and a high CD69 expression when stimulated with PHA-M (43.5%) were observed. Compared with negative control, TT specific CD4+ T cells produced more IFN-γ after 40 hours of co-culture (10 and 93pg/mL, respectively, Figure S4A); CD4+ T cells stimulated with 2d-stDCs pulsed with PHA-M, showed an IFN-γ production more than five hundred times greater than the control (5400 pg/mL, Figure S4A). These results evidence the capacity of 2d-stDCs to activate and induce proliferation of CD4+ T cells in response to specific stimuli.

Similarly, to TT assay, we use a Mixed Leukocyte Reaction (MLR) in the presence of 2d-stDCs as a model to study the ability of 2d-stDCs to initiate primary immune responses and to induce the proliferation and CD69 expression in purified CD4+ T cells in an in vitro assay using 2d-stDCs co-cultured with syngeneic CD4+ T cells (control) or with CD4+ T cells from different individuals (allogenic) previously labeled with CFSE, during 6 days. As could be shown by FC [FigureS3B], allogenic CD4+ T cells proliferated around 30% in response to allogeneic stimulation, compared to control, which showed only a 0.7% of cells growing. The percentage of CD69 positive cells was higher in the allogeneic co-culture (6-9%), compared to 2.7% in the control culture [Figure S3B]. In cultured 2d-stDCs with autologous CD4+ T cells no production of IFN-γ was evidenced, contrary to cultures containing heterologous lymphocytes with a production of 566, 338 and 430 pg/mL of cytokine on each of the tested individuals [Figure S4B]. Altogether the results of proliferation, expression of activation markers and cytokine secretion in cells stimulated with 2d-stDCs let us to argue for a high capacity of these cells to stimulate CD4+ T cells.

To assess the ability of 2d-stDC to induce an immune response in CD8+ T cells specific for tumor associated antigens (TAA), the expansion of CD8+ T cell precursors labeled by fluorescent tetramers was assessed in healthy donors that express the HLA-A*02:01 allele (determined by SSP-PCR). To do this, TAA peptides restricted to HLA-A2 (Melan-A and NY-ESO 1 at 10μM) were used to pulse 2d-stDCs that were co-cultured with autologous CD8+ T cells for 10 days in the presence of IL-2 and IL-7. After 10 days of cell culture, the T cells were restimulated with 2d-stDC pulsed with peptide for an additional 72 hours and stained with tetramers to identify populations of CD8+ T cells specific for the corresponding TAA. As shown in Figure 4A, the 2d-stDCs pulsed with peptides Melan-A and NY-ESO1 were

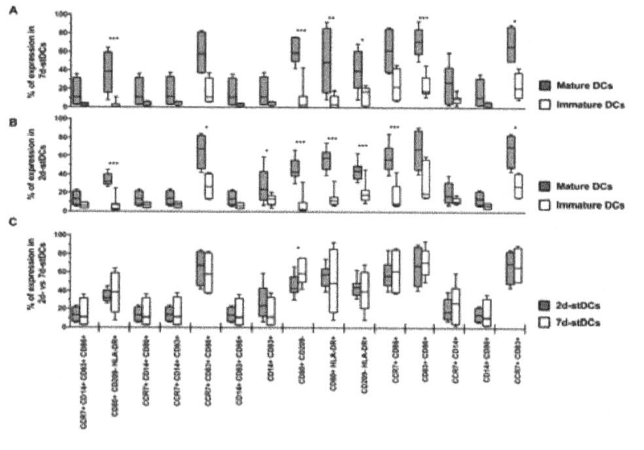

Figure 3: Use of multiple maturation markers for dendritic cells confirms that 2d-stDCs and 7d-stDCs have a similar phenotype. Analysis of multiple markers for characterization of dendritic cell maturation (two to four markers) and 2d-stDCs and 7d-stDCs monocyte-derived was analyzed by multiparametric flow cytometry. The statistical analysis of mature (gray box) and immature (white box) cells in 7d-stDCs (A) and 2d-stDCs (B). C. Comparison between mature 7d-stDCs (gray box) and 2d-stDCs (white box) derived from adherent and cryopreserved monocytes. The data showed the box and whisker (95% and 5%), Mann-Whitney * p <0.05, ** p <0.01, *** p <0.001.

efficient in the expansion of specific CD8+ precursor cells against these tumor antigens (11 and 5.15% respectively) compared to control (unstimulated cells, 0.067%). Subsequently the cytotoxic capacity of CTLs against tumor antigens were evaluated for CD8+ T cells cultured with 2d-stDCs pulsed with TAA-A2 and co-cultured with DCs pulsed with the corresponding peptide, as a control non-specific peptide of *Leishmania major* (L. major) and DCs pulsed with nopeptide were used in three healthy donors. To quantitate cytotoxicity the cells were incubated for 5 hours in the presence of anti-CD107a, CD107b and analyzed by flow cytometry. As shown in Figure 4B, a cytotoxic response (CD107a and CD107b positive cells) increased in cells specific against Melan-A antigen (8.36%) compared to control culture (1.22%) and culture with peptide *L. major* (1.31%). Subsequently, the cytotoxicity (CD107a) of CD8+ T cells in response to different TAAs such as Her-2/neu, NY-ESO1, telomerase and Melan-A in healthy donors expressing HLA-A2 was evaluated; it was observed an increase in positive CD107a cells (1.3, 4.45, 1.46, and 2.74% respectively) compared to the control culture (0.43%) [Figure 4C]. Finally, it was assessed expression of multiple intracellular cytokines (IFN-γ, TNF-α, and IL-2) in CD8+ T cells, after 20 hours of re-stimulation of T cells with mature 2d-stDC pulsed or not with a cocktail of TAA peptides were determined by flow cytometry multifunctional [Figure 4D] Compared to the control CD8+ T cells and DCs stimulated without peptide, DCs stimulated with peptide cocktail showed increased expression of cells expressing at least one cytokine is 3.5% vs 2.2%, however, the frequency of T cells that produce two or more cytokines is higher in cells stimulated with peptides compared to control [Figure 4E], the total percentage of cytokine-producing cells stimulated with peptide that are multifunctional is 3.4% compared to 0.58% of unstimulated T cells. These results support the idea that 2d-stDCs have the adequate functional capacity to stimulate CD4+ and CD8+ T cells.

Type I alpha DCs that produce IL-12 can be induced and maturated in two days

Type I alpha DCs, have been recently described by Mailliard et al.[23] using a specific combination of pro-inflammatory cytokines, type I interferons and a TLR3 ligand. So far, the needed time to induce mature alpha DCs (aDCs) has been 7d similar to 7d-stDCs, with few descriptions done in 5 days. To characterize the immune phenotype in 2d-aDCs we performed a side-by-side comparison of the immune phenotype and functional characteristics of immature DCs (iDCs) after treatment for 24 hours with either standard or alpha DC maturation cocktails (2d-stDCs and 2d-aDCs, respectively) based on the methodology of Dauer et al.[21, 28]. Purified monocytes from healthy donors were cultured for 24 hours in the presence of IL-4 (1000 UI/mL) and GM-CSF (750 UI/mL). Subsequently, to induce the maturation of 2d-aDCs or 2d-stDCs, iDCs were cultured in maturation cocktails containing IL-1β, IL-6, TNF-α, IFN-α, IFN-γ and Poly I:C [23] or IL-1β, IL-6, TNF-α and PGE2 [10], respectively. As shown in Figure 5A, we observed changes in the cell morphology and number of dendrites in response to the maturation stimuli. Consistently in several experiments performed, after 24 hours of maturation the round-shaped iDCs transformed into elongated 2d-aDCs which showed a higher number of thin dendrites than 2d-stDCs. Subsequently, we used multi-parametric Flow Cytometry (FC) to compare the expression of CD83, CD209, CD14, CCR7, CD80, HLA-DR, and CD86 in iDCs, 2d-aDCs, and 2d-stDCs. The standard cytokine cocktail for 2d-stDCs clearly induces a maturation phenotype in iDCs [Figure 5B], evidenced by the increased expression of CD80, CD83, CD86, CCR7 and HLA-DR and the decreased expression of CD209 and CD14; this phenotype was consistent with the maturation found in Figure 1C and standard studies describing 7d-stDCs [29, 30]. Notably, treating iDCs with maturation stimuli for 24 hours generates mature 2d-aDCs with a maturation phenotype similar to that of 2d-stDCs in terms of increased HLA-DR, CD80, CCR7 and CD83 expression and decreased CD209 and CD14 expression in both cell types [Figure 5B]. One of the main characteristics of Type I alpha DCs is the secretion of IL12p70 (IL-12), a key cytokine for the enhancement of tumor-specific T cell responses in mouse models [31]. To assess IL-12 secretion levels, we used the CBA kit (BD Biosciences) to quantify the pro-inflammatory cytokines in the supernatants of immature and mature DCs. The results revealed an important difference in the secretion of IL-12, which was only produced in 2d-aDCs (294 pg/mL) compared with the background levels induced in iDCs and the low levels of 2d-stDCs (34.5 and 35.1 pg/mL, respectively; Figure 5C). These results show that the induction of mature aDCs in two days is possible and have a compatible phenotype of a mature DCs similar to the phenotype of 7d-stDCs and 2d-stDCs with the clear advantage of the high production of IL-12.

2d-aDCs activates CD4+ T cells with a stem-like phenotype that produce IFN-γ

Recently Gattinoni et al. described a human stem cell-like memory (scm) T cell population characterized by a naïve phenotype (CD45RA+ CCR7+ CD62-L+ CD28+ CD27+) and the expression of CD95, IL-2Rβ, and CXCR3. This population exhibits a superior capacity for proliferation and antitumor activity compared to other memory subsets[32]. Thus, we evaluated the presence and response of scm T cells after stimulation with either 2d-aDCs or 2d-stDCs using an in vitro model for infectious diseases. We pulsed each DCs with a combination of *Mycobacterium tuberculosis* (Mtb) peptides described before[33] that are responsible for a specific in vitro response in tuberculosis-infected donors. For this purpose, we use PBMCs from a latent TB-infected donor (determined by a positive Tuberculin Skin Test – TST> 10mm, and a positive QuantiFERON TB Gold® test - Cellestis). We induced 2d-aDCs or 2d-stDCs pulsed with or without Mtb peptides. After stimulation for 6 days, the difference in the expansion of naïve, scm, memory or effector CD4+ T cellsub - populations were analyzed on each cell culture through FC subtracting the percentage of each population in peptide pulsed DCs minus the unpulsed DCs [Figure 6A] The IFN-γ production in each sub-population (percentage of IFN-γ+ CD4+ T cells pulsed minus unpulsed DCs) demonstrated that the scm compared to central memory (CM), Effector Memory (EM) or terminal effector (EF) T cell subpopulations produced higher amount of this cytokine only when the culture was stimulated with 2d-aDCs but not upon stimulation with 2d-stDCs [Figure 6B].

IL-12 produced by 2d-aDCs favors the expansion of circulating T follicular helper cells (TFH)

In the process of differentiation of TFH, the secretion of cytokines such as IL-23, TGF-β, and IL-12 by APC such as DCs, plays an important role in the differentiation process of naïve T cells into pre-TFH and from there to GC-TFH or resident memory TFH, with their function as helper cells in antibody production by B lymphocytes[34, 35]. Based on the high production of IL-12 by the 2d-aDCs [Figure 5C], we evaluated the ability of these DCs to induce expansion and activation of circulating TFH. To do this, we established an in vitro assay using 2d-aDCs and 2d-stDCs derived in PBMCs from healthy donors to characterize by FC the frequency of TFH (CD4+ CXCR5+ PD1+) in CD4+ T cells. Strikingly, we observed in PBMCs from four healthy donors a significant expansion of TFH favored by 2d-aDCs (mean 2.54%) compared with immature DCs (1.29%) and even 2d-stDCs (1.64%) as shown in figure 6C. We also evaluated in the TFH the expression of ICOS and IL-21 as functional markers of these cells in response to stimulation by DCs in combination with TCR activation by beads coupled with anti-CD3/CD28/CD2. In PBMCs from healthy donors, we assessed the expression of ICOS in TFH (CD4+ CXCR5+ PD1+) in response to stimulation of the TCR and we observed a significant increase in the presence of DCs, however 2d-aDCs induce an increased expression of ICOS in TFH (mean 77%) compared to 2d-iDCs (mean 61%) and 2d-stDCs (mean 58%) [Figure 6D left panel]. Finally, we evaluate the intracellular production of IL-21 in the TFH, and as expected, we observe a significant increase in IL-21 MFI in TFH cells when stimulated with 2d-aDCs (mean 31.28) compared to 2d-iDCs (mean -5.31) and 2d-stDCs (mean 18.7) [Figure 6D right panel]. Together, these results show that 2d-aDCs can induce IFN-γ producing CD4+ T cells and they can be used as a tool to evidence the activity of TFH circulating in peripheral blood probably due to its capacity to produce IL-12 in just 48 hours of culture.

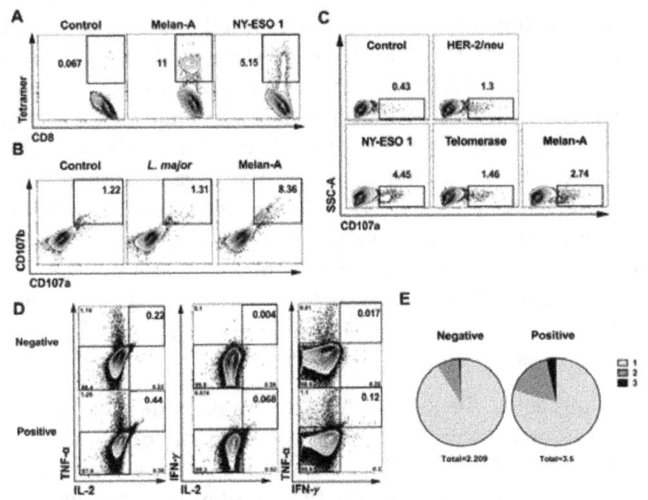

Figure 4: 2d-stDCs expands and activates cytotoxic and multifunctional specific Tumor Associated Antigens (TAAs) CD8+ T cells. A. Tetramer positive CD8+ T cells after 10-day cell culture with Melan-A or NY-ESO 1 peptide-pulsed 2d-stDCs and additional three days with pulsed 2d-stDCs from an HLA-A*02:01 healthy donors. B. Cytotoxicity (CD107a+ CD107b+) of CD8+ T cells one representative plot from 3 different HLA-A*02:01 healthy donors in response to 2d-stDCs pulsed with Melan-A peptide or controls (unpulsed DCs or non-related peptide L. major). C. Cytotoxicity (CD107a+) of different specific TAAs CD8+ T cells (Melan-A, Her-2/neu, NY-ESO1, and Telomerase) after 10-day stimulation with 2d-stDCs from an HLA-A*02:01 healthy donors. Numbers inside the plots represent the percentage of positive cells. D. Multifunctional CD8+ T cells induced by 2d-stDCs in response to TAAs peptide pool. CD8+ T cells were stimulated forten days with unpulsed or TAAs peptide pool pulsed 2d-stDCs with a 20 hour boosting with 2d-stDCs. D. Contour plots of intracellular staining with anti TNF-α, IL-2, and IFN-γ in negative (Top row) or peptide pulsed (bottom row). E. Pie chart of boolean analysis for cytokine production in negative or TAA peptide pool pulsed 2d-stDCs.

Two-day derived DCs induce expansion and activation of tumor-specific T cells in vitro in a breast cancer patient

In order to characterize antigen specific T cells present in peripheral blood of cancer patients, it is necessary to employ in vitro systems that allow amplifying TAA specific cells that are probably anergic and present at very low-frequencies. There are different methodologies to increase the number of cells, however, most methodologies require a large volume of blood to obtain the required cells; to address this we use two in vitro systems that not only can expand the antigen specific T cells with a restricted blood volume, but also can evaluate the behavior of naïve repertoire and the generation of memory cells specific to known TAA antigens. The first in vitro system is based on the method described by Martinuzzi et al.[25]. Briefly, iDCs were induced from PBMCs using IL-4 and GM-CSF for 24 hours in the presence of HER2/neu$_{369\text{-}377}$ peptide (KIFGSLAFL) and matured for an additional 24 hours using 2d-aDCs or 2d-stDCs maturation cocktails in the presence of additional HER2/neu peptide (5 µM), followed by stimulation for 6 days [Figure S1A]. The expansion of HER2-specific CD8+ T cells was analyzed through FC using the HER2 dextramer (Immudex), and the memory phenotype (CD45RO vs. CD62-L) was compared between T cells stimulated with iDCs, 2d-aDCs, and 2d-stDCs. We observed an increase in HER2-specific CD8+ T cells in response to stimulation with 2d-stDCs (3.6%) or 2d-aDCs (3.91%) compared with unpulsed DCs (0.94%) [Figure 7A], suggesting that stimulation with 2d-stDCs and 2d-aDCs induces the expansion of HER2-specific CD8+ T cells more than 3-fold compared with unpulsed DCs in the breast cancer patient.

Using naïve T-cells cells from the same patient sample, we used a second in vitro system based on the methodology of Moser et al.[27]. Briefly, naïve CD4+ and CD8+ T cells were sorted and cultured for 14 days with 2d-stDCs or 2d-aDCs in the presence of HER2/neu KIFGSLAFL peptide or the 20-mer HER2/neu peptide pool, followed by boosting with 2d-stDCs or 2d-aDCs for an additional 6 days [Figure S1B]. After stimulation, HER2-specific CD8+ T cells showed a 3 to 5-fold expansion [Figure 7B] compared with unpulsed DCs (from 0.38 to 1.39% with 2d-aDCs and 0.41 to 1.98% with 2d-stDCs, respectively). To analyze the activation of CD4+ T cells stimulated with 20-mer HER2/neu peptides, we measured CD154 (CD40-L) expression through FC [Figure 7C] and observed that only 2d-aDCs were able to induce high CD154 expression (MFI 8434 vs. 5190, pulsed vs. unpulsed DCs) compared with CD4+ T cells stimulated with 2d-stDCs (MFI 723 vs. 3543, pulsed vs. unpulsed DCs). We quantified the secretion of IFN-γ from the supernatants of CD4+ and CD8+ T cells pulsed and unpulsed with 2d-aDCs or 2d-stDCs using a CBA kit (BD biosciences). The results showed higher levels (>2 logs) of IFN-γ in CD4+ and CD8+ T cells stimulated with 2d-aDCs (delta

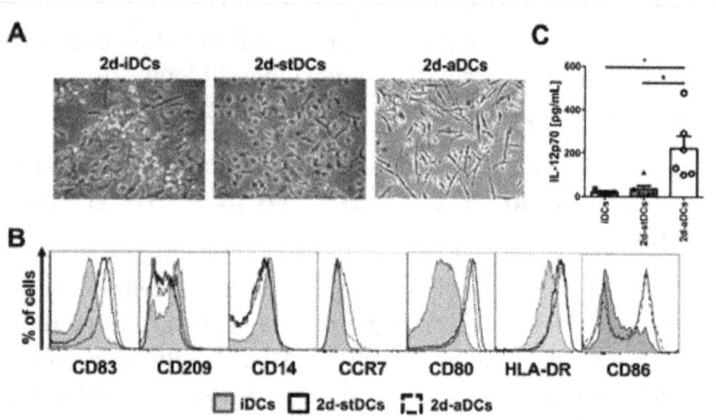

Figure 5: Two-day monocyte-derived dendritic cells using alpha cocktail have a phenotype of mature DCs and secrete high amounts of IL-12p70. A. Microscopic view (40X) of the monocyte-derived DCs 2d-iDCs, 2d-stDCs, and 2d-aDCs (left to right respectively). C. Histograms of the expression of maturation markers CD83, CD209, CD14, CCR7, CD80, HLA-DR, and CD86 of immature dendritic cells (gray histogram), 2d-stDCs (empty histogram), and 2d-aDCs (histogram with dashed line). D. The supernatant quantification of IL-12p70 secreted by 2d-iDCS, 2d-stDCs, and 2d-aDCs. IL-12 secretion is shown in bars (+/- SEM), Mann-Whitney * $p < 0.05$.

between pulsed minus unpulsed DCs) compared with CD4+ and CD8+ T cells stimulated with 2d-stDCs [Figure 7D]. These results led us to conclude that 2d-aDCs have a higher capacity than 2d-stDCs to expand and activate specific CD4+ and CD8+ T cells in response to HER2 peptides in a breast cancer patient before treatment.

Discussion

Since 1973 when Steinman and Cohn gave the name to the cells described by Paul Langerhans "dendritic cells" these cells have been used in different clinical trials as adjuvants to induce specific immune responses against different tumor antigens [1]. From the first study published in 1996 [6], where patients with B-cell lymphoma were vaccinated with autologous DCs (generated in 7 days know as Standard-DCs 7d-stDCs), these cells have been traditionally used as APCs in numerous clinical vaccination protocols [16] with a recently FDA approved therapeutic use for prostate cancer. In 2003, Dauer et al.[21] obtained in only 48 hours of in vitro culture, mature DCs from peripheral blood monocytes of healthy donors, called Fast-DCs (named here 2d-stDCs). Such cells, upon activation by pro-inflammatory mediators, expressed surface markers of mature DCs, and secrete IL-12p70 when stimulated only with CD40L and IFN-γ. Additionally, 2d-stDCs were capable of stimulate antigen specific Th1 immune responses [21]. Since then, the role of 2d-stDCs as APCs has been used in various in vitro assays [36-40] and in some clinical trials with breast cancer patients, who received 4 weekly vaccinations of 2d-stDCs pulsed with Her-2/neuropeptides [41]; in this study, seven of eleven patients showed a marked decrease in the expression of Her-2/neu in tumor specimens. This study demonstrated the potential of 2d-stDCs in the treatment of early breast cancer.

In order to determine possible changes in the phenotype of the cells in this study we derived 7 and 2 days DCs from monocytes and matured with pro-inflammatory cocktail from of fresh or frozen peripheral blood monocytes, purified by adherence to plastic or by negative selection (rosette formation with red blood cells and CD14 negative cells). In all methods used for their differentiation, DCs (2d and 7d) stimulated with standard pro-inflammatory cytokines cocktail showed significantly increased expression of CCR7, CD83 and CD80, surface markers important molecules for their performance as professional APCs as compared to unstimulated cells with the cocktail, we even compare the effect of several cytokine combinations including standard cocktail, IFN-γ, and IL-15 in the maturation phenotype of DCs [Figure S5].

The expression of CD83, a member of the immunoglobulin superfamily, is described as a characteristic marker for mature DCs and correlates with optimal mediated activation of T cells by DCs[42]. Also, studies in mice have shown that deficiencies in the expression of CD83 reduce the expression of MHC class II on the surface of 25 to 50% of splenic B cells, DCs, thymic epithelial cells, and peritoneal macrophages[43]. The interaction of the co-stimulatory molecules CD80 and CD86 (B7.1 and B7.2) with CD28 on T cells, provides a substantial activation and survival of T cells, together with binding MHC/peptide - TCR. Different studies have demonstrated that CD80 is more efficient than CD86 in the activation and proliferation of CD8+ T cells and the production of IL-12 by DCs [44-46] but the basis of these associations are unknown. Based on the results obtained in this work is interesting to note that CD80 acted as a differentiation marker of mature DCs compared to CD86 which, like HLA-DR, showed a high percentage of expression in mature and immature DCs (both 2d and 7d-stDCs).

CD14 expression was significantly increased in immature and mature cells derived in 2 days compared to 7 days DCs. This result contrasts with the observations of Dauer et al.[21] and could be explained by an internalization of CD14 antibody induced by the anti-CD14 beads used in Dauer's study for isolation of monocytes (positive selection). The expression of CD14 in 2d-stDCs suggest that these DCs are similar to dermal interstitial DCs as suggested by the work of Klechevsky et al., who have identified special properties of DCs expressing CD14 which preferentially initiate humoral immunity [47, 48].We observed a significant decrease in CD209 (DC-SIGN) in 2d-stDCs in response to maturation cocktail and independently of the purification or preservation process of monocytes. This molecule is often described as a phagocytic receptor with an important role in the acquisition and processing of the antigen on immature DCs,

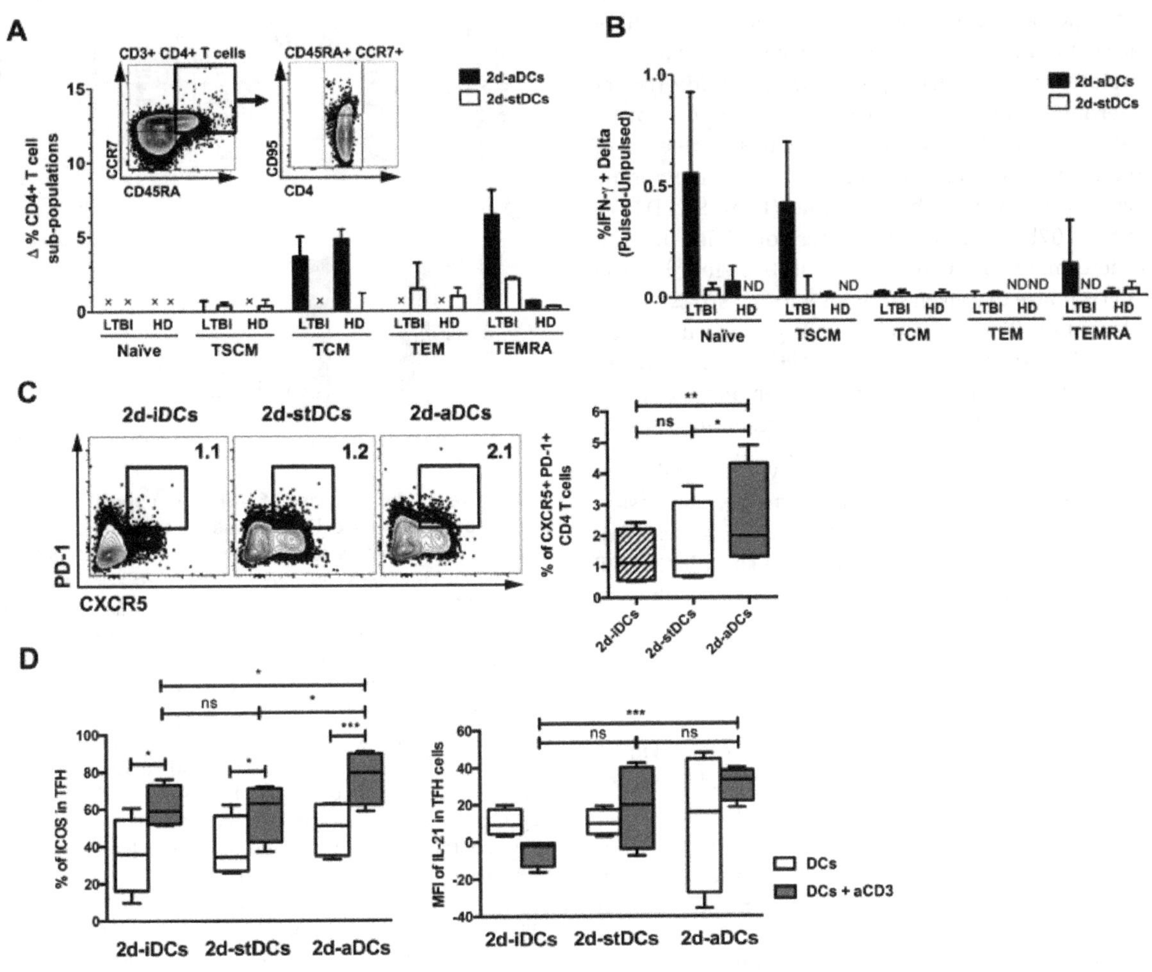

Figure 6: 2d-aDCs induce greater differentiation of memory T cells and expansion of TFH compared with 2d-stDCS. Phenotype characterization of CD4+ T cells stimulated with 2d-aDCs (black bars) or 2d-stDCs (white bars) pulsed with *Mtb* peptides minus unpulsed DCs (delta of the percentage) for two days in PBMCs from healthy donors (n=4) or latent tuberculosis infected (LTBI) donors (n=4), the insert represents contour plots of memory subpopulations (CD45RA vs. CCR7) and CD95+ in naïve cells for TSCM. B. Delta of the percentage of IFN-γ producing cells in each memory sub-population quantified in panel A stimulated by 2d-aDCs (black bars) or 2d-stDCs (white bars). C. Quantification of CD4+ T follicular helper (Tfh) cells (CXCR5+ PD1+) in response to 2d-iDCs, 2d-stDCs or 2d-aDCs (left to right panels respectively) and analysis of the quantification of Tfh percentage in healthy donors stimulated with corresponding DCs (n=4). D. Quantification of ICOS + in Tfh cells in response to stimulation with beads coupled to antibodies against CD3 / CD28 / CD2 in combination with DCs. Quantification of Tfh expansion in response to *Mtb* antigens (ESAT-6, QTF peptide pool or H37Rv sonicate) in healthy donors and LTBI donors. Nonparametric *t* test analysis and paired t-testwere done, * p<0.05, **** p<0.0001.

characterized with a large capacity of endo-phagocytosis [49, 50]. Since initially described in Human Immunodeficiency Virus (HIV), DC-SIGN has an important role in the immune modulatory response by this lectin in T cells and it has been demonstrated in multiple models of host-pathogen interaction[51, 52], suggesting that some pathogens have immune-evasive properties or immunosuppressant through DC-SIGN.

Overall, the use of fresh or cryopreserved monocytes to derive DCs does not affect the phenotype of the mature DCs derived in two or seven days. These results are of particular importance for immunotherapy considering the possibility of cryopreserving the cells usually obtained in large quantities by leukapheresis, and then used for later generations of DCs.

Besides achieving the expansion of TAA specific T cell precursors, is important to determine the functionality of these cells, for which CD8+ T cells generated in a specific line of peptide were re-stimulated with 2d-stDCs pulsed with peptide (target cells) and determined the fusion of the granules with the plasma membrane of the CD8+ T cells that results in the release of perforin and granzyme and surface expression of the membrane glycoproteins associated with the lysosome (LAMPS) CD107a (LAMP-1) and CD107b (LAMP-2). The expression of CD107a and CD107b was determined by FC in CD8+ T cells from four healthy individuals, in response to a given stimulus with 2d-stDCs pulsed with Melan-A peptide. Similarly, CD8+ T cells generated peptide specific lines NY-ESO1, Her-2/neu, and Telomerase showed expression of CD107a on the surface by FC, demonstrating the functional capabilities of 2d-stDCs as antigen-presenting cells.

Mature 2d-stDCs pulsed with peptides were capable to stimulate peptide-specific CD8+ CTLs (specific T cell expansion and cytotoxic functionality by CD107a/b expression). Furthermore, substantial production of IFN-γ by specific CD4+ T cell in response to Tetanus Toxoid (TT) and in a Mixed Leukocyte Reaction (MLR), suggests that antigen presentation by 2d-stDCs to T cells can induce the secretion of IL-12p70 by the DCs necessary to facilitate the stimulation of Th1 immune responses [Figure S4]. This is important if we consider that IL-12p70 production by DCs has become a key marker of competent DCs that will be used with immunotherapy purposes [53]. Recently, Mailliard et al.[23], described a type I alpha DCs, using a modification of the standard cocktail by adding Type I/II interferons and Poly I:C in order to obtain mature DCs capable of producing high amounts of IL-12p70 and activate Th1 cells responses. But similar to 7d-stDCs, the type I alpha DCs were produced after seven days of cell culture. These 7d-aDCS cells are able to load tumor specific antigens and activate breast cancer specific CTL responses [54]. In this work we take the advantage of producing mature DCs in 48 hours and analyzed the functional capability of Type I alpha DCs derived in two days (2d-aDCs) for activation of CD4 and CD8 antigen specific T cells.

The IL-12p70 production in culture supernatants of mature DCs was determined by CBA. As expected, and according to previous studies where the use of PGE2 in combination with the cocktail of pro-inflammatory cytokines inhibit the production of IL-12[21, 55], even after additional activation with IFN-γ [56], in DCs matured in 2 days with cocktail of pro-inflammatory cytokines produced no substantial amounts of IL-12p70, in contrast to the high production of IL-12 by 2d-aDCs [Figure 5C]. Taking altogether our results, the use of 2d-aDCs reinforces the benefits of this cytokine combination for differentiation and activation of T cells. We evaluated the functional capacity of 2d-aDCs and 2d-stDCs for both CD4 and CD8 T cell activation. For Th1 T cell differentiation, we used Mtb. peptides previously

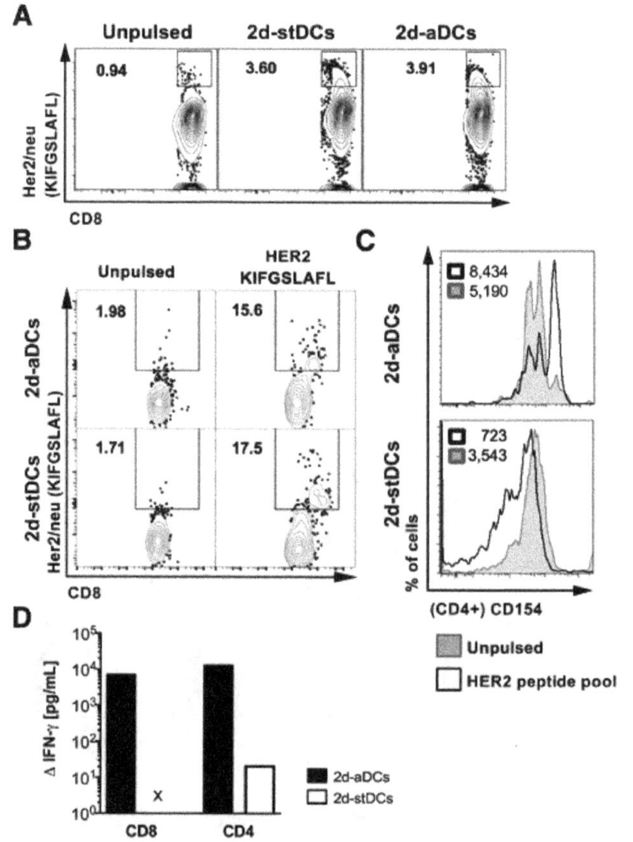

Figure 7: The increase of HER2/neu-specific CD8+ T cells and activation of CD4+ T cells in response to stimulation with 2d-aDCs in a breast cancer patient. A. Percentage of CD8+ tetramer-positive T cells in total PBMCs induced with iDCs, 2d-aDCs or 2d-stDCs pulsed with HER2 peptide pool for seven days (plot gated on single cells and CD3+ T cells) in a breast cancer patient before any treatment. B. Percentage of HER2 tetramer-positive CD8+ T cells after co-culture with 2d-aDCs or 2d-stDCs pulsed with HER2 peptide for 14 days and a boosted for six days with 2d-aDCs or 2d-aDCs pulsed with HER2 peptide (Unpulsed DCs tinted histogram, HER2 pulsed DCs empty histogram. Gated in singles cells, CD3+/CD8+ T cells). C. MFI of CD154 (CD40-L) on purified CD4+ T cells in response to 2d-aDCs or 2d-stDCs pulsed with HER2 peptides for 14 days and boosted with 2d-aDCs or 2d-aDCs pulsed with HER2 peptides for 6 days (Unpulsed DCs tinted histogram, HER2 peptide-pulsed DCs, empty histogram. Gated in singles cells, CD3+/CD4+ T cells). D. Analysis of IFN-γ secretion using a CBA kit (delta of the IFN-γ concentration of peptide-pulsed DCs minus unpulsed DCs) in CD4+ and CD8+ T cell supernatant stimulated with 2d-aDCs or 2d-stDCs.

described to induce IFN-γ secretion in immune TB donors [26, 57, 58]. In the present study, we compared the capacity of peptide-pulsed DCs to induce a Th1 phenotype with that of latent-infected donor cells cultured with purified CD4 T cells. After incubation for 6 days, we observed the increased production and secretion of IFN-γ, TNF-α and IL-6 compared with unpulsed DCs (data not shown). The analysis of the phenotype of responding cells showed that only stimulation with 2d-aDCs induced an increase in IFN-γ production in scm CD4+ T cells [Figure 6] and not in other memory, naïve or effector sub-populations. This result suggest the interesting possibility that the scm CD4+ T cell memory population is responsible of the IFN-γ production in individuals in which Mtb latency has been diagnosed with one of the common IGRA tests available (TB ELISPOT or QuantiFERON TB®).

Another important role of IL-12 is the stimulation of T follicular helper cells, a specific population of T cells involved in the antibody production by plasma cells by the release of IL-21. Recent studies show that IL-12 produced by DCs is important to expand and activate TFH cells [35]. We evaluate this possibility using our 2d-aDCs, and as expected we found a significant increase in the percentage of circulating CD4+ TFH cells (CXCR5+ and PD1+) in healthy donors, and more importantly, stimulating these cells with anti-CD3/CD28/CD2 beads we were able to induce the activation of TFH determined by the increase ICOS and IL-21 in TFH cells. With these results, we propose that the "in situ" induction of 2d-aDCs in PBMCs it may became a useful tool to study the phenotype and behavior of memory TFH present in peripheral blood and furthermore to evaluate the relationship between the detection of these cells in peripheral blood and the production of antigen specific high antibodies titers by B cells.

The functional capacity of 2d-aDCs to induce the expansion of CD8 T cells was evaluated in samples from breast cancer patients before chemotherapy to determine the degree of expansion and activation of specific T cells that recognize different TAAs. The results suggested that 2d-aDCs generated from total PBMCs (as described by Martinuzzi et al. [25]) or from purified monocytes as described by Moser et al. [27], [Figure 7A and 7B, respectively] induce the expansion of tumor antigen-specific CD8+ T cells against HER2 peptide and activate the effector function of these cells, evidenced by HER2-dependent IFN-γ secretion [Figure 7D] and the expression of CD154 in CD4 T cells obtained from breast cancer patients [Figure 7C]. These results suggest that 2d-aDCs induce the specific response and expansion of T cells in breast cancer similar to 2d-stDCs but with an increased Th1 cytokine production. Taken together, the data in the present study suggest that 2d-stDCs and 2d-aDCs are efficient short-term APCs and useful in vitro tools to induce TAA specific CD8+ and CD4+ T cell responses (TFH and stem cell like memory T cells) that are suitable for immunotherapy settings, reducing time and cost of DCs development. Further studies should evaluate the combination of 2d-stDCs and 2d-aDCs for cancer immunotherapy.

Acknowledgments

This study was supported through funding from the Universidad Nacional de Colombia. DIB, Vicedecanatura de Investigación Universidad Nacional Medical School; funds from a joint grant between Fundación Salud de los Andes, Univesidad Nacional and COLCIENCIAS. The authors express their gratitude to Dr. Fabio Méndez CEO at the Fundación Salud de los Andes (FSA) and to FSA for their generous support. The authors would also like to thank Dr. Bernardo Camacho and to personnel at the Hemocentro Distrital for their kind assistance in obtaining blood buffy coats from volunteers. We are very thankful to Dr. Larry Stern, Dr. Mauricio Calvo-Calle and personnel at their laboratories at the University of Massachusetts for their generous support in providing reagents, infrastructure and knowledge required for building up HLA-A2 tetramers. We thank Dr. Manuel Patarroyo and personnel at the FundaciónInstituto de Inmunología de Colombia (FIDIC) for assistance with peptide synthesis. Finally, our deepest gratitude to patients (especially to the patient FLB001) and healthy volunteers for their generous denotation of blood samples used in this study.

Funding

This work was funded through Dirección de Investigación de Bogotá (DIB)-HERMES Grants (Numbers 13245, 12543, 33317, 33290, 32181 and 14976) from the Universidad Nacional de Colombia and funds from the joint grant between Fundación Salud de los Andes, Universidad Nacional and COLCIENCIAS (Contract No. 110150227509). DBE was supported by the Fundación Salud de los Andes, Bogotá-Colombia South America.

Author contribution

Experimental work: DB, and DRTM. Data analysis: DB, DTRM and CP. Wrote the manuscript: DB, DRTM and CP.

References

1. Steinman, R.M. and Z.A. Cohn, *Identification of a novel cell type in peripheral lymphoid organs of mice. I. Morphology, quantitation, tissue distribution.* J Exp Med. 1973;**137**(5):1142-1162.
2. Palucka, K. and J. Banchereau, *Cancer immunotherapy via dendritic cells.* Nat Rev Cancer. 2012;**12**(4): 265-277.
3. Draube, A., et al., *Dendritic cell based tumor vaccination in prostate and renal cell cancer: a systematic review and meta-analysis.* PLoS One. 2011;**6**(4). doi.org/10.1371/journal.pone.0018801.
4. Avigan, D., *Dendritic cells: development, function and potential use for cancer immunotherapy.* Blood Rev. 1999;**13**(1): 51-64.
5. Banchereau, J, et al., *Immune and clinical responses in patients with metastatic melanoma to CD34(+) progenitor-derived dendritic cell vaccine.* Cancer Res. 2001;**61**(17): 6451-6458.
6. Hsu, F.J., et al., *Vaccination of patients with B-cell lymphoma using autologous antigen-pulsed dendritic cells.* Nat Med. 1996;**2**(1):52-58. doi:10.1038/nm0196-52.
7. Melief, C.J, *Cancer immunotherapy by dendritic cells.* Immunity. 2008;**29**(3):372-383.
8. Motta,M.R.,etal.,*Generation of dendritic cells from CD14+ monocytes positively selected by immunomagnetic adsorption for multiple myeloma patients enrolled in a clinical trial of anti-idiotype vaccination.* Br J Haematol. 2003;**121**(2):240-250.
9. Osada, T., et al., *Dendritic cell-based immunotherapy.* Int Rev Immunol. 2006;**25**(5-6):377-413.

10. Schuler, G., B. Schuler-Thurner, and R.M. Steinman, *The use of dendritic cells in cancer immunotherapy.* Curr Opin Immunol. 2003;**15**(2):138-147.

11. Lenahan, C. and D. Avigan, *Dendritic cell defects in patients with cancer: mechanisms and significance.* Breast Cancer Res. 2006;**8**(1):101. doi:10.1186/bcr1375.

12. Pinzon-Charry, A., et al., *Numerical and functional defects of blood dendritic cells in early- and late-stage breast cancer.* Br J Cancer. 2007;**97**(9):1251-1259. doi:10.1038/sj.bjc.6604018.

13. Curiel, T.J.,et al., *Dendritic cell subsets differentially regulate angiogenesis in human ovarian cancer.* Cancer Res. 2004;**64**(16):5535-5538.

14. Fricke, I.and DI. Gabrilovich, *Dendritic cells and tumor microenvironment: a dangerous liaison.* Immunol Invest. 2006;**35**(3-4):459-483.

15. Gabrilovich, D., *Mechanisms and functional significance of tumour-induced dendritic-cell defects.* Nat Rev Immunol. 2004;**4**(12):941-952.

16. Bloy, N., et al., *Trial watch: Dendritic cell-based anticancer therapy.* Oncoimmunology. 2014;**3**(11): e963424.

17. Berntsen, A., et al., *Therapeutic dendritic cell vaccination of patients with metastatic renal cell carcinoma:a clinical phase 1/2 trial.* J Immunother. 2008;**31**(8): 771-780. doi:10.1097/CJI.0b013e3181833818.

18. Kavanagh, B., et al., *Vaccination of metastatic colorectal cancer patients with matured dendritic cells loaded with multiple major histocompatibility complex class I peptides.* J Immunother. 2007;**30**(7): 762-772.

19. Wierecky, J., et al., *Immunologic and clinical responses after vaccinations with peptide-pulsed dendritic cells in metastatic renal cancer patients.* Cancer Res. 2006;**66**(11): 5910-5918.

20. Sallusto, F. and A. Lanzavecchia, *Efficient presentation of soluble antigen by cultured human dendritic cells is maintained by granulocyte/macrophage colony-stimulating factor plus interleukin 4 and downregulated by tumor necrosis factor alpha.* J Exp Med. 1994;**179**(4): 1109-1118.

21. Dauer, M., et al., *Mature dendritic cells derived from human monocytes within 48 hours: a novel strategy for dendritic cell differentiation from blood precursors.* J Immunol. 2003;**170**(8):4069-4076.

22. Dauer, M., et al., *FastDC derived from human monocytes within 48 h effectively prime tumor antigen-specific cytotoxic T cells.* J ImmunolMethods. 2005;**302**(1-2):145-155.

23. Mailliard, R.B., *alpha-Type-1 Polarized Dendritic Cells: A Novel Immunization Tool with Optimized CTL-inducing Activity.* Cancer Research. 2004;**64**(17):5934-5937.

24. Jonuleit, H., et al., *Pro-inflammatory cytokines and prostaglandins induce maturation of potent immunostimulatory dendritic cells under fetal calf serum-free conditions.* Eur J Immunol. 1997;**27**(12):3135-3142.

25. Martinuzzi, E., et al., *acDCs enhance human antigen-specific T-cell responses.* Blood. 2011;**118**(8): 2128-2137. doi:10.1182/blood-2010-12-326231.

26. Goletti, D., et al., *Region of difference 1 antigen-specific CD4+ memory T cells correlate with a favorable outcome of tuberculosis.* J Infect Dis. 2006;**194**(7):984-992.

27. Moser, J.M., et al., *Optimization of a dendritic cell-based assay for the in vitro priming of naive human CD4+ T cells.* J Immunol Methods. 2010;**353**(1-2):8-19. doi:10.1016/j.jim.2009.11.006.

28. Dauer, M., et al., *FastDC derived from human monocytes within 48 h effectively prime tumor antigen-specific cytotoxic T cells.* Journal of Immunological Methods. 2005;**302**(1-2):145-155.

29. Sallusto, F. and A. Lanzavecchia, *Understanding dendritic cell and T-lymphocyte traffic through the analysis of chemokine receptor expression.* Immunol Rev. 2000;**177**:134-140.

30. Obermaier, B., et al., *Development of a new protocol for 2-day generation of mature dendritic cells from human monocytes.* Biol Proced Online. 2003;**5**:197-203. doi:10.1251/bpo62.

31. Nishioka, Y., et al., *Induction of systemic and therapeutic antitumor immunity using intratumoral injection of dendritic cells genetically modified to express interleukin 12.* Cancer Res. 1999;**59**(16): 4035-4041.

32. Gattinoni, L., et al., *A human memory T cell subset with stem cell-like properties.* Nat Med. 2011;**17**(10): 1290-1297. doi: 10.1038/nm.2446.

33. Goletti, D., et al., *Selected RD1 peptides for active tuberculosis diagnosis: comparison of a gamma interferon whole-blood enzyme-linked immunosorbent assay and an enzyme-linked immunospot assay.* Clin Diagn Lab Immunol. 2005;**12**(11): 1311-1316. doi:10.1128/CDLI.12.11.1311-1316.2005.

34. Vinuesa, C.G., et al.,*Follicular Helper T Cells.* Annu Rev Immunol. 2016;**34**:335-368. doi:10.1146/annurev-immunol-041015-055605.

35. Schmitt, N., et al., *Human dendritic cells induce the differentiation of interleukin-21-producing T follicular helper-like cells through interleukin-12.* Immunity. 2009;**31**(1):158-169. doi:10.1016/j.immuni.2009.04.016.

36. Jarnjak-Jankovic, S., et al., *A full scale comparative study of methods for generation of functional Dendritic cells for use as cancer vaccines.* BMC Cancer. 2007;**7**:119.

37. Ramadan, G., et al., *Generation of cytotoxic T cell responses directed to human leucocyte antigen Class I restricted epitopes from the Aspergillus f16 allergen.* Clin Exp Immunol. 2005;**140**(1): 81-91.

38. Dauer, M., et al., *Combined use of toll-like receptor agonists and prostaglandin E(2) in the FastDC model: rapid generation of human monocyte-derived dendritic cells capable of migration and IL-12p70 production.* J Immunol Methods. 2008;**337**(2):97-105. doi:10.1016/j.jim.2008.07.003.

39. Zhu, F., et al., *Stimulation by means of dendritic cells followed by Epstein-Barr virus-transformed B cells as antigen-presenting cells is more efficient than dendritic cells alone in inducing Aspergillus f16-specific cytotoxic T cell responses.* Clin Exp Immunol. 2008;**151**(2):284-296.

40. Guan, X., et al., *A novel, rapid strategy to form dendritomas from human dendritic cells and hepatocellular carcinoma cell line HCCLM3 cells using mature dendritic cells derived from human peripheral blood CD14+ monocytes within 48 hours of in vitro culture.* World J Gastroenterol. 2004;**10**(24):3564-3568.

41. Czerniecki, B.J., et al., *Targeting HER-2/neu in early breast cancer development using dendritic cells with staged interleukin-12 burst secretion.* Cancer Res. 2007;**67**(4):1842-1852.

42. Kruse, M., et al., *Inhibition of CD83 cell surface expression during dendritic cell maturation by interference with nuclear export of CD83 mRNA.* J Exp Med. 2000;**191**(9):1581-1590.

43. Kuwano, Y., et al., *CD83 influences cell-surface MHC class II expression on B cells and other antigen-presenting cells.* Int Immunol. 2007;**19**(8):977-992.

44. Fields, P.E., et al., *B7.1 is a quantitatively stronger costimulus than B7.2 in the activation of naive CD8+ TCR-transgenic T cells.* J Immunol. 1998;**161**(10):5268-5275.

45. Gajewski, T.F., *B7-1 but not B7-2 efficiently costimulates CD8+ T lymphocytes in the P815 tumor system in vitro.* J Immunol. 1996;**156**(2): 465-472.

46. Pejawar-Gaddy, S. and M.A. Alexander-Miller, *Ligation of CD80 is critical for high-level CD25 expression on CD8+ T lymphocytes.* J Immunol. 2006;**177**(7): 4495-4502. doi:10.4049/jimmunol.177.7.4495.

47. Klechevsky, E., et al., *Understanding human myeloid dendritic cell subsets for the rational design of novel vaccines.* Hum Immunol. 2009;**70**(5):281-288. doi:10.1016/j.humimm.2009.02.004.

48. Klechevsky, E., et al., *Functional specializations of human epidermal Langerhans cells and CD14+ dermal dendritic cells.* Immunity. 2008;**29**(3):497-510. doi:10.1016/j.immuni.2008.07.013.

49. Geijtenbeek, T.B., et al., *Self- and nonself-recognition by C-type lectins on dendritic cells.* Annu Rev Immunol. 2004;**22**:33-54.

50. Kerrigan, A.M. and G.D. Brown, *C-type lectins and phagocytosis.* Immunobiology. 2009;214(7):562-575. doi: 10.1016/j.imbio.2008.11.003.

51. den Dunnen, J., S.I. Gringhuis, and T.B. Geijtenbeek, *Innate signaling by the C-type lectin DC-SIGN dictates immune responses.* Cancer Immunol Immunother. 2009;**58**(7):1149-57.doi:10.1007/s00262-008-0615-1.

52. Wu, L. and V.N. KewalRamani, *Dendritic-cell interactions with HIV: infection and viral dissemination.* Nat Rev Immunol. 2006;**6**(11):859-868.

53. DeBenedette, M.A., et al., *Priming of a novel subset of CD28+ rapidly expanding high-avidity effector memory CTL by post maturation electroporation-CD40L dendritic cells is IL-12 dependent.* J Immunol. 2008;**181**(8):5296-5305.

54. Park, M.H., et al., *Alpha-Type 1 Polarized Dendritic Cells Loaded with Apoptotic Allogeneic Breast Cancer Cells Can Induce Potent Cytotoxic T Lymphocytes against Breast Cancer.* Cancer Res Treat. 2011;**43**(1): 56-66. doi:10.4143/crt.2011.43.1.56.

55. van der Pouw Kraan, T.C., et al., *Prostaglandin-E2 is a potent inhibitor of human interleukin 12 production.* J Exp Med. 1995;**181**(2):775-779.

56. Dauer, M., et al., *IFN-alpha promotes definitive maturation of dendritic cells generated by short-term culture of monocytes with GM-CSF and IL-4.* J Leukoc Biol. 2006;**80**(2):278-286.

57. Petruccioli, E., et al., *IFNgamma/TNFalpha specific-cells and effector memory phenotype associate with active tuberculosis.* J Infect. 2013;**66**(6): 475-486. doi:10.1016/j.jinf.2013.02.004.

58. Goletti, D., et al., *Accuracy of an immune diagnostic assay based on RD1 selected epitopes for active tuberculosis in a clinical setting: a pilot study.* Clin Microbiol Infect. 2006;**12**(6):544-550.DOI: doi.org/10.1111/j.1469-0691.2006.01391.x.

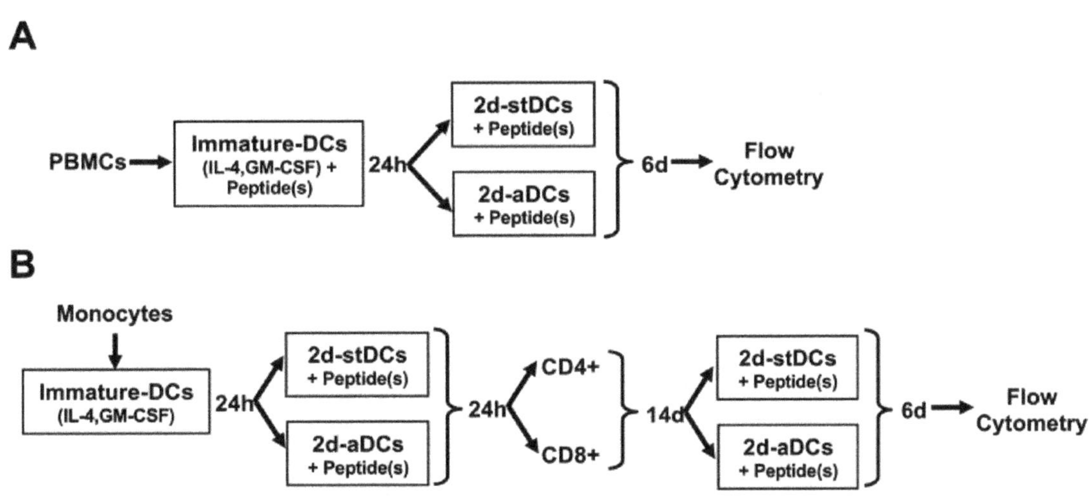

Figure S1: Flowchart of the methodology used for cell culture with 2d-aDCs and 2d-stDCs. A. Methodology for 2d-aDC and 2d-stDC induction in total PBMCs (based on the method of Martinuzzi et al.[25]). For flow cytometry, 10^6 PBMCs were stimulatedduring 24 h with IL-4 and GM-CSF (750 IU/mL and 1000 IU/mL, respectively) and with pro inflammatory cytokines and TLR ligands for the corresponding DCs with or without peptide(s) for six days. B. T cell stimulation scheme based on the methodology of Moser et al.[27]. Briefly 2d-aDCs and 2d-stDCs were derived from monocytes and cultured with purified CD4+ or CD8+ T cells for 14 days and re-stimulated for 6 days with peptide (5 μM)-pulsed 2d-aDCs or 2d-stDCs. Flow cytometry was performed using T cells for immuno-phenotyping and supernatants for cytokine quantification by CBA.

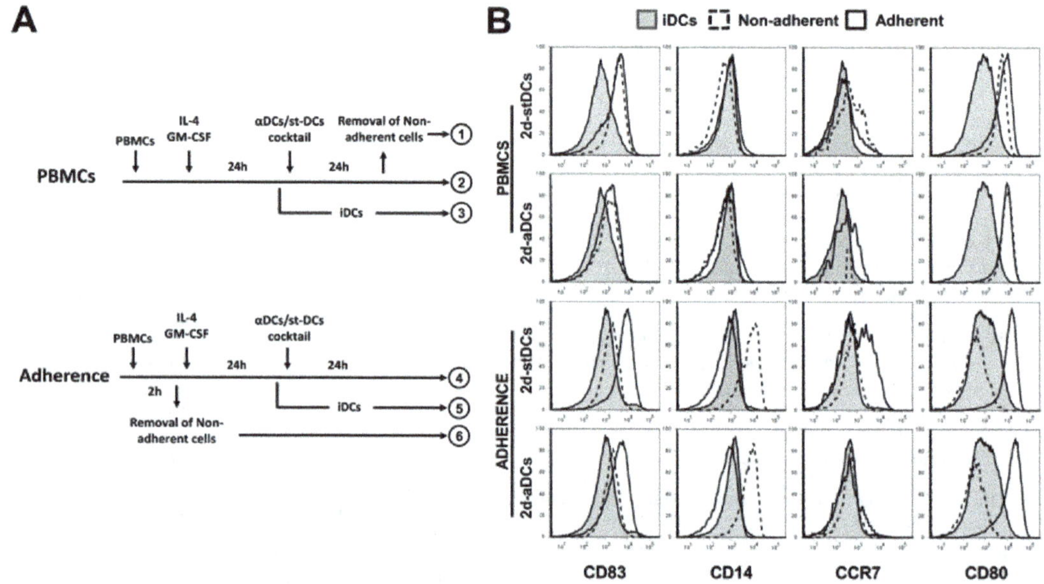

Figure S2: Maturation profile of DCs obtained through plastic adhesion or direct maturation in PBMCs. A. In vitro DCs generation scheme. The top scheme (PBMCs) DCs were induced in total cells with either aDC or stDC cytokine cocktails, (1) non-adherent cells exposed to maturation cocktail, (2) mature DCs adhered to plastic and (3) immature DCs. The bottom scheme represents the adherent method to induce mature DCs, (4) mature 2d-aDCs or 2d-stDCs derived from adherent cells, (5) iDCs from adherent cells (exposed only to IL-4 and GM-CSF) and (6) non-adherent cells without cytokine exposure. B. Expression profile (CD83, CD14, CCR7, and CD80) in iDCs (tinted histogram), adherent (empty histograms) and non-adherent cells (dashed histograms) derived from total PBMCs (top row) as described in panel A or adherence method (bottom rows) with aDCs or stDCs cytokine cocktail. All histograms were gated on single cells (FSC-A vs. FSC-H) and DCs based on FSC-A vs. SSC-A (CD11c-positive cells).

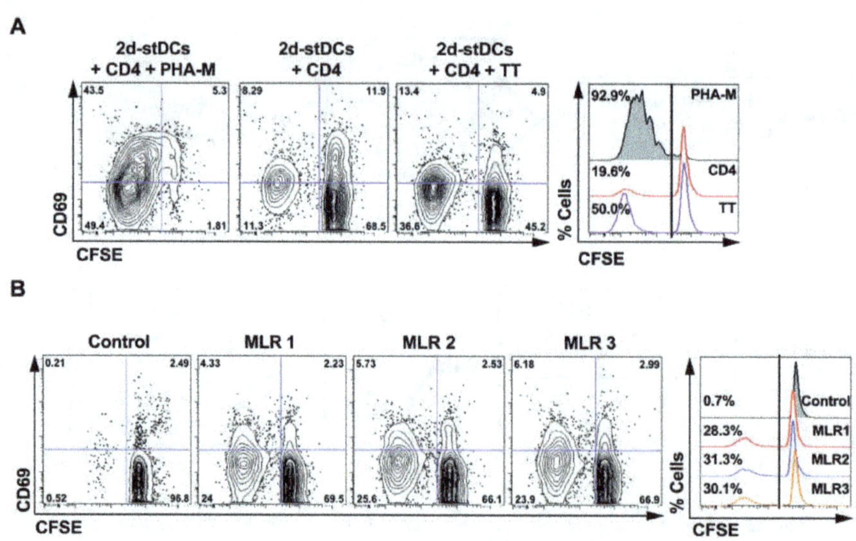

Figure S3: 2d-stDCs properly induce CD4+ T cell activation and proliferation. A. CFSE labeled CD4+ T cells were stimulated with unpulsed or pulsed 2d-stDCs with tetanus toxoid (TT) or PHA-M during 5 days. Contour plots of CFSE vs CD69 in CD4+ T cells (left panels) and overlapping histograms of CFSE dilution in CD4+ T cells (right panel). Proliferating and activated CD4+ T cells (CFSE- CD69+) in presence of PHA-M (positive control 43.5%), unpulsed (negative control 8.29%) and TT (13.4%) pulsed 2d-stDCs. Histograms (right panel) show dilution of CFSE in CD4+ T cells with PHA-M (92.9%), unpulsed (19.6%) and TT pulsed 2d-stDCs (50%). B. Activation and proliferation analysis of CD4+ T cells in a Mixed Leukocyte Reaction (MLR) with 3 different healthy donors in co-culture with 2d-stDCs, contour plots of CFSE vs CD69 in CD4+ T cells (left panels) and overlapping histograms of CFSE dilution in CD4+ T cells. Proliferating and activated CD4+ T cells (CFSE- CD69+) in presence of autologous CD4+ T cells (0.21%) or 3 different donors (4.33, 5.73 and 6.18%) and proliferation with autologous CD4+ T cells (0.7%) or with 3 different donors (28.3, 31.3 and 30.1%).

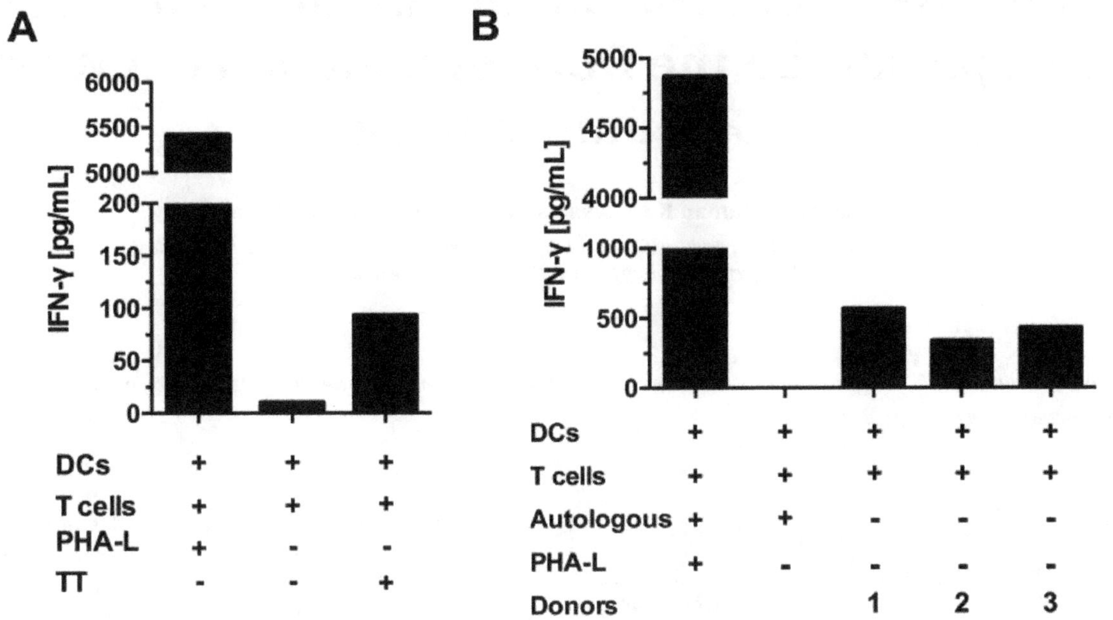

Figure S4: Production of IFN-γ after stimulation of CD4+ T cells with 2d-stDCs. IFN-γ production was determined by ELISA in culture supernatants after 40 hours of in vitro cell culture. A) The culture of CD4+ T cells stimulated with DCs pulsed with 2d-stDCs PHA-L (5μM), alone or with ten μg/mL of TT. B) Mixed Leukocyte Reaction (MLR) with 2d-stDCs in coculture with CD4+ T cells from 3 different healthy individuals, as positive control autologous CD4+ T cells, stimulated with PHA-L (5μM) or without PHA-L as a negative control.

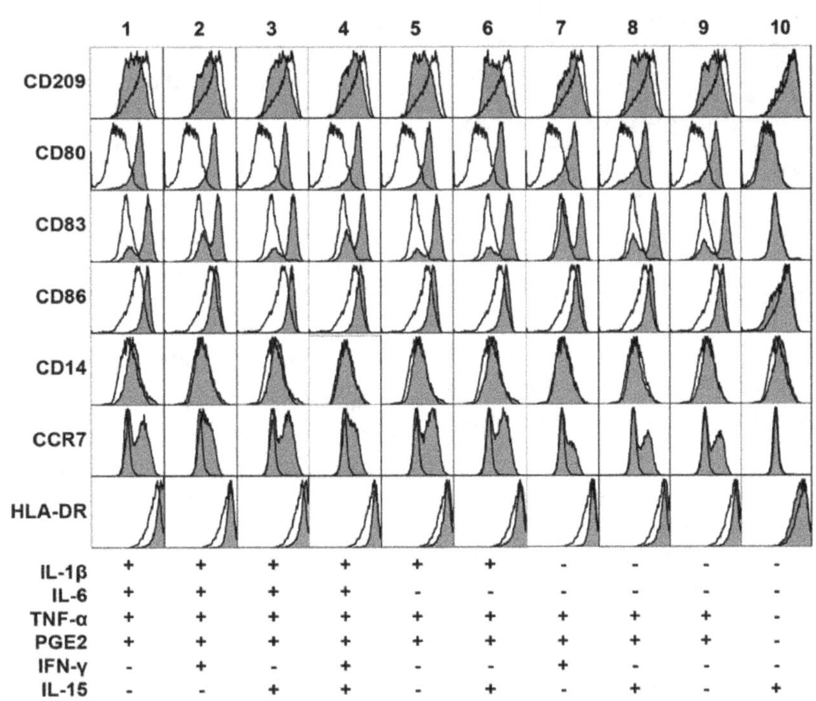

Figure S5: Different cytokine combinations can alter the phenotype of mature DCs. Ten different cytokine combinations were analyzed using Il-1β, IL-6, TNF-α, PGE-2, IFN-γ and IL-15. Phenotype analyses were done using flow cytometry to determine the expression of CD209, CD80, CD83, CD86, CD14, CCR7 and HLA-DR, after two days of maturation culture. Empty histograms represent immature DCs (IL-4 and GM-CSF) and filled histograms represent mature DCs with corresponding cytokine combination (1-10).

14

A Genotyping and Phenotyping Study Concerning the Possible Effects of Some Inflammatory Cytokine Gene Polymorphisms on the Development of Coal Workers' Pneumoconiosis

Ilker Ates[1]*, Asuman Karakaya[1], Sinan H. Suzen[1], Berran Yucesoy[2]

[1]*Department of Toxicology, Faculty of Pharmacy, Ankara University, 06100 Tandogan, Ankara, Turkey*

[2]*Health Effects Laboratory Division, CDC/NIOSH, Morgantown, WV, USA*

***Corresponding author:** Ilker ATES, Department of Toxicology, Faculty of Pharmacy, Ankara University, 06100 Tandogan, Ankara, Turkey, E-mail: ilkerates976@gmail.com*

Abstract

Cytokines are important for playing a major role in several inflammatory reactions resulting in development of several diseases as well as Coal Worker's Pneumoconiosis (CWP). Coal dust exposure stimulates inflammatory response leading to enhanced cytokine release from monocytes such as TNF-alpha and IL1. These released cytokines are the key points in the pathogenesis of CWP. The present study aimed to seek the cytokine gene profiles of Turkish coal workers by genotyping and phenotyping analysis of important CWP-related proinflammatory cytokines; TNF-alpha, IL1-alpha and IL1-beta. According to the genotyping results, TNFA -238 gene polymorphism was appeared to be a risk factor in development of CWP (OR=3.79) and regarding to the phenotyping analysis, both TNF-alpha and IL1 cytokine releases from the monocytes in CWP patients were enhanced significantly compared to the healthy workers. Therewithal, LPS and coal dust stimulated TNF-alpha release were higher significantly in allele 2 carriers than allele 1 carriers in both of the groups. These data propose that coal dust-induced TNF-alpha release from monocytes may be a valuable biomarker of CWP.

Introduction

Similarly to several developing countries, coal mining is still important in Turkey contrary to the shutdown of many coal mines in developed countries. Due to the official data of the Turkey Hard Coal Enterprise, in 2015 there were 17,014 registered coal miners in whole country (Turkey Hard Coal Enterprise, 2016).

Coal worker's pneumoconiosis (CWP) is an occupational lung disease signalized by fibrotic nodular lesions following the inhalation of coal dust. CWP is defined as the accumulation of the coal dust in lungs and the tissue's reaction to its presence (Morgan and Seaton, 1975). Cytokines are crucial mediators of toxic and pathogenic effects seen in mineral dust exposure of humans and playing a major role in several important biological processes particularly in inflammation and immune response (Schins and Borm, 1999). TNF-alpha and IL1 cytokines are called as proinflammatory cytokines and their key roles in CWP and silicosis development are endorsed by plenty of experimental researches (Zhai et al., 1998; Corbett et al., 2002; Nadif et al., 2003; Fan et al., 2010; Lee et al., 2010; Li et al., 2011, Wang et al., 2012; Kurniawidjaja 2014; Jiang PR et al., 2015). Latest studies have shown an association of TNF alpha gene polymorphisms with sensitivity to lung diseases including asthma, chronic bronchitis and fibrosing alveolitis (Huang et al., 1997; Whyte et al., 2000, Witte et al., 2002; Mukhopadhyay S et al., 2006; Yang et al., 2015; Makamure et al., 2016).

Because cytokines are key regulators of homeostatic mechanisms, possible variations in their levels or their structures may be related with the disease process (Ollier, 2004). The polymorphism commonly originates as a Single Nucleotide Polymorphism (SNP) in the network of cytokines and a vast number of SNPs have been recognized in regulatory regions of genes acting on their synthesis or degradation rates. Epidemiological researches have indicated that both pro- and anti-inflammatory cytokine SNPs are related with immune-mediated diseases (Zhai et al., 1998; Francis et al., 1999; Franceschi et al., 2001; Yucesoy et al., 2001; Kim et al., 2002; Ohtsuka et al., 2002; Yucesoy et al., 2005; Jonth et al., 2007; Fan et al., 2010; Li et al., 2011; Wang et al., 2012; Makamure et al., 2016).

Subsequent to coal dust exposure, lung tissue responds by starting three types of phenomena: an accumulation and activation of inflammatory cells in lower respiratory tract, fibroblast proliferation and an increased synthesis of extracellular matrix components (Vanhee et al., 1995). Macrophage-derived cytokines are the key points in coal dust inflammation. Presence of persistent stimulus and chronic release of cytokines may conclude in autoimmune and inflammatory diseases as well as CWP or silicosis. Recent records have displayed that following the coal dust exposure of alveolar macrophages, a significant release of TNF-alpha and IL1 from blood monocytes occurs (Lasalle et al., 1990; Jorna et al., 1994; Zhai et al., 1998; Kim et al., 1999; Borm

***Corresponding author email:** ilkerates976@gmail.com*

and Schins, 2001; Wang et al., 2005; Zhang et al., 2015).

Several researches have pointed out that TNF-alpha release from blood monocytes differentiates coal workers with respiratory and pulmonary effects of coal dust (Jorna et al., 1994; Zhai et al., 1998). According to the data of these studies, release of monocytes TNF-alpha is a valid means to CWP prognosis beneath history and X-rays when carefully applied. Comprehension of the underlying mechanisms of the CWP aetiology was fundamental in offering numerous biomarkers evaluated to appraise the effects following exposure to coal dust. Human studies exhibited some of these offered biomarkers for CWP including TNF-alpha released from monocytes and TNF-alpha polymorphism (blood cellular DNA) and also they were defined as biomarkers of sensitivity (Gulumian et al., 2006).

In this study, we aimed to follow two assumptions: First, CWP-associated pro-inflammatory cytokines (TNF-alpha, IL1-alpha and IL1-beta) gene polymorphisms related with CWP in a group of Turkish coal workers. Second, these gene polymorphisms have an important role on the release of these cytokines from blood monocytes due to the coal dust exposure in homozygous and minor variant allele carriers. To check over these assumptions; we gathered coal workers having a heavy dust exposure background and searched the stated cytokines releases from blood monocytes by genotyping and phenotyping analyses.

Materials and Methods

All the individuals were gathered from Kozlu/Zonguldak region of Turkey where coal is densely mined. For genotyping analyses, subjects were categorized into two groups: 67 retired coal worker with CWP as patients and 92 workers without any CWP evidence as controls. All active coal workers were chosen from the mining part of the mine (except the coal face or stope) having a mean coal dust exposure (0.07-1.25 mg/m^3 respirable coal dust fraction). Mean coal chemical properties of Kozlu mines were as follows: 55% carbon, 28% volatile substances, 10% ash, 7% moisture and also about 47% of the ash came out of crystalline silica. After defining the genotype profiles of the all individuals, their profiles were inspected cautiously for studied cytokine gene polymorphisms. Patients frequently (at least 3 of the 5) bearing 1/2+2/2 or 2/2 alleles (n=19) and healthy workers frequently (at least 3 of the 5) bearing homozygous 1/1 allele (n=19) a total of 38 subjects were chosen among the study groups to consider the possible effects of gene polymorphisms on the release of the studied cytokines from blood monocytes. All of the subjects belonging to study groups replied a brief questionnaire providing information about past medical and occupational background, smoking, lifestyle and drug usage. Informed consent was achieved from all individuals in consonant to the Helsinki declaration (1983) of the World Medical Association. We had approval of the Ankara University Faculty of Medicine Ethic Committee for this scientific research and posses informed consents of all persons involved in the study.

DNA preparation and genotyping

10 ml whole bloods of all studied individuals were collected by venipuncture in tubes containing heparin. Afterwards, DNA isolation were performed by Promega Wizard Genomic DNA Purification kit due to the manufacturer's instructions and all the DNA samples were stored at -20^{0}C till analyses. Genotyping of TNF-alpha -308G/A, TNF-alpha -238G/A, IL1-alpha +4845C/T and IL1-beta +3953C/T gene polymorphisms were carried out using a Polymerase Chain Reaction-Restriction Fragment Length Polymorphism (PCR-RFLP) method due to the formerly mentioned protocols with minor modifications from Ates et al., (2008). PCR reaction products were departed by gel electrophoresis on a 10% polyacrilamide-TBE gel at 200 V for 40 min and then viewed via an UV illumination following staining with ethidium bromide solution.

Monocyte isolation and measurement of cytokine release

The isolation of monocytes from the peripheral blood was performed due to the technique of Strober (1994). The cytokine release levels from adherent monocytes supernatants (5 x 10^6) were calculated by Enzyme-linked Immunosorbent Assay (ELISA) without a stimulant (spontaneous) or afterwards 18 h of incubation with coal dust (5 mg/ml) and LPS (1000 ng/ml) at 37^{0}C and 5% CO_2.

Statistical analysis

For whole of the statistical analyses, we exerted the version no 15 of the SPPS programme (SPSS Inc., Chicago, IL, USA). Possible significant associations between polymorphic variants and disease situation were commentated by χ^2 test and Fischer exact test where suitable. The χ^2 test was performed to check against allele and genotype frequencies between controls and cases. Odds ratios (OR) and incorporated 95% Confidence Intervals (CI) were evaluated by logistic regression. A p-value <0.05 was regarded as significant.

Results

(Table 1) demonstrates the demographic output of all the studied individuals. No interactions were obtained between the CWP and the parameters of age and smoking following the multiple linear regression method. On the other hand, work period was found to be associated with the development of the disease.

(Table 2) indicates the distribution of TNF-alpha and IL1 genotypes. Whole SNPs in both controls and cases were in the Hardy-Weinberg equilibrium. Entire risk of disease was significantly higher only with TNF-alpha (-238) variant (OR= 3.79, 95% CI: 1.37-10.46). In regards of TNF-alpha (-308) variant, an enhancement risk for CWP was observed however it was not significant (OR= 1.37: 95% CI, 0.73-2.58). No significant relationship was seen between neither the IL1-alpha nor IL1-beta variants with CWP.

(Table 3) represents the allelic distribution of the group of CWP patients matched with controls. The development of the disease was ascended in allele 2 carriers than the ones hauling allele 1. The basal, the LPS and the coal dust-caused cytokine release were seized up for appraising the *in vitro* cytokine release from the blood lymphocytes. As obviously seen from the data given in Table 4, there was a significant association between the patients hauling allele 2 compared to those of controls ($p< 0.05$).

In (Table 4), TNF-alpha and IL1 cytokine releases from blood monocytes of controls and CWP patients were demonstrated. The release order of studied cytokines was as spontaneous <coal dust-induced <LPS-induced. The release of these cytokines in CWP patients were significantly enhanced in contrast to the controls. Conversely, the coal dust-induced release of TNF-alpha in allele 2 carriers was significantly increased compared to those hauling allele 1 in both study groups. In despite of an increase of the IL1 cytokines release in allele 2 carriers in all subjects, no significant distinction was observed. Smoking was also considered as a confounding factor but no relationship was found of the possible effects of smoking on the release of cytokines.

Table 1: The demographic data of the subjects

	Control group	CWP Patients
No. of subjects	92	67
No. of subjects with simple pneumoconiosis (SP)	-	41
No. of subjects with progressive massive fibrosis (PMF)	-	26
Age (± S.E)	48.75 ± 5.21	58.60 ± 4.41
Work period (year) (± S.E)	13.21 ± 6.93[a]	22.39 ± 6.05
Smoking status (%)		
Yes	49 (%53.3)	-
Ex-smoker	29 (%31.5)	46 (%68.7)
No	14 (%15.2)	21 (%31.3)

[a] *P<0.05* compared with CWP patients

Table 2: The genotypic distribution of TNF-alpha and IL1.

Genotypes	TNFA (-308)		TNFA (-238)	
	Control n (%)	CWP patients n (%)	Control n (%)	CWP patients n (%)
1/1 alleles	47 (51.1)	29 (43.3)	86 (93.5)	53 (79.1)
1/2 + 2/2 alleles	45 (48.9)	38 (56.7)	6 (6.5)	14 (20.9)
OR* (CI)	1.00	1.37 (0.73-2.58)	1.00	3.79 (1.37-10.46)[a]
	IL1A (+4845)		IL1B (+3953)	
1/1 alleles	62 (67.4)	44 (65.7)	18 (19.6)	14 (20.9)
1/2 + 2/2 alleles	30 (32.6)	23 (34.3)	74 (80.4)	53 (79.1)
OR* (CI)	1.00	1.08 (0.55-2.10)	1.00	0.92 (0.42-2.01)

[a] *P<0,001* compared with controls, * Odds ratios (OR) and associated 95% confidence intervals (CI)

Table 3: Allelic distribution of patient group matching with controls

	Control group	Allele Freq. (%)	CWP patients	Allele Freq. (%)
Allele 1 carriers*	11	71.1	7	45.9
Allele 2 carriers*	8	28.9	12	54.1[a]
Total	19	100	19	100

[a] significantly different from the allele 2 carriers in control group, $p < 0.05$
* individuals bearing 1/1 or 1/2 were grouped as allele 1 carriers, patients bearing 1/2 or 2/2 were grouped as allele 2 carriers

Table 4: *In vitro* release of TNF-alpha from monocytes induced by LPS and coal dust among 19 healthy workers and 19 CWP patients

Monocyte TNF-alpha release (ng/ml)*			
	Unstimulated	Coal dust (5 mg/ml)	LPS (1.000 ng/ml)
Healthy workers			
Allele 1 carriers (n=11)	1.23 (0.11)	3.87 (1.55)	4.75 (1.89)
Allele 2 carriers (n=8)	1.26 (0.14)	5.52 (1.97)[a]	8.04 (2.26)[b]
CWP patients			
Allele 1 carriers (n=7)	4.51 (1.23)	10.29 (2.24)	13.75 (4.77)
Allele 2 carriers (n=12)	4.97 (1.44)	13.52 (3.85)[a]	21.84 (5.11)[b]
Monocyte IL1-alpha release (ng/ml)*			
Healthy workers			
Allele 1 carriers (n=11)	0.41 (0.19)	1.17 (0.57)	2.09 (1.42)
Allele 2 carriers (n=8)	0.59 (0.23)	1.34 (0.63)	2.51 (1.71)
CWP patients			
Allele 1 carriers (n=7)	2.51 (1.12)	5.62 (1.79)	7.81 (2.87)
Allele 2 carriers (n=12)	4.82 (2.27)	7.05 (3.06)	9.86 (3.88)
Monocyte IL1-alpha release (ng/ml)*			
Healthy workers			
Allele 1 carriers (n=11)	0.55 (0.23)	1.59 (0.74)	2.01 (0.94)
Allele 2 carriers (n=8)	0.89 (0.51)	1.80 (0.97)	2.73 (1.15)
CWP patients			
Allele 1 carriers (n=7)	3.25 (1.23)	5.59 (2.01)	6.72 (2.65)
Allele 2 carriers (n=12)	Allele 2 carriers (n=12)	7.49 (2.63)	8.73 (3.23)

*Values are mean and standard errors (in parenthesis)
a significantly differenat from the allele 1 carriers. $p <0.05$ and b $p <0.001$ (Student t-test)

Discussion

In this research, we gathered a group of coal workers with CWP who had a common history to heavy dust exposure. According to the data, we sighted significantly higher allele frequencies of -238A in patients contrast to controls. Moreover, it seems that TNF-alpha is an important parameter in development of CWP more than IL1. TNF-alpha -238A variant should be kept in mind as a risk parameter in development of the disease and there is an enhancing effect of TNF-alpha cytokine gene polymorphism on release of it from the blood monocytes.

Even though its pathophysiological mechanisms have not been fully identified, coal dust has been known as the most important factor for the development of CWP. Chronic diseases induced by inflammation including CWP are multi-factorial ones directly affected by both genetic and environmental parameters. Polymorphisms such as genetic factors can switch the size or severity of the disease. In regards to this, SNPs are important affecting the amount of inflammatory cytokines which have been related with too many different chronic inflammatory and autoimmune diseases (Zhai et al., 1998; Franceschi et al., 2001; Yucesoy et al., 2001; Ates et al., 2008; Wang et al., 2012; Yang et al., 2015; Makamure et al., 2016).

TNF-alpha is an major cytokine relative to CWP's inflammatory and fibrotic processes (Schins and Borm, 1999). Variations at -238 and -308 have been suggested to be responsible for the modification of TNF-alpha production (Fan et al., 2010). A relationship between -308 variant and CWP in Europeans (Zhai et al., 1998), in South Africans (Corbett et al., 2002), in Japans (Wang et al., 2005) and in Chinese (Fan et al., 2010) has been observed. In these researches, the possible effect of TNF-alpha -238 variant was also studied and due do the data, only Corbett et al., (2002) and Ates et al., (2008) have obtained distinct relationship between this variant with the development of CWP. In respect to our research, according to the genotyping data, we found -238 variant to be a powerful risk factor for CWP. On the contrary, we observed an enhancement risk factor for -308 variant in patients compared to controls but it was not significant.

IL-1 is one of the most studied proinflammatory cytokines as similar as TNF-alpha. There are many researches related with the possible effects of IL1 gene polymorphisms on several pulmonary and autoimmune diseases except CWP. In our study, we did not observe any significant relationship with IL1-alpha variants and the development of CWP. For each polymorphism, the odds ratio of disease and associated 95% confidence intervals for individuals carrying alleles 2 were reported for the CWP patients, using logistic regression models with adjustment for age, years of exposure and smoking.

Latest *in vitro* release researches have indicated that pro-inflammatory cytokine levels of TNF-alpha and IL1 were triggered in CWP (Schins and Borm, 1995; Zhai et al., 1998; Kim et al., 1999). According to our data, we found an enhanced release of TNF-alpha from blood monocytes after coal dust and LPS stimulation in contrast to spontaneous release in both controls and patients. The level of released TNF-alpha was significantly increased in allele 2 carriers than allele 1 hauliers. These results are alike to the data of Zhai and co-workers (1998) and Kim and co-workers (2002). On the other hand, even though there was a triggering effect in the release of IL1-alpha and IL1-beta, no significant differences were found between both allele1 and 2 hauliers. It was obviously observed that in patients, the levels of cytokine release from monocytes were about 3-4 fold higher than controls. These findings indicate an association between coal dust exposure and cytokine releases in the development of CWP.

In this study, smoking did not point out any significant contribution to the development of CWP.

Due to the limited sample size of the studied groups, we need to comment these findings cautiously as precursors. Further researches performed with sufficient "n" number are needed to verify these results.

In conclusion, due to our genotyping findings, TNFA –238 gene polymorphism was appeared to be a risk factor in development of CWP (OR=3.79) and regarding to the phenotyping analysis, both TNF-alpha and IL1 cytokine releases from the monocytes in CWP patients were enhanced significantly compared to the healthy workers. Therewithal, LPS and coal dust stimulated TNF-alpha release were higher significantly in allele 2 carriers than allele 1 carriers in both of the groups. These data propose that coal dust-induced TNF-alpha release from monocytes may be a valuable biomarker of CWP. Further intended comprehensive researches with a vast number of individuals should be performed to illuminate the underlying mechanisms and possible effects of these cytokine gene polymorphisms.

Acknowledgements

This study was supported by Research Fund of Ankara University

References

1. Ates I, Suzen HS, Yucesoy B, Tekin, IO, Karakaya A. Association of cytokine gene polymorphisms in CWP and its severity in Turkish coal workers. Am J Ind Med. 2008;51(10):741-747.
2. Borm PJA, Schins RPF: Genotype and phenotype in susceptibility to coal workers' pneumoconiosis. The use of cytokines in perspective. Eur Respir J. 2001;18:127-133.
3. Corbett EL, Mozzato-Chamay N, Butterworth AE, De Cock KM, Williams BG, Churchyard GJ, et al. Polymorphisms in the tumor necrosis factor-α gene promoter may predispose to severe silicosis in black South African miners. Am J Respir Crit Care Med. 2002;165(5):690-693.
4. Fan HM, Wang Z, Feng FM, Zhang KL, Yuan JX, Sui H, et al. Association of TNF-alpha-238 G/A and 308 G/A gene polymorphism with pulmonary tuberculosis among patients with coal workers' pneumoconiosis. Biomed Environ Sci 2010; 23(2):137-145.
5. Franceschi C, Valensin S, Lescai F, Olivieri F, Licastro F, Grimaldi LM, et al. Neuroinflammation and the genetics of Alzheimer's disease: the search for a pro-inflammatory phenotype. Aging Clin Exp Res. 2001;13(3):163-170.
6. Francis SE, Camp NJ, Dewberry RM, Gunn J, Syrris P, Carter ND, et al. Interleukin-1 receptor antagonist gene polymorphism and coronary artery disease. Circulation. 1999; 99(7):861-866.
7. Gulumian M, Borm PJA, Vallyathan V, Castranova V, Donaldson K, Nelson G, et al. Mechanistically identified suitable biomarkers of exposure, effect, and susceptibility for silicosis and Coal Worker's Pneumoconiosis: A comprehensive review. J Toxicol Environ Health B Crit Rev. 2006; 9(5):357-395.

8. Huang SL, Su CH, Chang SC. Tumor necrosis factor-α gene polymorphism in chronic bronchitis. Am J Respir Crit Care Med. 1997;156(5):1436-1439.

9. Jiang PR, Cao Z, Qiu ZL, Pan JW, Zhang N, Wu YF. Plasma levels of TNF-α and MMP-9 in patients with silicosis. Eur Rev Med Pharmacol Sci. 2015;19(9):1716-1720.

10. Jonth AC, Silveira L, Fingerlin TE, Sato H, Luby JC, Welsh KI, et al. TGF-β1 variants in chronic beryllium disease and sarcoidosis. J Immunol. 2007;179(6):4255-4262. doi:10.4049/jimmunol.179.6.4255.

11. Jorna THJM, Schins RPF, Lenaerts L, Derhaag TJJM, Wouters EFM, Borm PJA. Airflow obstruction and monocyte TNF release in coal workers. Exp Lung Res. 1994;20(5):421-431.

12. Kim KA, Lim Y, Kim JH, Kim EK, Chang HS, Park YM, et al. Potential biomarker of coal workers' pneumoconiosis. Toxicol Lett. 1999; 108:297-302.

13. Kim KA, Cho YY, Cho JS, Yang KH, Lee WK, Lee KH, Kim YS, Lim Y. Tumor necrosis factor-α gene promoter polymorphism in coal workers' pneumoconiosis. Mol Cell Biol. 2002;234/235(1):205-209.

14. Kurniawidjaja LM. Silicosis and its progress influenced by genetic variation on TNF-alpha locus-308, TNF-alpha and IL-10 cytokine on cement factory workers in Indonesia. Pak J Biol Sci. 2014;17:419-423.

15. Lassalle P, Gosset P, Arets C, Fournier E, Lafitte JJ, Degreef JM, et al. Abnormal secretion of interleukin-1 and tumor necrosis factor alpha by alveolar macrophages in coal workers' pneumoconiosis: comparison between simple pneumoconiosis and progressive massive fibrosis. Exp Lung Res. 1990;16(1):73-80.

16. Lee JS, Shin JH, Lee JO, Lee KM, Kim JH, Choi BS. Serum levels of interleukin-8 and tumor necrosis factor-alpha in coal workers' pneumoconiosis: One-year follow-up study. Saf Health Work. 2010;1(1):69-79.

17. Li J, Hao CF, Yao W, Yun YX, Gao B, Wang YX. A meta analysis of susceptibility to pneumoconiosis and polymorphism of tumor necrosis factor-α 308 and 238 locus. Zhonghua Yu Fang Yi Xue Za Zhi. 2011;45(6):547-552.

18. Makamure MT, Reddy P, Chutturgoon A, Naidoo RN, Menta G, Batterman S, et al. Tumour necrosis factor α polymorphism (TNF-308α G/A) in association with asthma related phenotypes and air pollutants among children in KwaZulu-Natal. Asian Pac J Allergy Immunol. 2016;34:217-222.

19. Morgan WK, Seaton A. Occupational lung diseases. WB Saunders Co. 1975;149-210.

20. Mukhopadhyay S, Hoidal JR, Mukherjee TK. Role of TNF alpha in pulmonary pathophysiology. Respir Res. 2006;11:7-125.

21. Nadif R, Jedlicka A, Mintz M, Bertrand JP, Kleeberger S, Kauffmann F. Effect of TNF and LTA polymorphisms on biological markers of response to oxidative stimuli in coal miners: a model of gene-environment interaction. J Med Genet. 2003;40:96-103. doi:10.1136/jmg.40.2.96.

22. Ohtsuka T, Yamakage A, Yamazaki S. The polymorphism of transforming growth factor-β1 gene in Japanese patients with systemic sclerosis. British J Dermatol. 2002;147(3):458-463.

23. Ollier WER. Cytokine genes and disease susceptibility. Cytokine. 2004;28:174-178.

24. Schins RPF, Borm PJA. Plasma levels of soluble tumour necrosis factor receptors are increased in coal miners with pneumoconiosis. Eur Respir J. 1995;8:1658-1663.

25. Schins RPF, Borm PJA. Mechanisms and mediators in coal dust induced toxicity: a review. Ann Occup Hyg. 1999;43(1):7-33. doi:10.1093/annhyg/43.1.7.

26. Strober W. Depletion of monocytes/macrophages from mononuclear cells using adherence method. In: Colligan JE, Kruisbeck M, Margulies DH, Shevach EM, Strober, W, Coico R, editors. Current Protocols in Immunology. USA: John Wiley &Sons Inc.; USA 1994. Section:7.1.3.

27. Turkey Herd Coal Enterprise. Turkey hard Coal Enterprise Annual Activity Report 2015. http://www.tki.gov.tr.

28. Vanhee D, Gosset P, Boitelle A, Wallaert B, Tonnel AB. Cytokines and cytokine network in silicosis and coal workers' pneumoconiosis. Eur Respir J. 1995;8:834-842.

29. Wang XT, Ohtsuka Y, Kimura K, Muroi M, Ishida T, Saito J, Munakata M. Antithetical effect of tumor necrosis factor-α gene polymorphism on coal workers' pneumoconiosis (CWP). Am J Ind Med. 2005;48:24-29.

30. Wang YW, Lan JY, Yang LY, Wang De J, Kuang J. TNF-α and IL-1RA polymorphisms and silicosis susceptibility in Chinese workers exposed to silica particles: a-case control study. Biomed Environ Sci. 2012;25(5):517-525.

31. Whyte M, Hubbard R, Meliconi R, Whidborne M, Eaton V, Bingle C, et al. Increased risk of fibrosing alveolitis associated with interleukin-1 receptor antagonist and tumor necrosis factor-α gene polymorphisms. Am J Respir Crit Care Med. 2000;162(2):755-758.

32. Witte JS, Palmer U, O'Connor RD, Hopkins PJ, Hall LM. Relation between tumor necrosis factor polymorphism TNFa-308 and risk of asthma. Eur. J Hum Genet. 2002;10:82-85.

33. Yang YH, Liu YQ, Zhang L, Li H, Li XB, Ouyang Q, et al. Genetic polymorphisms of the TNF-α-308 G/A are associated with metabolic syndrome in asthmatic patients from Hebei province, China. Int J Clin Exp Pathol. 2015;8(10):13739-13746.

34. Yucesoy B, Vallyathan V, Landsittel DP, Sharp DS, Weston A, Burleson GR, et al. Association of tumor necrosis factor-α and interleukin-1 gene polymorphisms with silicosis. Toxicol Applied Pharmacol. 2001;172(1):75-82.

35. Yucesoy B, Peila R, White LR, Wu KM, Johnson VJ, Kashon ML, et al. Association of interleukin-1 gene polymorphisms with dementia in a community-based sample: the Honolulu-Asia aging study. Neurobiol Aging. 2005;27(2):211-217.

36. Zhai R, Jetten M, Schins RPF, Franssen H, Borm PJA. Polymorphisms in the promoter of the tumor necrosis factor-α gene in coal miners. Am J Ind Med. 1998;34(4):318-324.

37. Zhang Y, Wang F, Lan Y, Zhou D, Ren X, Zhao L, et al. Roles of microRNA-146a and microRNA-181b in regulating the secretion of tumor necrosis factor-α and interleukin 1-β in silicon dioxide-induced NR8383 rat macrophages. Mol Med Rep. 2015;12(4):5587-5593.

15

A T cell equation as a conceptual model of T cell responses for maximizing the efficacy of cancer immunotherapy

Haidong Dong[1,2]*, Yiyi Yan[3,4], Roxana S. Dronca[3], and Svetomir N. Markovic[1,4]

[1]*Department of Immunology, Mayo Clinic, Rochester, MN, USA*
[2]*Department of Urology, Mayo Clinic, Rochester, MN, USA*
[3]*Division of Medical Oncology, Mayo Clinic, Rochester, MN, USA*
[4]*Division of Hematology, Mayo Clinic, Rochester, MN, USA*

***Corresponding author:** *Haidong Dong, Department of Immunology and Urology, Mayo Clinic, Rochester, MN, USA, E-mail: dong.haidong@mayo.edu*

Abstract

Following antigen stimulation, the net outcomes of a T cell response are shaped by integrated signals from both positive co-stimulatory and negative regulatory molecules. Recently, the blockade of negative regulatory molecules (i.e. immune checkpoint signals) demonstrates promising therapeutic effects in treatment of human cancers, but only in a fraction of cancer patients. Since this therapy is aimed to enhance T cell responses to cancers, here we devised a conceptual model by integrating both positive and negative signals in addition to antigen stimulation that can evaluate strategies to enhance T cell responses. A digital range of adjustment of each signal is formulated in our model for prediction of a final T cell response. Our model provides a rational combination strategy for maximizing the therapeutic effects of cancer immunotherapy.

Introduction

T cell response to antigen stimulation is a tightly controlled process. Recent clinical trials have demonstrated that unleashing T cell responses to cancers could be an effective approach in the treatment of human malignancies [1-3]. The identification of immune checkpoint molecules in tumor immune evasion greatly contributes to the development of therapeutics aimed to block immunosuppressive mechanisms in order to enhance antitumor T cell immunity [4,5]. Although the clinical outcomes of immune checkpoint blockade are promising, the low efficiency and potential adverse effects remain as major challenges. Combination therapy among different immune checkpoint targets or other therapies (i.e. chemotherapy, targeting therapy or radiotherapy, etc.) is speculated to increase the efficacy of cancer immunotherapy [6,7]; however, the rationale of optimized combination is still lacking. Given the complex responses of T cells that are regulated by a battery of signals at different stages of activation and differentiation [8], a conceptual model is needed to design an effective combined therapy that would maximize the therapeutic effects of each components of a regimen. Here, we devised a conceptual model of T cell responses that can be used to predict an outcome of a T cell response according to changes of positive or negative regulatory signals. Importantly, based on this equation, our model provides a rationale for synergistic treatment combination aimed to decrease resistance and maximize T cell responses against cancers.

Two signal theory in shaping T cell responses

Antigen stimulation initiates a T cell response through the T Cell Receptor (TCR); however, the net outcome of a T cell response (activation, anergy, or tolerance) to this antigen is regulated by two additional signals, i.e. costimulation (CD28) or checkpoint (CTLA-4 or PD-1, etc.),which are integrated with the TCR signaling pathway [9]. To represent a T cell response that is initiated by antigen via TCR engagement and regulated by integrated positive or negative signals, we present an equation as below:

$$R = \frac{P*T}{N*T+1}$$

In this equation, R is for Response; T is for TCR signal; P is for Positive costimulation signals (like CD28 or signal 3 cytokines IL-2 or IL-12); N is for Negative checkpoint signals (like CTLA-4, PD-1 or IL-10). According to this equation, we defined that when R=1, a T cell response is turned on; when R=0, a T cell response is turned off; when R>1, a T cell response is enhanced; when R<1, a T cell response is deterred or in a tolerogenic status. In the following sections, we will give several examples of different outcomes of a T cell response based on the integration of TCR signals along with positive or negative regulatory signals in this mathematical model to see how our equation would predict a T cell response. The relationship of N or P with T is defined as N*T or P*T, rather than N+T or P+T according to the proximal signaling integration of N or P with TCR signals.

Besides checkpoint molecules that are directly integrated within TCR signaling pathway, other immune regulatory systems work in parallel with TCR signals to control T cell responses. These other regulatory mechanisms include, but are not limited to, regulatory T cells (Treg) or myeloid-derived suppressor cells

***Corresponding author email:** *dong.haidong@mayo.edu*

(MDSCs) [10]. To include these parallel rather than proximal signals in regulation of TCR signals in addition to N, we used a constant number "1" as an inclusive and independent parameter in our T cell equation, and defined their relation with TCR using "+" to present the additional regulations.

T cell response is dependent on antigen stimulation though the TCR

First of all, this equation should be able to predict the fundamental role of antigen stimulation of TCR in the T cell response. Actually, in the absence of antigen stimulation and without TCR signals, i.e. T=0, then R will be 0, and there is no T cell response (**Calculation 1**).

$$Calculation1: R = \frac{P*0}{N*0+1} = \frac{0}{0+1} = 0$$

This equation also explains why positive or negative signals *alone* do not have any effects on T cell response in the absence of TCR stimulation, since when T=0, R will always be 0 whether P or N is 1 or not.

T cell response is dependent on costimulation and regulated by immune checkpoint signals

It has been established that a full activation (response) of T cells is dependent on the presence of costimulation, i.e. CD28 engagement [11,12]. In the absence of costimulation (when P=0), R will always be 0, though there is a TCR stimulation (T=1) (**Calculation 2**). The outcome of calculation 2 explains T cell anergy [13], i.e. a mere TCR stimulation is not able to initiate a full T cell response.

$$Calculation2: R = \frac{0*1}{0*1+1} = \frac{0}{0+1} = 0$$

When there is a costimulation signal (P=1), a full T cell response will be generated (R=1) (**Calculation 3**), as long as negative signals are absent (N=0).

$$Calculation3: R = \frac{1*1}{0*1+1} = \frac{1}{0+1} = 1$$

If a negative signal is present (N=1), R will be 0.5, which is less than 1, indicating a deferred T cell response or T cell tolerance (**Calculation 4**).

$$Calculation4: R = \frac{1*1}{1*1+1} = \frac{1}{1+1} = 0.5 < 1$$

According to Calculation 4, T cell tolerance is established by negative signals via immune checkpoint molecules. Calculation 4 also suggests that although both TCR stimulation and positive regulatory signals (costimulation) are present, there is no guarantee that a full T cell response can be generated due to immune regulatory mechanisms (N+1). Thus, our equation demonstrates a critical role of negative signals (immune checkpoints) in restraining the T cell response, which could be crucial in order to prevent pathology caused by any ongoing or unlimited T cell responses. We acknowledge that our T cell equation reflects how naïve T cells are primed to become activation, while the re-activation of memory T cells could be independent of co-stimulation (i.e. P signal).

Breaking T cell tolerance and enhancing a T cell response

As shown in Calculation 4, the presence of negative signals or immune checkpoints significantly compromises the generation of a full T cell response initiated by TCR stimulation in the presence of costimulation. In order to enhance T cell response or break a T cell tolerance, we have to increase the strength of either costimulation or TCR stimulation. To that end, if we increase costimulation P to 2, and keep the others at the same levels (T=1, N=1), we will have R = 1 (**Calculation 5**), suggesting a T cell response can be restored through the increase of costimulation. This calculation is in line with an early observation that introduction of CD28 costimulation (positive signals) enhances T cell response [11,12] or introduction of B7 (CD80/CD86) molecules into tumor cells results in a strong antitumor response in vivo [14].

$$Calculation5: R = \frac{2*1}{1*1+1} = \frac{2}{1+1} = 1$$

However, a mere increase of TCR stimulation (let T=2) cannot restore a T cell response in the presence of negative signals (when N=1) (**Calculation 6**). This outcome may explain some preclinical and clinical observations showing strong antigenicity (e.g. high affinity antigen peptides) alone did not initiate a strong T cell response and failed to generate a protective T cell immunity. Occasionally, some high affinity antigen peptides may overcome the requirement of co-stimulation in activation of T cells , however, most of the cases involved activation of memory or memory-like T cells.

$$Calculation6: R = \frac{1*2}{1*2+1} = \frac{2}{2+1} = 0.67 < 1$$

Next we examined to what degree a reduction of negative signals would be required to restore or enhance a T cell response. According to **Calculation 7**, when N is in a range of 0.1 to 0.9, R will always be less than 1, suggesting partial reduction of negative signals is not enough to restore a T cell response. As indicated from calculation 3, only a complete blockade or absence of negative signals, i.e. when N=0, a full T cell response can be achieved. This result underscores the strategy currently used in treatment of human cancer by a complete blockade of immune checkpoints (PD-1 or CTLA-4) in order to achieve objective clinical responses. Actually, the combination of anti-PD-1 and anti-CTLA-4 treatment achieved higher response rather than either alone [15].

$$Calculation7: R = \frac{1*1}{0.1\,or\,0.9*1+1} = \frac{1}{0.1\,or\,0.9+1} = 0.9\,or\,0.5 < 1$$

However, since a complete blockade of negative signals only can be achieved in a fraction of cancer patients, and in most situations negative signals can only be partially reduced, additional approaches are needed to restore or increase a T cell response. To that end, if negative signals are partially reduced (let N=0.5), our equation suggests a partial increase of costimulation (when P=1.5) will be able to restore a T cell response **(Calculation 8)**.

$$Calculation8: R = \frac{1.5*1}{0.5*1+1} = \frac{1.5}{0.5+1} = 1$$

In order to enhance T cell response (i.e. to let R>1), a double increase of costimulation (let P=2) is needed as shown in **Calculation 9** if the negative signals are also partially reduced (N=0.5).

$$Calculation9: R = \frac{2*1}{0.5*1+1} = \frac{2}{0.5+1} = 1.3 > 1$$

It could be very challenging, if not impossible, to have a double increase of costimulation in order to enhance a T cell responses, for example, to have poorly immunogenic tumor cells express B7 costimulatory ligand [14] or to provide additional costimulation signals directly to T cells (e.g. 41BB stimulation) [16]. Alternatively, a combination of partially increased TCR stimulation and costimulation (T=1.5; P=1.5) with partially decreased negative signals (N=0.5) will be able to give an enhanced T cell response(R=1.29 >1) **(Calculation 10)**. This calculation indicates that a synergistic combination can be achieved by integrating a suboptimal increase of TCR and costimulation and suboptimal decrease of negative signals (e.g. partial immune checkpoint blockade) in order to enhance T cell responses.

$$Calculation10: R = \frac{1.5*1.5}{0.5*1.5+1} = \frac{2.25}{0.75+1} = 1.29 > 1$$

Discussion

Here we present a T cell equation as a conceptual model that can be used to predict the net outcomes of T cell responses by integrating both regulatory and stimulatory signals along with antigen stimulation. The T cell equation(R=P*T/[N*T+1]) gives a digital range (0.1-0.9 or 1-2) of adjustment for each regulatory or stimulatory signal T cells may receive during antigen stimulation. As predicted from Calculation 10, a synergistic combination is generated by integrating each signal when each signal can only be adjusted in a suboptimal condition due to practical limitations. The predication of our equation underscores the significance of current clinical efforts in seeking synergistic combination treatment of human cancers in order to decrease drug resistance and to increase the efficacy of cancer immunotherapy.

Our model predicts that simply increasing TCR stimulation is not enough to increase T cell responses due to the regulation of immune checkpoints at least in priming naïve T cells. In line with this predication, objective cancer responses have not been achieved in clinical trials with several tumor antigens that have strong antigenicity [17]. As predicted by our equation, if we combine tumor antigen peptides with immune adjuvants that are used to increase costimulation, i.e. to increase the expression of stimulatory molecules by antigen presenting cells, such tumor vaccine formulations are able to generate tumor antigen specific T cell responses [18]. However, since some adjuvants have the potential to increase the expression of immune checkpoint molecules [19], the therapeutic effects of tumor antigen vaccine are compromised due to immune regulatory mechanisms. To maximize the therapeutic effects of tumor vaccine, our calculations 9 and 10 suggest that components capable of increasing costimulation or decreasing negative regulatory signals, or both, should be integrated in an optimal formulation for tumor vaccines [20].

Our T cell equation clearly presented the significance of the immune checkpoint in controlling the T cell response. This prediction is echoed by recent successful treatments of some human cancers with the immune checkpoint blockade strategy (CTLA-4 or PD-1) that aimed to restore or enhance antitumor T cell immunity [21,22]. Interestingly, to achieve more efficient reduction of negative signals as shown in our equation to reduce N value as close as possible to 0, a combined therapy of both PD-1 and CTLA-4 therapy has been approved by FDA to gain a synergy effect in treatment of metastatic melanoma [15]. However, this combined blockade of immune checkpoints might increase the risk of enhanced adverse effects in some patients. As predicted by the T cell equation, adjustment of other stimulatory or regulatory signals should be considered in order to gain a safer and stronger antitumor T cell response. The field of cancer immunotherapy has moved from an era of empirical combinations to one of rational design by considering the compatibility of each regulatory or stimulatory mechanisms [7]. To that end, the predication of our conceptual model of T cell response provides a rationale to design a synergistic combination that takes into count each of the major factors that work together to affect the net outcome of a T cell response. As predicted by Calculation 10, a synergistic combination can be achieved by integrating sub optimally adjusted stimulatory or regulatory signals in order to enhance T cell responses in cancer patients.

Our T cell equation, as a conceptual model, is not designed for dose estimation or calculation in application of a particular regimen of cancer immunotherapy, but rather our model may predict a final outcome based on the signal strength a regimen may bring in. Since no defined dose-response has been established in cancer immunotherapy, i.e. the highest dose is not always the optimal dose, our model provides a way to evaluate how signals (positive or negative) are integrated for achieving a maximal effect in promoting T cell responses. Based on our T cell equation, a level of signal strength can be determined (for example to set N=0.5). Accordingly, the actual dose (concentration) of a regimen (antibody used to block immune checkpoint) will be determined by selecting a dose that would lead to 50 percent reduction of negative signals. The actual effects of 50 percent reduction of negative signals in cancer treatment eventually will be evaluated by objective biomarkers or clinical responses.

Some chemotherapy drugs cause immunogenic cell death (ICD) in tumor cells, such as doxorubicin, mitoxantrone, oxaliplatin, and cyclophosphamide. Accumulating clinical data indicate that activation of adaptive immune responses induced by immunogenic cell death is associated with improved disease outcome in cancer patients [23]. According to our T cell equation, the ICD of tumors likely contribute to the increase of T (TCR) signals by releasing more immunogenic tumor antigens, and to the increase of P (costimulation) signals by releasing a series of immunostimulatory damage-associated molecular patterns (DAMPs), so called natural adjuvants [24], that promotes antigen presentation and T cell priming [25]. In combination with immune checkpoint inhibitors (like anti-PD-1) that aim to reduce N (negative) signals, it is predictable that chemotherapy drugs that cause ICD can achieve additive or synergistic clinical activity by coincidently increasing P and T and decreasing N according to our T cell equation.

Taken together, our T cell equation provides a conceptual model of T cell responses for designing synergistic treatment combinations aimed to defuse resistance and maximize T cell responses against cancers. Our T cell equation indicates that a combined therapeutic formula should include approaches capable of increasing tumor antigen stimulation and costimulation, and at the same time, reducing or blocking immune checkpoint signals.

Author contribution

H.D. conceived the concept, designed the key T cell equation, and wrote the main manuscript text. Y.Y., R.D., and S.M contributed to the discussion of clinical implications of this equation and revised versions of some equations. All authors reviewed the manuscript.

Acknowledgements

Authors have no competing financial interests. Financial Support: Mayo Foundation (Dong), and in part by NCI R21 CA197878 (Dong/Dronca) and NIH/NIAID R01 AI095239 (Dong).

References

1. Topalian SL, Drake CG, Pardoll DM. Targeting the PD-1/B7-H1(PD-L1) pathway to activate anti-tumor immunity. Curr Opin Immunol. 2012;24(2):207-212. Doi:10.1016/j.coi.2011.12.009
2. Sznol M, Chen L. Antagonist antibodies to PD-1 and B7-H1 (PD-L1) in the treatment of advanced human cancer--response. Clin Cancer Res. 2013;19(5):1021-1034. Doi:10.1158/1078-0432.CCR-12-2063
3. Pardoll DM. The blockade of immune checkpoints in cancer immunotherapy. Nature Reviews Cancer. 2012;12:252-264. Doi:10.1038/nrc3239
4. Dong H, Strome SE, Salomao DR et al. Tumor-associated B7-H1 promotes T-cell apoptosis: a potential mechanism of immune evasion. Nat Med. 2002;8:793-800. Doi:10.1038/nm730 nm730
5. Dong H, Chen L. B7-H1 pathway and its role in the evasion of tumor immunity. J Mol Med. 2003;81(5):281-287. Doi:10.1007/s00109-003-0430-2
6. Dronca RS, Dong H. Immunomodulatory antibody therapy of cancer: the closer, the better. Clin Cancer Res. 2015;21(5): 944-946. Doi:10.1158/1078-0432.CCR-14-2111
7. Melero I, Berman DM, Aznar MA, Korman AJ, Gracia JL, Haanen J. Evolving synergistic combinations of targeted immunotherapies to combat cancer. Nature Reviews Cancer. 2015;15:457-472. Doi:10.1038/nrc3973
8. Zang X, Allison JP. The B7 family and cancer therapy: costimulation and coinhibition. Clin Cancer Res. 2007;13(18):5271-5279. Doi:10.1158/1078-0432.CCR-07-1030
9. Chen L, Flies DB. Molecular mechanisms of T cell co-stimulation and co-inhibition. Nature Reviews Immunology. 2013;13:227-242. Doi:10.1038/nri3405
10. Zou W. Immunosuppressive networks in the tumour environment and their therapeutic relevance. Nature Reviews Cancer. 2005;5:263-274. Doi:10.1038/nrc1586
11. Krummel MF, Allison JP. CD28 and CTLA-4 have opposing effects on the response of T cells to stimulation. J Exp Med. 1995;182(2):459-465.
12. Linsley PS, Ledbetter JA. The role of the CD28 receptor during T cell responses to antigen. Annu Rev Immunol. 1993;11:191-212. Doi:10.1146/annurev.iy.11.040193.001203
13. Schwartz, R.H. T cell anergy. *Annu Rev Immunol.* 2003;21:305-334.
14. Chen L, Ashe S, Brady WA, Hellstrom I, Hellstrom KE, Ledbetter JA, et al. Costimulation of antitumor immunity by the B7 counterreceptor for the T lymphocyte molecules CD28 and CTLA-4. Cell. 1992;71(7):1093-1102.
15. Larkin J, Chiarion-Sileni V, Gonzalez R et al. Combined Nivolumab and Ipilimumab or Monotherapy in Untreated Melanoma. N Engl J Med. 2015;373:23-34. Doi:10.1056/NEJMoa1504030
16. Ascierto PA, Simeone E, Sznol M, Fu YX, Melero I. Clinical experiences with anti-CD137 and anti-PD1 therapeutic antibodies. Seminars in oncology. 2010;37(5):508-516. Doi:10.1053/j.seminoncol.2010.09.008
17. Thomas A, Giaccone G. Why has active immunotherapy not worked in lung cancer? Annals of oncology: official journal of the European Society for Medical Oncology / ESMO. 2015;26(11):2213-2220 Doi:10.1093/annonc/mdv323
18. Mehta NK, Moynihan KD, Irvine DJ. Engineering New Approaches to Cancer Vaccines. Cancer immunology research. 2015;3(8):836-843. Doi:10.1158/2326-6066.CIR-15-0112
19. Pulko V, Liu X, Krco CJ, Harris KJ, Frigola X, Kwon ED ,et al. TLR3-stimulated dendritic cells up-regulate B7-H1 expression and influence the magnitude ofCD8 T cell responses to tumor vaccination. J Immunol. 2009;183(6):3634-3641. Doi:jimmunol.0900974 [pii]10.4049/jimmunol.0900974

20. Webster WS, Thompson RH, Harris KJ, Frigola X, Kuntz S, Inman BA, et al. Targeting molecular and cellular inhibitory mechanisms for improvement of antitumor memory responses reactivated by tumor cell vaccine. J Immunol. 2007;179(5):2860-2869. DOI:10.4049/jimmunol.179.5.2860

21. Topalian SL, Hodi FS, Brahmer JR et al. Safety, activity, and immune correlates of anti-PD-1 antibody in cancer. N Engl J Med. 2012;366:2443-2454. doi:10.1056/NEJMoa1200690

22. Hamid O, Robert C, Daud A et al. Safety and tumor responses with lambrolizumab (anti-PD-1) in melanoma. N Engl J Med. 2013;369:134-144. Doi:10.1056/NEJMoa1305133

23. Bezu L, Gomes-de-Silva LC, Dewitte H, Breckpot K, Fucikova J, Spisek R, et al. Combinatorial strategies for the induction of immunogenic cell death. Frontiers in immunology. 2015;6:187. Doi:10.3389/fimmu.2015.00187

24. Rock KL, Hearn A, Chen CJ, Shi Y. Natural endogenous adjuvants. Springer seminars in immunopathology. 2005;26(5):231-246. Doi:10.1007/s00281-004-0173-3

25. Kepp O, Senovilla L, Vitale I et al. Consensus guidelines for the detection of immunogenic cell death. Oncoimmunology. 2014;3(9):e955691. Doi:10.4161/21624011.2014.955691

Detection of Estrogen and IL-17A Levels in Serum of Premenopausal and Postmenopausal Women in Kirkuk City Iraq

Najdat Ali Al-Kadhi[1], Fatima Mustafa Mohammed[2], Ektifa Abdulhameed[3] and Chateen I Ali Pambuk[2*]

[1]*Kirkuk Technical College, Northern Technical University, Iraq*

[2]*College of Dentistry, Tikrit University, Iraq*

[3]*College of Education for Women, Tikrit University, Iraq*

***Corresponding author:** Dr. Chateen I. Ali Pambuk, Assistant Professor, College of Dentistry University Of Tikrit, PhD Medical Immunology, E-mail: dr.chatin2@yahoo.com*

Abstract

The study aimed to the detection of estrogen & IL-17A levels in the serum of apparently healthy postmenopausal women. The study carried out on the 190 women, they were divided randomly to two groups: the first group included 90 postmenopausal women & the second group was included 100 premenopausal women. From each subject (5 ml) of venous blood were drawn in plain tube & centrifuged by centrifuge for 5 minutes at 3000 rpm. The serum was used for evaluation of IL-17A By Enzyme Linked Immunosorbent Assay (ELISA), while the level of estrogen was determined by the minividas. The results of the study were recorded a significant decline ($p < 0.01$) in the concentration of estrogen & significant decrease ($p < 0.05$) in the level of IL-17A in the serum of postmenopausal women when compared with the premenopausal women. From the results of the study, we concluded a significant decline in the levels of IL-17A, which may be contributed to the significant decrease in the level of estrogen in the Serum of postmenopausal women. In conclusion, the postmenopausal women recorded a significant decline in the serum concentration of IL-17A particularly in the presence of Lower estrogen level.

Keywords: Pre and Postmenopausal; Interleukin-17A; Estrogen

Introduction

The state of Menopause in women life considered as an important physiological process, characterized by the dramatic cessation of menstruation due to the loss of ovarian follicular activity [1]. The onset of menopause associated with the decline in the levels of estrogen and consequently alteration of lipid fractions in the serum of menopausal women [2,3]. Deficiency of some hormones during menopause associated with the lower systemic inflammatory status ,which inflammation manifested by the increased serum levels of key proinflammatory cytokines IL-1, IL-6 or tumor necrosis factor alpha TNF-)[4].

The incidence of cardiovascular events increases when there are several circulating inflammatory markers including hs-CRP & IL-6[5]. Recently, approved that the elevation of serum levels of IL-8 in premenopausal ,perimenopausal, postmenopausal women & oophorectomized women with sever hot flushes were significantly higher than those in women without hot flushes[6]. In chronic inflammatory disease, a new candidate IL-17A produced by T helper 17 (Th17) lymphocytes & effects on neutrophil recruitment & granulopoieses [7]. Estrogen deficiency in postmenopausal women recorded significant increase in IL-17A level which highlighted to its role in chronic inflammatory events & increases their susceptibility to the chronic inflammatory disease [8].

Significant decrease of estrogen concentration in menopausal women regarded to the acceleration or aggravation of immune & inflammatory events which leads for the libration of proinflammatory cytokines such as IL-1, IL-6, TNF- & IL-17 resulting an immune dysfunction [9,10]. The relationship between hormonal decline associated with the menopause & increased serum levels of proinflammatory cytokines is not yet fully understood. So this study aimed to investigate the level of serum estrogen & IL-17A in healthy women at the premenopausal & postmenopausal status.

Materials & Methods

Study Population

For detection of estrogen level, the study carried out on 190 healthy premenopausal (mean age 36 years) & postmenopausal (mean age 36 years) women. The 100 premenopausal women were divided randomly to two subgroups, the first subgroup included 63 women (mean age 30.5 years) & second subgroup included 37 women (mean age 40 years). While 90 healthy postmenopausal women were divided to two subgroups, the first subgroup included 67 women (mean age 54.5 years) & second subgroup included 23 women (mean age 63 years).

For detection of serum IL-17A level, the study carried out on 74 women were 40 of them at premenopausal state while 34 of them at the postmenopausal state. Blood samples were collected on fasting for 10-14 hrs. Serum were obtained & kept at deep freeze (- 20 C^0) till use.

***Corresponding author email:** dr.chatin2@yahoo.com*

Detection of Estrogen Concentration

Total serum estrogen level quantitatively determined according to the Vidas estradiol 11 test by vidas instrument which used Enzyme Linked Fluorescently Assay (ELFA) techniques. (According to the technical specification provided by the manufacturer).

Detection of Interleukin 17A Level

Serum IL-17A level were detected by Human IL-17 Elisa kit from (Bioassay technology laboratory / USA), applied the method of Enzyme Linked Immunosorbent Assay (ELISA) from (Biotek-elx800, USA). (According to the technical specification provided by the manufacturer).

Statistical Analysis

The data were analyzed by the statistical program by Minitab according to T-test. The data displayed as Mean ± SD. P- value (p`< 0.05) refers to the statistically significant differences and (p < 0.01) refers to highly significant differences.

Results

The results of the study revealed in the Table 1 , that the serum of estrogen concentration recorded a significant decrease (p`< 0.01) in postmenopausal women (38.1 + 11.58) ng /L when compared with the premenopausal women (192.24 + 41.13) ng/L ,while there was no significant differences (p> 0.05) between subgroups (age groups) , as in premenopausal subgroups existed (196 ± 42.13),(186.81 ± 35.08)Ng/l & in postmenopausal subgroups existed (46.53 ± 11.70), (31.20 ± 9.52) ng/l.

Table 1: Concentration of estrogen (ng/l) in serum of premenopausal and Postmenopausal women.

Subjects	Total No.	Study Groups	Estrogen Level (Ng/L) Mean ± SD	Total
Premenopausal Women	100	27 – 34 years	196.27 ± 42.13	192.24±41.13
		N = 63		
		35 – 45 years	187.81 ± 35.08	
		N = 37		
Premenopausal Women	90	51 - 58 years	46.53 ± 11.70	38.10 ± 11.58
		N = 67		
		59 – 67 years	31.20 ± 9.52	
		N = 23		

Fig.1 showed the results of IL-17A level in the serum, which recorded a significant decrease (p< 0.05) in postmenopausal women (22.16 + 3.1) ng / L when compared with the premenopausal women (30.87 + 6.15) ng / L. ** Highly significant p< 0.01

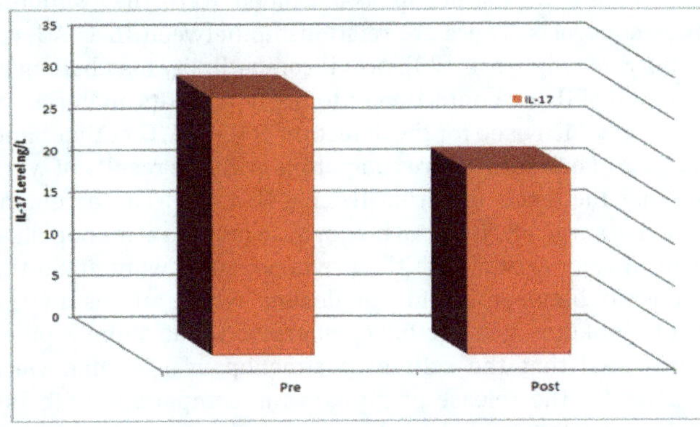

Figure 1: Reveals concentration of IL-17A (ng/l) in serum of premenopausal and postmenopausal women.

Discussion

The onset of menopause associates with the onset of alterations in the hormonal & systemic inflammatory status. The process of the inflammation characterized by increased key proinflammatory cytokine levels in the serum. Cytokines are signaling protein molecules with important role in intercellular communication. Their sources from all cells & tissues, which differentiates them from hormones [11,12].

From the results of the study which revealed a significant decrease (p< 0.01) in the levels of estrogen in the serum of postmenopausal women as shown in Table 1 & Fig.1, that this finding were compatible with many studies. Kilm & Chandala 213 found that the level of estrogen were low in postmenopausal women [13]. The level of estrogen decrement in postmenopausal women was considered normal state [14]. Cessation of the menstruation is a common event in menopause which results from cessation of the activity of the ovarian follicles, due to hormonal deficiency, which is a contributory factor for the increased incidence of inflammatory disease [15].

Also, the disturbances in the level of lipid & lipoprotein fractions in the serum, in addition to the negative alterations in the levels of glucose, insulin metabolism, fat distribution, blood coagulation, fibrinolysis & endothelial defects results from an abrupt decline in the level of estrogen hormone [16]. Moreover, estrogen plays a significant role for heart (protective role) by increases release of nitrous oxide, which is important for stabilizing of endothelial cells & synergize the antioxidant effects with the changes of fibrinolytic proteins which all of them considered from cardio-protective mechanism which lost with the onset of menopause [17,18]. While a significant decline of estrogen level in postmenopausal women have many undesirable changes which influences on the health of women, because there are an association between estrogen decrement in postmenopausal women with fat deposition in the body & an increase the incidence of many disease, and alteration of HDL, LDL levels, leads to great impact for the development of cardiovascular disease [19-22].

The results of the study also revealed a significant decrease (p< 0.05) in the levels of IL-17A for postmenopausal women. This result were incompatible with the results of [7] found the increment of IL-17A level associated highly with the age & period of menopause in postmenopausal women were they suffered from osteoporosis. There are relationship between IL-17A level in serum & osteoporosis [23]. Also incompatibility may belong to the method of IL-17A detection or to the type of kits, materials & instrument which used for the detection of IL-17A. On other hand the results of our study were compatible with the results of were found that the levels of serum IL-17A, IL-4 & IL-10 decrement recorded in surgically induced menopause or natural menopausal women in comparison with the premenopausal women. So the relationship between hormonal decline & alterations in the level of cytokines not yet fully understood, but some studies demonstrated that the cells of postmenopausal women were less active for the release of cytokines in comparison with the premenopausal women [12,14].

The study concluded a significant decline in the levels of IL-17A which contribute to the significant decrease in the concentration of estrogen hormone in the serum of postmenopausal women.

References

1. Andrei M, Mihu D, Costin N, Mihu C. Proinflimmatory and anti-inflammatory cytokine changes related to menopause. Prezeglad Menopauzalny J. 2014;13 (3) : 162-168; Doi :10.5114/pm.2014.43818
2. Kilim S.R, Chandala S. R. A comparative study of lipid profile and oestradiol in pre& postmenopausal women. J Clin Diagn Res. 2013; 7 (8):1596–1598; Doi:10.7860/JCDR/2013/6162.3234
3. Bade G, Shah S, Nahar P, Vaidya S. Effect of menopause on lipid profile in relation to body mass index. Chronicles of Young Sci. 2014; 5 (1): 20–24
4. Charles C, Yuskavage J, Carlson O, John M, Tagalicud AS, Maggio M, et al. Effect of high dose isoflavones on metabolic & inflammatory markers in healthy Postmenopausal women. Menopause. 2009;16(2):395–400;Doi: 10.1097/gme.0b013e3181857979
5. Ridker PM, Hennekens CH, Buring JE, Rifai N. C-reactive protein and other markers of inflammatory in prediction of CVD in women. N. Engl J Med. 2000;342:836–843; Doi :10.1056/NEJM200003233421202
6. Yasui T, Uremura H, Tomita J, Miyatani Y, Yamada M, Kuwahara A, et al. Association of interleukin-8 with hot flushes in premenopausal, perimemopausal and postmenopausal women and bilateral oophorectomized women. J Clin Endocrinol Metab. 2006;91(12):4805–4808;Doi:10.1210/jc.2006-1100
7. Molnar I. The role of IL-17A in postmenopausal inflammatory events, Such as in osteoporosis. 4th international Conference & exhibition on Immunology. OMICS;2015
8. Molner I, Bohaty I, Somogyine-Vari E. IL-17A mediated RANK ligand elevation involved in postmenopausal osteoporosis. Osteoporosis International. 2014;25(2):783–786; Doi:10.1007/s00198-013-2548-6
9. Teitelbaum SL. Postmenopausal osteoporosis, T cell, and immune dysfunction. Proc Natl Acad Sci.2014;101(48): 16711 – 16712; Doi: 10.1073/pnas.0407335101
10. Tyagi AM, Srivastava K, Mansoori MN, Trivedi R, Chattopadhyay N, Singh D. Estrogen deficiency induces the differentiation of IL-17 secreting Th 17 cells: a new candidate in the pathogenesis of osteoporosis of osteoporosis. Plos one.2012;7(9):445–452;Doi: 10.1371/journal.pone.0044552
11. Cannon JG. Inflammatory cytokines in non pathological state. News Physiol Sci. 2000;15:298–303
12. Malutan A.M, Costin N, Ciortea R, Mihu D. Variation of anti-inflammatory cytokines in relationship with menopause. Applied Med Informatics.2013;32(2):30-38
13. Kilim R.K, Chandala R.C. A comparative study of conventional risk factors in patients with coronary heart disease. JAMA. 2013; 290(7):898–904
14. Sultan N, Nawaz M, Sultan A, Fayaz M, Baseer A. Effect of menopause on serum HDL- cholesterol level. J. Ayub Med Coll Abbottabad. 2003;15(3)24–26
15. Su HI, Freeman EW. Hormone changes associated with the menopausal transition. Minerva Ginecol.2009;61(6):483–489
16. Bales AC. In search of lipid balance in order women. New studies raise questions about what works best. Postgrad Med. 2000;108(7):57–72; Doi :10.3810/pgm.2000.12.1312
17. Igweh JC, Nwagha IU, Karo JM. The effects of menopause on the serum lipid profile of normal females of south east Nigeria. Niger J Physiol Sci. 2005;20(1-2):48-53
18. Taddei S, Virdis A, Ghiadoni L, Mattei P, Sudano I, Bernini G, et al. Menopause is associated with endothelial dysfunction in women. Hypertension. 1996;28(4):576–582
19. Pisani M. Aging & lung disease. A clinical guides Springer. 2012; P: 157
20. Pollard M. Western Disease: An evolution perspective. Cambridge University press. 2008;P: 92
21. Goroll A., Mulley A. Primary care medicine: Office evaluation & management of the adult patient. Lippincott Williams & Wilkins. 2012;P:20–24
22. Davis SR, Lambrinoudalei I, Lumsden M, Mishra GD, Pal L, Rees M, et al. Menopause. Nature review, Disease primers. 2015;1:1–19
23. Korayem HK, Rezk MM, Hassan MM, El-Tawab SS, Elsaid SS. Relation between serum IL-17 level & risk of osteoporotic fracture in premenopausal rheumatoid arthritis patient: Clinical, radiological, laboratory studies. The Egyptian Rheumatology; Available online J. 2015

17

Assessment of Genetic Mutations DMD, DYSF, EMD, LMNA, DUX4, DMPK, ZNF9, PABPN1 Genes Induction Duchenne Muscular Dystrophy

Shahin Asadi[1*], Vida Vahdani Kia[2], Rana Bagheri[2], Mahsa Jamali[3] and Samaneh Sadeh Dell[3]

[1]*student of Molecular Genetics, Director of the Division of Medical Genetics and Molecular Research, Molecular Genetics-IRAN-TABRIZ*

[2]*Master of Molecular Biology-Genetics, Director of the Division of Medical Genetics and Molecular Research, Molecular Genetics-IRAN-TABRIZ*

[3]*Master of Molecular Genetics, Director of the Division of Medical Genetics and Molecular Research, Molecular Genetics-IRAN-TABRIZ*

***Corresponding author:** Shahin Asadi, Director of the Division of Medical Genetics and Molecular Research, Molecular Genetics-IRAN-TABRIZ, E-mail: shahin.asadi1985@gmail.com*

Abstract

Background Importance and Aim: Our aim is to investigate the genetic and genetic patterns of Duchenne muscular dystrophy.

Methodology: In this study we have analyzed 20 people. 10 patients Duchenne muscular dystrophy disease and 10 persons control group. The genes DMD, DYSF, EMD, LMNA, DUX4, DMPK, ZNF9, PABPN1 analyzed in terms of genetic mutations made. In this study, people who have genetic mutations were targeted, with nervous disorders, Duchenne muscular dystrophy disease.

Results and Conclusion: In fact, of all people with Duchenne muscular dystrophy disease. 10 patients Duchenne muscular dystrophy disease had a genetic mutation in the genes DMD, DYSF, EMD, LMNA, DUX4, DMPK, ZNF9, PABPN1 Duchenne muscular dystrophy disease. Any genetic mutations in the target genes control group did not show.

Keywords: Genetic study; Duchenne muscular dystrophy disease; Mutations The genes DMD, DYSF, EMD, LMNA, DUX4, DMPK, ZNF9, PABPN1

Introduction

Duchenne muscular dystrophy (DMD) is a severe type of muscular dystrophy.[1] The symptom of muscle weakness usually begin around the age of four in boys and worsens quickly.[2] Typically muscle loss occurs first in the upper legs and pelvis followed by those of the upper arms. This can result in trouble standing up.[1] Most are unable to walk by the age of 12.[2] Affected muscles may look larger due to increased fat content. Scoliosis is also common. Some may have intellectual disability. Females with a single copy of the defective gene may show mild symptoms.[1]

The disorder is X-linked recessive. About two thirds of cases are inherited from a person's parents, while one third of cases are due to a new mutation. It is caused by a mutation in the gene for the protein dystrophin. Dystrophin is important to maintain the muscle fiber cell membrane. Genetic testing can often make the diagnosis at birth. Those affected also have a high level of creatine kinase in their blood.[1]

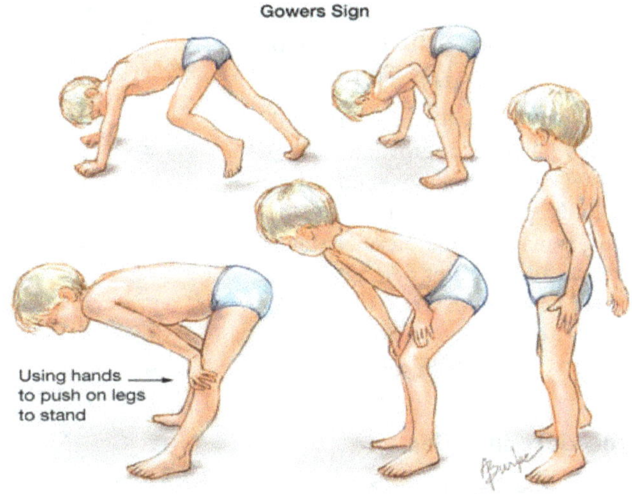

Figure 1: Schematic View of Child with Duchenne Muscular Dystrophy.

***Corresponding author email:** shahin.asadi1985@gmail.com*

No cure for muscular dystrophy is known. Physical therapy, braces, and corrective surgery may help with some symptoms.[2] Assisted ventilation may be required in those with weakness of breathing muscles.[1] Medications used include steroids to slow muscle degeneration, anticonvulsants to control seizures and some muscle activity, and immunosuppressant's to delay damage to dying muscle cells.[3,4]

DMD affects about one in 5,000 males at birth. It is the most common type of muscular dystrophy.[1] The average life expectancy is 26; however, with excellent care, some may live into their 30s or 40s. [5-8] Gene therapy, as a treatment, is in the early stages of study in humans.[9,10][Figure 1]

Materials and Methods

In this study, 10 patients with Duchenne muscular dystrophy disease, and 10 persons control group were studied. Peripheral blood samples from patients and parents with written permission control were prepared. After separation of serum, using Real Time-PCR technique of tRNA molecules was collected. To isolate Neuroglial cells erythrocytes were precipitated from hydroxyethyl starch (HES) was used. At this stage, HES solution in ratio of 1to5with the peripheral blood of patients and controls were mixed. After 60 minutes of incubation at room temperature, the supernatant was removed and centrifuged for 14 min at 400 Gera. The cell sediment with PBS (phosphate buffered saline), pipetazh and slowly soluble carbohydrate ratio of 1to2 onficole (Ficol) was poured in the 480G was centrifuged for 34 minutes. Mono nuclear Neuroglial cells also are included, has a lower density than ficole and soon which they are based. The remaining erythrocytes have a molecular weight greater than Ficol and deposited in test tubes. [11-18]

The supernatant, which contained the mono nuclear cells, was removed, and the 400 Gera was centrifuged for 12 minutes. Finally, the sediment cell, the antibody and Neuroglial cells was added after 34 minutes incubation at 5 °C, the cell mixture was passed from pillar LSMACS. Then the cells were washed with PBS and attached to the column LSMACSS pam Stem cell culture medium containing the transcription genes DMD, DYSF, EMD, LMNA, DUX4, DMPK, ZNF9, PABPN1, and were kept. [19]

To determine the purity of Neuroglial cells are extracted, flow cytometry was used. For this purpose, approximately 4-5 × 10^3 Neuroglial cells were transfer red to1.5ml Eppendorf tube and then were centrifuged at 2000 rpm for 7minutes at time. Remove the supernatant culture medium and there maining sediment, 100µl of PBS buffer was added. After adding 5-10µl PE monoclonal anti body to the cell suspension for 60 min at 4°C incubated and read immediately by flow cytometry. For example, rather than control anti body Neuroglial cells PE, igg1 negative control solution was used. [20]

Total mRNA Extraction Procedure Includes

1. 1ml solution spilled Qiazolon cells, and slowly and carefully mixed and incubated at room temperature for 5 minutes. Then 200µl chloroform solution to target mix, and then transfer the micro tubes was added, and the shaker well was mixed for 15 seconds. The present mix for 4 minutes at room temperature and then incubated for 20 min at 4°C and was centrifuged at 13200 rpm era. Remove the upper phase product was transfer reductase new micro tube and to the one times the volume of cold ethanol was added. The resulting mixture for 24 hours at -20°C was incubated. [21-27]

2. Then for 45min at 4°C and was centrifuged at 12000 rpm era. Remove the supernatant and the white precipitate, 1ml of cold 75% ethanol was added to separate the sediment from micro tubes were vortex well. The resulting mixture for 20 min at 4°C and by the time we were centrifuged 12000 rpm. Ethanol and the sediment was removed and placed at room temperature until completely dry deposition. The precipitate was dissolved in 20µl sterile water and at a later stage, the concentration of extracted mRNA was determined. [28]

To assessment the quality of mi-rna, the RT-PCR technique was used. The cDNA synthesis in reverse transcription reaction (RT) kit (Fermentase K1622) and 1µl oligoprimers 18 (DT) was performed. Following the PCR reaction 2µm dNTP, 1µg cDNA, ferment as PCR buffer 1X, 0 / 75µm mgcl2, 1.25 U / µL Tag DNA at 95°C for 4min, 95°C for 30s, annealing temperature 58°C for 30s, and72 °C for 30 seconds, 35 cycles were performed. Then 1.5% agarose gel, the PCR product was dumped in wells after electrophores is with ethidium bromide staining and color was evaluated. [29-30]

Results

The results of the PCR and RT-PCR reaction for Duchenne disease target genes are as follows:[Figure 2-8]

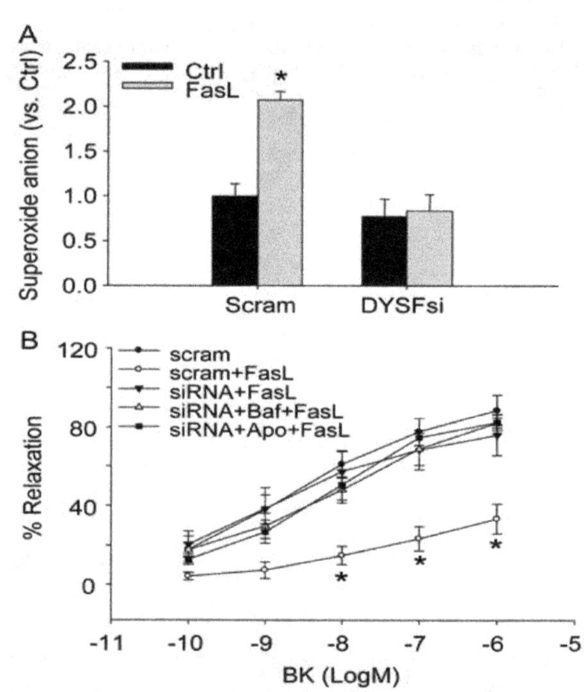

Figure 2: Diagnosis of DYSF gene expression in patients with Duchenne Muscular Dystrophy.

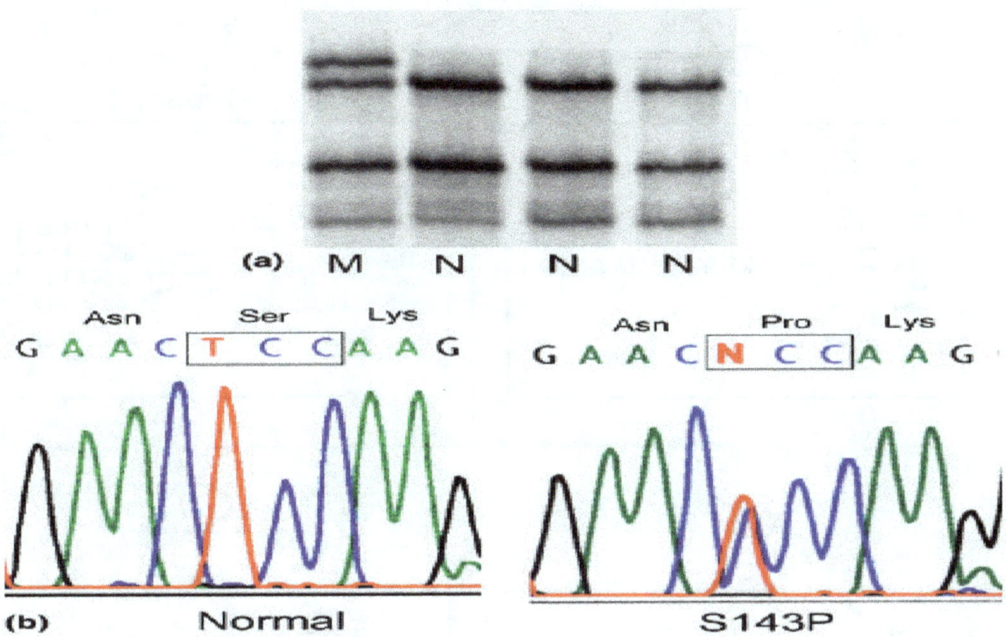

Figure 3: Schematic view of the nucleotide sequence of target mutated genes in patients with Duchenne scrotum dystrophy compared to normal group.

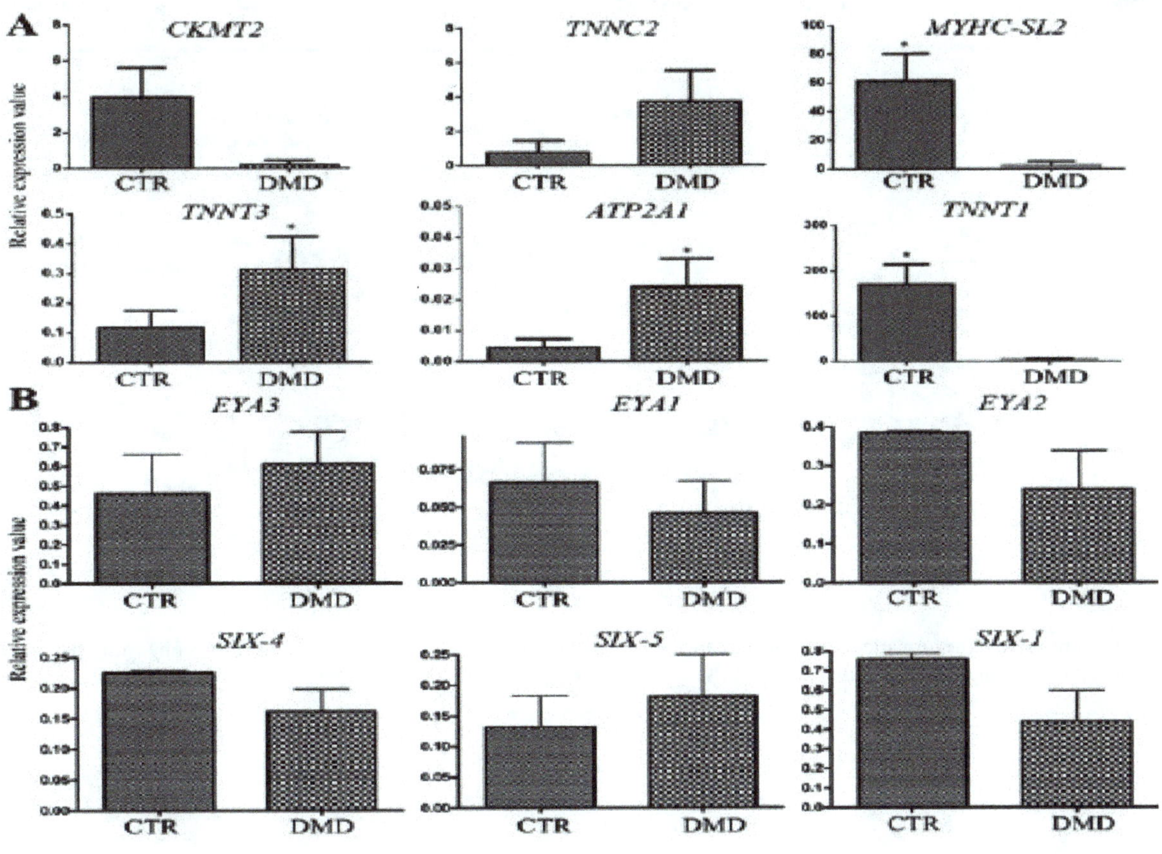

Figure 4: Schematic representation of the expression mutated genes expression diagram in patients with Duchenne scrotum dystrophy with MHC expression changes compared to normal group.

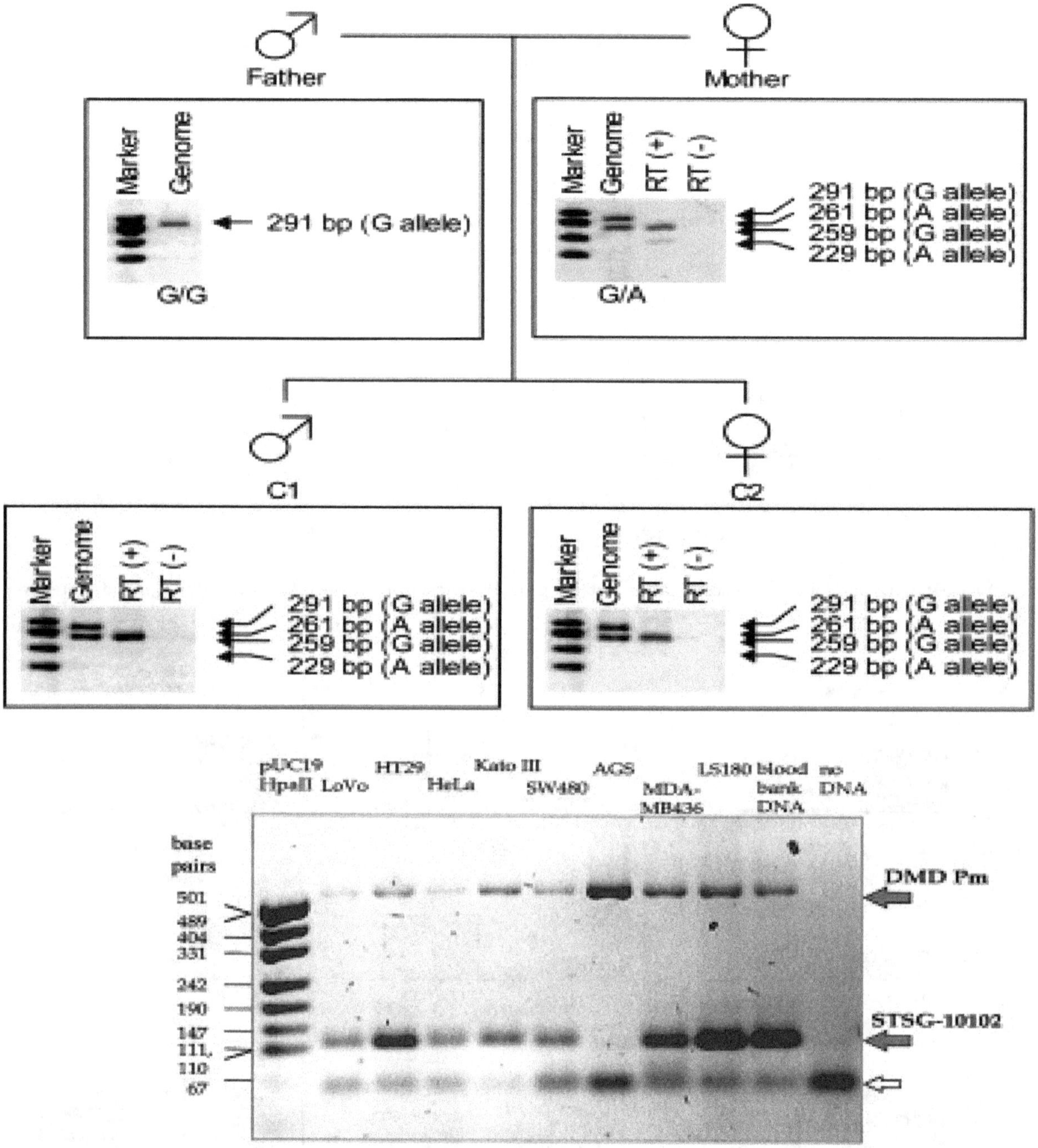

Figure 5: Schematic view of the formation of a target mutated genes band in patients with Duchenne scrotum dystrophy with a pattern of bond formation in parent genes carrying Duchene muscle dystrophy target mutations.

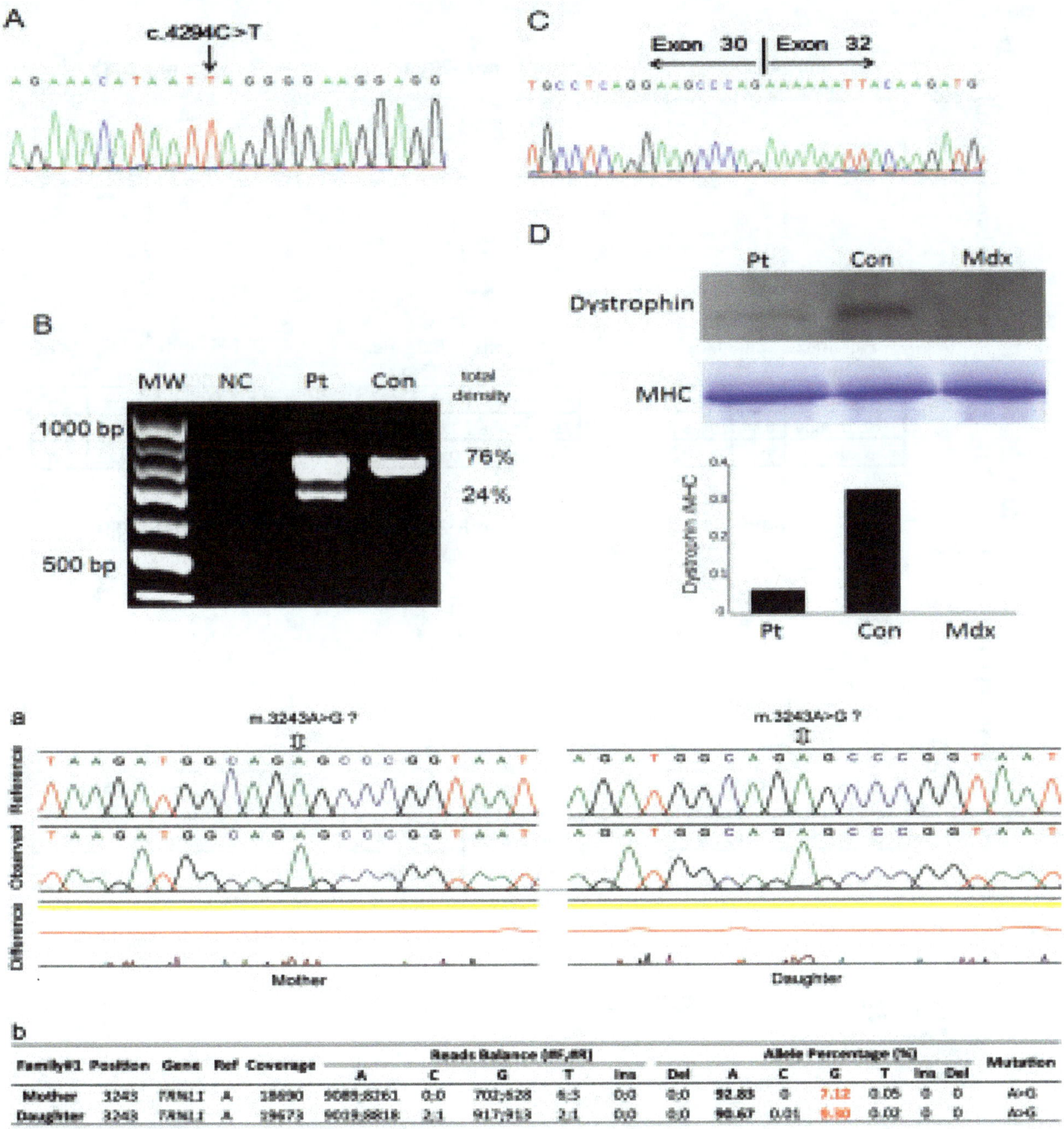

Figure 6: Schematic View of the Formation of the Dystrophin-mutated Gene Genes in Patients with Duchenne Sciaticular Dystrophy with the Bond Formation in the Genes of the Mothers and Daughters of the Duchenne Muscular Dystrophy Mutations.

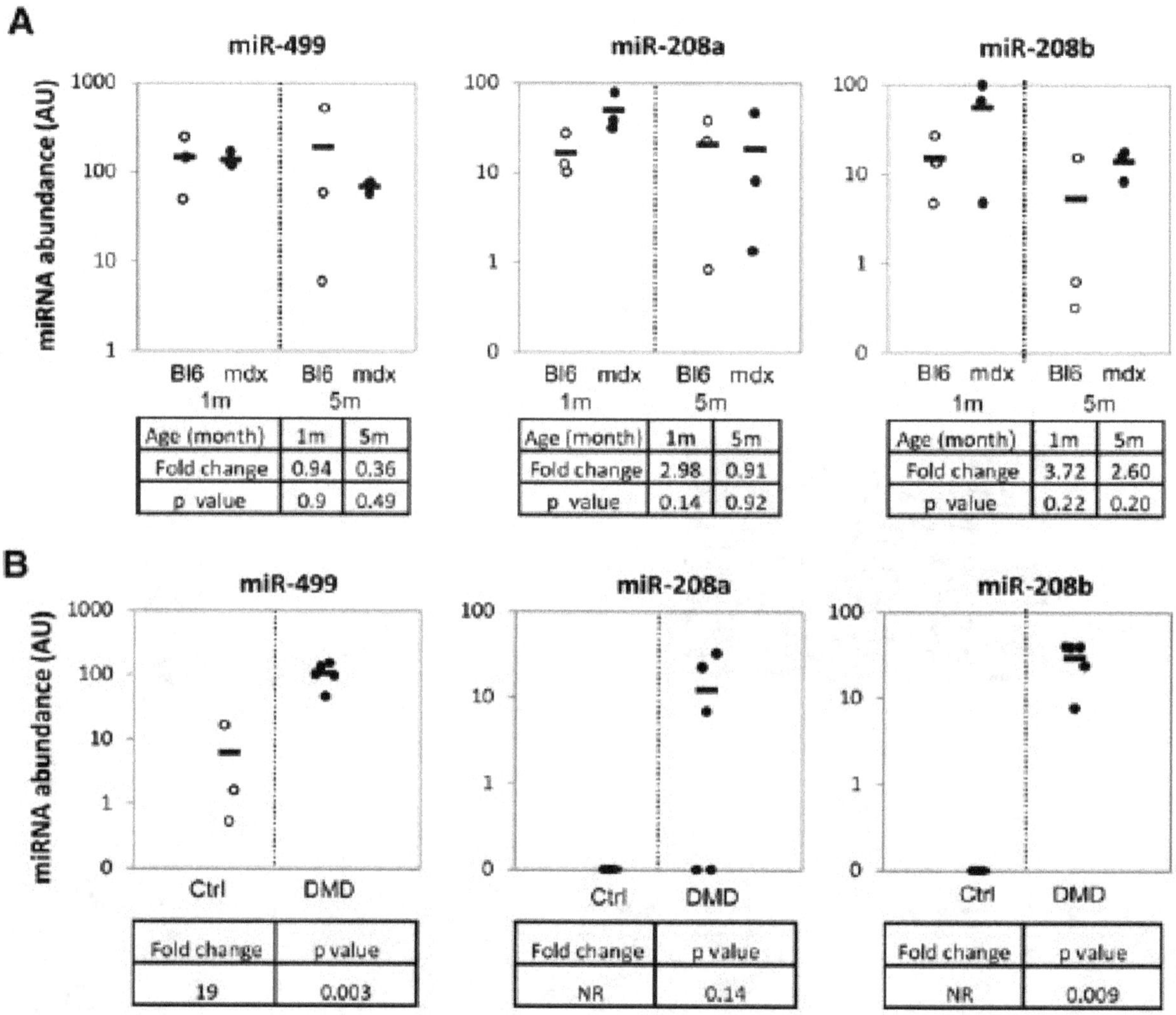

Figure 7: Schematic view of the pattern of altered mirna expression of dystrophin in patients with Duchenne Muscular Dystrophy.

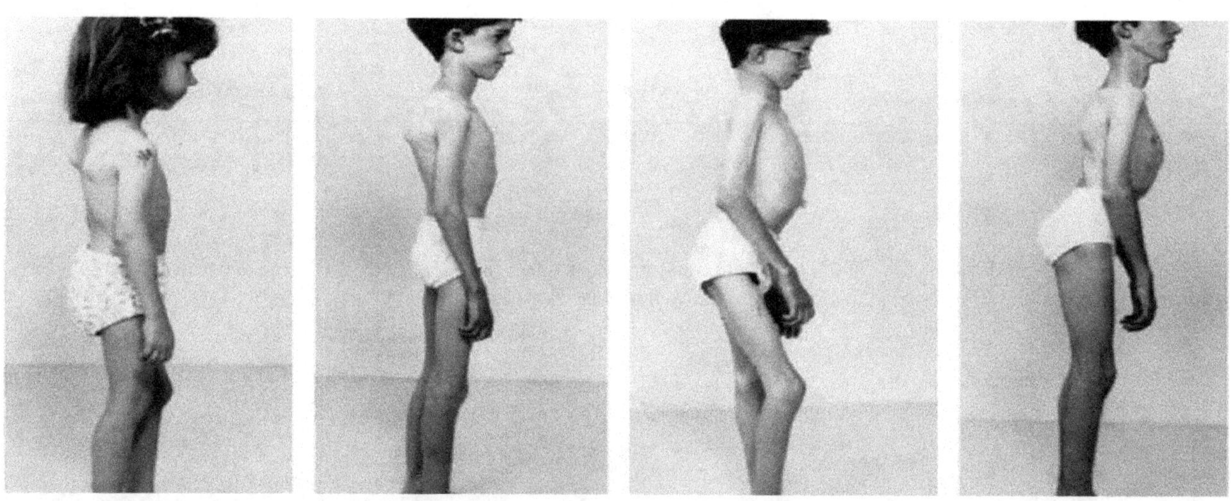

Figure 8: Images of Children with Duchenne Muscular Dystrophy.

Discussion

According to studies by Goemans NM and colleagues in 2011, similar results were obtained with the results of this study in the study of genetic mutations effective in Duchenne disease. Since this study was the first in Iran, so the need for further research into the Duchenne disease should be created by other scholars of the world for different human races in the world.[Figure 9]

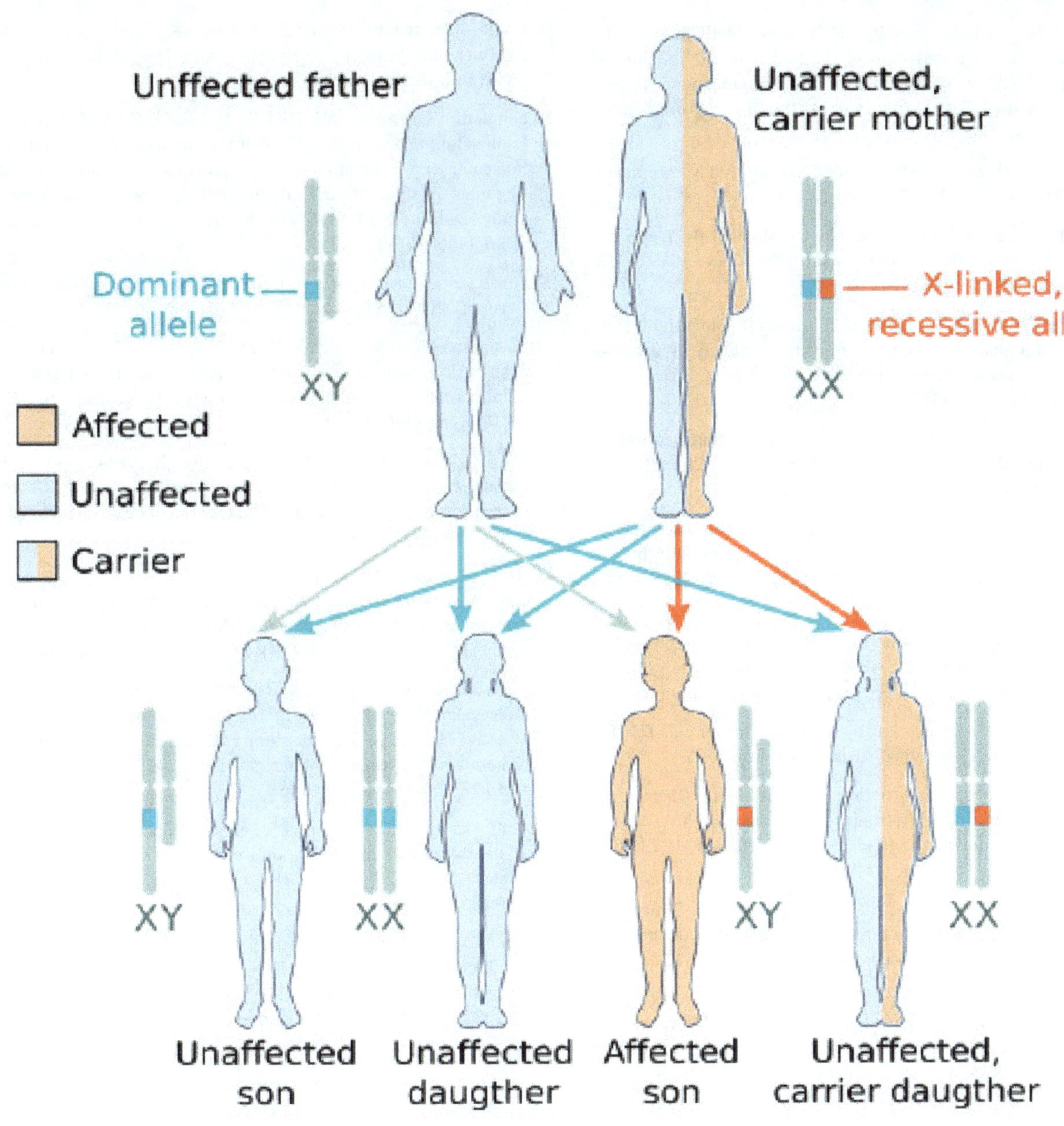

Figure 9: schematic view of an X Linked-recessive hereditary pattern that Duchenne muscular dystrophy also follows this pattern.

Conclusion

According to the results of sequencing the genome of patients with Duchenne muscular dystrophy disease, and the genetic mutations DMD, DYSF, EMD, LMNA, DUX4, DMPK, ZNF9, PABPN1genes found that about 100% of patients with Duchenne muscular dystrophy disease, they have this genetic mutations. Patients with Duchenne muscular dystrophy disease, unusual and frightening images in the process of Duchenne muscular dystrophy disease, experience. Lot epigenetic factors involved in Duchenne muscular dystrophy disease. But the most prominent factor to induce Duchenne muscular dystrophy disease, mutations is DMD, DYSF, EMD, LMNA, DUX4, DMPK, ZNF9, PABPN1 genes. This gene can induce the birth and can also be induced in the adulthood.

References

1. Muscular dystrophy: hope through research. NINDS. 2016
2. Rowland, L. P. Clinical perspective: phenotypic expression in muscular dhystrophy. In R. C. Strohman & S. Wolf (eds.), gene expression in muscle. New york, NY: Plenum Press. 1985;3-5
3. Nicholson LV, Johnson MA, Bushby KM, Gardner-Medwin D, Curtis A, Ginjaar IB, et al. Integrated study of 100 patients with Xp21 linked muscular dystrophy using clinical, genetic, immunochemical, and histopathological data. Part 1. Trends across the clinical groups. J Med Genet. 1993; 30(9): 728-736.
4. Muntoni F. Is a muscle biopsy in Duchenne dystrophy really necessary?. Neurology. 2001; 57(4): 574–575.
5. Flanigan KM, Andrew VN, Dunn DM, Alder J, Mendell JR, Weiss RB. Rapid Direct Sequence Analysis of the Dystrophin Gene. AJHG. 2003; 72(4): 931-939; Doi: 10.1086/374176
6. Dany A, Barbe C, Rapin A, Réveillère C, Hardouin JB, Morrone I, et al. Construction of a quality of life questionnaire for slowly progressive neuromuscular disease. Quality Life Res.2015; 24 (11): 2615–2623; Doi: 10.1007/s11136-015-1013-8
7. Falzarano MS, Scotton C, Passarelli C, Ferlini A. Duchenne muscular dystrophy: from diagnosis to therapy. Molecules. 2015;20(10): 18168–18184; Doi: 10.3390/molecules201018168
8. Marques MJ, Ferretti R, Vomero VU, Minatel E, Neto HS. Intrinsic laryngeal muscles are spared from myonecrosis in the mdx mouse model of Duchenne muscular dystrophy. Muscle nerve. 2017;35(3): 349–353; Doi: 10.1002/mus.20697
9. Ferretti R, Marques MJ, Khurana TS, Santo Neto H. Expression of calcium-buffering proteins in rat intrinsic laryngeal muscles. Physiol Rep. 2015;3(6): e12409; Doi: 10.14814/phy2.12409
10. De rosa G. Da conte a Duchenne [by conte in Duchenne]. Dm (in Italian). Unione italiana lotta alla distrofia muscolare.2005.
11. Dunckley MG, Manoharan M, Villiet P, Eperon IC, Dickson G. Modification of splicing in the dystrophin gene in cultured mdx muscle cells by antisense oligoribonucleotides. Hum Mol Genet. 1998;7(7): 1083–1090
12. Goemans NM, Tulinius M, van den Akker JT, Burm BE, Ekhart PF, Heuvelmans N, et al. Systemic administration of PRO051 in Duchenne's muscular dystrophy". N Engl J Med. 2011; 364(16):1513-1522; Doi: 10.1056/NEJMoa1011367
13. England SB, Nicholson LV, Johnson MA, Forrest SM, Love DR, Zubrzycka-Gaarn EE, et al. Very mild muscular dystrophy associated with the deletion of 46% of dystrophin. Nature.1990;343(6254): 180-182; Doi: 10.1038/343180a0
14. Wilton SD, Lloyd F, Carville K, Fletcher S, Honeyman K, Agrawal S, et al. Specific removal of the nonsense mutation from the mdx dystrophin mRNA using antisense oligonucleotides. Neuromuscul disord. 1999; 9(5): 330-338.
15. Goyenvalle A, Vulin A, Fougerousse F, Leturcq F, Kaplan JC, Garcia L, et al. Rescue of dystrophic muscle through u7 snRNA-mediated exon skipping. Science. 2004; 306(5702):1796–1799; Doi: 10.1126/science.1104297
16. Morgan JE, Pagel CN, Sherratt T, Partridge TA. Long-term persistence and migration of myogenic cells injected into pre-irradiated muscles of mdx mice. J Neurol Sci. 2003;115(2):191–200; Doi: 10.1016/0022-510X(93)90224-M
17. Dellavalle A, sampaolesi M, Tonlorenzi R, Tagliafico E, Sacchetti B, Perani L, et al. Pericytes of human skeletal muscle are myogenic precursors distinct from satellite cells. Nature Cell biology. 2007; 9(3): 255-267; Doi: 10.1038/ncb1542
18. Rodino-Klapac LR, Chicoine LG, Kaspar BK, Mendell JR. Gene therapy for Duchenne muscular dystrophy. Arch Neurol. 2007;64(9):1236–1241; Doi: 10.1001/archneur.64.9.1236
19. Sebahattin C, Virginia AG, Michela G, Lucy F, Silvia T, Karen A, et al. Exon skipping and dystrophin restoration in patients with Duchenne muscular dystrophy after systemic phosphorodiamidate morpholino oligomer treatment: an open-label, phase 2, dose-escalation study. Lancet. 2011; 378(9791): 595–605; Doi: 10.1016/S0140-6736(11)60756-3
20. Spreitzer H. Rimeporide. Österreichische apothekerzeitung (in German). 2015; 69(11):12.
21. Cavazzana-Calvo M, Payen E, Negre O, Wang G, Hehir K, Fusil F, et al. Transfusion independence and HMGA2 activation after gene therapy of human β-thalassaemia. Nature. 2010; 467(7313):318-322; Doi: 10.1038/nature09328
22. Yang ZJ, Zhang YR, Chen B, Zhang SL, Jia EZ, Wang LS, et al. Phase I clinical trial on intracoronary administration of Ad-hHGF treating severe coronary artery disease. Mol Biol Rep. 2009; 36(6): 1323-1329; Doi: 10.1007/s11033-008-9315-3
23. Hahn W, Pyun WB, Kim DS, Yoo WS, Lee SD, Won JH, et al. Enhanced cardioprotective effects by coexpression of two isoforms of hepatocyte growth factor from naked plasmid DNA in a rat ischemic heart disease model. J Gene Med. 2011;13(10): 549-555; Doi: 10.1002/jgm.1603
24. Deev RV, Bozo IY, Mzhavanadze ND, Voronov DA, Gavrilenko AV, Chervyakov YV, et al. pCMV-vegf165 Intramuscular Gene Transfer is an Effective Method of Treatment for Patients with Chronic Lower Limb Ischemia. J Cardiovasc PharmacolTher.2015;20(5): 473-482;Doi: 10.1177/1074248415574336
25. Aiuti A, Biasco L, Scaramuzza S, Ferrua F, Cicalese MP, Baricordi C, et al. Lentiviral hematopoietic stem cell gene therapy in patients with Wiskott-Aldrich syndrome. Science. 2013; 341(6148) :1233151; Doi: 10.1126/science.1233151
26. Lee TW, Southern KW. Topical Cystic Fibrosis Transmembrane Conductance Regulator Gene Replacement for Cystic Fibrosis-Related Lung Disease. Cochrane Database Syst Rev. 2013;(11): CD005599; Doi: 10.1002/14651858.CD005599.pub4
27. MacLaren RE, Groppe M, Barnard AR, Cottriall CL, Tolmachova T, Seymour L, et al. Retinal Gene Therapy in Patients with Choroideremia : Initial Findings from a Phase 1/2 Clinical Trial. Lancet. 2014; 383(9923):1129–1137; Doi: 10.1016/S0140-6736(13)62117-0
28. Tebas P, Stein D, Tang WW, Frank I, Wang SQ, Lee G, et al. Gene editing of CCR5 in autologous CD4 T cells of persons infected with HIV. N Engl J Med. 2014; 370(10):901-910; Doi: 10.1056/NEJMoa1300662
29. Thrasher AJ, Gaspar HB, Baum C, Modlich U, Schambach A, Candotti F, et al. Gene Therapy: X-SCID Transgene Leukaemogenicity. Nature. 2006; 443 (7109): E5-E6; Doi: 10.1038/nature05219
30. Blaese RM, Culver KW, Miller AD, Carter CS, Fleisher T, Clerici M, et al. T lymphocyte-directed gene therapy for ADA- SCID: initial trial results after 4 years. Science. 1995; 270(5235): 475–480.

A Review On Nk^{46} the Immunogenetic Predictive Candidate Gene of Unexplained Recurrent Miscarriages

S Daniel Chukwuemaka Ogbuabor[1], Peter U Achukwu[2], Alphosus O Ogbabor[1] and Nosa Terry Omorodion[2,3*]

[1] Department of Health Management, Faculty of Health Sciences and Technology, University of Nigeria, Nigeria

[2] Department of Medical Laboratory Sciences, Faculty of Health Sciences and Technology, University of Nigeria, Nigeria

[3] Health Services Department, University of Benin, Nigeria

***Corresponding author:** Omorodion Nosa Terry, Health Services Department University of Benin, Benin city, Edo state, Nigeria, E-mail: terry.omorodion@uniben.edu*

Abstract

Recurrent miscarriages of unknown etiology present a challenge to the clinician with a profound socio-psychological impact in addition to pathophysiological burden for affected couples. Currently, there exists a plethora of articles dealing with genetic polymorphisms associated with recurrent miscarriages with the polymorphic analysis being concentrated on a limited numbers of genes. The NKP^{46} gene a major regulator of natural killer cell function has a strong biological literature which suggests a putative role in the immunogenetics of unexplained recurrent miscarriages. Therefore plausible to propose the NKp^{46} gene a candidate gene in the etiology of unexplained recurrent miscarriages.

Keywords: NK Cells; NKP^{46}; Genes; Immunogenetics;

Introduction

In humans pregnancy may be interrupted at various stages of development. Receptive interruptions of pregnancy resulting in loss are referred to as Recurrent Miscarriages. Miscarriage is defined by the World Health Organization as a loss of a viable pregnancy before 24 weeks of gestation [1]. Miscarriages remain the most common pregnancy complications [1]. The exact frequency of miscarriages is, however unknown as miscarriages frequently occur before the woman is aware of her pregnancy. It is estimated that more pregnancies are lost spontaneously than actually carried to term [2,3]. Suggested risk factors for recurrent miscarriages include: Genetics, immunologic, anatomic, endocrinilogical and environmental factors [3]. However, in about 50% of cases, the etiology is unknown. Since a proportion of recurrent miscarriages are thought to be immunologic and genetic origin, investigations of genes involved in biological mechanisms within the feto-maternal interface becomes suggestive [2]. Sequence variations have been located in genes involved in immunologic mechanisms.

The CTL-A gene is a regulator of T-cell activation and an A/G polymorphism in exon 1 of this gene has been reported to be associated with recurrent miscarriages in the Chinese population [3]. Polymorphisms in the HLA –G promoter region, the HLA-G *0104 and *0105 alleles and the HLA-DRBI *1505 allele have been equally associated with the immunopathogenesis of recurrent miscarriages but polymorphisms has not yet been studied extensively on many other genes including the NKP^{46} gene [4-6]. Studies have also shown higher levels of nucleotide polymorphisms in many other populations [7]. Currently, research is focused on cellular constituents or processes to explain the pathophysiologic mechanisms underlying recurrent miscarriages [7]. Natural killer cells have been extensively evaluated as probable contributory factors [8]. They are the predominant leucocytes populations present in the endometrium during the period of implantations and early pregnancy, functioning as local immunomodulators in the regulation of trophoblast and placental growth [8]. Being the predorminant leucocytes present in implantation sites during the first trimester. , a role for this particular interaction regarding materno-fetal tolerance is suggestive in recurrent miscarriages , the fetus being semi allogenic in nature [9,10]. Previous studies have reported that reduction in the number of peripheral natural killer cells enhances the progression of normal pregnancy and that an increased natural killer cell populations is associated with miscarriages [8]. Polymorphisms in some natural killer cell regulatory genes as a result may be an unexplainable factor underlying recurrent miscarriages in some individuals. The genetic origin of disease either partly or wholly due to abnormalities within the genetic code. Genetic polymorphisms are variations in DNA that are observed in 1% or more of the population [11]. Most common polymorphisms are potential regulatory polymorphism located in noncoding regions and may have np discernable effect on the protein products [11,12]. However, some polymorphisms may alter protein structure and function through a single nucleotide base substitution in a gene's coding region and may increase or decrease gene expression either by affecting mRNA stability [11]. Such polymorphism could as a result present apathophysiological conditions. The

***Corresponding author email:** terry.omorodion@uniben.edu*

study of genetic polymorphisms helps define pathophysiologic mechanisms, identify individuals at risk for disease and suggest novel targets requiring only access to a polymerase chain reaction machine, funding for reagents and DNA and samples from cases and controls [11,12].

NK Cells

Natural killer cells are the most recently describes of the innate immunocompetent cells. There is vivid increase in the number of these cells during the secretory phase of the menstrual cycle as progesterone levels increase. These numbers rise further if the pregnancy ensues. Since progesterone levels drop at the end of the menstrual cycle and NK cells also undergo death, it has been proposed that their survival depends on progesterone level [13]. NK cells become granulated and come in contact with vessels and endometrial glands [13]. They are largely granular lymphocytes that populate the materno- foetal interface during pregnancy, protecting the mother and foetus. The materno-foetal interface during pregnancy represents an immunological battle-field characterized by direct contact between foetal antigens and maternal immunocompetent cells from theuterus, the foetus being semi allogenic in nature. Therefore, for successful pregnancy to occur the immune response by immunocompetent cells such as NK cells is modulated at the materno-foetal interface throughout gestation and recurrent miscarriages maybe a result of a failure of such modulations in some individuals. Phenotypically, natural killers cells are characterized by the expression of CD56 and CD16 receptors on their cells surface. Based on the concentration of CD56 antigen, they can be divided into two subpopulations, CD56dim, and CD56 bright. CD56dim cells demonstrate high toxicity invitro whereas CD56 bright cellsexhibit low cytotoxic capacity; however the CD56bright produces important immunoregulatoy cytokines particularly interferon gamma (IFN-Y). Approximately 90% of peripheral natural killer cells are CD56dim and express high level of CD16 while the remaining 10% are CD56bright with minimal or no expression of CD16. Several studies have demonstrated that peripheral natural killer cells do not change during menstrual cycle; however numerical and functional decrease has been observed during pregnancy suggesting the failure of their regulation as a major factor in recurrent miscarriages [14]. The human endometrium contains a substantial population of natural killer cells referred to as uterine natural killer cells (uNkcells). Atlthough there is no concensus about the origin of uterine natural killer cells, it is believed that mature peripheral natural killer cells or immature precursors may migrate into the endometrium from the blood possibly in response to chemokines produced by cells within the endometrium. The production of CXCL-12 by extravillous trophoblast (EVT) cells has been linked to be able to attract natural killer cells into the individual in pregnancy while transforming growth factor beta 1(TGF-BETA 1) has equally been implicated to modify peripheral blood natural killer cells to uterine natural killer cells. Studies have shown that endometrial natural killer cells participate in the materno-foetal interactions during implantation, trophoblast invasion, placentation, organogenesis and foetal development which represent a new perspective in the field of reproductive immunology [14]. The prevalence of NK cells in women with RM is unclear. A recent study had concluded that women with recurrent miscarriages had a significantly higher NK percentage than controls [15].

NKp^{46} Gene

The NKp^{46} gene (NCRI, CD335) is a member of natural cytotoxicity receptors; activating receptors capable of inducing natural killer cell mediated cytotoxicity. It is stably expressed and specifically present on all resting and activated human natural killer cells [16]. NKp^{46} is a 46KDa type-1 transmembrane glycoprotein that is not only a member of the natural cytotoxic receptors (NCRs) family but also a member of the immunoglobulin superfamily [16]. It is a 9 exon count protein encoded by the NCR1 gene located on the human chromosome close to the highly polymorphic Leucocyte Receptor Complex. NKp^{46} gene is currently considered the most reliable identifying marker for natural for natural killer cells across species [16]. Natural Killer cells are effector regulatory lymphocytes of the innate immune system that contribute to tumor surveillance hematopoietic allograft rejection, control of microbial infections and pregnancy [17]. They have also been suggested to provide a link between innate and acquired immunity through production of cytokines and interaction with antigen –presenting cells [17]. Since the NKp^{46} gene is thought to be the main natural killer cell activating receptor and also proven to display functional cross talk with other natural cytotoxic receptors and natural killer cell activating receptors, we propose a strong involvement of this gene in the pathogenesis of unexplained recurrent miscarries suggesting it a candidate gene for the condition [18].

The Candidate Gene Approach

The candidate gene is a gene of documented biological activity involved in the pathophysiology or biological pathways of a given medical condition with polymorphic activity. This condition is a major factor of inter-individual variability [19]. New effective methods for genetic screening together with the information offered by the Human Genome project have made the candidate gene approach a commonly used method to search for disease genes as information on physical locations and sequence arrangement of many genes are available. The candidate gene approach can be applied when the biochemical or physiological background of the defect of interest is known, when a chromosomal region has been linked to a disease or if an animal model of a disease has established [20,21]. Interaction partners of identified proteins defective on a disease could as well be considered as a candidate gene for disease of interest. The mouse is widely used model organisms for studying mammalian gene functions and genes causing a phenotype in a mouse can be used as a candidate gene in human disease studies. The NKp^{46} gene has been extensively studied in the mouse and proven to excite same qualities in human. This is however speculative.

Conclusion

Progress in natural killer cell receptor genetics may likely prove their significance in pathological conditions such as recurrent miscarriages. We propose a genome wide association study of theNKP^{46} gene as a predictive candidate gene in the etiology of unexplained recurrent miscarriages. It is however, the unlimited potentials of genetics to help predict who will get a disease and /or who once diagnosed with a disease will have an unfavorable prognosis that inspired review.

References

1. Lara M, shillito J, Tang T. Preventing recurrent miscarriages of unknown etiology. Obstet Gynecol. 2013;15(2):99-105.
2. Rai R, Regan L. Recurrent miscarriage. Lancet. 2006;368(9535):601-611. doi:http://dx.doi.org/10.1016/S0140-6736(06)69204-0
3. Stephenson M, Kutteh W. Evaluation and management of Recurrent Early pregnancy loss. Clinical obstet Gynecol. 2007;50(1):132-145.
4. Aldrich CL, Stephenson MD, Karrison T, Odem RR, Branch DW, Scott JR, et al. HLA-G Genotype and pregnancy outcome in couples with unexplained Recurrent Miscarriages. Mol Hum Reprod. 2001;7(12):1167-1172.
5. Ober C, Aldrich CL, Chervoneva I, Billsrand C, Rahimov F, Gray HL, et al. Variation in the HLA-G promoter region influences miscarriage rate. Am J Hum Genet. 2003;72(6):1425-1435. doi:10.1086/375501
6. Takakuwa K, Adachi H, Hataya I, Ishii k, Tamura M, Tanaka M. Molecular Genetics Studies of HLA-DRBI Alleles in patients with unexplained Recurrent Abortion in the Japanese population. Human Reprod. 2003;18(4):728-733.
7. Cambell MC, Tishkoff SA. Africa Genetic Diversity: Implications for Human Demographic History, Modern Human Origins and complex Disease Mapping. Ann Rev Genomics Human Genetics. 2008;9:403-433 doi:10.1146/annurev.genom.9.081307.164258
8. Diejomaoh MF. Recurrent spontaneous miscarriages is still a challenging diagnostic and therapeutic quaqmire. Med Princ Pract. 24(suppl):38-55. doi:10.1159/000365973
9. Garcia CA, Robbinson J, Alejandro Madrigal J, Marsh SGE. Natural Killer Cell Receptors: Functional Roles. Immunologia. 2003;22(2):190-202.
10. Caligiuri AM. Human Natural Killer Cells. Blood. 2008;112:461-469. doi:https://doi.org/10.1182/blood-2007-09-077438
11. Michael CL, Maliarik M, Rybicki B. Genetic polymorphisms in lung diseases: bandwagon or breakthrough? Respir Res. 2002;3(1):15. doi:10.1186/rr164
12. Albert RP. What is a functional genetic polymorphism? Defining classes of functionality. J Psychiatry neurosci. (2011); 36(6):363-365. doi:10.1503/jpn.110137
13. Manaster I, Mandelboim O. The unique properties of uterine NK cells. Am J ReprodImmunol. 2010;63(6):434-444.
14. Rodriques MN, Favaron PO, Dombrowski JG, Souza RMD, Miglino MA. Role of Natural Killer (NK) Cells during pregnancy; A review. Open Journal of Animal Sciences. 2013;3(2):138-144. Doi:10.4236/ojas.2013.32021
15. King K, Smith S, Chapman M, Sacks G. Detailed analysis of peripheral blood natural killer (NK) cells in women with recurrent miscarriage. Hum Reprod. 2010;25(1):52-58. doi:10.1093/humrep/dep349
16. Noronha EL, Harman MR, Wagner B, Antczack FD. Generation and characterization of Monoclonal antibodies to equine NKP^{46} . Vet Immunol Immunopathol. (2012); 147(1-2):60-68. doi:10.1016/j.vetimm.2012.04.003
17. Storset KA, Kulberg S, Berg I, Boysen P, Hope CJ, Dissen E. NKP46 defines a subset of bovine leucocytes with natural killer cell characteristics. Eur J Immunol. (2004); 34(3):669-676. doi:10.1002/eji.200324504
18. Mrosikiewicz AS. Significance of genetic polymorphism investigations in pregnancy complications. Archives of perianal medicine. 2013;19(1):7-11.
19. Collins FS. Positional cloning: Lets not call it reverse anymore. Nat genet. 1992;1(1):3-6. doi:10.1038/ng0492-3
20. Ballabio A. The rise and fall of positional cloning? Nat genet. 1993;3(4):277-279.
21. Zhu M, Zhao S. Candidate gene identification approach: Progress and Challenges. Int J Biol Sci. 2007;3(7):420-427.

Antibody Response in Broiler Chickens Infected with Different Developmental Stages of *Eimeria Tenella*

*Kaze Paul Davou[1], Idris Lawal[2], Ajanusi Joseph[2], Saidu Lawal[3], Dogo Goni Abraham[1]

[1]Department of Veterinary Parasitology and Entomology, Faculty of Veterinary Medicine, University of Jos, Nigeria.

[2]Department of Veterinary Parasitology and Entomology, Faculty of Veterinary Medicine, Ahmadu Bello University, Zaria, Nigeria.

[3]Department of Veterinary medicine, Faculty of Veterinary Medicine, Ahmadu Bello University, Zaria, Nigeria.

***Corresponding author:** *Kaze Paul Davou, Department of Veterinary Parasitology and Entomology, Faculty of Veterinary Medicine, University of Jos, Nigeria, Email: paulkaze@yahoo.com*

Abstract

Antibodies (IgG or IgY) titre values were higher in broilers sera infected with sporulated oocyst and merozoites reaching a peak on day 10 of post primary and secondary infections and day 5 post tertiary infection in sera of broilers (treated and non- treated). At tertiary infection, antibodies increases at day 5, 7, 11 and 14 indicating that antibodies increases in broilers infected with the invasive or zoite stages, (sporozoite and merozoite) of the parasite.. There was a significant difference in the antibody output between the sera of the broiler groups ($p<0.05$).

Introduction

Coccidiosis caused by protozoans of genus *Eimeria tenella* is a chicken parasitic disease of great economical importance globally characterized by haemorrhage leading to mortality. Conventional disease control strategies depend on vaccination and prophylactic use of anticoccidial drugs. Research has been carried out worldwide to try to elucidate the mechanism of protective immunity against coccidiosis. It was concluded from early studies that cellular immunity is the key to protection against *Eimeria*, whereas humoral immunity plays a very minor role in resistance against infection. By contrast, other studies have pointed towards the ability of antibody to block parasite invasion, development and transmission and to provide passive and maternal immunity against challenge infection. Herein, recent results demonstrate the ability of antibodies (raised by live immunization or against purified stage-specific *Eimeria* antigens) to inhibit parasite development *in vitro* and *in vivo* and readdress the question of the role of antibody in protection against coccidiosis [1]. Enzyme-linked immunosorbent assay has already been used extensively to measure anti-*Eimeria* antibodies in chickens [2]. The aim of the study was to determine the antibody response in Broilers infected with different developmental stages of *Eimeria* tenella.

Materials and Methods

Study Area

The experimental settings were at the PETCA building, Anguldi, 5 kilometers from the National Veterinary Research Institute, Vom, Jos Plateau State, Nigeria, where the laboratory work was carried out. The Jos Plateau lies on the pre-cambian from the cambian to jurasic northern Nigeria crystalline complex in central Nigeria. Its average elevation is about 1,250 m above mean sea level. The state is bounded on the north and west by Kaduna plains (on the average of 600 m above mean sea level) and on the south by Benue plains (on the average of 700 m above mean sea level), [3]. Geographically, the Jos Plateau is located between latitude 08°24'N and longitude 008°32' and 010°38' east. The land surface of Jos Plateau consists of plains, hills, depressions and todes of various forms, shapes and sizes. It is a major tourist centre in Nigeria with agriculture as the main occupation of the people. The high altitude confers on the Plateau lower temperature than those encountered elsewhere in Nigeria except the Obudu and Mambilla Plateau. The dry season is determined by the north easterly tropical continental air masses known as harmattan (from October – April) and the wet season is the most tropical maritime air masses from May – September. The average annual rainfall is about 1,100 mm and is evenly distributed. Another element of climate is temperature December and January experience temperatures below 15^{0}C. During February and March, the temperature rises again about 25^{0}C. Most of the human activities are mining and agriculture involving rearing of chickens in both the rural and urban areas for subsistency and income [3].

Experimental Birds

Four hundred (400) day-old broilers (marshal breed) were

***Corresponding author email:** *paulkaze@yahoo.com*

purchased from ECWA farms, Jos, brooded and used for the study. The birds were randomly distributed into six different groups of 40 each, in a clean wire cage (n = 40). At two weeks old, each group was again subdivided into two, treated and non treated, of twenty broilers (n =20) each. The birds were kept in a clean building, and the legs banded or labelled under strict biosecurity measures. Feed (Broiler starter, Grand cereals and oil mills, PLC, Zawan, Jos-Plateau, Nigeria) and water were provided *adlibitum* The birds were vaccinated with Newcastle disease vaccine (La-Sota) at day 21 and Gomboro disease vaccines at days 14 and 28.

Experimental infection of broilers with infective materials and monitoring

The experimental birds, except the control were orally given primary and secondary challenge infections with the various developmental stages of *Eimeria* tenella, respectively at week 2 and $3^{1/2}$ while at week 5 of age, all birds were infected the sporulated oocyst of the parasite (Table 1). Each group was subdivided into Treated (n = 20) and Non- Treated (n = 20). In each infected group, birds in one of the subdivisions were treated

Table 1: Experimental infection of broilers with developmental stages of *Eimeria tenella*.

Group	Treatment and No. of birds	Infection type/ Age of bird		3o /wk 2 challenge with virulent E. tenella
		Io /wk 2	2o /wk 3	
I	T(n=20)	10^5 USO	10^5 USO	10^5 SO
	NT(n=20)	10^5 USO	10^5 USO	10^5 SO
II	T(n=20)	10^5 SO	10^5 SO	10^5 SO
	NT(n=20)	10^5 SO	10^5 SO	10^5 SO
III	T(n=20)	10^5 SCZ	10^5 SCZ	10^5 SO
	NT(n=20)	10^5 SCZ	10^5 SCZ	10^5 SO
Iv	T(n=20)	10^5 MRZ	10^5 MRZ	10^5 SO
	NT(n=20)	10^5 MRZ	10^5 MRZ	10^5 SO
V	T(n=20)	10^5 GMT	10^5 GMT	10^5 SO
	NT(n=20)	10^5 GMT	10^5 GMT	10^5 SO
VI		0	0	0

KEY: USO-Unsporulated oocyst; SO- Sporulated oocyst; SCZ- Schizoites; 1o- primary infection; 2o- Secondary infection; 3o- Tertiary infection; WK-Week; T –Treated; NT – Non treated; MRZ- Merozoites; GMT- Gametocytes;

with amprolium 250 WSP[R] Holland was administered in drinking water at a concentration of 250 mg/1 (0.025%) for a period of 5 days as prescribed by the Manufacturer at the appearance of visible clinical signs.

To obtain serum, blood samples were collected from the experimental birds using the method described by Talebi and Mulcahy [4]. Briefly, 1 ml of blood sample was obtained from the wing vein of each bird using 20 gauge needle (Becton Dickson co., Plymouth, UK) into a 2 ml vacutainer. Samples were obtained on days 2, 4, 6, 8, and 10 after primary and secondary infections, and on days 5, 7, 11, 14, 17, 20 and 24 after tertiary infection [5]. The blood which had been allowed to clot for 1 hour at room temperature was left over night at 4°C and then centrifuge at 800g for 5 minutes. The serum samples were thereafter heated at 56°C for 30 minutes to inactivate the compliment before storage at -20°C. All sera were analyzed with the developed ELISA Triplicate.

Enzyme-linked Immunosorbent Assay

Hay Dottom Nune certified microtiter plates (Roskilde Denmark) were coated with 50 μ1 of soluble *E. tenella* antigen (sporozoites from characterized sporulated *Eimeria tenella* oocysts)/web at a concentration of 5 μ g/ml carbonate buffer (pH 9.6) for 1 hour at 39°C. The plates were rinsed five times with saline/tween (S/T), and treated with 75 μ l of PBS containing 3% BSA, 1% rabbit serum and 0.05% sodium azide for 1 hr at room temperature to block non-specific adsorption. The plates were washed five times with saline/tween (S/T). A 50 μ l test serum sample, diluted 1 : 1000 in PBS-T (including 1% rabbit serum and 0.05% sodium azide) was added to each well and incubated for 2 hours at room temperature. The plates were washed 5 times with S/T and 50 μ l of 1:1000 dilution of rabbit anti chicken immunoglobulin peroxidase (Pelfreeze Rogers, Arkansas) in PBS-T was added. After 2 hours incubation, plates were washed five times with S/T and freshly prepared substrate solution (2mM OPD 6.15Mm H_2O_2 in 0.1M citrate buffer pH 6.0) was added per well. The enzyme-substrate reaction was stopped after 30 minutes by addition of 100 μ l to each well of 2N H_2SO_4. Absorbance were measured at 492 nM (A_{492}) in a Biotele ELISA Reader (Ref S1118170, Multiskan Ex, USA). All serum samples from the experiment were analyzed on a single day.

Enzyme linked Immunospot Assay

The spleen was crushed by pressing on fine mesh Petri dishes containing RPMI-1640 (Sigma, Aldrich Cheme, GmbH, Germany). The suspension was then passed through nylon cell strainer (70μm; Becton, Dickson, Lincoln Park, NJ). The filtrate was centrifuged at 250 g for 10 minutes at 4oC and the sediment collected. Lysis buffer (1ml/ spleen) was added for erythrocyte lysis and placed on ice for 2 minutes. The suspension was passed through the cell strainer again and was centrifuged again 250 g for 10 minutes at 4 °C to collect the sediment. The cell suspension (10μl) was mixed with same amount of trypan blue (Sigma, Aldrich Cheme GmbH, Germany) and the number of cells was counted in a haemocytometer. The centrifuged ion was adjusted to 10^6 cells / 100 μl with RPMI-1640. Nitro cellulose – microlitre plates (96 wells, Millipode multiscreen MAHA) were used in the ELISPOT assay. Individual wells of the plate were filled with 100 μl of goat anti-chick 1g (H+L) – UNLB (primary antibody) at a final concentration of 2 μl/ml and were allowed to stand overnight at 4°C in a humid chamber. Unadsorbed antibodies were removed by three successive washings with PBS. Wells were immediately filled with 100 μl RPMI – 1640 to saturate the remaining finding sites and incubated at 37°C for 2hrs. The medium was discarded and the plates were dried with absorbent paper incubation of Ig-secreting cells. A 100 μl cell suspension containing 10^6 cells was dispensed into each well in duplicate and they were incubated undisturbed at 37°C for 4 hours. The plates were rinsed twice by immersion in PBS containing 0.05% Tween 20 (PBST) for 2-3 minutes. The wash buffer was removed from the plates and the outer surfaces of the plates were dried carefully. A 100 μl of PBST containing Goat anti-chick IgG-AP (1000-fold dilution) were added to each well and the plates incubated at 4°C over night. The plates were then rinsed three times by immersion in PBST and dried. Each well was then filled with 100 μl BCIP/NBT solution; prepared by adding 66μl of NBT (containing 50 μl/ml nitroblue tetrazolium in 70% N, N-dimethylformamide) in alkaline phosphate buffer (containing 5.8 g Nacl, 0.1 g $Mgcl_2$ 12.1 g Tris). The plates were thoroughly washed with running tap water and air-dried for 24 hours. Blue spots showing fuzzy borders were considered positive for immunoglobulin G (IgG).

Results

Antibodies (IgG or IgY) titre values were higher in sera from broilers infected with sporulated oocyst and merozoites reaching a peak on day 10 of post primary and secondary infections and day 5 post tertiary infection in both broilers treated and non treated, (Figure 1,2,3). The antibodies values were relatively low in broilers infected with unsporulated oocysts, schizonts and gametocytes at primary and secondary infection in both treated and non treated broilers at day 10 (Figure 1,2). At tertiary infection, antibodies increases at day 5, 7, 11 and 14 (Figure 3)

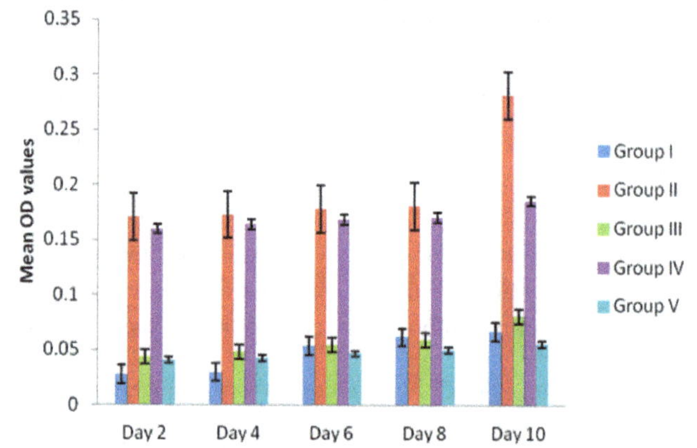

Figure 1: Antibodies level in sera of the experimentally infected broilers with the different stages (unsporulated oocyst, sporulated oocyst, schizonts, merozoites and gametocytes) of *Eimeria tenella* at optical density (O.D) or absorbance of 492 nm) at primary infection.

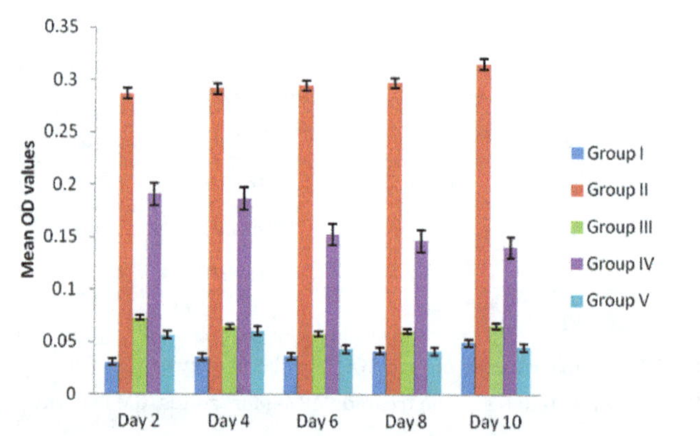

Figure 2: Antibodies level in sera of the experimentally infected broilers with the different stages (unsporulated oocyst, sporulated oocyst, schizonts, merozoites and gametocytes) of *Eimeria tenella* at optical density (O.D) or absorbance of 492 nm) at secondary infection.

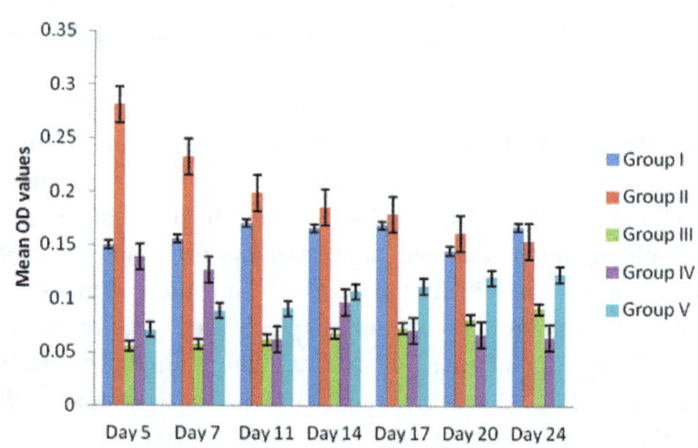

Figure 3: Antibodies level in sera of the experimentally infected broilers with the different stages (unsporulated oocyst, sporulated oocyst, schizonts, merozoites and gametocytes) of *Eimeria tenella* at optical density (O.D) or absorbance of 492 nm) at tertiary infection.

respectively. Generally, antibodies levels of sera from the infected broilers with the different developmental stages of the parasite, treated and non treated increased post inoculation and after reaching peak levels, they began to decline (Figure 3). The control birds show no antibodies in the sera. The study demonstrates a non significant difference in the antibody titre values of the treated and non treated sera of the infected broilers groups (II and IV), $p<0.05$.

Discussion

Serum antibody levels increased rapidly on day 10 in the broilers at primary and secondary infections and day 5 at tertiary infection. This varied from the reports Bumstead, et al. who recorded a peak of humoral immune response between day 14 and 21 post coccidial infection in birds [6]. This may be due to differences in the immunogenic potential of the isolate, age, environment and genetic background of the birds. Antibodies remain significantly high in broilers infected with sporulated oocyst (sporozoite) and merozoite at the end of each infection period, suggesting that the level of antibodies appears to be related to the severity of the developmental stage of the parasite. This is in concordance with the reports of Constantinoiu et al. who reported high antibodies levels persistence in commercial flock after natural exposure to Eimeria or following infection with live vaccine [7]. The present study revealed that there was no significant difference in the antibody titre values in the treated and non treated broilers. This is consistent with the results of Kiani and Farhang, but is inconsistent with the reports of Kurkure et al. who stated that chicks treated with coxynil showed higher antibody titre values than those maintained on feed without coxynil [8,9,10]. There are still debates on antibodies inducing protective immunity. Dalloul and Lillehoj, stated that antibodies play a minor role as cell mediated immunity (CMI) [11]. Gilbert et al., reported that the levels of serum antibodies following infection do not correlate with protection or oocyst output and antibody levels in chickens [4,12]. This variation may be due to the age, dose and strain of the parasite as well as the genetic background of the broilers. However, the study agrees with the findings of Rose who showed that antibody could have deleterious effects, including agglutination, lysis, neutralization of infectivity and morphological changes on various developmental stages of Eimeria if they come in close contact with the parasite [13]. The first subunit vaccine (CoxAbicR) is based on transfer of protective antibodies from immunized hens to embryo indicating that antibodies do play an important role in immunity [14].

Conclusion

The following can be concluded from the results obtained:

1. The sporozoites and merozoites showed strong infectivity and elicited stronger antibody titre values in infected broilers with sporulated oocysts and merozoites in infected birds at primary-secondary-tertiary infections, indicating that they might be potential vaccines candidates against avian coccidiosis.

2. The immunoglobulins were IgG or IgY

References

1. Wallach M. Role of antibody in immunity and control of chicken coccidiosis. Trend in Parasitology. 2010;28:382-387.
2. Smith N, Wallach M, Miller CMD, Morgenstern R, Braun R, Eckert J. Maternal transmission of immunity to Eimeria maxima. Infection and Immunity. 1994;62(11):1348-1352.
3. Plateau Agricultural Development Programme (PADP). (2002). Annual Report 2002.
4. Talebi A, Mulcahy J. Correlation between immune responses and oocyst production in chickens monospecifically infected with Eimeria maxima. Avian Pathology. 1995;24(3):485-495.
5. Rose ME, Hesketh P. Immunity to Coccidia in Chickens:Adaptive transfer with peripheral blood lymphocytes and spleen cells. Parasite Immunology. 1982;4(3):171-185.
6. Bumstead JM, Bunstead N, Rothwell L, Tomley FM. Comparision of immune responses in bred lines of chickens to Eimeria maxima and Eimeria tenella. Parasitology, 1995;III:143-151.
7. Constantinoiu C, Molloy J, Jorgensen W, Coleman G. Development and validation of an ELISA for detecting antibodies to Eimeria tenella in chickens. 2017;150(4):306-313.
8. Conway D, McKenzie M. Poultry Coccidiosis Diagnostic and Testing Procedures. 3rd Edn, 2121 state Avneu, Ames, lowa, USA. Veterinary Parasitology, 2007;150:306-313.
9. Kiani R, Farhang HH. Development of an ELISA test for serological diagnosis of coccidial infections and studying of resistance against coccidiostats based on flock history. Asian Journal of Biological Sciences. 2008;1:77-83.
10. Kurkure NV, Kolte SW, Bhandarkar AG, Kalorey DR. Evaluation of herbal coccidiostat 'Coxynil' in broiler. Indian Journal of Experimental Biology. 2006;44(9):740-744.
11. Dalloul RA, Lillehoj HS. Recent advances in immunomodulation and vaccination strategies against coccidiosis. Avian Diseases. 2005;49(1):1-8.
12. Gilbert JM, Bhanushali J, McDougald LR. An Enzyme Linked Immunosotbent Assay (ELISA) for coccidiosis in chickens: Correlation of Antibody levels with Prior Exposure to Coccidia in the laboratory and in the field. Parasite Immunology. 1998;30:97-100.
13. Rose M. Immunity to Eimeria infections. Veterinary Immunology and Immunopathology Journal. 1987;17(1-4):333-343.
14. Belli SI, Mai K, Skene CD, Gleeson MT, Witcombe DM, Katrib M, et al. Characterization of the antigenic and immunogenic properties of bacterially expressed, sexual stage antigens of the coccidian parasite, Eimeria maxima. Vaccine. 2004;22:4316-4325.

20

Some Attempted Strategies towards the Control of Avian Coccidiosis

Kaze Paul Davou[1*], Idris Lawal[2], Ajanusi Joseph[2] and Saidu Lawal[3]

[1]*Department of Veterinary Parasitology and Entomology, Faculty of Veterinary Medicine, University of Jos, Nigeria.*

[2]*Department of Veterinary Parasitology and Entomology, Faculty of Veterinary Medicine, Ahmadu Bello University, Zaria, Nigeria.*

[3]*Department of Veterinary Medicine, Faculty of Veterinary Medicine, Ahmadu Bello University, Zaria, Nigeria.*

***Corresponding author:** *Kaze Paul Davou, Department of Veterinary Parasitology and Entomology, Faculty of Veterinary Medicine, University of Jos, Nigeria. E-mail: paulkaze@yahoo.com*

Abstract

his paper focuses on anticoccidial drugs and resistance, poultry management, alternatives for anticoccidial drugs including dietary modulation, natural additives and herbs comprising of botanicals to coccidia. The paper also viewed at the treatment programme for coccidiosis control as well as its potential implications in meat tissue to man.

Keywords: Anticoccidial drugs; Avian Coccidiosis; Coccidiostats

Introduction

Anticoccidial feed additives have been used for more than 50 years to remedy or treat coccidiosis in poultry [1] and the aim of the paper was to review some attempted strategies employed towards the control of avian coccidiosis. Coocidiosis causes annual losses of US $ 2.4 billion to the poultry industry worldwide [2] in both the layer and broiler industries. Conventional disease control strategies depend on vaccination and proplylactic use of anticoccidial drugs. However, resistance against anticoccidial compounds is widely spread and coccidiostats as feed additives was banned in Europe by the year 2012 [Regulation [EC] No 1831/2003 of the European parliament and of the council of 22 September, 2003 on additives for use in animal nutrition]

Anticoccidial Drugs

These are synthesized drugs, which include variant groups of completely different chemical classes:

1. Amprolium is good against *E. tenella* but is not very effective against *E.acervulina* and *E. maxima.*

2. Nicarbazin is a broad-spectrum anticoccidial, it is used in colder seasons or climatic areas and the drugs should not be used in birds older than 20 days because the possibility of strong growth depression. - Robendine is safe broad-spectrum anticoccidial but it must be used with caution because of it's potential fast resistance build up.

3. Halofuginone and Lerbek effects on *E. tenella* are coccidiostatic activity and no coccidiocidal effect, but good for control of *E. acevurlina.*

4. Clinacox [Dicluzuril]. This has a broad-spectrum activity against all Eimeria species. The potential of *Eimeria* species, especially *Eimeria tenella* and *Eimeria maxima* to develop resistance to the drug is low. It is also used for "clean-up" program after the use of ionophore [3].

Sulfonamide Products

The drug exerted a major impact on the worldwide production of poultry meat [4]. Veterinarians and Animal scientist regularly use sulfonamides for therapeutic and prophylactic. Sulfadimethoxine, and sulfaquinoxaline are mainly used for prevention or treatment of poultry coccidiosis, and are generally co-administered in feed. The treatment of hens with sulfonamides-supplemented feed may result in sulfonamides residues being present in market eggs if these drugs have been improperly administered or if the withdrawal time for the treated hens has not been observed. To assure the food safety for consumers, the European Union has set a maximum residue limit for sulfonamides in foods of animal origin such as meat, milk, and eggs [5]. Misuse of these veterinary drugs in laying hens is of great concerns because the drug residues are turning up in eggs, which is an indispensable food for the consumers because it is highly nutritious, cheap and readily available.

A strong residue monitoring of sulfonamides in eggs is thus an important specific activity to guarantee the food safety. Removing the waste of organic solvents is also a serious problem on the world scale. From the view point of the effect of organic solvents to environments and analysts, analytical methods for the monitoring should avoid the use of organic solvents [6, 7].The feeding of 2,500 parts per million [ppm] sulfaquinoxaline causes a severe anemia in chickens with hemorrhages on the legs, breast muscle, and in abdominal organs [8]. Toxicity is more likely to be observed when medication is given in the water during hot weather. Feeding 300-ppm sulfaquinoxaline to growing chickens for 8 weeks reduced the weight gain of female birds but adverse, effects were not observed when sulfaquinoxaline was administered to growing chickens at 300-ppm in various feeding schedules. Continuous feeding of 125-ppm sulfaquinoxaline was highly efficacious in preventing naturally acquired caecal and intestinal coccidiosis. The total efficacy benefits of sulfaquinoxaline in comparison with other sulfonamides were associated to the fact that it is more readily absorbed than other sulfonamides when given in the feed.

Ionophore Products

Ionophores are the major group of poultry feed additives the polyether antibiotics commonly called Ionophores. Six compounds have become available [Monensin, Laslocid, Salinomycin , Narasin , Maduramycin and Semduramycin], the mechanism of action of all ionophores is very similar since they mediate the transport of mono and divalent cations throw the membrane of the parasite, resulting in disturbance of its osmotic balance. Ionphores can be divided into three groups according to the precise of action and chemical structure; monovalent [Monensin, Narasin and Salinomycin], monovalent glycoside [Maduramycin and Semduramycin] and divalent [Laslocid]. Laslocid and Maduramycin are more effective against E. tenella than Monensin, Narasin and Salinomycin [3].

Polyether Ionophors

They are produced by fermentation of *Streptomyces* or *Actinomadura* and they are the most widely used agents, such as salinomycin, monensin, lasalocid and narasin. They act through a general mechanism of changing ion transport and disrupting osmotic balance in the parasite.

Mode of Action of Anticoccidial Drugs

The biochemical effects of anticoccidials are numerous, but each class of chemical compound is unique in the type of action exerted on the parasite and its development stage. Different modes of action have been observed and this can be divided into different broad categories, according to Chapman [1997] [9] and McDougald [2003].

Drugs that Affect Cofactor Synthesis

Several drugs affect biochemical pathways that are dependent upon an important cofactor. For instance, amprolium competitively inhibits the uptake of thiamine by the parasite.

Drugs that Affect Mitochondrial Function

These drugs inhibit energy metabolism in the cytochrome system of the *Eimeria*. For instance, quinolones and clopidol inhibit electron transport in the parasite mitochondrion, but by different pathways.

Drugs that Affect Membrane Function

Ionophores in common have the ability to form lipophylic complexes with alkaline metal cations [Na^+, K^+, and Ca^{++}] and transport these cations through the cell membrane and then affect a range of processes that depend upon ion transport, such as influx of sodium ions thus, causing severe osmotic damage. These drugs act against the extracellular stages of the life cycle of the *Eimeria.*

Resistance to Anticoccidial Drugs

In 1963, the World Health Organization [WHO] defined resistance as "ability of a parasite strain to multiply or to survive in the presence of concentrations of a drug that normally destroy parasites of the same species or prevents their multiplication". Resistance may be relative [increasing doses of the drug being tolerated by the host] or complete [maximum doses being tolerated by the host] [10]. Anticoccidial drugs added to the feed are a good preventive measure and are well adapted to large-scale use, but continuous use of these drugs leads inevitably to the emergence of *Eimeria* strains that are resistant to all anticoccidial drugs, including ionophores [1]. Resistance can develop quickly, as in the case of quinolones and clopidol, or it may take several years for the Coccidia to become tolerant, as in the case of polyether ionophores [11].

Origin of Resistance Anticoccidial Drugs

There are three key factors contributing to drug resistance in commercial poultry production [12]:

- The intense and the continuous use of anticoccidial drugs in the poultry industry providing the basis for changing gene frequency through genetic selection.
- Coccidia are ubiquitous in poultry facilities and the large reproductive potential forms a large reservoir of genetic variation, which leads to the development of drug resistance.
- The life cycle of *Eimeria* is complex and involves a period of asexual and sexual stages. The nuclei of the asexual stage of *Eimeria* contain haploid complement chromosomes. Most drugs are active against this haploid stage, resulting in the removal of the most sensitive ones. This enables the more resistant ones to increase and thus rapidly becoming the dominant phenotype that spreads through the parasite population.

Poultry House Management

The high standard of flock hygiene, sanitation and poultry farm management helps in achieving optimal benefit from the use of anticoccidial drugs in preventing coccidiosis [9]. However, the sanitary practice alone is inadequate for complete removal of coccidial oocysts. This is because of the following:

- There have been too many failures in sanitary programs
- Oocysts are extremely resistant to common disinfectants
- House sterilization is never complete
- An oocyst-sterile environment for floor-maintained birds could prevent early establishment of immunity and thus allow late outbreaks [11].

Alternative for Anticoccidial Drugs

The constant and extensive use of the anticoccidial drugs for prevention and control of coccidiosis in poultry has been a major factor in the success of the industry. This beneficial use of anticoccidial drugs is associated with a widespread drug resistance of Coccidia in the United States, South America and Europe [11]. The first step of defense against development of resistance is the use of shuttle or dual programs [two or more drugs employed within a single flock] and frequent rotation of drugs [rotation of different compounds between flocks] [11]. The awareness by the consumers to avoid chemotherapeutics, the high development costs and low profits have not encouraged the pharmaceutical industry to develop new anticoccidial products [9]. Thus, alternatives progressively and currently been sought.

Dietary Modulation of Coccidia

The study of the interactions between diet composition and Coccidia is of great interest. Before the availability of effective anticoccidial drugs, recommendations for coccidial control included the formulation of diets that were considered capable of reducing the severity of infection such as diets containing skimmed milk, buttermilk, or whey [13]. But due to the development of the efficient, low-cost anticoccidial drugs caused lesser interest in dietary modulation. However, with the appearance of resistance to coccidiostats, the consumers' concern, and the expected regulations to ban the coccidiostats in the future, the possible role of nutrition has recently attracted interest [14].

Vitamins and minerals

Many vitamins change the immune status and the resistance of the host against Eimeria infections. Many works reported that vitamin A deficiency depresses T-lymphocyte response to mutagens [15] and reduces specific antibody production to protein antigens. Recently, [16] reported that vitamin A deficiency in chickens caused alteration in the IEL subpopulation, reduced the local cell-mediated immunity, and lowered the ability of birds to resist *E. acervulina* infection. Vitamin E and selenium generally improve resistance to coccidiosis, improve weight gain [El-Boushy, 1988], and reduce mortality due to E. tenella infection [17].

Vitamin C is known to possess immunity-enhancing effects in chickens and positive effect on birds' performance during coccidial challenge has been observed [18], but it had no effect on the lesion scores due to *E. tenella* or *E. acervulina* infection [19] found that feeding a diet with extra vitamins A, C, D_3, K, and selenium had no beneficial effects on the performance of chickens with subclinical infection caused by *E. maxima*, and *E. tenella*. Additional, the authors reported that performance in the birds supplemented with vitamins was even poorer than in birds fed the control diet. These results are inconsistent with previous work of [17] who fed 0.025 or 0.50 mg Se/kg of diet, noted a reduced mortality, an increase in body weight, and improved resistance against E. tenella.

Products Rich in N-3 Fatty Acid

The n-3 fatty acids are polyunsaturated fatty acids, the major fatty acids being eicosapentaenoic acid [EPA] and docosahexaenoic acid [DHA], found abundantly in fish oil, and alpha-linolenic acid [ALA], being a major component of flaxseed oil. Allen et al. [20-22], they reported that fish-liver oil exerts favorable control on the course of coccidiosis. They also worked on a series of experiments using fish oil, flaxseed oil and flaxseed in diets fed to male chickens from day 1 of age through 3 weeks of age and challenged with *E. tenella* at 2 weeks of age. The researchers reported a significant reduction in caecal lesion scores and in the histological examination, a significant reduction in the degree of parasitaization and retarded development of the *E. tenella* parasite was observed. The suggested mode of action is that the n-3 fatty acids infiltrate the tissues of the parasite, which in turn become more susceptible to oxidative attack by phagocytic cells.

Additionally, n-3 fatty acids have been shown to enhance the immune response in birds infected with E. tenella. However, little if any response was seen in the birds' performance, which is of most importance in poultry production. The n-3 fatty acids were proven ineffective against moderate or severe infection with E. maxima, and did not counteract reduced body-weight gain and lesion scores. The reason for the differences in response between these two Eimeria species to dietary n-3 fatty acids is not yet known [23].

Betaine Supplementation

Betaine supplementation has been shown to have positive effects on the water balance of broiler chicks stressed by high ambient temperature or coccidiosis [24], and to protect the cells from osmotic stress, allowing them to continue regular metabolic activities under conditions that would normally inactivate the cell [25]. [26] reported that betaine, in combination with the ionophore and salinomycin had a significant positive effect on the performance of chickens infected with *E. acervulina*, *E. maxima*, and *E. tenella*, the effect being greater than that mediated by betaine or salinomycin alone. Moreover, the combination resulted in a slight decrease in development and invasion of the epithelium by *E. acervulina*, while there was an increase in the invasion of *E. tenella*.

However, the diet supplemented with betaine alone decreased the invasion of E. acervulina and E. tenella as indicated by the number of sporozoites present in the intestinal epithelium after the challenge. Klasing *et al.* [2002] later clarified this effect when they found that chickens fed betaine had more lymphocytes in the epithelium and in the lamina propria during *E. acervulina* infection than those fed the diet without betaine. This effect of betaine could result in more effective clearance of sporozoites that explain the decreased numbers in the epithelium as reported by [26], while [27] found that betaine as a single feed supplement significantly improved chickens' body weight and tended to reduce the feed conversion ratio during coccidiosis infection. When betaine was used in the combination with the ionophore narasine, betaine showed no effects on birds' performance when *Eimeria tenella* was the major pathogenic species. The exact action of betaine is not fully understood. [26] suggested that betaine might increase performance in chickens infected by coccidiosis by inhibition of coccidial invasion and indirectly by supporting intestinal structure and function that could enhance the ability of the infected chickens to with stand coccidial infection.

Whole Wheat

The use of whole grains in broiler feeds is a frequent practice in Europe [28]. Many works indicated that offering broilers a whole cereal grains and balanced pellets greatly reduced the severity of infection with *Eimeria* as judged based on the reduction in output of oocysts [29-34] investigated the effects of whole wheat inclusion in broiler feeds with or without access to grit, and they observed no significant differences in faecal oocyst yields, lesion scores, or performance in birds infected with *E. tenella* or *E. maxima.* They concluded that the decrease in output of oocysts as caused by inclusion of whole cereals in the diet, and observed in the previous experiments, was not due to the increase in the viscosity of the digesta or the crushing of oocysts by an active gizzard and that whole wheat addition to the diet of broiler chickens provides no control of coccidiosis.

Exogenous enzymes

The use of exogenous enzymes in food processing started as early as 1900 and the majority of the enzymes have been derived from fermentation by microorganisms [35]. When broilers fed diet rich in wheat, barley, oat, or rye, the presence of non-starch polysaccharides [arabinoxylans and β-glucans] can give rise to high viscosity in the small intestine thereby decreasing the contact of endogenous digestive enzymes and its substrates. This results in a decrease in absorption and broilers' performance, and increase in the size of the GIT, pancreas, and the liver [36,37] reported an improvement in broilers' performance, a reduction in the size of digestive organs and the GIT size, and an increase in the total volatile fatty acids in the caecum, when a wheat-based diet was supplemented with the 200 mg exogenous enzymes xylanase or β-glucanase per kg feed. Addition of exogenous xylanase has been found to improve the performance and to reduce ileal digesta viscosity in *Eimeria*-infected birds [38]. It was concluded that intestinal viscosity and the size of the gizzard might affect the severity of the *Eimeria* infection. However, others did not observe effects of increased intestinal digesta viscosity on the severity of the *Eimeria* infection, when a large increase in viscosity was being induced by the inclusion of carboxymethyl cellulose in the feed [19, 34].

Electromagnetic Fields

Electromagnetic fields [EMF] have been in use as therapeutic modalities for at least 40 years. It is well known that selected electromagnetic fields [EMF] can have beneficial effects on bones, joints, and neurological disorders, as well as wound healing [39]. Anti-inflammatory aspects of EMF exposure have been reported to be due to the activation of A_2A adenosine receptors in human neutrophils [40]. Generally, inflammation is characterized by massive infiltration of T lymphocytes, neutrophils and macrophages into the damaged tissue [41].

In earlier studies, it has been reported that EMF mediate positive effects on wound healing, controlling the proliferation of inflammatory lymphocytes, and therefore demonstrating beneficial effects on inflammatory disease [42]. Many authors [42- 44] have discussed the effects initiated by various EMF signals and stated that EMF causes stress at the cellular level and that this leads to production of cytokines and consequently a biological response, including an immune response. Recently, [45] reported that exposure of broiler chickens to EMF antagonized the effects of coccidial infection in birds infected with a mixture of sporulated oocysts containing *E. acervulina, E. maxima,* and *E. tenella*. It was found that the severity of the intestinal lesions mediated by *E. acervulina* and *E. maxima* were reduced in the EMF-treated birds.

Natural Additive and Herbs

A number of natural herbs have been tested as anticoccidial dietary additives. Artemisinin isolated from Artemisia *annua,* is a naturally occurring endoperoxide with antimalarial properties. It has been found effective in reducing oocyst output from both *E. acervulina* and *E. tenella* infections when fed at levels of 8.5 and 17 ppm in starter diets [22]. The mode of action is thought to involve oxidative stress. Extracts from 15 Asian herbs were tested for anticoccidial activity against *E. tenella* and the test criteria were survival rate, bloody diarrhoea symptoms, lesion scores, oocyst output, and technical performance. Practical applications of these findings, such as the use of the products in starter rations or combinations of them with current anticoccidials or vaccines, appear possible and need to be investigated [1]. Therefore far, chemoprophylaxis and anticoccidial feed additives have controlled the disease but the situation has been complicated by the emergence of drug resistance [46] and their potentially toxic effects on the animal health [47].

Furthermore, drug or antibiotic residues in poultry products may be potentially hazardous to consumers. Another approach for coccidiosis control is the vaccination of birds with live *Eimeria* oocysts, but, in cases of poor management, these vaccines can trigger severe reactions that may affect the performance of flocks, mainly in broilers because of their rearing period [48]. As a result of this drawback of live vaccines, attenuated vaccines [with reduced pathogenicity] have been developed, but these are expensive to produce.

Botanicals and Coccidiosis

Cost effective alternative strategies are being tried for more effective and safer control of avian coccidiosis [49]. The use of botanicals has played a strong role in the control of avian coccidiosis, as they are not only natural products but may include new therapeutic molecules to which immunity has not yet developed. The use of botanicals as anticoccidial reduces, therefore, holds possible as an alternative in the control of coccidiosis.

Aloe Species

Aloes are believed to have several medicinal properties and are used to treat various ailments. There are more than 360 known Aloe species, but the most recommended type of Aloe in controlling coccidiosis is Aloe *excelsa* [50] revealed that the anticoccidial effects of *A. excelsa* were comparable with sulphachlopyrazine sodium monohydrate in terms of improved live weight gains and reduction in oocyst output in infected broiler chickens. Other species of *Aloe* plant such as *Aloe vera* have also been reported to have anticoccidial activities.

Aloe vera treatments show toxic effects on the intestinal tract by benefiting microflora and reducing bowel putrefaction as well as reducing inflammation [51]. An in vitro study was undertaken to determine the effect of three concentrations [15%, 30%, and 45%] of *A. Vera* and *A. spicata* on the inhibition of the sporulation of avian coccidia oocysts [52]. The two extracts showed a concentration-dependant anticoccidial effect; however, *A. spicata* inhibited sporulation to a greater extent than *A. vera*. In another study [37] dietary supplementation of *A. Vera* resulted in significantly lower gut lesion scores and reduced faecal oocyst shedding of *E. maxima* in broiler chickens. These authors [37] suggested that reduced faecal oocyst shedding, a protective role against *Eimeria* infection, in *Aloe*-based chicken diets could be associated more with cell-mediated responses than antibody responses.

Artemisia Species

The most common species is *Artemisia annua* which has been reported for its antiparasitic activities. *A. annua* is a common type of wormwood botanical anticoccidials: Abbas *et al.* [2004] and [53] conducted the first experimental trial to evaluate the anticoccidial activity of *A. annua* extracts against *E. tenella* in chickens. *A. annua* extracts showed the anticoccidial activity in terms of improved weight gain, improved feed conversion ratio and reduced lesion scores in infected chickens. Later, [23] reported a significant anticoccidial effect of *A. annua* against *E. tenella*, measured as reduced lesion scores, when fed to broiler chickens for three weeks as dried leaves at a dietary concentration of 5% [equivalent to 17 ppm pure artemisinin].

The pure form of artemisinin, fed for a period of 4 weeks at levels of 2, 8.5 and 17 ppm, significantly decreased oocyst output from single and dual species infection with *E. tenella* and *E. acervulina*. Moreover, artemisinin isolated from *A. sieberi* was also found to be effective against *E. tenella* and *E. acervulina* but not against *E. maxima* [54]. So far, a limited amount of work has been carried out to determine the anticoccidial effect of Artemisia spp. in layer chickens. [55] Studied the effect of feeding 20% dried pulverized *A. annua* leaves against *E. tenella* both in broiler and layer chickens. The anticoccidial effects of diets containing *A. annua* leaves were almost equal to the commercial anticoccidials both in broiler and layer chickens. The proposed mechanism of action of artemisinin involves cleavage of endoperoxide bridges by iron producing free radicals [hypervalent iron-oxo species, epoxides, aldehydes, and dicarbonyle compounds] which damage biological macromolecules causing oxidative stress in the cells of the parasite [56].

Azadirachta Indica [Neem] Plant

Azadirachta indica [neem] plant is commonly available in Asian and African countries and is well known in the therapy of a number of infectious diseases including coccidiosis. Neem fruit, at a concentration of 150 g/50 kg feed, has been found to have anticoccidial effects against *E. tenella* infection by reducing oocyst excretion and mortality in broiler chickens [57]. In addition to the anticoccidial effect of neem fruit, some reports have shown the anticoccidial activity of an aqueous extract of neem leaves against E. tenella alone [58] as well as in a mixed infection [Biu *et al.*, 2006], which was comparable to the commercial anticoccidials amprolium and baycox.

The exact mechanism of action of neem against coccidian parasites is unknown, but a report by the National Research Council [1992] [59] suggested that aqueous neem leaf extract, when taken orally, produces an increase in red cells, white blood cells and lymphocyte counts thus enhancing the cellular immune response, increasing antibody production and so most pathogens can be removed before they cause the symptoms associated with disease. Further study is needed to determine the maximum safe levels of neem supplementation because the higher doses, due to its bitterness, may show adverse effects on feed intake which will change the performance parameters of birds.

Beta Vulgaris

The beneficial effects of incorporating sugar beet [Beta vulgaris] solids in animal feeds on livestock growth and overall performance have been known for a long time. One of the active ingredients is betaine which protects cells against osmotic stress by stabilizing cell membranes through the maintenance of osmotic pressure in the cells.

Curcuma Longa

Curcuma longa L. [Zingiberaceae], commonly known as turmeric, is a medicinal plant widely used and cultivated in the tropical regions. In developing countries like Pakistan, poultry farmers provide turmeric powder as a feed additive for the control of coccidiosis in broilers [60]. The active compound of turmeric is the phenolic compound curcumin, which has been shown to have antioxidative, anti-inflammatory and immunomodulatory properties [56]. In an experimental study, the anticoccidial effect of dietary supplementation of 1% curcumin was observed in chickens after infection of *E. maxima* and *E. tenella species.* Improved weight gain, reduced lesion scores and oocyst counts were shown only against *E. maxima*. A significant reduction of plasma $NO2^-$ and $NO3^-$ was found only in *E. maxima*-infected and curcumin-treated birds, and hence provides a possible explanation for the difference in anticoccidial activity found for both *Eimeria species* [56]. Later [60] reported that dietary supplementation with 3% *C. longa* powder was effective against a mild infection of *E. tenella.*

The proposed mechanism of action of *C. longa* [curcumin] involves the induction of oxidative stress against coccidia. Further researches are required to determine the possible anticoccidial activity of different concentrations of whole *C. longa* and its active ingredient curcumin against different *Eimeria species* in poultry.

Echinacea Purpure

Echinacea and its different preparations contain a variety of active substances such as flavonoids, polysaccharides, glycoproteins, alkamides, cinnamic acids, essential oils and phenolic compounds [61; 62] which are effective in treatment of various ailments and are proven to be beneficial in promoting immunity [Bauer, 1999]. This plant is known to have anti-inflammatory, antioxidant and immunomodulating properties that may be linked to its anticoccidial effects [62]. In an experimental trial [56...], ground root preparations of *E. purpurea* [0.1% -0.5%] were offered to broilers for two weeks which ameliorated weight gain reduction and birds had fewer coccidial lesions after a mixed challenge infection with E. *acervulina, E. maxima, E. tenella* and *E. necatrix.* The exact mechanism of action is still unknown, but because of its antioxidant properties *Echinacea* therapy may induce a state of oxidative stress against *Eimeria species.*

Origanum Vulgare

The essential oils of *Origanum vulgare* are well known for their antiprotozoal activity [63, 64] carried out a study to examine the effect of dietary supplementation of *O. oregano* [*O.* vulgare] essential oil on performance of broiler chickens experimentally infected with *E. tenella.* It was concluded that *O. oregano* essential oils, mainly carvacrol and thymol, had anticoccidial effects against *E. tenella.* Some studies suggest that vaccination against coccidiosis, in combination with *O. oregano* containing compounds, may be an alternative control method for intestinal health in chickens [Waldenstedt, 2000a]. In addition, some works suggested the use of dried oregano leaves as a natural herbal growth promoter for early maturing of birds [65]. The dietary supplementation of *O. oregano* containing plants like *O. vulgare,* thus, seems equally effective for maintaining the performance and reducing pathogenic parameters in infected birds.

Saccharum Officinarum

Sugar cane [*Saccharum officinarum*] extract [SCE], a well known natural immunostimulant, is reported to have protective effects against E. tenella infection in chickens [66]. Some studies [67] showed a significant increase in the number of IgM- and IgG plaque-forming cell responses of peripheral blood leukocytes [PBL], intestinal leukocytes, splenocytes, in addition to significantly higher phagocytic activity of PBL and antibody responses in chickens that had been orally administered with either sugar cane extract [SCE] or the polyphenol-rich fraction [PRF]. Most recently, [68] reported the immunotherapeutic effects of sugar cane extract against mixed *Eimeria species* in broiler chickens. The results of these researches suggested that sugar cane extract has an immunostimulating effect in chickens and their administration may augment protective immunity against coccidiosis.

Triticum Aestivum

The *supplementation* of whole *Triticum aestivum* [wheat] grains in broiler feeds is common practice in Europe [28] because dietary fibre anti- oxidants may actually quench the soluble radicals that are continuously formed in the intestinal tract [69]. Many reports [56] have noted the protective effects of whole cereal grains against coccidiosis in broiler chickens measured as a reduction of oocyst output. However,[32] and [34] demonstrated the effects of whole wheat inclusion in broiler feeds with or without access to grit, and observed no significant differences in oocyst counts of mixed *Eimeria species.* They concluded that the reduction in output of oocysts by supplementation with whole cereals in the diet was not a result of the crushing of oocysts by an active gizzard or the increase in the viscosity of the digesta. Furthermore, they concluded that the whole wheat supplementation provided no control of coccidiosis in broiler chickens.

Yucca Schidigera

Plant extracts with high saponin content are a good source of natural antimicrobial compounds. *Yucca schidigera* is a major source of natural saponins that cause the inhibition of protozoan development by interacting with the cholesterol present on the parasite cell membrane, thus resulting in parasite death [70]. Several studies have shown a beneficial and synergistic effect between the coccidiosis vaccine and the *Y. schidigera* extract in improving weight gains, feed conversion ratio and maintaining the integrity of the intestinal villi in chickens [71]. These improvements in the performance parameters of birds may be the result of the potential of saponins [extracted from the *Y. schidigera*] to improve the absorption of nutrients by the intestinal mucosal surface [72]. These saponins are steroidal glycosides with strong surfactant activity, reducing the superficial tension of fluids and allowing better absorption of nutrients by the intestinal epithelium.

Treatment Programme for Coccidiosis Control

Shuttle or Dual Program

The use of one product in the starter and another in the grower feed is called a shuttle program in the US and a dual program in other countries. The shuttle program usually is intended to manage coccidiosis control. Intensive use of the polyether ionophore drugs for many years produced strains of coccidia in the field that have reduced sensitivity to the ionophores. It is a common practice to use another drug such as nicarbazin or halofuginone in the starter or grower feed to bolster the anticoccidial control and take some pressure off the ionophore. The use of shuttle programs is thought to reduce buildup of drug resistance. In 1988, approximately 80% of the US producers used some type of shuttle program [73], in which two compounds usually a synthetic agent[such as Incarbazin] and Ionophore [such as Salinomycin] are employed successively in single flock. During 1999 in the US, shuttles involving synthetic drugs followed by Ionophores were employed by approximately 25% of broiler complexes [74].

Future Hazards of Anticoccidial Residues in Broilers Meat Tissues to Man

Anticoccidial drugs play an important role in animal production, especially in intensive broiler production. They are used for disease prevention and therapy, as well as for their growth-stimulating effect. These drugs add to the recovery of animals from protozoal endoparasites, increase breeding productivity and decrease economic losses caused by coccidiosis. However, mass and long-term administration of these substances has brought problems connected to the occurrence of unfavorable residues in animal products for human consumption.

The residues of anticoccidial drugs represent a potential risk to human health. Proper administration of these substances will ensure minimal content in animal products that will minimize health risks. To protect the health of consumers against the entry of residues of anticoccidial drugs into the food chain, it is necessary to monitor drug residues in animals for food production and for valid veterinary hygienic legislation to pay appropriate attention to this group of drugs [75]. Some anticoccdial drugs such as ionophores are not used in human medicine due to their potent cardiovascular effects. Ensure that recommended withdrawal periods are observed, it has been suggested that residues of ionophores in food could cause adverse health effects in humans as a result of their cardiovascular toxicity. Since poultry litter is extensively applied to land as manure ionophores and their degradation products may readily enter the soil and water environment.

Some studies have been published regarding the environmental fate of ionophores and thus it is difficult to assess their potential impact. Biodegradation studies have indicated that monensin is degradable under aerobic conditions with or without manure and in manure piles within 33 days. Degradation in manure piles under anaerobic conditions was less extensive. It should be assumed that the microbiological activity of soil will be affected, at least initially following application of ionophore containing manure and this may affect nutrient release.

Direct effects on plants are not expected except that an inhibitory effect on apple pollen has been reported for monensin. Ionophores may cause irritation and allergic reaction in humans and protective clothing and dust masks should be used whenever there is a risk of exposure. Alarming human health hazards, the emergence of resistant strains of bacteria in birds and passage of these or other resistant factors via food chain from birds to human beings. Use of antibiotics at sub-therapeutic levels in broiler feeds may lead to the development of resistant strains of bacteria in the bird. While consuming the meat containing residues of antibiotics over protracted period may lead to emergence of resistant gut flora and pathogens in human beings such as E. Coli and Salmonella spp. Production of harmful effects from direct toxicity or from the allergic reactions [hypersensitivity reactions] in persons already sensitized to them.

Certain drugs and or their metabolites possess carcinogenic potential e.g.sulphamethazine residues containing meat preserved with sodium nitrate may develop a triazine complex that has a considerable carcinogenic potential. Prolonged ingestion of tetracycline present in the broiler meat has detrimental effects on teeth and bones in growing children. Some tetracyclines, most therapeutic antibiotics are relatively heat stable and resist both pasteurization and cooking process [76]. Adverse effects on the cartilage development in children may result if the broiler meat contains quinolone residues. Drug residues may destroy the useful micro flora of gastrointestinal tract, especially in children and hence lead to enteritis [diarrhoea, dysentery] like problems. Super infections that refer to as fresh invasion or re40 infection added to an already existing infection. Candidiasis caused by Candida albicans is a classical example of the unhealthy consequence of the use of antibiotics. Residues of chloramphenicol are known to cause bone marrow depression and problems like anaemia in consumers [76].

In addition, there are many safe veterinary drugs and none withdrawal period like, amprolium [77]. Factors that leading to the occurrence of antibiotics residues in animal products are; failure to observe drug withdrawal period, extended usage or excessive dosages of antibiotics, non-existence of restrictive legislation or their inadequate enforcement, poor records of treatment, failure to identify treated animals, lack of advice on withdrawal periods, off-label use of antibiotics, availability of antibiotics to lay persons as over the counter drugs in the developing countries, the addition of antibiotics as milk preservatives during hauling from the centre of production[villages] to the centers of consumption [cities or factories] and lack of consumer awareness about the magnitude and human health hazards associated with antibiotic residues in the food of animal origin [76].

Anticoccidial Testing in Birds

Three types of tests are generally used to study anticoccidial drugs in broiler birds. These are; Battery tests: Done 7–14 days, tests with birds in wire cages, Standard grow-out test: Done 6–8 weeks tests on birds in floor pens and Full-scale tests which is done in commercial facilities. Each type has a different objective and value to the investigator for example; the battery test is used most effectively to measure the efficacy of an anticoccidial drug against a variety of field isolates of Coccidia. This is an efficient and relatively inexpensive testing procedure. The floor- pen test is an intermediate testing procedure with a primary goal of providing statistically useful performance data under controlled conditions. Individually, the predictive value of each test is limited. One cannot, for example, confidently extrapolate performance data in a seven-day battery test to market weight, nor can one predict from a few commercial trials the efficacy of an anticoccidial agent in preventing the lesions of major species of Coccidia. As a whole, when properly conducted, the tests complement one another by providing a comprehensive picture of the efficacy, safety and economic value of an anticoccidial agen.

Conclusion

Treatment and control of the disease are beset with several problems prominent of which is the poor understanding of the immune response. Another factor is the increasing incidence of drug resistance in field strains of *Eimeria*. Furthermore, due to health awareness there is increasing concern regarding drug residues in poultry products and growing pressure from Government and consumer on the production of drug-free poultry products [78].

References

1. Allen PC, Fetterer RH. Recent advances in biology and immunobiology of Eimeria species and in diagnosis and control of infection with these coccidian parasites of poultry. Clin Microbiol Rev. 2002; 15(1):58-65.
2. Shirley MW1, Smith AL, Tomley FM. The biology of avian Eimeria with emphasis an emphasis on their control by vaccination. Adv Parasitol. 2005; 60:285-330. DOI:10.1016/S0065-308X(05)60005-X
3. Leu S. Planning your attack against coccidiosis. Poultry Worid.1999; 15: 111-113.
4. Chapman HD, Cherry TE, Danforth HD, Richards G, Shirley MW, Williams RB. Sustainable coccidiosis control in poultry production: The role of live vaccines. Int J Parasitol. 2002; 32(5):617-29.
5. Brussels. The rules governing medical products in the European Community IV, Commission of the European Community. 1991
6. Fang GZ , He JX , Wang S. Multiwalled carbon nanotubes as sorbent for on-line coupling of solid-phase extraction to high-performance liquid chromatography for simultaneous determination of 10 sulfonamides in eggs and pork. Journal of Chromatography. A. 2006; 1127(1-2):12-17. DOI: 10.1016/j.chroma.2006.06.024.
7. Zheng MM , Zhang MY , Peng GY , Feng YQ. Monitoring of sulfonamide antibacterial residues in milk and egg by polymer monolith microextraction coupled to hydrophilic interaction chromatography/ mass spectrometry. Analytica Chimica Acta. 2008; 625(2):160-172. DOI: 10.1016/j.aca.2008.07.033.
8. Khan MA, Younas M, Khan I, Abbas RZ, Ali M. Comparative efficacy of some herbal and homeopathic preparations against coccidiosis in broilers. International Journal of Agriculture and Biology. 2008; 10(3): 358-360.
9. Chapman, H. D. Biochemical, genetic and applied aspects of drug resistance in Eimeria parasite of the fowl. Avian Pathol. 1997; 26(2):221-244. DOI:10.1080/03079459708419208.
10. Chapman, H. D. Anticoccidial Drug Resistance. The Biology of the Coccidia. P. L. Long, ed. University Park Press: Baltimore. 1982.
11. McDougald LR. Coccidiosis. in Poultry Diseases. 2003.
12. Jeurissen S H M, Veldman B. The interaction between feed (components) and Eimeria infection in poultry health. in Nutrition and Health of the Gastrointestinal Tract. M. C. Blok, H.A. Vahl, L. de Braak, G. Hemke, and M. Hessing, eds. Wageningen Academic Publisher, Wageningen, The Netherlands. 2002.
13. Hussain Abo Alqomsan. Prevalence of caecal coccidiosis among broilers in Gaza strip. A Thesis submitted in Partial Fulfillment of the Requirement for the Degree of Master of Biological Sciences.2010.
14. Gabriel I, Mallet S, Leconte M. Effect of whole wheat feeding on the development of coccidial infection in broiler chickens until market-age. Animal Feed Science and Technology. 2006; 129: 179-303. DOI: 10.1016/j.anifeedsci.2006.01.004

15. Aharon Friedman, David Sklan. Impaired T lymphocytes immune response in vitamin A depleted rats and chickens. British Journal of Nutrition. 1989; 62(2): 39-449. DOI:10.1079/BJN19890044
16. Dalloul RA, Lillehoj HS, Shellem TA, Doerr JA. Effect of vitamin A deficiency on host intestinal immune response to Eimeria acervulina in broiler chickens. Poult Sci. 2002; 81(10):1509-1515. DOI:10.1093/ps/81.10.1509
17. Colnago GL, Jensen LS, Long PL.. Effect of selenium and vitamin E on the development of immunity to coccidiosis in chickens. Poult Sci. 1984; 63(6):1136-1143. DOI:10.3382/ps.0631136
18. Attia M E, Fathy I M, Attia A M N. The effect of dietary vitamin C on the severity of coccidiosis in Fayomi chicks. Veterinary Medical Journal.1978;20: 65-74.
19. Waldenstedt L, Elwinger K, Lundén A, Thebo P, Bedford MR, Uggla A. Intestinal digesta viscosity decreases during coccidial infection in broilers. Br Poult Sci. 2000;41(4):459-64 DOI:10.1080/71365495 9.
20. Allen PC, Danforth HD, Morris VC, Levander OA. Association of lower plasma carotenoid with protection against caecal coccidiosis by diet high in n-3 fatty acids. Poult Sci. 1996;75: 966-972.
21. Allen PC, Lydon J, Danforth HD. Effects of components of Artemisia annua on coccidia infections in chickens. Poult Sci. 1997; 76(8): 1156-63. DOI:10.1093/ps/76.8.1156
22. Allen PC, Lydin J, Danforth HD. Effects of components of Artemisia annua on coccidia infections in chickens. Poultry Science, 76: 1156-1163.
23. Augustine PC, Danforth H. Influence of betaine and salinomycin on the intestinal absorption of methionine and glucose and on the ultrastructure of intestinal cells and parasite development stages in chicks infected with Eimeria acervulina. Avian Dis. 1999; 43(1): 89-97.
24. R Ko, L T Smith, G M Smith (1994). Glycine betaine confers enhance osmotolerance and cryotolerance on listeria monocytogenes. J Bacteriol. 1994; 176: 426-431.
25. Augustine PC, McNaughton JL, Virtanen E, Rosi L. Effect of betaine on the growth performance of chicks inoculated with mixed cultures of Avian Eimeria species and on invasion and development of Eimeria tenella and Eimeria acervulina in vivo and in vitro. Poult Sci. 1997; 76(6): 802-9. DOI:10.1093/ps/76.6.802
26. L Waldenstedt, K Elwinger, P Thebo, A Uggla. Effect of betaine supplement on broilers performance during an experimental coccidial infection. Poultry Science. 78(2): 182-189. DOI:10.1093/ps/78.2.182
27. Banfield M, Forbes J. Feed content and structure effects on coccidiosis: In proceedings of the 12th European Symposium on Poultry Nutrition, WPSA, Veldhoven, The Netherlands. 1999.
28. Cumming R B. The effect of dietary fibre and choice feeding on coccidiosis in chickens. Proceedings 4th AAAP. Animal Science Congregation.1987.
29. Cumming R B. The biological control of coccidiosis. 19th World Poultry Congregation; 1992; 2: 425-428.
30. Cumming R B. Opportunities for whole wheat grain feeding. 9 thEuropean Poultry Conference. 1994; 2: 219-222.
31. Waldenstedt L, Elwinger K, Hooshmand-Rad P, Thebo P, Uggla A. (1998). Comparison between effects of standard feed and whole wheat supplemented diet on experimental Eimeria tenella and Eimeria maxima infections in broiler chickens. Acta Vet Scand. 1998;39(4):461-71.
32. M J Banfield, R P Kwakkel, M Groeneveld, R A H Ten Doeschate J M Forbes. Effects of whole wheat substitution in broiler diets and viscosity on a coccidial infection in broilers. British Poultry Science, (1999);40: 58-60.
33. Banfield MJ, Kwakkel RP, Forbes JM. Effects of wheat structure and viscosity on coccidiosis in broiler chickens. Animal Feed Science and Technology, 2002; 98: 37-48. DOI:10.1016/S0377-8401(02)00032-9
34. Clarkson K, Jones B, Bott R, Bower B, Chotani G, Becker T. Enzymes: screening, expression, design, and production. Pages 325-352, in Enzymes in Farm Animal Nutrition. M. R. Bedford, ed. CABI Publication. Wilsshire.2001. DOI : 10.1079/9780851993935.0315
35. Wang ZR, Qiao SY, Lu WQ, Li DF. Effects of enzyme supplementation on performance, nutrient digestibility, gastrointestinal morphology, and volatile acid profiles in the hindgut of broilers fed wheat-based diets. Poult Sci. 2005; 84(6): 875-881. DOI:10.1093/ps/84.6.875
36. Yim D, Kang SS, Kim DW, Kim SH, Lillehoj HS, Min W. Protective effects of Aloe vera-based diets in Eimeria maxima-infected broiler chickens. Exp Parasitol. 2011; 127(1): 322-325. DOI: 10.1016/j.exppara.2010.08.010
37. Morgan A J, Bedford M J. Advances in the development and application of feed enzymes. Australia Poultry Symposium. 1995; 7: 109-115.
38. Montesinos MC, Yap JS, Desai A, Posadas I, McCrary CT, Cronstein BN. Reversal of the anti-inflammatory effects of methotrexate by the non selective adenosine receptor antagonist theophylline and caffeine: evidence that the anti-inflammatory effects of methotrexate are mediated via multiple adenosine receptors in rat adjuvant arthritis. Arthritis Rheum. 2000; 43(3): 656-663. DOI: 10.1002/1529-0131(200003)43:3<656::AID-ANR23>3.0.CO;2-H.
39. Vallbona C, Richard T. Evolution of magnetic therapy from alternative to traditional medicine. Physiological Medical Research. (1999); 10: 729-754. DOI: 10.1016/S1047-9651(18)30190-6
40. Gessi S, Varani K, Merighi S, Ongini E, Borea PA. A_2 adenosine receptors in human peripheral blood cells. British Journal Pharmacology, Br J Pharmacol. 2000; 129(1): 2-11. DOI:10.1038/sj.bjp.0703045
41. Blank M, O Khorkova, Goodman R. Changes in polypeptide distribution stimulated by different levels of electromagnetic and thermal stress. Bioelectrochemical Bioenergy.1994; 33(2): 109-114.

42. Goodman R, Blank M, Lin H, Dai R, Khorkova O, Soo L, Weisbrot D. Henderson. A Increased levels of hsp70 transcripts induced when cells are exposed to low frequency electromagnetic fields. Bioelectrochemical Bioenergy. 1994; 33(2): 115-120. DOI:10.1016/0302-4598(94)85002-X.

43. Mevissen M, Häussler M, Szamel M, Emmendörffer A, Thun-Battersby S, Löscher W. Complex effects of long term 50 Hz magnetic field exposure in vivo on immune functions in female Sprague-dawley rats depend on duration of exposure. Bioelectromagnetics. 1998; 19(4):259-270.

44. Elmusharaf M A, Bautista V, Nollet L, Beynen A C. Effect of a mannanoligosaccharide preparation on Eimeria tenella infection in broiler chickens. Int.ernational Journal of Poultry Science.2006; 5: 583-588. DOI: 10.3923/ijps.2006.583.588

45. Abbas RZ, Iqbal Z, Blake D, Khan MN, Saleemi MK. Anticoccidial drug resistance in fowl coccidia: the state of play revisited. World's Poultry Science Journal,2011; 67(2): 337-350.

46. Nogueira VA, Franca TN, Peixoto PV. Ionophore poisoning in animals. Pesquisa Veterinária Brasileira, 2009; 29: 191-197.

47. Chapman HD. Practical use of vaccines for the control of coccidiosis in the chicken. World's Poultry Science Journal.2000; 56(1): 7-20. DOI:10.1079/WPS20000002

48. Abbas RZ, Munawar SH, Manzoor ZA, Iqbal Z, Khan MN, Saleemi MK, Zia MA, Yousaf A. Anticoccidial effects of acetic acid on performance and pathogenic parameters in broiler chickens challenged with Eimeria tenella. Pesq. Vet. Bras. 2011; 31(2): 99-103. DOI:10.1590/S0100-736X2011000200001

49. Gadzirayi CT, Mupangwa J F, Mutandwa E (2005). Effectiveness of Aloe excelsa in controlling coccidiosis in broilers. Journal of Sustainable Development in Africa. 2005; 7(1): 10-14.

50. Jeffrey S Bland, Ir M Menco. Effect of orally consumed Aloe vera juice on gastrointestinal function in normal humans. Preventive Medicine.2006; 14: 152- 154.

51. Marizvikuru M, Evison B, Michael C, Tiniyoko EH. The in vitro studies on the effect of Aloe vera ((L.) Webb. and Berth.) and Aloe spicata (L. f.) on the control of coccidiosis in chickens. International Journal of Applied Research in Veterinary Medicine. 2006; 4: 128-133.

52. Oh HG, Youn HG, Noh HJ, Jang JW, Kang YB. Anticoccidial effects of artemisin on the Eimeria tenella. Korean Journal Veterinary Research, 1995; 35: 123-130.

53. Arab HA, Rahbari S, Rassouli A, Moslemi MH, Khosravirad F. Determination of artemisinin in Artemisia sieberi and anticoccidial effects of the plant extracts in broiler chickens. Trop Anim Health Prod. 2006; 38(6): 497-503.

54. EA Brisibe, UE Umoren, PU Owai, F Brisibe. Dietary inclusion of dried Artemisia annua leaves for management of coccidiosis and growth enhancement in chickens. African Journal of Biotechnology. 2008; 7: 4083-4092.

55. Allen PC. Dietary supplementation with Echinacea and development of immunity to challenge infection with coccidia. Parasitol Res. 2003; 91(1): 74-8. DOI: 10.1007/s00436-003-0938-y

56. Murtaza Ali Tipu , TN Pasha, Zulfiqar Ali. Comparative efficacy of salinomycin sodium and neem fruit (Azadirachta indica) as feed additive anticoccidials in broilers. International Journal of Poultry Sciences. 2002; 1(4): 91-93. DOI: 10.3923/ijps.2002.91.93

57. Takagi M, Kuriyagawa T, Hirose J, Ryuno T, Imura Y, Okamoto K, Omata Y, Yasui T, Deguchi E. (2006). Anticoccidial Efficacy of Natural Herbs Extracts in calves. Journal of Animal and Veterinary Advances, 2006; 5(12):1096-1100.

58. National Research Council . Neem: a tree for solving global problems. pp.115 (National Academy Press, Washington D.C.). NIDAULLAH, H., DURRANI, F.R., AHMAD, S., JAN, I.U. and GUL, S. (2010). Aqueous extract from different medicinal plants as anticoccidial, growth promotive and immunostimulant in broilers. ARPN Journal of Agricultural and Biological Science. 1992; 5: 53-59.

59. Abbas RZ, Iqbal Z, Khan MN, Zafar MA, Zia MA. Brazilian Archives of Biology and Technology. 2010; 53: 63-67.

60. Liu YC, Zeng JG, Chen B, Yao SZ. Investigation of phenolic constituents in Echinacea purpurea grown in China. Planta Med. 2007;73(15):1600-5. DOI: 10.1055/s-2007-993742

61. Zili Zhai, Yi Liu, Lankun Wu, David S. Senchina, Eve S. Wurtele, Patricia A. Murphy, Marian L. Kohut, and Joan E. Cunnick. Enhancement of innate and adaptive immune functions by multiple Echinacea species. J Med Food. 2007; 10(3): 423–434. DOI: 10.1089/jmf.2006.257

62. Milhau G, Valentin A, Benoit F, Mallie M, Bastide JM, Pelissier Y, et al. In vitro antimalarial activity of eight essential oils. Journal of Essential Oil Research, 1997;9(3): 329-333. DOI: 10.1080/10412905.1997.10554252

63. Giannenas I, Florou-Paneri P, Papazahariadou M, Christaki E, Botsoglou NA, Spais AB. Effect of dietary supplementation with oregano essential oil on performance of broilers after experimental infection with Eimeria tenella. Arch Tierernahr. 2003; 57(2): 99-106.

64. Bampidis VA, Christodoulou V, Florou-Paneri P, Christaki E, Chatzopoulou PS, Tsiligianni T, Spais AB. Effect of dietary dried oregano leaves on growth performance, carcass characteristics and serum cholesterol of female early maturing turkeys. Br Poult Sci. 2005; 46(5): 595-601. DOI: 10.1080/00071660500256057.

65. El-Abasy M1, Motobu M, Na KJ, Shimura K, Nakamura K, Koge K, Onodera T, Hirota Y. Protective effects of sugar cane extracts (SCE) on Eimeria tenella infection in chickens. J Vet Med Sci. 2003; 65(8): 865-871.

66. Hikosaka K, El-Abasy, M, Koyama Y, Motobu M, Koge K, Isobe T, Kang CB, Hayashidani H, Onodera T, Wang PC, Matsumura M, Hirota Y.. Immunostimulating effects of the polyphenol-rich fraction of sugar

cane (Saccharum officinarum L.) extract in chickens. Phytother Res. 2007;21(2):120-125.

67. Awais MM, Akhter M, Muhammed F, Haq AU, Anwar MI. Immunotherapeutic effects of some sugar cane (Saccharum officinarum L.) extracts against coccidiosis in industrial broiler chickens. Exp Parasitol. 2011; 128(2):104-10.DOI: 10.1016.
68. YM Bao, M Choct .Dietary NSP nutrition and intestinal immune system for broiler chickens. World's Poultry Science Journal. 2010; 66(3): 511-518. DOI:10.1017/S0043933910000577
69. Wang Y, Mcallister T A, Newbold C J, Rode L M, Cheeke P R, Cheng K J. Effects of Yucca schidigera extract on fermentation and degradation of steroidal saponins in the rumen simulation technique (RUSITEC). Animal Feed Science and Technology,1998; 74: 143-153. DOI: 10.1016/S0377-8401(98)00137-0
70. D M Alfaro, A V F Silva, S A Borges, F A Maiorka, S Vargas, E Santin. Use of Yucca schidigera extract in broiler diets and its effects on performance results obtained with different coccidiosis control methods. Journal of Applied Poultry Research. 2007; 16(2): 248-254. DOI: 10.1093/japr/16.2.248
71. Mcallister TA, Wang Y, Hristov AN, Olson ME, Cheeke PR. Applications of Yucca schidigera in livestock production. Eimeria tenella Proceedings of 33rd Pacific Northwest Animal Nutrition Conference, Canada, pp. 1998;109-119.
72. Saif Y, Barnes A, Glisson F, Mcdougald L, Swaye D. Diseases of Poultry. 12th Edn, lowa state press, USA. 2003
73. Chapman, H.D. Drug program and immunity applications for drug withdrawal. Avian Pathology. 1998; 28 : 521-535.
74. Jevinova P, Laciakova A, Pipova M, Mate D Kozaroval. Legislative treatment of anticoccidial drugs and their residues in poultry personal. Slovakia University Veterinary Journal., 73, 041 81. [87] HACCP manual monensin and salinomycin in veterinary medicine – overview: http://www.bioagrimix.com/haccp, 12/1/2010.2010
75. Javaid A, Awan A, Athar M. Rational use of drugs in broiler meat production. International Agricultural and Biological Journal, 2000; 02(3): 269–272.
76. Donald C, Pharm D. Veterinary drug hand book. 3rd Edn, White Bear Lake, Minnesota, USA.1999
77. Williams RB. Anticoccidial vaccines for broiler chickens: Pathway to success. Avian Pathol. 2002 ; 31(4): 317-353. DOI:10.1080/03079450220148988

21

The Role of Inflammation and Immune Activation in Non-AIDS related Co-Morbidities in HIV infection: Determinants and Outcomes

Chidozie E Agu[1*], Ikenna K Uchendu[2], Chigozie N Okwuosa[2] and Peter U Achukwu[2]

[1]*Prime Health Response Initiative (PHRI)–Sub-recipients of Global Fund Impact HIV Project, Ilorin, Kwara State, Nigeria*

[2]*Department of Medical Laboratory Science, Health Sciences and Technology, College of Medicine, University of Nigeria Enugu Campus*

***Corresponding author:** Chidozie Elochukwu Agu, Prime Health Response Initiative (PHRI) – Sub-recipients of Global Fund Impact HIV Project, Ilorin, Kwara State, Nigeria; E-mail: chidozieagu@gmail.com, cagu@phrinitiative.com*

Abstract

Individuals living with HIV infection on antiretroviral therapy are at an increased risk of developing non-AIDS related diseases and experience incomplete immune restoration. The path physiology has been linked to HIV-specific mechanisms as well as non-specific generalized responses to infection which are thought to contribute to the ongoing activation of the immune system. Factors such as the early loss of gastrointestinal (GI) tract mucosal integrity, the pro-inflammatory cytokine milieu, co-infections and marked destruction of lymph node architecture all contribute to the ongoing immune activation as well as deficient immune recovery. Intensive studies on HIV are gradually aiding us understand the processes that link HIV infection to the onset of immunodeficiency. CD4+ T cells exhaustion represents the most fundamental events in HIV infection. Also, HIV-infected individuals show a strong association with individuals of old age: their immune systems are marked by a loss of regenerative capacity and an aggregation of ageing T cells. This review discusses the reason for the development of immune activation and inflammation in the early stages of HIV infection and the long-term effect of these processes to the immune system and health. The three major aspects of HIV disease pathogenesis: reduction of CD4+ T cells, immune activation and depletion of regenerative capacity shall be linked to this process

Keywords: Inflammation; Immune Activation; HIV Infection; CD4+ T Cells; Antiretroviral Therapies

Introduction

Inflammation

Inflammation is a broad term that represents the processes that takes place in the body when the immune system is stimulated to respond to a threat. Though an active immune system is important to maintain good health, in some cases persistent immune activation and inflammation due to an ongoing disease like HIV-infection can result in health-related challenges through-out the body [1].Inflammation explains the complex cascade of events that takes place during immune recognition of antigens and goes into action, including movement and initiation of different kinds of white blood cells (leukocytes) and release of chemical messengers referred to as cytokines [1]. The immediate immune response to infection or acute injury is often referred to as "inflammation" [2]. Macrophages and leukocytes resident in the tissues are stimulated when microorganism gain entry into the body (via a cut) through release of toxins and other signals from injured cells and blood vessels. Nuclear factor kappa-B (NF-kB) is cellular protein is produced which turns on the genes required for immune response. Newly activated macrophages release pro-inflammatory cytokines, including interleukin 1 (IL-1), interleukin-6 (IL-6), and tumor necrosis factor-alpha (TNF-alpha). Soon, neutrophils and other immune cells move to the site and ingest the pathogens through phagocytosis or destroy them by releasing toxic substance [3].

The production of prostaglandin is stimulated by these "first-line defenders" within cells. They also stimulate an acute-phase response (APR), stimulating the liver to secrete acute-phase proteins like C-reactive protein (CRP), fibrinogen, and plasminogen. These chemical initiates physiological changes locally such as vasodilation and elevated permeability resulting to redness, swelling, heat and pain which are the classic signs of inflammation [4]. They also play a role in coagulation (blood clotting) and tissue repair [3]. "Systemically, pro-inflammatory signals act on the brain and elsewhere in the body, causing fever, loss of appetite, fatigue, and other flu-like symptoms. An extreme version of this reaction, known as a "cytokine storm," has proven fatal in clinical trials of experimental therapies and has been proposed as an explanation for the high death rate during the 1918 influenza pandemic. First-line defenders release additional cytokines, including interferon-gamma, that promote longer-term immune activation mediated by the lymphocytes: T-cells, B-cells, and natural killer cells (NK cells). Antigen-presenting

***Corresponding author email:** chidozieagu@gmail.com, cagu@phrinitiative.com*

cells such as macrophages capture pathogens and display pieces of them (antigens) on their surface [4]. Lymphocytes interact with these cells and learn to recognize and directly target those pathogens. The inflammatory response, therefore, is the result of a complex interplay of many different types of immune cells that use hundreds of chemical messengers to communicate among themselves, forming cascades and feedback loops" [5].

Under normal circumstances, the immune response is self-limiting and turns itself off when the threat is no more- for example, when a wound heals, or a bout of infection resolves. But inflammation can become chronic if the trigger persists or if suppressive control mechanisms do not work properly. Pro-inflammatory prostaglandins and acute-phase proteins are short-lived, and their effects are temporary unless there is an ongoing signal to produce more [5]. And just as some chemical messengers promote immune activation, opposing signals act to inhibit responses. Anti-inflammatory cytokines include IL-4, IL-10, and transforming growth factor-beta (TGF-beta). This fine-tuned system can go awry when the immune system is faced with a threat it cannot overcome [2]. This occurs, for example, during persistent infection. Other causes of chronic inflammation include autoimmune conditions, obesity, chronic stress, and exposure to toxins such as tobacco smoke. Numerous pathogens, including HIV, hepatitis B and C viruses, and herpes viruses, can remain in the body over the long term. Though the immune system may respond by producing antibodies and activating killer T-cells; however, this response is not always enough to clear infection. For some pathogens, such as Hepatitis CVirus, a few people can clear the infection either spontaneously or with treatment [1]. Others, like HIV, appear to always persist for life. In contrast with localized acute inflammatory responses, chronic inflammation may be systemic, affecting the entire body. The overall effect is persistent immune activation, but it is more accurately thought of as immune dysregulation, characterized by a shift in leukocyte activity [3]. During chronic inflammation, neutrophils become less active, while T-cells and other lymphocytes take on a larger role. Persistent activation of T-cells accelerates their maturation and progression through the cell cycle of growth and division. Eventually, T-cells burn out prematurely and may undergo apoptosis or lose their ability to divide. Long-term immune activation and sustained high levels of pro-inflammatory cytokines can induce damage through-out the body, and chronic inflammation is increasingly recognized as a common denominator underlying a host of progressive and age-related diseases" [4].

Chronic inflammation and immune activation in HIV infection

"Following the identification of HIV in the early 1980s, it was recognized that HIV-infection consists of a complex interaction between immunodeficiency, chronic inflammation and immune activation [6]. These disturbances were found to activate essentially all cellular compartments of both the innate and adaptive immune systems, including monocytes/ macrophages, NK cells, B cells, and both HIV-specific and non-HIV specific CD4+ and CD8+ T lymphocytes [7,8]. In several seminal studies, Giorgi et al. [9] demonstrated that T cell activation levels, as measured by increased expression of CD38 on CD8+ T cells, added to the predictive value of very low CD4+ T cell counts and were more prognostic of clinical progression and shorter survival than plasma viral load in people with very low CD4 T cell counts. Studies of Simian immunodeficiency virus (SIV) have provided further evidence linking inflammation and dysregulated immune activation with progressive HIV pathogen city. Like HIV-infection, SIV- infection is characterized by high levels of viral replication and the rapid destruction of infected CD4+ T cells in sooty mangabeys and rhesus macaques. Upon infection, sooty mangabeys, which are the natural hosts of SIV and do not experience progressive immunodeficiency despite high viraemia, exhibit relatively restrained levels of immune activation in the early phases of disease in contrast to rhesus macaques, which experience high levels of T cell activation and succumb to SIV-mediated pathogenesis, much like their HIV-infected human counterparts [10]. This suggests that sustained, uncontrolled levels of inflammation and immune activation play a determinant role in distinguishing between pathogenic vs. non-pathogenic models of SIV infection. Similarly, HIV-2, which presents a milder, less pathogenic disease-course than HIV-1, is characterized by considerably lower levels of immune activation than those observed in HIV-1-infected individuals [11]. It has been hypothesized that persistently heightened levels of inflammation and immune activation manifest in the ongoing proliferation, expansion, and destruction of T cells, which leads to the exhaustion of the regenerative capacity of the immune system and ultimately immunodeficiency" [8,12]

Causes of Immune activation and inflammation in HIV-infected individuals

"During HIV-1 infection, the establishment of immune activation and inflammation involve several mechanisms that are either directly or indirectly related to viral replication [13], (Figure 1). The common cause of T cell activation during an infection is antigenic stimulation by the virus, which is the foundation of the adaptive immune response. During primary infection, HIV-1 stimulates strong T cell responses, in particular CD8+ T cells, which can persist during the chronic infection phase due to the continuous replication of the virus: around 20% of circulating CD8+ T cells can be HIV-specific in untreated chronically infected patients [14, 15]. HIV- specific CD4+ T cell responses are usually present at a lower magnitude (i.e. up to 3% of circulating CD4+ T cells), which may be related to their preferential depletion by the virus [7].

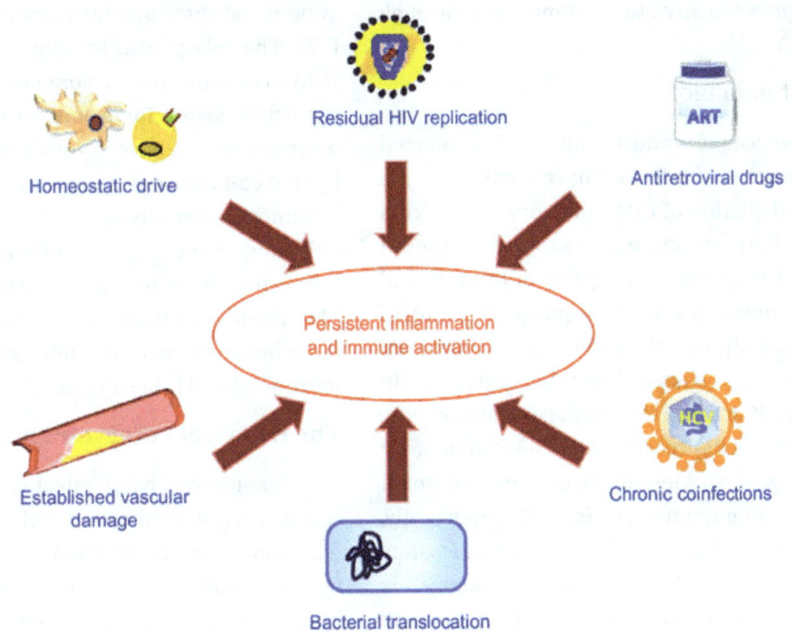

Figure 1: Factors implicated in persistent inflammation and immune activation in HIV patients

Nonetheless, the degree of activation during HIV-1 infection is such that induction with HIV antigens alone cannot account for the complete phenomenon of immune activation observed. Although the physiological impact is not yet known, *in vitro* studies suggest that HIV gene products can induce directly the activation of lymphocytes and macrophages, and the production of proinflammatory cytokines and chemokines. For instance, the envelope protein gp120 may be able to activate cells or to enhance their responsiveness to activation, even in absence of direct infection, through binding to CD4 or co-receptors [16]. The accessory protein Nef is also able to lead to lymphocyte activation either directly [17,18], or through the infection of macrophages [18].The major drivers of immune activation in HIV-1 infection are viral replication, viral proteins, microbial translocation (from GI tract "damage"), co-infection, viral reservoirs and time of ART initiation. Of these, GI tract damage and associated stimulation of cells of the innate and acquired immune systems are possibly the most important. This is supported by the fact that the natural hosts such as SMs do not show evidence of epithelial barrier breakdown nor microbial translocation and consequently, no pathological immune activation [20]. Also, in cart treated individuals, the contribution of viral replication is minimized; therefore, ongoing immune activation is primarily due to non-viral replication factors".

HIV-infected patients have a persistent state of inflammation and immune activation in spite of the suppression of HIV replication via ART.

Various factors might be implicated:

1) homeostatic drive: after reaching an immune/inflammatory set point, the immunological and inflammatory response persist in spite of eliminating the initial stimulus
2) Residual non-detected HIV replication
3) Proinflammatory effects of certain antiretroviral drugs
4) Translocation of bacterial products through damaged intestinal mucosa
5) Coexistence of chronic HCV or herpes virus infection, common in the HIV population
6) Established vascular lesions. Abbreviations: HIV, human immune deficiency syndrome; ART, antiretroviral therapy; HCV, hepatitis C virus. [Adopted from Luis et al. [13]]

HIV Viral proteins

"Although the physiological impact is not yet known, in vitro studies suggest that HIV gene products can stimulate directly the activation of lymphocytes and macrophages, and the production of pro-inflammatory cytokines and chemokines. For instance, the envelope protein gp120 may be able to activate cells or to enhance their responsiveness to activation, even in absence of direct infection, through binding to CD4 or co-receptors [16].HIV gene products, such as Env, Tat, and Nef, have been proposed to be involved in HIV-induced immune activation. The Nef protein of HIV-1 has lost the ability to down modulate the CD3-TCR complex from the surface of infected T cells [21]. Consequently, HIV-1Nef may directly contribute to immune activation by rendering infected CD4+ T cells highly sensitive to re-stimulation through the T-cell receptor (TCR)".

Gastrointestinal tract damage

Studies in the SIV-macaque model and in HIV-infected patients during the acute stage of infection have highlighted the massive and irreversible depletion of CD4+ memory T cells from gut mucosal tissue [22]. Early infection and rapid depletion of these cells are associated with loss of integrity of the mucosal barrier which in turn becomes a source for ongoing activation of the innate immune system [23]. Translocation of gastrointestinal (GI) tract microbes and other bacterial products such as lip polysaccharide (LPS) directly activate macrophages and dendritic cells via toll-like receptors (TLRs) to produce pro-inflammatory cytokines and reactive oxygen species (ROS), causing additional generalized activation of the immune system [23]. Importantly, the loss of memory CD4+ T cells lining the GI tract mucosa is not reversed by cart [24] and therefore, the GI "damage" is likely to remain a significant contributing factor to ongoing activation of the immune system. In addition, recent studies have demonstrated that markers of innate immune activation may be stronger predictors of death during cart than T cell activation [25]. Furthermore, plasma levels of soluble CD14, a marker of microbial translocation and monocyte activation; were shown to predict mortality in HIV infection independently" [26].

Co-infections

Cytomegalovirus (CMV)

"Another important contributing factor to persistent activation of the immune system is the reactivation of other latent viral infections, particularly cytomegalovirus (CMV) [27]. Earlier studies had shown that patients in the sexually exposed group (as opposed to the hemophiliac or intravenous drug user groups) who were CMV antibody (Ab) positive had more rapid HIV disease progression than those who were CMVAb negative [28]. Since then, with effective cart, it has become evident that plasma CMV DNA levels are associated with progression to non-CMV AIDS-defining events [29] and that CMV-specific T cell responses persist at very high levels even during long-term cart [27]. Importantly, it has since been demonstrated that anti-CMV treatment with valganciclovir significantly reduced T cell activation levels and this effect persisted even after stopping the drug [30]. It was suggested that up to 25% of abnormal T cell activation during treated HIV disease may be due to CMV [30]. It seems reasonable to say that larger studies will be important to determine the clinical benefit of treatment of CMV".

Other co-infections

The significant contribution of co-infections to morbidity and mortality in HIV infection has been well documented [31, 32]. Many co-infections non-specifically activate the host immune system and some organisms can directly facilitate HIV replication [27]. The role played by some co-infections such as hepatitis C (HCV) remains unclear. Some have reported HCV to be a relevant predictive factor for a lack of immune recovery on cart [33]; whereas others have shown that CD4 recovery is not affected by this co-infection. Many studies have focused on the effects of treatment of various co-infections on levels of HIV viral load; with only modest reductions in plasma HIV RNA levels [33]. However, it is also relevant to investigate the impact of treatment on levels of immune activation. Co-infections have been shown to activate the cellular arm of the immune system, thereby effectively adding more activated CD4+ T cells for HIV infection and replication [31].

The HIV viral reservoir

"Despite receiving effective ART, HIV patients may have residual viral replication below the detection limits of the techniques commonly used, and/or they may have episodes of transient viral replication. There are conflicting reports in the literature regarding the relative contribution of the HIV reservoir to levels of immune activation. Residual viral replication has been associated with higher CD4+ and CD8+ T cell activation levels [34]. To address this question further, studies have considered the additional benefit of treatment intensification with the viral integrase inhibitor raltegravir. Some have demonstrated an impact on immune activation levels [34], whereas others have shown that treatment intensification has no effect on CD8+ T cell activation [35, 36]. The importance of distinguishing between effects on the CD4+ as opposed to the CD8+ T cell compartment was highlighted by the study of Massanella *et al.* [37], which demonstrated that in a long-term trial of raltegravir intensification therapy, CD38 levels on CD8+ T cells were significantly reduced; however, no effect on CD4+ T cell counts, or activation levels were detected.

Homeostatic drive and time to ART initiation

"Sustained viral replication may promote an immune/inflammatory response that cannot be reversed after a certain point. Therefore, starting ART before reaching this immune/inflammatory set point may prevent the state of persistent inflammation and immune activation. Supporting this hypothesis, the results of several studies have demonstrated that starting ART with low CD4 counts and/or a lowest CD4 nadir was associated with worse immunologic outcomes, even if patients achieved effective viral suppression [38,39].Treating HIV infection in the acute phase significantly reduces the proportion of activated CD38+HLA-DR+CD8 T-cells when compared to non-treated patients [39,40]. In the Options Project, patients who started ART in the first 6 months after infection had a lower proportion of activated CD8 T-cells than those who initiated treatment 2 years or more after the infection [41]. In another interesting study,

Burdo et al. [42] demonstrated that patients with chronic HIV infection experienced a decrease of sCD163 levels after 3 months of ART; however, the sCD163 plasma concentration remained elevated as compared with controls. By contrast, in patients with early HIV infection (1-year post-infection), the sCD163 levels at 3 months of treatment were similar to those of controls. It seems reasonable to conclude that these studies suggest that early ART could result in the decreased activation of CD8 T-cells and monocytes–macrophages, and they support the hypothesis that the activation of these cells could be reversed to normal levels with early ART.

ART-dependent effects

Some antiretroviral drugs can induce endothelial dysfunction and oxidative stress and promote an inflammatory response. Drugs such as ritonavir, indinavir, lopinavir, zidovudine, and abacavir have been associated with these deleterious effects. It has been suggested that other anti-retroviral drugs, such as raltegravir, may have an additional beneficial effect on these processes, independently of the suppression of viral replication. Some studies have shown that substituting a PI or non-nucleoside/nucleotide reverse transcriptase inhibitors with raltegravir reduces the levels of circulating inflammatory markers such as IL-6, hsCRP, or D-dimer.153–155

The consequences of immune activation and inflammation in HIV-1 infection

The initiation of this state of immune activation and inflammation and its long-term establishment due to persistence of the virus have extensive and detrimental effects on the immune system and human health [43], (Figure 2).

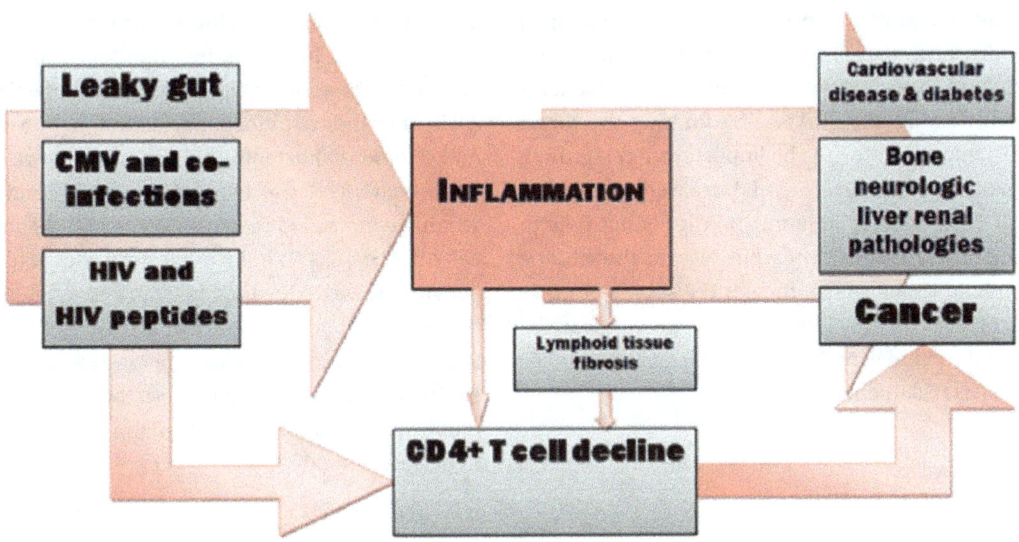

Figure 2: Schematic representation of some of the etiologies and consequences of inflammation in HIV-1 infection

CD4+ T cell depletion

The hallmark depletion of CD4+ T cells is central to the pathogenesis of HIV-infection and AIDS. Direct and indirect mechanisms are delineated involving both the virus itself and non-specific responses to infection and immune activation [44]. Early studies showed that CD4+ T lymphocytes from HIV-infected persons had an enhanced propensity to apoptosis. Furthermore, this was directly related to the degree of lymphocyte activation and correlated with disease progression [44, 45]. To limit the potentially harmful effects of ongoing cell proliferation, well-coordinated death signals are unregulated coincident upon repetitive T cell receptor (TCR) engagement by antigen [46]. Therefore, persistent exposure to antigen, whether in the form of HIV peptides or other foreign antigens, will elicit the signal for T cells to undergo the physiological form of death, apoptosis. This process is termed activation-induced cell death (AICD) and is mediated in CD4+ T cells predominantly via the engagement of the death receptor Fas (CD95) by its ligand (FasL) [46]. This is relevant in the context of HIV infection in that levels of both Fas and FasL have been shown to be increased and further, that CD4+ T cells have an enhanced propensity to AICD [47].In addition to this, HIV peptides such as Nef and Tat, without direct infection of CD4+ T cells, have been shown to induce the up-regulated expression of Fas and FasL [48]. The cytokine environment and particularly the T cell growth factor, interleukin-2 (IL-2), can also modulate Fas-mediated apoptosis [49]. Thus, it is pertinent to note that the very process inducing CD4+ T cells to proliferate is simultaneously predisposing them to death.

Lymphoid tissue pathology

"Chronic antigen stimulation and inflammation result in lymphoid tissue hyperplasia and ultimately in the diffuse effacement of lymph node architecture [50]. Even with effective ART, HIV structural proteins have been shown to persist in the germinal centers (GC) of lymph nodes, providing a source for ongoing immune activation. In addition, there is an accumulation of T follicular helper cells (TFH) within the GC [51] which promote the skewing of B cell subsets and efficiently support ongoing viral replication even during negative plasma viremia [52]. However, recently it was shown that immune activation (as measured by sCD14) rather than direct infection with SIV or HIV was associated with in vivo accumulation of TFH cells within GC [51]. The pro-inflammatory cytokine IL-6 drives the accumulation of T cells in the GC with concomitant up-regulation of bcl-6 (a transcription factor required for TFH formation and B cell help), and this was associated with hyper secretion of IgG1 [53]. Ongoing inflammation promotes the deposition of collagen which disrupts the fibro-reticular network within lymphoid tissues [54]. This impairs homeostatic mechanisms required for T cell homing and survival by limiting the direct contact of naıve T lymphocytes with the important T cell growth cytokine, interleukin-7 (IL-7), resulting in decreased naıve T cell proliferation [54]. Thus, chronic inflammation leads not only to an increased demand for CD4+ T cells but also to the fibrotic "scarring" of lymphoid tissue which impairs the supply of CD4+ T cells during cart [55]. The possibility that this process may be partially reversible if treated early bodes well for the use of anti-fibrotic agents such as Pirfenidone".

HIV-related lymphomas

Such activity and pathology in the lymph nodes as described above would predispose HIV-infected individuals to significantly higher incidences of B-cell non-Hodgkin lymphomas (NHL). All sub-types of NHL show a 60--200-fold increase in HIV-infected persons. Importantly, HIV itself is not classified as a direct carcinogen [56]; but rather facilitates the development of lymphomas via indirect mechanisms; including chronic inflammation and immunodeficiency. In addition, the reactivation or acquisition of other directly oncogenic viruses such as Epstein–Barr virus (EBV) and human herpes virus 8 (HHV-8) provide the "second hit" required for the development of lymphomas [57]. The introduction of ART saw a significant decrease in incidence in HIV-related lymphomas in first world countries; however, this does not appear to have been the case in resource-limited countries [58]. It is possible that in these settings, late presentation of the lymphomas with higher levels of general antigenic exposure and inflammation may preclude the beneficial effects seen from ART in resource- rich countries.

Inflammatory-associated aging

Diseases traditionally associated with aging include liver and kidney disease; cardiovascular, bone loss with associated fractures, cognitive impairment and cancers. As ART has improved the survival of patients with HIV, they are living to older age and the shift has moved from typical HIV-associated diseases and opportunistic infections to disorders associated with aging. However, these are occurring at a younger age in these persons [59]. Inflammation and immune dysfunction may contribute to the increased prevalence of age-related disorders in this population. Higher levels of inflammatory markers such as IL-6 and C-reactive protein (CRP), as well as the coagulation marker, D-dimer; have been associated with increased mortality [60]. The adaptive immune system has an important role in normal aging. It becomes less competent resulting to a state of chronic inflammation [61]. As described above, HIV is also associated with a chronic inflammatory state. HIV-infected subjects older than 50 years show a slower response to treatment. Although ART generally improves immune function, its effects may also contribute to accelerated aging in HIV-infected persons. Certain ART's may induce mitochondrial dysfunction and oxidative stress and telomere shortening [62]. Nucleoside reverse transcriptase inhibitor (NRTI) use may lead to mitochondrial toxicity which results in increased reactive oxygen species and slowing down of cell division [62]. Tenofovir has been shown to inhibit telomerase activity in vitro leading to shortening of telomere length in peripheral blood mononuclear cells [63]. Protease inhibitors can induce senescence markers, oxidative stress and inflammation in human coronary artery endothelial cells i*n vitro* [64]. The implications hereof with regards to patient management remain to be determined".

HIV and neuro cognitive disorders

"The neurocognitive disorders associated with HIV range from mild neuropsychological impairment to HIV-associated dementia [65]. Although the prevalence of severe dementia associated with HIV has decreased owing to the introduction of ART, neurological effects persist. These are thought to be a result of the increased incidence of age-associated disorders in HIV as described above, the ongoing inflammation, increased use of drugs of abuse in HIV-infected individuals and the effects of ART [65]. The inflammation that has been described to occur with HIV-infection also occurs locally within the central nervous system resulting in macrophage activation and the severity of neurological disorders in HIV correlates with the amount of microglia activation [66]. Some ART's may have neurotoxic side-effects and exacerbate central nervous system disorders in HIV-infected individuals [66]. As described above, HIV-infection is associated with the same immune system disorders as aging which may be particularly relevant for the development of dementia. In addition, HIV is associated with an increased prevalence of atherosclerosis, coagulopathy, cardiovascular disease, hypertension, diabetes and the metabolic syndrome. These are all risk factors associated with the development of cerebrovascular accidents (CVAs) which have a higher incidence in HIV infection" [66].

Cardiovascular disease (CVD)

The risk of cardiovascular disease (CVD) is significantly increased in HIV infection and is likely to be the combined result of traditional risk factors, ART-induced cardio-toxicity and HIV-related immune dysfunction. In addition, an increased incidence of traditional CVD risk factors such as smoking, dyslipidemia, diabetes, hypertension and central obesity has been found in HIV-infected individuals [67]. Atherosclerosis is increased in HIV infection. Immune activation induces coronary artery endothelial cells to produce chemokines and adhesion molecules which enhance the development of atherosclerosis [68]. The HIV virus itself via its tat protein can directly activate endothelial cells leading to the up-regulated expression of adhesion molecules such as E-selectin [69]. The HIV envelope glycoprotein gp120 can also increase T cell adherence. In addition, both these proteins have been shown to induce endothelial cell apoptosis. Levels of soluble immune activation markers such as IL-6, adhesion molecules and D-dimer correlate with endothelial dysfunction in HIV infection [61]. It is postulated that the persistent immune activation in HIV infection may result in atherosclerosis and thickened carotid intimal thickness [70].The risk of acute myocardial infarction (AMI) has also been found to be higher in HIV-infected individuals with a recent study showing a 50% increased risk [71].

Another important contributing factor to the risk of CVD is CMV co-infection. CMV has been shown to infect endothelial cells resulting in endothelial damage and accelerated atherosclerosis. Microbial translocation with increased levels of LPS may also perpetuate chronic inflammation and endothelial dysfunction, thereby facilitating accelerated atherosclerosis. The important causal link between microbial translocation and the development of both atherosclerosis and thrombotic disease came from the study of Pandrea *et al.* [72]; their data demonstrated that pigtail macaques developed both atherosclerosis and thrombotic diseases during chronic pathogenic SIV-infection, mimicking the vascular pathology observed in HIV-infected individuals. In addition, vascular pathology in the pigtail macaque SIV model was associated with increases in systemic monocyte/ macrophage activation and coagulation markers" [72].

Inflammation and thrombosis

"An important paradigm is the link between inflammation and thrombosis: inflammation promotes thrombosis, and thrombosis can amplify inflammation [73]. HIV-infection is associated with an increased risk of thrombosis which may be worsened by certain ART regimens [73]. Inflammation induces an imbalance between endothelial pro-coagulant and anti-coagulant properties and this is facilitated by the interaction of leukocytes, endothelial cells and platelets [74]. The marker of coagulation, the D-dimer, has been shown consistently to be a valuable marker of risk for adverse events. A study in untreated late stage HIV-infection found that D-dimer levels were elevated and were strongly associated with mortality after initiation of ART. It was suggested that increased D-dimer levels may be useful to identify those who may need aggressive clinical monitoring after the initiation of ART [74]. Another study highlighted monocyte as an important cell that provides the link between inflammation and thrombosis [75]. It was demonstrated that monocytes from HIV-infected patients showed significantly up-regulated expression of the pro-coagulant tissue factor and that this correlated with markers of immune activation and soluble levels of CD14, the receptor for LPS [75]. The up-regulated expression of this tissue factor promotes the synthesis of thrombin which directly activates platelets, further predisposing patients to thrombus formation".

Diabetes mellitus

"HIV infection is associated with lip dystrophy which involves mitochondrial dysfunction, adipose tissue redistribution, altered differentiation of adipocytes, increased adipocyte lipolysis and apoptosis [76]. This leads to altered adipokine secretion and the release of pro-inflammatory cytokines and free fatty acids which exacerbate chronic inflammation, dyslipidemia and insulin resistance (76). Lip dystrophy may be worsened by ART as mitochondrial toxicity, described with thymidine-based NRTIs, may combine with a direct role of HIV-1 infection via Vpr and tat proteins [77]. The atherogenic lipid profile found in HIV- infected individuals is also referred to as the "diabetic dyslipidemia" and is strongly associated with diabetes [77]. HIV inflammation may lead to insulin resistance due to pro-inflammatory cytokines such as IFN-α and the disturbed secretion of adipokines [77]. Study such as the Swiss HIV cohort study is along-term study which showed that HIV-infected individuals receiving ART had a higher prevalence of diabetes [78]. An important consideration is that diabetes, especially when poorly controlled, may lead to an increased risk of infection and reduced immunity which has been found to increase the prevalence of tuberculosis (TB) infection in some cohorts"[79].

Bone disease

"Bone disease in HIV includes osteoporosis, osteonecrosis and osteomalacia [80]. This is likely due to the combination of traditional risk factors, direct effects of HIV-1 infection (peptides and inflammation) and effects of ART (80). Various studies have shown that the prevalence of osteopenia in HIV is 22–71% and the prevalence of osteoporosis in HIV is 3–33% [78]. HIV-infected individuals tend to have a higher incidence of traditional risk factors such as smoking and alcohol use, they tend to weigh less, use medication that may affect bone mineral density such as selective serotonin re-uptake inhibitor (SSRI) anti-depressants and steroids, and have more diarrhea and malabsorption1 [81]. High viral load has been associated with a decreased bone mineral density and low CD4+ T cell count is an independent risk factor. Both these factors have been associated with increased frailty [80]. Certain viral proteins such as Vpr and gp120 may stimulate osteoblast activity leading to increased bone resorption [82]. Gp120 can also stimulate apoptosis of osteoblasts and shift the differentiation of mesenchymal cells from osteoblasts to adipocytes. P55-gag suppresses osteoblast activity and leads to increased osteoblast apoptosis [82].

Osteoblasts are derived from precursors of monocyte-macrophage lineage and have a membrane surface receptor known as receptor activator NF-kB (RANK). Its ligand is known as RANKL and has important immunological functions including the regulating of T-cell growth and dendritic cell function [83]. In addition, RANK-RANKL interaction stimulates the formation of osteoclasts. The inflammatory cytokines associated with HIV infection can activate RANKL, stimulate osteoclast formation and induce the apoptosis of osteoblasts [84]. LPS from microbial translocation can also directly stimulate osteoclast synthesis by producing inflammatory cytokines and RANKL [83]. The effects of ART are also important in the development of bone pathology. Most bone loss occurs early after initiation of cart and stabilizes in a year or two [84]. Some ARTs lead to mitochondrial toxicity, which elevates lactic acid levels and the bone is resorbed to act as a buffer in this situation. Other ARTs can inactivate vitamin D or cause its catabolism by stimulating cytochrome p450 enzymes [84]. Tenofovir acts directly on the proximal renal tubules and can induce renal phosphate wasting with increased parathyroid hormone levels thereby facilitating bone resorption"[79].

Renal and liver complications

Kidney disease in HIV-infected individuals may be due to direct effects of HIV, the chronic inflammation associated with HIV-infection, traditional risk factors of kidney disease and the toxic effects of certain ART [85]. After the introduction of ART in the 1990s, the prevalence of certain diseases traditionally associated with HIV-infection decreased; however others including kidney and liver disease, increased [86].A study in France has found that the following are risk factors for the development of renal complications in HIV-infection: female gender, older age, diabetes, hyperlipidaemia, low CD4 count and the use of tenofovir [87]. A study conducted in the US found similar risk factors but added African-American race and higher viral loads [88]. HIV-associated nephropathy (HIVAN), the typical kidney disorder associated with HIV-infection and presenting with severe proteinuria and progressive kidney damage, is found to be more prevalent in Blacks [87]. This disorder is caused by HIV-1 infection of the kidney itself [89], however, may also have an underlying genetic component [87]. Kidney disease may affect vitamin D metabolism leading to decreased activation of vitamin D and this may worsen the bone diseases associated with HIV-infection described above. HCV co-infection is also associated with kidney disease as reported in a study [89].

In addition, kidney disease is a CVD risk factor and may worsen the CVD risk in HIV-infected individuals described above. Liver disease is associated with the increased co-infection with hepatitis viruses in HIV-infected individuals [90]. Hepatitis Cvirus HCV is transmitted parenterally; therefore, co-infection with HIV is common, especially in intravenous drug users. Although there is now a vaccine available for hepatitis B virus (HBV), co-infection with HIV is common, especially amongst intra- venous drug users. The SMART study found that HIV immunodeficiency was exacerbated by HBV infection [91]. Other viral hepatitis, such as hepatitis D and E may also lead to liver disease in HIV-infection [91]. Certain ART may cause mitochondrial dysfunction promoting the development of liver diseases and others are associated with lipid disturbances like those found in the metabolic syndrome, which may result in non-alcoholic fatty liver disorder [92]. Importantly, all these factors described heighten the risk of developing hepatocellular carcinoma in HIV infection" [92].

Non-HIV-related malignancies

"In the early pre-ART era of HIV infection, AIDS-defining cancers (ADC) such as Kaposi's sarcoma, non-Hodgkin's lymphoma (NHL) and cervical cancer were highly documented [93]. However, since the early introduction of ART in 1996 and the introduction of the regimes presently used in 2002, there has been a decline in ADC's and an increased reporting on "non-AIDS-defining cancers" (NADC's) such as melanomas, Hodgkin's lymphoma, anal, prostate, hepato- cellular, lung and colorectal cancers [93,94]. The potential causes of NADC's include direct and indirect oncogenic effects of other viruses and HIV, immunosuppressant, chronic inflammation and immune activation, ART and the traditional risk factors [94]. As successful ART improves survival, age may also play an important role in the increased prevalence of NACD's. The immunosuppressant associated with HIV-1 infection has been directly associated with an increased risk of ADC [93]. CD4+ T cell counts are inversely associated with NACD risk, whereas the association is not as strong as for ADC [95]. Additionally, the HIV tat protein can block tumor suppressor genes, inhibit cell apoptosis and affect the cell cycle. HIV-infected individuals have also been shown to have impaired DNA repair ability [96].

Chronic inflammation and immune activation promote increased cell proliferation and the generation of potentially damaging reactive oxygen species [95]. Pro-carcinogenic cytokines and growth factors may also be stimulated [95]. The immune dysfunction associated with HIV infection may also result in impaired immune surveillance with the impaired ability to detect early tumor cells [97]. An increased incidence of traditional risk factors such as smoking, and alcohol consumption has been found in HIV-infected individuals and this may increase the risk of certain NACD's. Earlier initiation of ART has not been proven to decrease the incidence of cancer [98] and in fact some studies have found a higher incidence of cancer in those on ART [94]. As HIV-infected individuals are followed up regularly and may have more regular medical examinations than the public, this may facilitate earlier cancer detection and a perceived higher prevalence than in the general population. However, HIV-infected individuals tend to present at a younger age with more advanced aggressive disease; which is associated with a worse prognosis and more metastases" [94]

Approaches to "switching off" inflammation and immune activation in HIV infection

"Several approaches have attempted to "dampen" the activated state associated with HIV-infection with varying degrees of success. Early work using general immunosuppressant's such as hydroxyurea, prednisone and mycophenolic acid showed limited benefit or even harm in some cases, illustrating that blanket immunosuppressant is counter-productive [99]. A randomized trial that added cyclosporine A to the ART regimen in the chronic stage of the disease showed only minimal and transient increases in CD4+ T cell counts [100]. Selective anti-inflammatory approaches are more likely to be successful. Targeting LPS-induced monocyte/macrophage expression of the enzyme cyclo-oxygenase 2 with cox-2 inhibitors such as Celecoxib has shown promising results. A small randomized placebo-controlled trial (RPCT) in patients with stable viral loads on long-term ART significantly improved markers of immune activation [101]. Subsequently, for patients not yet on ART, the use of Celecoxib for 12 weeks was shown to reduce chronic immune activation with the additional evidence of improved T cell function in vivo [102]. In a similar approach targeting cells of the innate immune system; a study utilizing the anti-malarial agent hydroxychloroquine, which is known to inhibit endosomal TLR, was shown to decrease immune activation levels significantly in HIV-infected, ART-treated, immunological non-responders" [103].

"The lipid-lowering agents, the statins, have well-describedanti-inflammatory properties [104]. Results of a double-blind RPCT using Atorvastatin showed modest but significant reductions in the proportions of activated T lymphocytes. In addition, statins have been shown to reduce levels of CRP and improve patients' outcome as shown in the JUPITER study [105]. Future studies will be of value to determine the impact of these agents on both pro-inflammatory and T cell activation markers. In addition, it will be important to establish the degree to which these markers should be decreased for clinical benefit to be attained [106]. The value of aspirin in the context of HIV management remains contentious. Early studies did not appear to show any benefit; however, this was prior to the advent of ART. In view of the heightened risk for CVD in HIV-infection, the possibility of revisiting aspirin for these individuals has been explored with some promising results. However, another study showed that aspirin was poorly tolerated and failed to improve endothelial function in virologically suppressed HIV-infected adults" [107]

Conclusion

"Persistent immune activation and inflammation are key driving forces in the loss of CD4+ T cells, progression to AIDS and other complications, as shown in Figure 1. The ongoing priming of the immune system results in the release of pro-inflammatory cytokines and recruitment of immune cells to sites of inflammation or infection. An inability to switch into "anti-inflammatory mode" results in the constant erosion of immune protection. Immunotherapy's that limit ongoing immune activation or selectively "switch off" the attendant inflammation will be important avenues for future research. The field of inflammation in HIV-1 infection is extensive and further investigation is required to determine which of the causes impact to a greater or lesser extent on the various complications".

Acknowledgments

None

References

1. Dai G, Kaazempur-Mofrad MR, Natarajan S, Yuzhi Zhang, Saran Vaughn, Brett R Blackman, Roger D Kamm, et al. Distinct endothelial phenotypes evoked by arterial waveforms derived from atherosclerosis-susceptible and -resistant regions of human vasculature. Proc Natl Acad Sci. 2004;101(41):14871–14876.
2. Ridker PM, Cushman M, Stampfer MJ, Tracy RP, Hennekens CH. Inflammation, aspirin, and the risk of cardiovascular disease in apparently healthy men. N Engl J Med. 1997;336(14):973–979.
3. Hansson GK. Inflammation, atherosclerosis and coronary artery disease. N Engl J Med. 2005;352:1685–1695.
4. Ridker PM, Hennekens CH, Buring JE, Rifai N. C-reactive protein and other markers of inflammation in the prediction of cardiovascular disease in women. N Engl J Med. 2000;342(12):836–843.
5. Pai JK, Pischon T, Ma J, Manson JE, Hankinson SE, Joshipura K, Curhan GC, et al. Inflammatory markers and the risk of coronary heart disease in men and women. N Engl J Med. 2004;351(25):2599–2610.
6. Fauci A S. Multifactorial nature of human immunodeficiency virus disease: implications for therapy. Science. 1993;262(5136): 1011-1018.
7. Douek DC, Brenchley JM, Betts MR, Ambrozak DR, Hill BJ, Okamoto Y, Casazza JP, et al. HIV preferentially infects HIV-specific CD4+ T cells. Nature. 2002;417(6884):95–98.
8. Deeks SG, Kitchen CM, Liu L, Guo H, Gascon R, Narvaez AB, Hunt P, et al. Immune activation set point during early HIV infection predicts subsequent CD4+ T-cell changes independent of viral load. Blood. 2004;104(4):942-947.

9. Giorgi JV, Hultin LE, McKeating JA, Johnson TD, Owens B, Jacobson LP, Shih R, et al. Shorter survival in advanced human immunodeficiency virus type 1 infection is more closely associated with T lymphocyte activation than with plasma virus burden or virus chemokine coreceptor usage. The Journal of infectious diseases. 1999;179(4):859-870.
10. Silvestri G, Sodora DL, Koup RA, Paiardini M, O'Neil SP, McClure HM, Staprans SI. Nonpathogenic SIV infection of sooty mangabeys is characterized by limited bystander immunopathology despite chronic high-level viraemia. Immunity. 2003;18(3):441-452.
11. Sousa AE. CD4 T cell depletion is linked directly to immune activation in the pathogenesis of HIV-1 and HIV-2 but only indirectly to the viral load. Journal of immunology. 2002;169(6):3400-3406.
12. McCune JM. The dynamics of CD4+ T-cell depletion in HIV disease. Nature. 2001;410(6831):974-979.
13. Luis M, Alfonso RN, Juan MA, Jesús E, Juan G, Juan AM. Influence of immune activation and inflammatory response on cardiovascular risk associated with the human immunodeficiency virus. Vascular Health and Risk Management. 2015;11:35-48. Doi: 10.2147/VHRM. S65885
14. Betts MR, Ambrozak DR, Douek DC, Bonhoeffer S, Brench- ley JM, Casazza JP, Koup RA, et al. Analysis of total human immunodeficiency virus (HIV)-specific CD4(+) and CD8(+) T cell responses: relationship to viral load in untreated HIV infection. JVirol. 2001;75(24):11983-11991.
15. Papagno L, Appay V, Sutton J, Rostron T, Gillespie GM, Ogg GS, King A, et al. Comparison between HIV- and CMV-specific T cell responses in long-term HIV infected donors. Clin Exp Immunol. 2002;130(3):509–517.
16. Lee C, Liu QH, Tomkowicz B, Yi Y, Freedman BD, Coll- man RG. Macrophage activation through CCR5- and CXCR4- mediated gp120-elicited signaling pathways. J LeukocBiol. 2003;74(5):676-682.
17. Wang JK, Kiyokawa E, Verdin E, Trono D. The Nef protein of HIV-1 associates with rafts and primes T cells for activation. Proc Natl Acad Sci USA. 2000;97(1):394-399.
18. Simmons A, Aluvihare V, McMichael A. Nef triggers a transcriptional program in T cells imitating single-signal T cell activation and inducing HIV virulence mediators. Immunity. 2001;14(6):763-777.
19. SwinglerS, Mann A, Jacque J, Brichacek B, Sasseville VG, Williams K, Lackner AA, et al. HIV-1 Nef mediates lymphocyte chemo- taxis and activation by infected macrophages. Nat Med. 1999;5(9):103-997.
20. Estes JD, Harris LD, Klatt NR, Brian Tabb, Stefania Pittaluga, Mirko Paiardini, G. Robin Barclay, et al. Damaged intestinal epithelial integrity linked to microbial translocation in pathogenic simian immunodeficiency virus infections. PloS Pathogen. 2010;6:e1001052.
21. Schindler M, Münch J, Kutsch O, Li H, Santiago ML, Bibollet-Ruche F, Müller-Trutwin MC, et al. Nef-mediated suppression of T cell activation was lost in a lentiviral lineage that gave rise to HIV-1. Cell. 2006;125(6):1055-1067.
22. Paiardini M, Frank I, Pandrea I. Mucosal immune dysfunction in AIDS pathogenesis. AIDS Rev. 2008;10(1):36-46.
23. Brenchley JM, Price DA, Schacker TW, Tedi E Asher. Microbial translocation is a cause of systemic immune activation in chronic HIV infection. Nat Med. 2006;12(12):1365-1371.
24. Mehandru S, Poles MA, Tenner-Racz K, Jean-Pierre P, Manuelli V, Lopez P, Shet A, et al. Lack of mucosal immune reconstitution during prolonged treatment of acute and early HIV-1 infection. PLoS Med. 2006;3(12):17147468.
25. Tenorio A, Zheng E, Bosch R, Steven G. Deeks, Benigno Rodriguez, Supriya Krishnan, Peter W. Hunt, et al. Soluble markers of inflammation and coagulation, but not T cell activation, predict non-AIDS-defining events during suppressive ART. 2013.
26. Sandler NG, Wand H, Roque A, Law M, Nason MC, Nixon DE, Pedersen C, et al. Plasma levels of soluble CD14 independently predict mortality in HIV infection. J Infect Dis. 2011;203(6):780-790. Doi: 10.1093/infdis/jiq118
27. Naeger DM, Martin JN, Sinclair E, Peter W. Hunt, David R. Bangsberg, Frederick Hecht, Priscilla Hsue, et al. Cytomegalovirus- specific T cells persist at very high levels during long-term antiretroviral treatment of HIV disease. PLoS One. 2010;5:e8886.
28. Robain M, Boufassa F, Hubert JB, Dussaix E, Sadeg K, Meyer L. Is cytomegalovirus infection a co-factor in HIV-1 disease progression? Epidemiol Infect. 2000;125(2):415-420.
29. Deayton JR, Sabin CA, Johnson MA, Emery VC, Wilson P, Griffiths PD. Importance of cytomegalovirus viraemia in risk of disease progression and death in HIV-infected patients receiving highly active antiretroviral therapy. Lancet. 2004;363(9427):2116–2121.
30. Hunt PW, Martin JN, Sinclair E, Epling L, Teague J, Jacobson MA, Tracy RP, et al. Valganciclovir reduces T cell activation in HIV-infected individuals with incomplete CD4+ T cell recovery on antiretroviral therapy. J Infect Dis. 2011;203(10):1474–1483. Doi: 10.1093/infdis/jir060
31. Chang CC, Crane M, Zhou J, Michael Mina, Jeffrey J Post, Barbara A Cameron, Andrew R Lloyd, et al. HIV and co-infections. Immunol Rev. 2013;254(1):114-142.
32. Modjarrad K, VermundSH. Effect of treating co-infections on HIV-1 viral load: a systematic review. Lancet Infect Dis. 2010;10(7):455–463. Doi: 10.1016/S1473-3099(10)70093-1
33. Negredo E, Massanella M, Puig J, Pérez-Alvarez N, Gallego-Escuredo JM, Villarroya J, Villarroya F, et al. Nadir CD4 T cell count as predictor and high CD4 T cell intrinsic apoptosis as final mechanism of poor CD4 T cell recovery in virologically suppressed HIV-infected patients: clinical implications. Clin Infect Dis. 2010;50(9):1300-1308. Doi: 10.1086/651689
34. Buzo´n M, Massanella M, Llibre J, Esteve A, Dahl V, Puertas MC, Gatell JM, et al. HIV-1 replication and immune dynamics are affected by raltegravir intensification of HAART-suppressed subjects. Nat Med. 2010;16(4):460-465. Doi: 10.1038/nm.2111
35. Gandhi RT, Zheng L, Bosch RJ, Chan ES, Margolis DM, Read S, Kallungal B, et al. The effect of Raltegravir intensification on low-level residual viremia in HIV-infected patients on antiretroviral therapy: a randomized controlled trial. PLoS Med. 2010;7(8):e1000321. Doi: 10.1371/journal.pmed.1000321

36. Hatano H, Hayes TL, Dahl V, Sinclair E, Lee TH, Hoh R, Lampiris H, et al. A randomized, controlled trial of raltegravir intensification in antiretroviral-treated, HIV-infected patients with a suboptimal CD4þ T cell response. J Infect Dis. 2011;203(7):960–968. Doi: 10.1093/infdis/jiq138
37. Massanella M, Negredo E, Pe´rez-Alvarez N, Puig J, Ruiz-Hernández R, Bofill M, Clotet B, et al. CD4 T-cell hyperactivation and susceptibility to cell death determine poor CD4 T-cell recovery during suppressive HAART. AIDS. 2010;24(7):959-968. doi: 10.1097/QAD.0b013e328337b957
38. Ho JE, Scherzer R, Hecht FM, Maka K, Selby V, Martin JN, Ganz P, et al. The association of CD4+ T-cell counts and cardiovascular risk in treated HIV disease. AIDS. 2012;26(9):1115-1120. doi: 10.1097/QAD.0b013e328352ce54
39. Vinikoor MJ, Cope A, Gay CL, Ferrari G, McGee KS, Kuruc JD, Lennox JL, et al. Antiretroviral therapy initiated during acute HIV infection fails to prevent persistent T-cell activation. J Acquir Immune Defic Syndr. 2013;62(5):505-508. Doi: 10.1097/QAI.0b013e318285cd33
40. Chevalier MF, Petitjean G, Dunyach-Rémy C, Céline Didier, Pierre-Marie Girard, Maria Elena Manea, Pauline Campa, et al. The Th17/Treg ratio, IL-1RA and sCD14 levels in primary HIV infection predict the T-cell activation set point in the absence of systemic microbial translocation. PLoS Pathog. 2013;9(6):e1003453.
41. Jain V, Hartogensis W, Bacchetti P, Hunt PW, Hatano H, Sinclair E, Epling L, et al. Antiretroviral therapy initi- ated within 6 months of HIV infection is associated with lower T-cell activation and smaller HIV reservoir size. J Infect Dis. 2013;208(8):1202–1211. Doi: 10.1093/infdis/jit311
42. BurdoTH, Lentz MR, Autissier P, Krishnan A, Halpern E, Letendre S, Rosenberg ES, et al. Soluble CD163 made by monocyte/macrophages is a novel marker of HIV activity in early and chronic infection prior to and after anti-retroviral therapy. J Infect Dis. 2011;204(1):154-163. Doi: 10.1093/infdis/jir214
43. Hayley I, Annalise E Zemlin, Rajiv T Erasmus, Richard H. Glashoff. Role of inflammation in HIV-1 disease progression and prognosis. Critical Reviews in Clinical Laboratory Science. 2014;51(2):98-111. Doi: 10.3109/10408363.2013.865702
44. Badley AD, Pilon AA, Landay A, Lynch DH. Mechanisms of HIV-associated lymphocyte apoptosis. Blood. 2000;96(9):2951-2964.
45. Gougeon ML, Lecoeur H, Dulioust A, Enouf MG, Crouvoiser M, Goujard C, Debord T, et al. Programmed cell death in peripheral lymphocytes from HIV-infected persons: increased susceptibility to apoptosis of CD4 and CD8 T cells correlates with lymphocyte activation and with disease progression. J Immunol. 1996;156(9):3509-3520.
46. Van Parijs L, Refaeli Y, Lord JD. Uncoupling IL-2 signals that regulate T cell proliferation, survival, and Fas-mediated activation-induced cell death. Immunity. 1999;11(3):281-288.
47. Kaplan D, Sieg S. Role of the Fas/ Fas ligand apoptotic pathway in HIV type 1 disease. J Virol. 1998;72(8):6279-6282.
48. Campbell GR, Watkins JD, Esquieu D. The C terminus of HIV-1 Tat modulates the extent of CD178-mediated apoptosis of T cells. J Biol Chem. 2005;280(46):38376-38382.
49. Zheng L, Trageser CL, Willerford DM, Lenardo MJ. T cell growth cytokines cause the superinduction of molecules mediating antigen-induced T lymphocyte death. J Immunol. 1998;160(2):763-769.
50. Lederman MM, Margolis L. The lymph node in HIV pathogenesis. Semin Immunol. 2008;20(3):187-195. Doi: 10.1016/j.smim.2008.06.001
51. Petrovas C, Yamamoto T, Gerner MY, Boswell KL, Wloka K, Smith EC, Ambrozak DR, et al. CD4 T follicular helper cell dynamics during SIV infection. J Clin Invest. 2012;122(9):3281-3294. Doi: 10.1172/JCI63039
52. Vinuesa CG. HIV and T follicular helper cells: a dangerous relationship. J Clin Invest. 2012;122(9):3059-3062. Doi: 10.1172/JCI65175
53. Lindqvist M, van Lunzen J, Soghoian DZ, Kuhl BD, Ranasinghe S, Kranias G, Flanders MD, et al. Expansion of HIV-specific T follicular helper cells in chronic HIV infection. J Clin Invest. 2012;122(9):3271-3280. Doi: 10.1172/JCI64314
54. Zeng M, Smith AJ, Wietgrefe SW, Southern PJ, Schacker TW, Reilly CS, Estes JD, et al. Cumulative mechanisms of lymphoid tissue fibrosis and T cell depletion in HIV-1 and SIV infections. J Clin Invest. 2011;121(3):998-1008. Doi: 10.1172/JCI45157
55. Schacker T, Brenchley J, Beilman G, Reilly C, Pambuccian SE, Taylor J, Skarda D, et al. Lymphatic tissue fibrosis is associated with reduced numbers of naive CD4þ T cells in human immunodeficiency virus type 1 infection. Clin Vaccine Immunol. 2006;13(5):556-560.
56. Bouvard V, Baan R, Straif K, Grosse Y, Secretan B, El Ghissassi F, Benbrahim-Tallaa L, et al. A review of human carcinogens – Part B: biological agents. Lancet Oncol. 2009;10(4):321-322.
57. Carbone A, GloghiniA. AIDS-related lymphomas: from pathogenesis to pathology. Br J Haematol. 2005;130(5):662-670.
58. Ulrickson M, Oliver W, Casper C. Epidemiology, diagnosis, and treatment of HIV-associated non-Hodgkin lymphoma in resource-limited settings. Adv Hematol. 2012. Doi:10.1155/2012/932658
59. Deeks SG. HIV infection, inflammation, immunosenescence, and aging. Annu Rev Med. 2011;62:141-155. Doi: 10.1146/annurev-med-042909-093756
60. Kuller LH, Tracey R, Belloso W, Stephane De Wit, Fraser Drummond, H Clifford Lane, Bruno Ledergerber, et al. Activation of inflammatory and coagulation pathways is associated with mortality in patients with HIV infection. PLoS Med. 2008;5:e203.
61. Hsue PY, Deeks SG, Hunt PW. Immunologic basis of cardiovas- cular disease in HIV-infected adults. J Infect Dis. 2012;205(3): S375-382. Doi: 10.1093/infdis/jis200
62. Caron M, Auclairt M, Vissian A. Contribution of mitochondrial dysfunction and oxidative stress to cellular prema- ture senescence induced by antiretroviral thymidine analogues. AntivirTher. 2008;13(1):27-38.
63. Leensyah E, Cameron P, Solomon A, Tennakoon S, Velayudham P, Gouillou M, Spelman T, et al. Inhibition of telomerase activity by HIV nucleotide reverse transcriptase inhibitors: a potential factor contributing to HIV-associated accelerated aging. J Infect Dis. 2013;207(7):1157-1165. Doi: 10.1093/infdis/jit006

64. Lefe`vreC, Auclair M, Boccara F, Bastard JP, Capeau J, Vigouroux C, Caron-Debarle M, et al. Premature senescence of vascular cells is induced by HIV protease inhibitors: implication of prelamin A and reversion by statin. ArteriosclerThrombVascBiol. 2010;30(12):2611-2620. doi: 10.1161/ATVBAHA.110.213603
65. Gannon P, Khan MZ, Kolson DL. Current understanding of HIV-associated neurocognitive disorders pathogenesis. CurrOpinNeurol. 2011;24(3):275-283. doi: 10.1097/WCO.0b013e32834695fb
66. LetendreSL, Ellis RJ, AncesBM. Neurologic complications of HIV disease and their treatment. Top HIV Med. 2010;18(2):45-55.
67. Saves M, Chene G, Ducimetiere P, Leport C, Le Moal G, Amouyel P, Arveiler D, et al. Risk factors for coronary heart disease in patients treated for human immunodeficiency virus compared with the general population. Clin Infect Dis. 2003;37(2):292-298.
68. Zanni MV, Grinspoon SK. HIV-specific immune dysregulation and atherosclerosis. Curr HIV AIDS Rep. 2012;9(3):200–205. Doi: 10.1007/s11904-012-0123-y
69. Cota-Gomez A, Flores NC, Cruz C, Casullo A, Aw TY, Ichikawa H, Schaack J, et al. The human immuno- deficiency virus-1 Tat protein activates human umbilical vein endothelial cell E-selectin via an NF-kappa B-dependent mech- anism. J Biol Chem. 2002;277(17):14390-14399.
70. TincatiC, Bellistri GM, Casana M. CD8þ hyperactivation and senescence correlate with early carotid intima-media thick- ness in HIVþ patients with no cardiovascular disease. J Acquir Immune DeficSyndr. 2009;51:642-644.
71. Freiberg MS, Chang CH, Kuller LH, Skanderson M, Lowy E, Kraemer KL, Butt AA, et al. HIV infection and the risk of acute myocardial infarction. JAMA Intern Med. 2013;173(8):614-622. Doi: 10.1001/jamainternmed.2013.3728
72. PandreaI, CornellE, Wilson C, Ruy M Ribeiro, Dongzhu Ma, Jan Kristoff, Cuiling Xu, et al. Coagulation biomarkers predict disease progression in SIV-infected nonhuman primates. Blood. 2012;120(7):1357-1366. Doi: 10.1182/blood-2012-03-414706
73. Matta F, Yaekoub AY, Stein PD. Human immunodeficiency virus infection and risk of venous thromboembolism. Am J Med Sci. 2008;336(5):402–406. Doi: 10.1097/MAJ.0b013e31816dd2fd
74. Woollard KJ, Suhartoyo A, Harris EE, Eisenhardt SU, Jackson SP, Peter K, Dart AM, et al. Path physiological levels of soluble P-selectin mediate adhesion of leukocytes to the endothelium through Mac-1 activation. Circ Res. 2008;103(10):1128–1138. Doi: 10.1161/CIRCRESAHA.108.180273
75. Funderburg NT, Mayne E, Sieg SF, Asaad R, Jiang W, Kalinowska M, Luciano AA, et al. Increased tissue factor expression on circulating monocytes in chronic HIV infection: relationship to in vivo coagulation and immune activation. Blood. 2010;115(2):161-167. Doi: 10.1182/blood-2009-03-210179
76. Giralt M, Domingo P, Villarroya F. Adipose tissue biology and HIV-infection. Best Pract Res Clin Endocrinol. 2011;25(3):487-499. Doi: 10.1016/j.beem.2010.12.001
77. Gutierrez AD, Balasubramanyam A. Dysregulation of glucose metabolisms in HIV patients: epidemiology, mechanisms, and management. Endocrine. 2012;41(1):1-10. Doi: 10.1007/s12020-011-9565-z
78. Hasse B, Ledergerber B, Furrer H, Battegay M, Hirschel B, Cavassini M, Bertisch B, et al. Morbidity and aging in HIV-infected persons: the Swiss HIV cohort study. Clin Infect Dis. 2011;53(11):1130-1139. Doi: 10.1093/cid/cir626
79. Gupta S. Tenofovir-associated Fanconi syndrome: review of the FDA adverse event reporting system. AIDS Patient Care STDS. 2008;22(2):99-103. Doi: 10.1089/apc.2007.0052
80. Powderly WG. Osteoporosis and bone health in HIV. Curr HIV AIDS Rep 2012; 9(3):218-222. Doi: 10.1007/s11904-012-0119-7
81. Kanis JA, Borgstrom C, De Laet H, et al. Assessment of fracture risk. Osteoporos Int 2005;16(16):581-589. Doi: 10.1007/s00198-004-1780-5
82. Fakruddin JM, Laurence J. HIV envelope gp 120-mediated regulation of osteoclastogenesis via receptor activator of nuclear factor kappa B ligand (RANKL) secretion and its modulation by certain HIV protease inhibitors through interferon-gamma/ RANKLK cross-talk. J Biol Chem 2003;278(48):48251-48258. Doi: 10.1074/jbc.M304676200
83. Anderson DM, Maraskovsky E, Billingsley WL, et al. A homologue of the TNF receptor and its ligand enhance T-cell growth and dendritic-cell function. Nature 1997; 390(6656):175-179. Doi: 10.1038/36593
84. Ofotokun I, McIntosh E, Weitzmann MN. HIV: inflammation and bone. Curr HIV AIDS Rep 2012;9(1):16-25.Doi: 10.1007/s11904-011-0099-z
85. Maggi P, Bartolozzi D, Bonfanti P, et al. Renal complications in HIV disease: between present and future. AIDS Rev 2012; 14(1): 37-53.
86. Selick R, Byers R, Dworkin M. Trends in diseases reported on U.S. death certificates that mentioned HIV infection, 1987–1999. J Acqir Immune DeficSyndr2006; 43: 378-87
87. Atta M. Diagnosis and natural history of HIV-associated nephropathy. Adv Chronic Kidney Dis 2010;17(1):52-58. doi: 10.1053/j.ackd.2009.08.005
88. Fernando S, Finkelstein F, Moore B, Weissmann S. Prevalence of chronic kidney disease in an urban HIV infected population. Am J Med Sci 2008; 335(2):89-94. Doi: 10.1097/MAJ.0b013e31812e6b34
89. Marras D, Bruggeman LA, Gao F, et al. Replication and compartmentalization of HIV-1 in kidney epithelium of patients with HIV-associated nephropathy. Nat Med 2002; 8(5):522-526. Doi: 10.1038/nm0502-522
90. Soriano V, Puoti M, Sulkowski M, et al. Care of patients coinfected with HIV and HCV: 2007 updated recommendations from the HCV-HIV International Panel. AIDS 2007; 21(9):1073-1089. Doi: 10.1097/QAD.0b013e3281084e4d
91. Dore G, Soriano V, Rockstroh J, et al. Frequent HBV rebound among HIV-HBV coinfected patients following antiretroviral treatment interruption. AIDS 2010; 24(6):1479-1485. Doi: 10.1097/QAD.0b013e328334bddb
92. Lin W, Weinberg EJ, Chung RT. Pathogenesis of accelerated fibrosis in HIV/HCV co-infection. J Infect Dis 2013; 207:S13-18. J Infect Dis. 2013 Mar 15; 207(Suppl 1): S13–S18.doi: 10.1093/infdis/jis926
93. Casper C. The increasing burden of HIV-associated malignancies in resource-limited regions. Annu Rev Med 2011;62:157–70. Doi: 10.1146/annurev-med-050409-103711
94. Cutrell J, BedimoR. Non-AIDS defining cancers among HIV- infected

patients. Curr HIV AIDS Rep 2013; 10(3):207–216. Doi: 10.1007/s11904-013-0166-8

95. Dubrow R, Silverberg MJ, Park LS, et al. HIV infection, aging, and immune function: implications for cancer risk and prevention. CurrOpin Oncol 2012; 24(5):506-516. DOI: 10.1097/CCO.0b013e328355e131

96. Engels EA, Biggar RJ, Hall HI, Cross H, Crutchfield A, Finch JL, et al. Cancer risk in people infected with human immunodeficiency virus in the United States. Int J Cancer 2008; 123:187-194. Doi: 10.1002/ijc.23487

97. Tyerman Z, AboulafiaDM. Review of screening guidelines for non-AIDS-defining malignancies: evoking issues in the era of highly active antiretroviral therapy. AIDS Rev 2012; 14(1):3-16.

98. KitahataMM, Gange SJ, Abraham AG, et al. Effect of early versus deferred antiretroviral therapy for HIV on survival. N Engl J Med2009; 360(18):1815-1826. Doi: 10.1056/NEJMoa0807252

99. Hunt PW, DeeksSG. Immune-based therapy for HIV infection: are acute and chronic HIV infection different diseases? J Infect Dis2006;194:1632-1634.

100. Lederman MM, Smeaton L, Smith KY, Rodriguez B, Pu M, Wang H, et al. Cyclosporin A provides no sustained immunologic benefit to persons with chronic HIV-1 infection starting suppressive antiretroviral therapy: results of randomized, controlled trial of the AIDS Clinical Trials Group. J Infect Dis 2006; 194(12):1677-1685. Doi: 10.1086/509261

101. Kvale D, Ormaasen V, Kran AM, Johansson CC, Aukrust P, Aandahl EM et al. Immune modulatory effects of cyclooxygenase type 2 inhibitors in HIV patients on combination antiretroviral treatment. AIDS 2006(6); 20:813–820. Doi: 10.1097/01.aids.0000218544.54586.f1

102. Pettersen FO, Torheim EA, Dahm AEA, et al. Old drugs, new tricks – how some anti-inflammatory drugs can be used to treat HIV an exploratory trial of cyclooxygenase type 2 Inhibitor in HIV-1 infection: down regulated immune activation and improved T cell-dependent vaccine responses. J Virol2011;85:6557-6566

103. Piconi S, Parisotto S, Rizzardini G, et al. Hydroxychloroquine drastically reduces immune activation in HIV-infected, ART-treated, immunological non responders. Blood 2011;118: 3263–3272

104. Weitz-Schmidt G. Statins as anti-inflammatory agents. Trends Pharmacol Sci 2002; 23:482–486.

105. RidkerPM, Danielson E, Fonseca FAH, et al. Rosuvastatin to prevent vascular events in men and women with elevated C-reactive protein. N Engl J Med 2008; 359(21): 2195–207. Doi: 10.1056/NEJMoa0807646

106. CarrA. Statins as anti-inflammatory therapy in HIV disease? J Infect Dis 2011;203:751–752

107. Hileman CO, Carman TL, Gripshover BM, O'Riordan M, Storer NJ, Harrill DE et al. Salsalate is poorly tolerated and fails to improve endothelial function in virologically suppressed HIV-infected adults. AIDS 2010; 24: 1958–1961.Doi: 10.1097/QAD.0b013e32833c3251

Estrogen Receptor Alpha Binding to ERE is Required for Full Tlr7- and Tlr9-Induced Inflammation

Melissa A Cunningham[1,2]*, Jena R Wirth[1], Osama Naga[1], Jackie Eudaly[1] and Gary S Gilkeson[1]

[1]Division of Rheumatology and Immunology, Medical University of South Carolina, USA
[2]Ralph H. Johnson Veterans Affairs Hospital, Charleston, South Carolina, USA

***Corresponding author:** *Melissa A Cunningham, 96 Jonathan Lucas Street, Suite 814, MSC637, Charleston, SC 29425, USA, E-mail: cunnima@musc.edu*

Abstract

We previously found that a maximum innate inflammatory response induced by stimulation of Toll-like Receptors (TLRs) 3, 7 and 9 requires ERα, but does not require estrogen in multiple cell types from both control and lupus-prone mice. Given the estrogen-independence, we hypothesized that ERα mediates TLR signaling by tethering to, and enhancing, the activity of downstream transcription factors such as NFκB, rather than acting classically by binding EREs on target genes. To investigate the mechanism of ERα impact on TLR signaling, we utilized mice with a knock-in ERα mutant that is unable to bind ERE. After stimulation with TLR ligands, both *ex vivo* spleen cells and Bone Marrow-Derived Dendritic Cells (BM-DCs) isolated from mutant ERα ("KIKO") mice produced significantly less IL-6 compared with cells from Wild-Type (WT) littermates. These results suggest that ERα modulation of TLR signaling does indeed require ERE binding for its effect on the innate immune response.

Keywords: (ERα); (TLRs); (DCs)

Abbreviations

ERα: Estrogen receptor alpha; TLRs: Toll-like receptors; DCs: Dendritic cells

Introduction

One of the more profound features of lupus is that females have a 9:1 prevalence of disease over males. The cause of the sex bias in lupus is likely multifactorial, including differences in the sex chromosomes, sex hormones and their receptors. Although estrogen acts primarily via its receptors, estrogen receptor alpha and beta (ERα/ERβ), estrogen can also act through non-receptor mediated mechanisms. Interestingly, there is growing evidence that ERs can mediate physiologic functions independent of estrogen. We previously showed that an optimal inflammatory response by TLRs is dependent on ERα, but independent of estrogen in multiple cell types. B cells and DCs derived from both B6 ERαKO and lupus-prone ERαKO mice had a significantly blunted response to TLR 7 and 9 ligands [1]. These data suggest that ERα modulation of TLR signaling may play a role in lupus pathogenesis, and appears to be independent of estrogen effects.

Based on these findings, we hypothesized that ERα mediates TLR signaling by a genomic, but non-classical mechanism, i.e. by tethering to and enhancing the activity of downstream transcription factors such as NFκB, thereby altering the innate immune response and exacerbating inflammation. An alternative mechanism to explain this ligand independence is activation via kinase cascades including MAPKs (rapid signaling pathway). To investigate the mechanism of ERα impact on TLR signaling, we utilized mice with a knock-in ERα mutant ("KIKO") that is unable to bind ERE, but otherwise functions normally with regard to ligand binding, activation, etc [2]. Female mice that carry a single copy of this non-classical ERα knock-in mutation are infertile due to severe ovarian and uterine defects, [3] but their immune phenotype is not known.

In this communication we show that both *ex vivo* spleen cells and bone-marrow-derived dendritic cells (BM-DCs) from KIKO mice (ERα DNA-binding mutant mice) behave similarly to ERαKO mice in that TLR-stimulated endpoints are blunted. We previously showed that multiple TLR-induced cytokines are impacted by ERα IL-6, MCP-1, IL-23, IL-17 among others. In this study we looked at IL-6, which has multiple pro-inflammatory effects and is a potent activator of the NFκB pathway. IL-6 is known to play a critical role in the immunopathology of SLE in both humans and mice (contributes to B cell hyperactivity and differentiation of T cells into effector cells, including Th17 cells) [4-6]. Blocking IL-6 in mouse models of SLE significantly improves disease [7]. Herein we show that DNA binding of ERα is required for TLR-induced IL-6 production in murine immune cells. These findings suggest that ERE binding is indeed required for ERα modulation of TLR-induced inflammation, despite the lack of estrogen in the system, and future studies will confirm this result in a lupus mouse model.

Materials and Methods

Mice

Female NERKI (ERα DNA-binding mutant) and Ex3a (ERa-null) mice on the C57BL/6/129 background (kind gift of Ken Korach, NIEHS, NC) were crossed to obtain the "KIKO" mouse

***Corresponding author email:** *cunnima@musc.edu*

as previously described. All mice were maintained at the Ralph H. Johnson VAMC Animal Care Facility (Charleston, SC) using Institutional Animal Care and Use Committee approved protocols.

Generation of BMDCs and Spleen cells

Bone marrow-derived DCs were generated using a modified version of the protocol originally described by Inaba et al. [8], without lymphocyte depletion. Briefly, equal numbers of BM cells from WT and KIKO mice were suspended in complete RPMI supplemented with 20ng/mL murine GM-CSF and 20ng/mL murine IL-4 (R&D systems, Minneapolis, MN) and cultured in T75 flasks at 1×10^6 cells/ml ($\sim 20 \times 10^6$/flask) for 7 days. BMDCs were harvested from flasks, counted, and re-plated in 6-well plates at 1×10^6 cells/ml (4×10^6/well) for 18h. For spleen cells, mice were sacrificed and spleens harvested and kept in ice-cold RPMI. Spleens were processed and subjected to red blood cell lysis. Cells were washed twice in cold RPMI before being counted and cultured in 12- or 6-well plates at 1×10^6 cells/ml ($2\text{-}4 \times 10^6$/well) for 18h.

Treatment of DCs with TLR agonists

BM-DCs were harvested on day 7 from BM cultures as described above, seeded at 1×10^6 cells/ml into 6- or 12-well plates in estrogen-free phenol red-free RPMI with 10% charcoal-dextran-stripped FCS and treated with vehicle or TLR agonist: loxoribine (TLR7/8 agonist; 50-200µmol, Sigma–Aldrich, St. Louis, MO), or CpG DNA (TLR9 agonist; 1µg/ml, Hycult Biotech, Canton, MA) for 18h.

Cytokine Release Assay

Cytokine release by DCs and spleen cells was determined by culturing 2×10^6/mL cells with either vehicle or TLR agonist. After 18h, culture supernatants were harvested and cytokine (IL-6) concentrations were measured by sandwich ELISA as per the manufacturer's protocol (eBioscience, Inc., San Diego, CA) using a micro-plate luminometer (Thermo Scientific Multiskan Ascent).

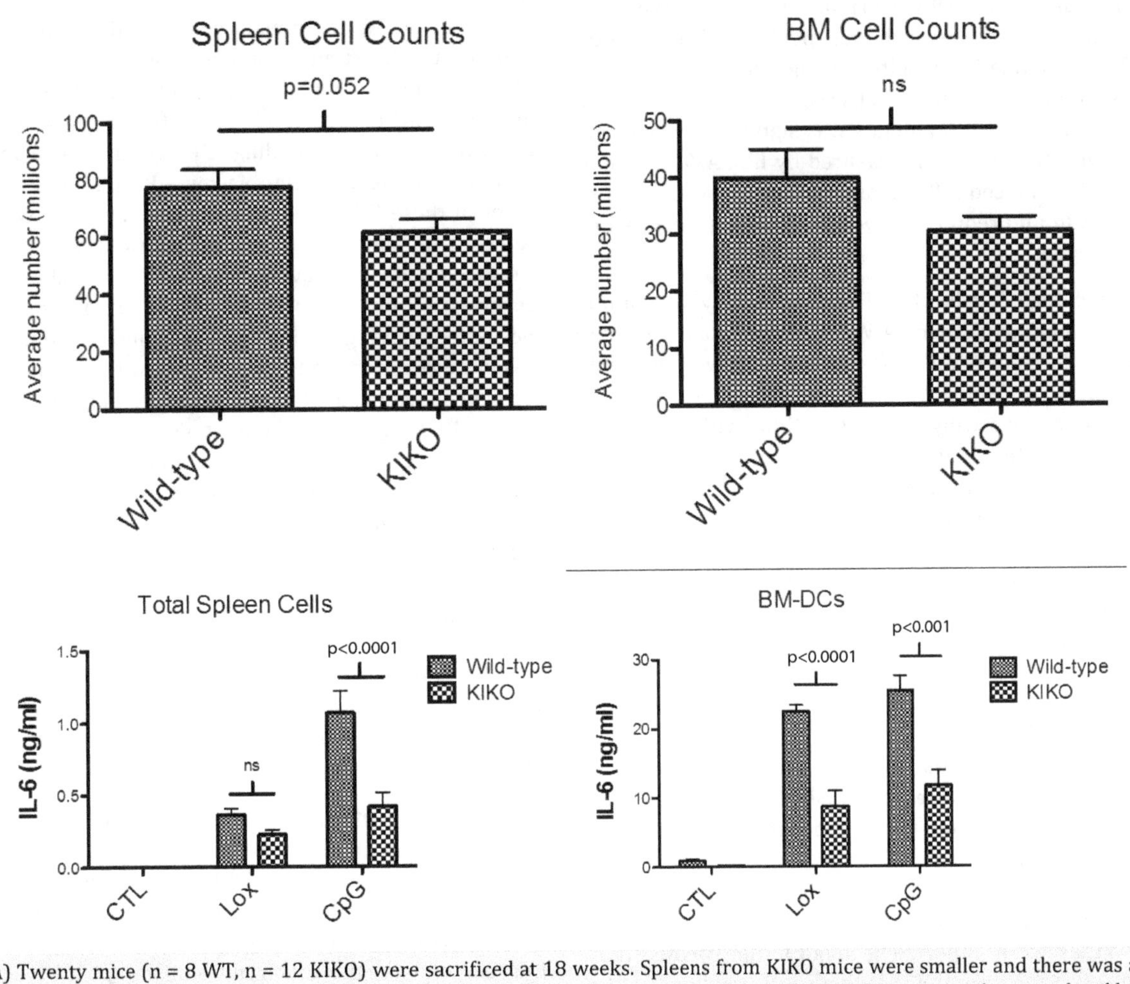

Figure 1: (A) Twenty mice (n = 8 WT, n = 12 KIKO) were sacrificed at 18 weeks. Spleens from KIKO mice were smaller and there was a significant difference in spleen cell counts between WT and KIKO mice. In a subset of animals (n = 4 WT, n = 5 KIKO), femurs were harvested and bone marrow cells were counted. There was a trend toward reduced numbers in the KIKO mice that did not reach significance. (B) *Ex vivo* spleen cells (n = 4 WT, n = 7 KIKO) and cultured BM-DCs (n = 4 WT, n = 5 KIKO) were stimulated overnight (18h) with 200µmol loxoribine (TLR 7 ligand) or 1µg/mol CpG DNA (TLR9 ligand) under estrogen-free media conditions. TLR stimulation resulted in increased IL-6 production that was significantly decreased in media from KIKO mice compared to wild-type mice.

Results

Spleen counts are significantly reduced in ERα DNA binding domain mutant mice ("KIKO") compared with wild-type mice

Twenty C57BL/6 mice (8 WT and 12 KIKO) were sacrificed at 18 weeks. Spleen cells were isolated and counted. Spleens from KIKO mice were smaller and there was a significant difference in spleen cell counts between WT and KIKO mice (Figure 1A). This result aligns with previously published studies demonstrating a critical role for estradiol and ERα in immune organ development (spleen, thymus) as well as murine DC development [9-13]. In a subset of animals (4 WT and 5 KIKO) femurs were harvested and bone marrow cells were counted. There was a trend towards reduced number of cells in the KIKO mice, however it did not reach significance.

TLR7- and 9-induced IL-6 production by ex vivo *spleen cells and BM-DCs is significantly reduced in KIKO mice*

We examined IL-6 production by spleen cells following TLR stimulation. Spleens were harvested and spleen cells were stimulated overnight (18h) with loxoribine (TLR 7 ligand) or CpG DNA (TLR9 ligand) under estrogen-free media conditions. Both TLR7 and TLR9 stimulation resulted in significantly increased levels of IL-6 production by spleen cells (Figure 1), however, IL-6 produced by cells from KIKO animals was significantly reduced compared to wild-type mice, as measured by ELISA. This result suggests that TLR-induced inflammatory cytokine production is modulated by ERα and requires direct ERE binding, despite the absence of estrogen.

DCs express high levels of TLRs and are key mediators of the innate immune response. We isolated bone marrow hematopoietic cells from WT or KIKO mice and derived DCs with selective/supplemented media. Following harvest on d7, DCs were stimulated under estrogen-free conditions with loxoribine or CpG DNA for 18h. In DC cultures derived from WT mice, both the TLR7 and TLR9 ligands stimulated robust IL-6 production, however, KIKO DC IL-6 production was significantly decreased. TLR stimulation increased IL-6 levels more than 20-fold in media from B6 WT DCs, with the stimulation index reduced by ~50% in KIKO animals. These data indicate that ERα significantly modulates TLR7 and 9 responses by DCs via direct ERE binding on target genes.

Discussion and Conclusion

We previously reported that lupus prone ERαKO mice had significantly reduced renal disease and significantly prolonged survival [14]. We subsequently demonstrated that ERα modulates TLR signaling in both C57BL/6 and lupus prone mice (NZM2410 and MRL/lpr) [1]. The mechanism of ERα effect on TLR-induced inflammatory endpoints is currently unknown and is the focus of the current report. The major finding in this study is the requirement for ligand-independent ERE binding by ERα for robust stimulation of TLR-induced inflammatory endpoints.

It is well known that estrogen can modulate IL-6 gene expression [15,16]. Classically, this would occur via ERα binding to an estrogen response element (ERE). There is a growing evidence, however, that the molecular mechanisms, by which ERα exerts its effects, on IL-6 and other target genes, are more complicated than the classic pathway of ligand-activated transcriptional activation. ERα also acts via multiple non-classical signaling pathways to regulate cellular responses. For example, ERα may bind to other transcription factors such as AP-1, C/EBPβ, and NFκB to regulate transcription of IL-6 and others [17-22]. It is also possible that ERα exerts some of its effects by differentially recruiting co-activators or co-repressors, such as p300 to the transcriptional complex to impact gene expression depending on the cell type and environment [23,24]. This study provides additional evidence for ligand-independent actions of ERα, since the experiments reported herein, were done under estrogen-free conditions. Our results were unexpected, however, in that we hypothesized the mechanism of ERα impact on TLR signaling would be both estrogen-independent and ERE-independent. We speculated that the effect would be genomic, by ERα tethering to other transcription factors, but would not require direct DNA binding.

Similar to ERαKO animals, however, IL-6 expression levels in response to TLR stimulation were significantly decreased in the setting of an ERα DNA binding mutant. This DNA binding domain mutant (NERKI) retains all other functions of ERα, including tethering and ligand binding [2]. The mutant also retains the rapid action effects of cytoplasmic ERα (ex. MAPK signaling). If either of these ERα mechanisms were involved, we would have expected to see no change in TLR-stimulated IL-6 production compared with WT levels. Again, all stimulation experiments were carried out under estrogen-free conditions, thus, despite the mutant having an intact ligand-binding domain, the effect was estrogen-independent. One caveat to this, however, is that estrogen is required for the normal development of immune cells such as DCs. This development also requires ERα; as has been demonstrated by our experiments and other experimental works in the field [1,13]. Thus, cells are exposed to estrogen *in vivo* and during initial culture. There may be some threshold or triggering event dependent on estrogen that impacts (imprints on) future signaling (i.e. a developmental effect). Our results suggest that if imprinting on immune cell development is the underlying mechanism, then ERE binding by ERα is also required for immune cell development. Further work is needed with conditional knockout animals and/or *in vitro* knockdown experiments to determine whether the estrogen-independent effect of ERα on TLR signaling is absolute.

This study provides further evidence for a ligand-independent effect of ERα on TLR-induced gene expression in spleen cells and bone marrow-derived dendritic cells, which we have shown is ERE-dependent. Additional work is needed to elucidate the specific mediators of TLR signaling upstream of ERE binding by ERα. Defining the molecular mechanism(s) of ERα effects on TLR signaling is critical to our understanding of female-predominant autoimmune diseases such as SLE and may lead to future therapies that target particular ERα actions and modulate innate immunity.

Acknowledgement

We have been supported with the grant by South Carolina Clinical and Translational Research (SCTR) Institute, NIH Grant number UL1 TR000062, KL2 TR000060.

References

1. Cunningham MA, Naga OS, Eudaly JG, Scott JL, Gilkeson GS (2012) Estrogen receptor alpha modulates Toll-like receptor signaling in murine lupus. Clin Immunol 144(1): 1-12.
2. Sinkevicius KW, Burdette JE, Woloszyn K, Hewitt SC, Hamilton K, et al. (2008) An estrogen receptor-alpha knock-in mutation provides evidence of ligand-independent signaling and allows modulation of ligand-induced pathways in vivo. Endocrinology 149(6): 2970-2979.
3. Jakacka M, Ito M, Martinson F, Ishikawa T, Lee EJ, et al. (2002) An estrogen receptor (ER) alpha deoxyribonucleic acid-binding domain knock-in mutation provides evidence for nonclassical ER pathway signaling in vivo. Mol Endocrinol 16(10): 2188-2201.
4. Richards HB, Satoh M, Shaw M, Libert C, Poli V, Reeves WH (1998) Interleukin 6 dependence of anti-DNA antibody production: evidence for two pathways of autoantibody formation in pristane-induced lupus. J Exp Med 188(5): 985-90.
5. Ryffel B, Car BD, Gunn H, Roman D, Hiestand P, et al. (1994) Interleukin-6 exacerbates glomerulonephritis in (NZB x NZW)F1 mice. Am J Pathol 144(5): 927-937.
6. Finck BK, Chan B, Wofsy D (1994) Interleukin 6 promotes murine lupus in NZB/NZW F1 mice. J Clin Invest 94(2): 585-591.
7. Mihara M, Takagi N, Takeda Y, Ohsugi Y (1998) IL-6 receptor blockage inhibits the onset of autoimmune kidney disease in NZB/W F1 mice. Clin Exp Immunol 112(3): 397-402.
8. Inaba K, Inaba M, Romani N, Aya H, Deguchi M, et al. (1992) Generation of large numbers of dendritic cells from mouse bone marrow cultures supplemented with granulocyte/macrophage colony-stimulating factor. J Exp Med 176(6): 1693-1702.
9. Carreras E, Turner S, Frank MB, Knowlton N, Osban J, et al. (2010) Estrogen receptor signaling promotes dendritic cell differentiation by increasing expression of the transcription factor IRF4. Blood 115(2): 238-246.
10. Carreras E, Turner S, Paharkova-Vatchkova V, Mao A, Dascher C, et al. (2008) acts directly on bone marrow myeloid progenitors to differentially regulate GM-CSF or Flt3 ligand-mediated dendritic cell differentiation. J Immunol 180(2): 727-738.
11. Kovats S, Carreras E (2008) Regulation of dendritic cell differentiation and function by estrogen receptor ligands. Cell Immunol 252(1-2): 81-90.
12. Mao A, Paharkova-Vatchkova V, Hardy J, Miller MM, Kovats S (2005) Estrogen selectively promotes the differentiation of dendritic cells with characteristics of Langerhans cells. J Immunol 175(8): 5146-5151.
13. Paharkova-Vatchkova V, Maldonado R, Kovats S (2004 Estrogen preferentially promotes the differentiation of CD11c+ CD11b(intermediate) dendritic cells from bone marrow precursors. J Immunol 172(3): 1426-1436.
14. Svenson JL, EuDaly J, Ruiz P, Korach KS, Gilkeson GS (2008) Impact of estrogen receptor deficiency on disease expression in the NZM2410 lupus prone mouse. Clin Immunol 128(2): 259-268.
15. Jilka RL, Hangoc G, Girasole G, Passeri G, Williams DC, et al. (1992) Increased osteoclast development after estrogen loss: mediation by interleukin-6. Science 257(5066): 88-91
16. Ray A, Prefontaine KE, Ray P (1994) Down-modulation of interleukin-6 gene expression by 17 beta-estradiol in the absence of high affinity DNA binding by the estrogen receptor. J Biol Chem 269(17): 12940-12946.
17. Biswas DK, Singh S, Shi Q, Pardee AB, Iglehart JD (2005) Crossroads of estrogen receptor and NF-kappaB signaling. Sci STKE 2005(288): pe27.
18. Jakacka M, Ito M, Weiss J, Chien PY, Gehm BD, Jameson JL (2001) Estrogen receptor binding to DNA is not required for its activity through the nonclassical AP1 pathway. J Biol Chem 276(17): 13615-13621.
19. Safe S (2001) Transcriptional activation of genes by 17 beta-estradiol through estrogen receptor-Sp1 interactions. Vitam Horm 62: 231-252.
20. Ray P, Ghosh SK, Zhang DH, Ray A (1997) Repression of interleukin-6 gene expression by 17 beta-estradiol: inhibition of the DNA-binding activity of the transcription factors NF-IL6 and NF-kappa B by the estrogen receptor. FEBS Lett 409(1): 79-85.
21. Kurebayashi S, Miyashita Y, Hirose T, Kasayama S, Akira S, et al. (1997) Characterization of mechanisms of interleukin-6 gene repression by estrogen receptor. J Steroid Biochem Mol Biol 60(1-2): 11-7.
22. Galien R, Garcia T (1997) Estrogen receptor impairs interleukin-6 expression by preventing protein binding on the NF-kappaB site. Nucleic Acids Res 25(12): 2424-2429.
23. Louthrenoo W, Kasitanon N, Sukitawut W, Wichainun R (2003) A clinical study of crystal-proven gouty arthritis in a university hospital. J Med Assoc Thai 86(9): 868-875.
24. Louthrenoo W, Kasitanon N, Mahanuphab P, Bhoopat L, Thongprasert S (2003) Kaposi's sarcoma in rheumatic diseases. Semin Arthritis Rheum 32(5): 326-333.

Interleukin-1β Production in Human Monocytes/ Macrophages is Differentially Regulated by Mek1 upon Sterile and Infectious Inflammatory Conditions

Rakel Carpintero[#], Karim J. Brandt[#], Lyssia Gruaz and Danielle Burger*

Division of Immunology and Allergy, Inflammation and Allergy research group, Hans Wilsdorf Laboratory, Department of Internal Medicine, University Hospital and Faculty of Medicine, University of Geneva, Switzerland

[#]Contributed equally to the study and to the preparation of this manuscript

***Corresponding author:** Danielle Burger, Hans Wilsdorf Laboratory, IARG, BB-4866, University Hospital, 4 rue Gabrielle Perret-Gentil, CH-1211 Geneva 14, Switzerland, Email: danielle.burger@hcuge.ch*

Abstract

Deregulation of the production of IL-1β and its natural inhibitor, the secreted form of IL-1 receptor antagonist (sIL-1Ra), plays an important role in the pathogenesis of chronic inflammatory diseases such as rheumatoid arthritis and multiple sclerosis. Relevant to the latter conditions direct cellular contact with stimulated T cells potently triggers cytokine production in human monocytes. Identification of signal transduction pathways specific to pathogenic induction of cytokines may lead to new therapeutic approaches. Two different stimuli were compared to investigate the implication of MEK1 and MEK2 in the control of IL-1β and sIL-1Ra production by human monocytes: (i) soluble extracts of plasma membranes from stimulated T cells (CE_{sHUT}), mimicking cellular contact with T cells, i.e., chronic/sterile inflammatory conditions; and (ii) LPS that is relevant to infectious inflammation. The ATP-noncompetitive MEK1/2 (U0126) and MEK1 (PD98059) specific inhibitors diminished the expression (protein and mRNA) of IL-1β in CE_{sHUT}-activated monocytes. In contrast, only the concomitant inhibition of MEK1 and MEK2 inhibited IL-1β production in LPS-activated monocytes, whereas the inhibition of MEK1 only did not affect IL-1β production. In CE_{sHUT}- and LPS-activated monocytes, MEK1 inhibition slightly affected sIL-1Ra production that was significantly inhibited by U0126. These results suggest that MEK1 and MEK2 are differentially involved in the regulation of the IL-1 system upon chronic/sterile and infectious inflammatory conditions. MEK1 which is dispensable to IL-1β production in LPS-activated monocytes represents a potential therapeutic target whose inhibition could participate in the restoration of IL-1β/sIL-1Ra balance in chronic/sterile inflammation without affecting regular responses to pathogens.

Keywords: Chronic/sterile inflammation; IL-1β; sIL-1Ra; Human monocytes; MAP kinases; Transduction signals; Cytokines; LPS

Abbreviations

IL-1β: Interleukin-1β; sIL-1Ra: Secreted form of IL-1 Receptor Antagonist; LPS: Lipopolysaccharides; CESHUT: CHAPS Extract of Membranes Isolated from Stimulated HUT-78 cells; ERK: Extra-Cellular-Signal-Regulated Kinase; MEK: Mitogen-Activated Protein Kinase; PI3K: Phosphoinositide 3-Kinases

Introduction

Interleukin-1β (IL-1β) is a major proinflammatory cytokine which signals through ligation of the IL-1 receptor I (IL-1RI) and IL-1 receptor accessory protein (IL-1RAcP) [1]. The production of IL-1β is tightly controlled at several levels and required two signals, one activating gene transcription and leading to the expression of cytoplasmic pro-IL-1β, the other activating an inflammasome that activates caspase-1 and in turn the cleavage of pro-IL-1β into its mature form IL-1β [2,3]. The mechanism by which IL-1β is secreted is different from the canonical secretory mechanisms and remains elusive. One of the main natural inhibitors controlling mature IL-1β activity is the secreted form of IL-1 receptor antagonist (sIL-1Ra), which binds IL-1RI without inducing signal transduction [3,4]. Deregulation of the production of IL-1β and its natural inhibitor, sIL-1Ra, plays an important role in various chronic inflammatory diseases [5] including multiple sclerosis [6] and rheumatoid arthritis [7]. Relevant to these conditions direct cellular contact with stimulated T cells potently triggers cytokine production in human monocytes [8,9]. Direct cellular contact with stimulated T cells is now recognized a major pathway for the production of cytokines (e.g. IL-1β and tumor necrosis factor - TNF -) in monocytes/macrophages under chronic/sterile conditions [10-20]. Indeed, contact-mediated activation of monocytes/macrophages by stimulated T lymphocytes is as potent as optimal doses of LPS to inducing IL-1β and TNF production in monocytes [8,21]. We therefore assume that this mechanism is highly relevant to the pathogenesis and persistence of chronic/sterile inflammation in diseases with autoimmune etiology such as multiple sclerosis and rheumatoid arthritis in which auto reactive T cells play a major part.

An important challenge in therapeutic approaches in inflammatory diseases is to dampen inflammation without affecting the regular inflammatory responses to pathogens. To identify therapeutic targets specific to chronic inflammation we undertook to characterize signaling pathways that trigger the production of IL-1β and sIL-1Ra in human monocytes activated upon either chronic/sterile or infectious conditions.

***Corresponding author email:** danielle.burger@hcuge.ch*

In vitro, chronic/sterile inflammation was mimicked by contact with stimulated T cells. To obviate the complications of having the simultaneous presence of at least two viable cell types in culture, we developed strategies allowing only interactions between stimulated T cell molecules and monocytes, consisting in to use soluble extracts of plasma membranes isolated from T cells [22]. In parallel, LPS was used as a prototypical infectious stimulus. Using these models, we previously demonstrated that phosphatidylinositide-3 kinase δ (PI3Kδ) regulate IL-1β and sIL-1Ra expression in opposite ways in human monocytes activated by LPS or contact with T cells, dampening the production of pro-inflammarory cytokines in LPS-activated monocytes, but inducing it in contact-activated monocytes [23,24]. Therefore, inhibition of PI3Ks could exacerbate inflammatory response to pathogens.

MEK1 and MEK2 are two Thr/Tyr dual mitogen-activated protein kinases (MAPK) whose activation is triggered by a wide variety of stimuli. Although their respective genes are located on different chromosomes, MEK1 and MEK2 display 80% sequence homology and are ubiquitously expressed in cells and tissues [25,26]. ERK1 and ERK2 are the only protein substrates of MEK1/2 that have been identified to date [25]. It is commonly assumed that MEK1 and MEK2 are functionally equivalent, whilst several lines of evidence indicate that they are regulated differentially and that each may fulfill non-redundant functions. Indeed, in contrast with MEK2$^{-/-}$ mice that show no phenotypic abnormalities [27], MEK1$^{-/-}$ mice display recessive lethality, their homozygous mutant embryos dying by day 10.5 of gestation [28]. More recently, it was shown that active MEK2 may function as a regulatory scaffold protein promoting a crosstalk between different transduction pathways [29], further demonstrating that MEK1 and MEK2 may display non-redundant functions in spite of identical downstream substrates. We also highlighted a non-redundant role of MEK1 and MEK2 in human monocytes activated by interferon-β. In the latter conditions, the expression of sIL-1Ra was controlled by a MEK2-PI3Kδ pathway, MEK1 being dispensable [30]. In contrast to other protein kinase inhibitors, inhibitors of MEK1 (PD98059) and MEK1/2 (U0126) do not target the ATP pocket (i.e. they are ATP-noncompetitive inhibitors), and thus are less prone to display off-target effects [31,32]. In the present study, by using U0126 and PD98059, we demonstrate that MEK1 is dispensable to IL-1β production and poorly involved in the induction of sIL-1Ra upon LPS-activation of monocytes whereas both MEK1 and MEK2 are required to IL-1β production under chronic/sterile inflammatory conditions. The present results identify MEK1 as a potential therapeutic target to dampen chronic/sterile inflammation without affecting inflammatory response to pathogens.

Materials and Methods

Materials

The following materials were purchased from the designated suppliers: FCS, streptomycin, penicillin, L-glutamine, RPMI-1640, and PBS free of Ca^{2+} and Mg^{2+} (Invitrogen); Lymphoprep (Axis-Shield); MEK1/2 inhibitor, U0126 and MEK1 specific inhibitor, PD98059 (LC Laboratories); lipopolysaccharides (LPS, Ultra-pure LPS-Ek, InvivoGen); Phaseolus vulgaris leucophytohemagglutinin (PHA, E-Y Laboratories Inc., San Mateo, CA); and phorbol myristate acetate (PMA, Sigma Chemicals Co., St. Louis, MO); Other reagents were of analytical grade or better.

Monocytes: Peripheral blood monocytes were isolated from buffy coats of blood of healthy volunteers provided by the Geneva Hospital Blood Transfusion Center (Switzerland) as previously described [33]. In accordance with the ethical committee of the Geneva Hospital, the blood bank obtained informed consent from the blood donors.

T Cells and Preparation of T cell Plasma Membranes: HUT-78, a human T cell line, was purchased from the American Type Culture Collection. HUT-78 cells were cultured and activated by PHA (1 μg/ml) and PMA (5 ng/ml) in RPMI 1640 medium supplemented with 10% heat activated FCS, 50 μg/ml streptomycin, 50 U/ml penicillin, 2 mM glutamine (complete RPMI medium) in a 5% CO_2-air humidified atmosphere at 37°C as described elsewhere [34]. Plasma membranes of stimulated HUT-78 cells were isolated and solubilized with CHAPS to obtain membrane CHAPS extract (CE_{sHUT}) as previously described [35]. The capacity of CE_{sHUT} to activate human monocytes was equivalent to living HUT-78 cells or primary human T lymphocytes (i.e., in cocultures), fixed T cells, or isolated membranes as previously determined [34-36]. Protein concentration was determined by the method of Bradford [37]. CE_{sHUT} was endotoxin-free as determined by the Endochrome-K LAL kit (Charles River Laboratories Inc.).

Western blot analysis

Human monocytes were resuspended at 6 x 10^6 cells/ml in complete RPMI medium and 500 μl was placed in 2-ml polypropylene tubes (Eppendorf) at 37°C for 1 h. Cells were preincubated for 45 min in the presence or absence of 5 μM of U0126 or PD98059 and then activated with 100 ng/ml LPS or 6 μg/ml CE_{sHUT}. At the indicated time, the activation was stopped by the addition of 800 μL of ice-cold PBS before centrifugation and cell lysis. Total cell lysates were prepared and subjected to Western blot analysis as described previously [38]. Nitrocellulose membranes were probed with rabbit anti-phospho-ERK1/2-p44/42 MAPK (Thr202/Tyr204), mouse anti-ERK1/2-p44/42 MAPK (Cell Signaling Technology), and mouse anti-β-tubulin (Sigma). Secondary IR700/800 conjugated goat anti-rabbit or goat anti-mouse antibodies (Rockland) were used, and antibody bound proteins were detected and quantified with an Odyssey system (Li-Cor).

Cytokine production

Isolated monocytes (5 x 10^4 cells/200μl/well) were preincubated for 45 min in the presence or absence of the indicated concentration of kinase inhibitor in complete RPMI medium and then activated for 24 h with 100 ng/ml of LPS or 6 μg/ml of CE_{sHUT}. All conditions were conducted in triplicate. After supernatant harvesting, cells were lysed in 200 μl/well PBS containing 1% Nonidet P40. Culture supernatants and/or cell lysates were tested for the production of IL-1β and sIL-1Ra by commercially available enzyme immunoassay kits (eBioscience).

mRNA quantification

Monocytes (2 x 10^6 cells/2ml/well) were cultured in 6-well plates for 45 min in the presence or absence of 5µM of U0126 or 5 µM of PD98059 and then activated by LPS (100 ng/ml) or CE_{sHUT} (6 µg/ml proteins) for 3 h. Preparation of total RNA was carried out with Nucleo Spin RNA II kit (Macherey-Nagel) and quantitative real-time duplex PCR analysis was conducted as described previously [38]. The levels of mRNA expression were normalized with the expression of a housekeeping gene (18S). Cytokines and 18S probes were purchase from Applied Biosystems. All measurements were carried out in triplicates.

Statistical analysis

When required, significance of differences between groups was evaluated using Student's *t*-test.

Results

Production of IL-1β and sIL-1Ra in human monocytes

Previous works demonstrated that CE_{sHUT} and LPS induce IL-1β and sIL-1Ra production in human monocytes [23,24]. Since the levels of cytokine production varied between different monocyte preparations (i.e., blood donors), the results below are presented as percentages of the cytokine production measured in the absence of inhibitor. Table 1 shows the mean production of IL-1β, pro-IL-1β and sIL-1Ra observed in the presented experiments. CE_{sHUT} induced lower IL-1β production but higher production of sIL-1Ra as compared with LPS. However, the total production of IL-1β and pro-IL-1β (i.e., extra- and intracellular IL-1β) triggered by CE_{sHUT} or LPS was similar reaching 3102 ± 1038 pg/ml and 3217 ± 1068 pg/ml, respectively, suggesting that LPS was more efficient in inducing IL-1β secretion. CE_{sHUT} was more potent than LPS to trigger sIL-1Ra production, confirming previous data [24].

MEK/ERK pathway activation by CE_{sHUT} and LPS

To assess MEK1/2 activation in monocytes upon different inflammatory conditions, monocytes were activated by CE_{sHUT} or LPS and phosphorylation of ERK1/2 (i.e., the canonical substrates of MEK1/2) was measured by Western blot. The maximum ERK1/2 phosphorylation in monocyte was observed after 15-20 min activation independently of the stimulus (Figure 1A). The phosphorylation of ERK1/2 was inhibited to a higher extent in the presence of U0126 (MEK1/2 inhibitor) than in the presence of PD98059 (MEK1 inhibitor) independently of the stimulus (Figure 1B). These results demonstrate that MEK inhibitors decreased ERK1/2 phosphorylation and that their efficiency was independent of the stimulus.

IL-1β and sIL-1Ra production is differentially regulated by MEK1 and/or MEK2 upon chronic/sterile and acute/infectious inflammatory conditions

To elucidate the implication of MEK1 and/or MEK2 in chronic/sterile and infectious inflammation, the effects of the MEK1 inhibitor, PD98059, and the MEK1/2 dual inhibitor, U0126, were tested in monocytes activated by CE_{sHUT} or LPS. In monocytes activated by CE_{sHUT}, the production of IL-1β and sIL-1Ra was inhibited by U0126 and PD98059 in a dose dependent manner (Figures 2A and 2B). As expected, the inhibition of both MEK1 and MEK2 (U0126) was more efficient to decrease IL-1β and sIL-1Ra production than the inhibition of MEK1 only (PD98059) suggesting that the activation of MEK1 and MEK2 was required for optimal induction of cytokine production. In contrast with the production of sIL-1Ra (Figure 2B), CE_{sHUT}-induced IL-1β production was exquisitely sensitive to U0126 which diminished the production by more than 60% at a concentration as low as 1 µM. Although inhibited, the secretion of sIL-1Ra was decreased to a lower extent than that of IL-1β in the presence of MEK inhibitors and reached only 44 ± 5% and 19 ± 13% at the highest concentration of kinase inhibitors used in this study (Figures 2A and 2B). These results indicate that MEK1 and MEK2 do not represent a major regulator of sIL-1Ra production in chronic/sterile inflammatory conditions. In LPS-activated human monocytes, the production of cytokines was differentially regulated by MEK1 and MEK2 (Figures 2C and 2D). IL-1β production was strongly inhibited in the presence of U0126 reaching a plateau at 62 ± 14% inhibition at 2.5 µM U0126. In contrast, IL-1β levels remained unaffected by PD98059 even at high concentrations (Figure 2C), suggesting that MEK1 was dispensable to IL-1β production upon LPS-activation of human monocytes. sIL-1Ra production was regulated by both MEK1 and MEK2, even thought it was less sensitive to PD98059 than to

Table 1: Production of IL-1β, pro-IL-1β and sIL-1Ra by human monocytes.

Stimulus	IL-1β (pg/ml)	Pro-IL-1β (pg/ml)	sIL-1Ra (pg/ml)
CE_{sHUT} (6 µg/ml)	654 ± 245	2,528 ± 1,469	18,376 ± 2,287
LPS (100 ng/ml)	1,367 ± 737	1,851 ± 534	6,058 ± 602

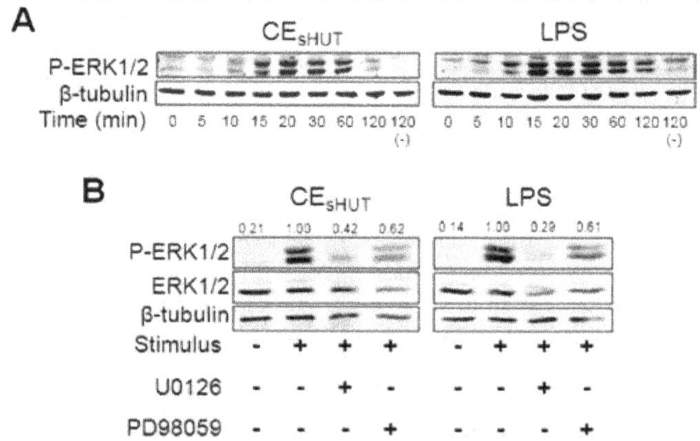

Figure 1: Induction of ERK1/2 phosphorylation by CE_{sHUT} and LPS and inhibition by U0126 and PD98059.

A) Isolated human monocytes (3 x 10^6 cells) were activated for the indicated time with either CE_{sHUT} or LPS. Cells were lysed and subjected to Western blot analysis as described in Materials and Methods. (-) monocytes incubated in the absence of stimulus for 120 min.

B) Monocytes were activated for 20 min in the presence or absence of 5 µM of the indicated inhibitor. Western blots were immunostained for phosphorylated ERK1/2 (P-ERK1/2), total ERK1/2 and β-tubulin as indicated. Immunoblotting data are representative of 3 separate experiments. ERK1/2 phosphorylation quantified by the Odyssey system (Li-Cor) is mentioned above the blot, the ratio of phosphorylated ERK1/2 to total ERK1/2 being considered as 1.0 in monocytes activated in the absence of inhibitor.

U0126 inhibition (Figure 2D). In the presence of PD98059, sIL-1Ra production was only slightly inhibited, reaching a maximum inhibition of 31 ± 6% at 10 μM PD98059. Together these results suggest that both MEK1 and MEK2 activation was required to IL-1β and sIL-1Ra optimal production in monocytes activated by CE_{sHUT}, i.e., upon sterile inflammatory conditions, but only MEK2 was required to IL-1β production in LPS-activated monocytes. Since only mature IL-1β is released in the extracellular space, we assessed whether MEK1 and/or MEK2 controlled IL-1β production at the transcriptional or the maturation/secretion level, and measured the effects of inhibitors on the production of pro-IL-1β, i.e., the production of intracellular IL-1β. As shown in Figures 2E and 2F, the inhibition of pro-IL-1β expression by U0126 in CE_{sHUT}- and LPS-activated monocytes was comparable to that observed for the production of IL-1β. These data suggest that MEK1/2 pathway control the expression of both IL-1β and, to a lower extent, sIL-1Ra upstream protein translation independently of the stimulus and that MEK1 is dispensable to IL-1β production in LPS-activated monocytes.

Cytokine transcript expression is differentially regulated by MEK1 and/or MEK2 upon chronic/sterile and acute/infectious inflammatory conditions

To determine whether MEK1 and/or MEK2 modulated the expression of IL-1β and sIL-1Ra mRNA as suggested by results of Figure 2, we determined the expression of IL-1β and sIL-1Ra transcripts in monocytes in the absence or presence of MEK inhibitors. As shown in Figure 3, the expression of IL-1β and sIL-1Ra mRNA reflected the results obtained at the protein level independently of the stimulus. Indeed, IL-1β mRNA and sIL-1Ra mRNA expression was inhibited by MEK1 and MEK1/2 inhibitors in CE_{sHUT}-activated monocytes, with higher inhibition observed with U0126. In LPS-activated monocytes sIL-1Ra mRNA was inhibited by U0126 and PD98059, whereas only U0126 inhibited IL-1β transcript expression (Figure 3B) further suggesting that MEK1 was dispensable to the induction of IL-1β expression upon infectious inflammation. Together these results demonstrate that the MEK-ERK1/2 pathway regulates the IL-1β system in both sterile and infectious conditions, but only MEK2 is required to optimally triggering IL-1β production in LPS-activated monocytes.

Discussion

The present results demonstrate that MEK1 activation is dispensable to signal transduction triggered by LPS that leads to IL-1β production. This contrasts with signal transduction triggered by CE_{sHUT} (i.e., direct cellular contact) which requires the activation of both MEK1 and MEK2 to optimally inducing IL-1β expression. The expression of sIL-1Ra is only slightly regulated by MEK1 and MEK2, independently of the stimulus. The low extent of sIL-1Ra inhibition observed with U0126 or PD98059 suggests that the MEK-ERK pathway does not play a major role in the regulation of sIL-1Ra production.

LPS signaling through ligation of the toll-like receptor 4 (TLR4) leads to NFκB activation and transcription of numerous genes including *IL1B* and *IL1RA*. It is well known that in addition

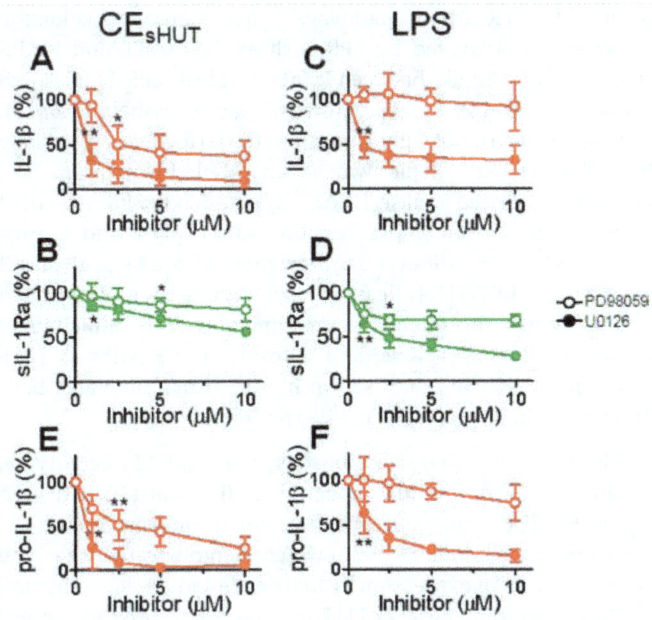

Figure 2: MEK1 and MEK2 differentially regulate IL-1β downstream CE_{sHUT}- and LPS-activation of monocytes. Isolated human monocytes were treated with the indicated concentration of U0126 (closed circles) or PD98059 (open circles) for 45 min before activation by CE_{sHUT} (A, B, and E) or LPS (C, D, and F). The production of IL-1β (A and C, red symbols), sIL-1Ra (B and D, green symbols) and pro-IL-1β (E and F, red symbols) was measured in cell lysates and supernatants harvested after 24h. Results obtained from at least 3 different donors are presented as mean ±SD of percentage of cytokine production observed in the absence of inhibitor. The lowest inhibitor concentration which induced a significant inhibition of cytokine production is indicated: (*) $p \leq 0.05$ and (**) $p \leq 0.01$ as determined by Student's *t*-test.

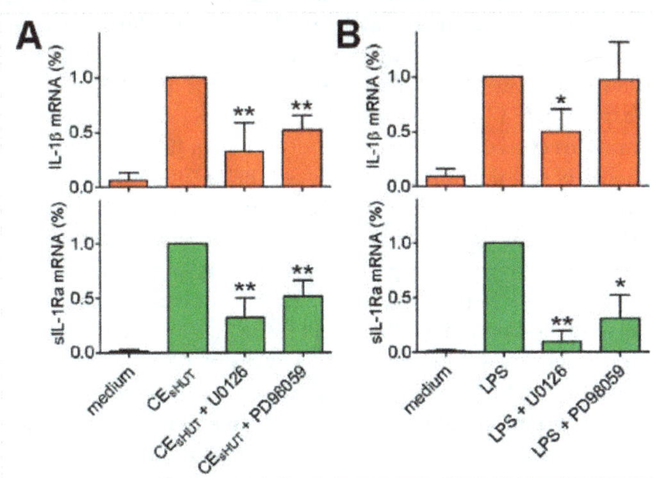

Figure 3: MEK1 is dispensable to IL-1β induction by LPS. Isolated human monocytes (2 x 10^6 cells) were activated for 3 h with CE_{sHUT} (A) or LPS (B) as described in Materials and Methods in the presence or absence of 5 μM of the indicated inhibitor. IL-1β (red) and IL-1Ra (green) transcripts were analyzed by quantitative real-time PCR. Results obtained from at least 3 different donors are presented as mean ± SD of percentage of cytokine production observed in the absence of inhibitor. Significance was assessed by Student's *t*-test (*) $p \leq 0.05$ and (**) $p \leq 0.01$.

to the LPS-TLR4-NFκB pathways other signal transduction pathways are activated by LPS such as PI3K-AKT and MAPK pathways. Cross-talk between the canonical LPS-TLR4-NFκB pathways and MAP kinase pathways was described, although mechanisms were not fully elucidated [39]. However, it is likely that MEK1/2-ERK1/2 pathway is activated downstream IL-1 receptor associated kinase (IRAK) 1/2 and IκBα kinase (IKK) β which activate tumor progression locus (Tpl) 2 and in turn MEK1/2 [40]. The differential activation of MEK1 and MEK2 downstream LPS-TLR4 ligation may represent a fine-tuning process controlling infectious inflammation with induction of dampening signals such as the PI3Kδ-AKT-GSK3 pathway [24]. This type of control is not set-up by CE_{sHUT} that activates both MEK1/2 leading to deregulated IL-1β production.

The present study establishes that MEK1 activity is dispensable to the optimal induction of IL-1β in LPS-activated monocytes. This is reminiscent of results in the murine monocyte/macrophage cell line J774A.1 which demonstrated that the induction of IL-1β expression by LPS depended on the activation of MAPKs including ERK1/2 [41]. In agreement with the present results, the latter study showed that IL-1β production was not inhibited in the presence of 10 μM PD98059 although ERK1/2 phosphorylation was inhibited by very high concentration of PD98059 (50 μM). Together with our results, this suggests that MEK1 was hardly involved in IL-1β induction by LPS in murine and human monocytes/macrophages.

The mechanism by which MEK2 controls the production of IL-1β downstream LPS stimulation remains to be determined. Although MEK1 and MEK2 activities are rarely distinguished in the literature, growing evidence indicates that MEK1 and MEK2 may be differentially regulated and exert non-redundant functions [42-51]. That MEK1 and MEK2 play different roles in the regulation of several cellular processes implies that either substrate other than ERK1/2 might be differentially phosphorylated by MEK1 and MEK2, or MEK1/2 functions that do not directly require kinase activity, or both [29,52]. However, ERK1/2 is still the only substrates of MEK1/2 identified to date. Noticeably, in studies revealing different functions of MEK1 and MEK2, MEK2 is often characterized as an important element in cross-talk between two pathways involving direct interaction between active MEK2 and other signaling molecules [29,30,53]. It has also been described that MEK2 may exert activity independently of ERK1/2 phosphorylation [29,30,53]. In the latter studies, MEK2 is likely to display protein scaffold activity promoting Pin1 binding to BPGAP1 and membrane localization of PI3Kδ, respectively. In the study by Pan et al. [29] the scaffold activity of MEK2 aimed at diminishing the activation of ERK1/2. The present study does not prove a role of regulatory scaffold for MEK2, it only establishes that there is a pathway initiated by MEK2 that leads to IL-1β production in LPS-activated human monocytes.

As shown in scheme of Figure 4, the triggering of IL-1β system in human monocytes is tightly controlled downstream LPS with induction of dampening signals such as the PI3Kδ-AKT-GSK3 pathway. This is not observed with the pathologic induction of IL-1β by CE_{sHUT} which triggers pathways that all converge to activate IL-1β production. In contrast with the inhibition of PI3K pathway that might have pernicious effects potentially leading to exacerbated inflammatory response to pathogens, the inhibition of MEK1 may restore a balanced production of IL-1β and sIL-1Ra in chronic/sterile inflammatory conditions without affecting responses to infectious pathogens.

Conclusion

Our results and previous studies [24,40] demonstrate that the induction of IL-1β expression upon chronic/sterile inflammatory conditions escape regulatory mechanisms that usually control it upon inflammatory response to infectious agents represented here by LPS (Figure 4). This further identifies contact with stimulated T cells as an uncontrolled, deregulated mechanism triggering pathogenic inflammation. That MEK1 is dispensable to induction of IL-1β by LPS designates this kinase as a potential therapeutic target to dampen detrimental inflammation without affecting protective inflammatory response to pathogens. There is currently no cure for autoimmune diseases with chronic/sterile inflammation. Patients are usually treated with immunosuppressive drugs aiming at diminishing overall inflammatory responses. Usage of protein kinase inhibitors has become an attractive class of drugs among which MEK1 and

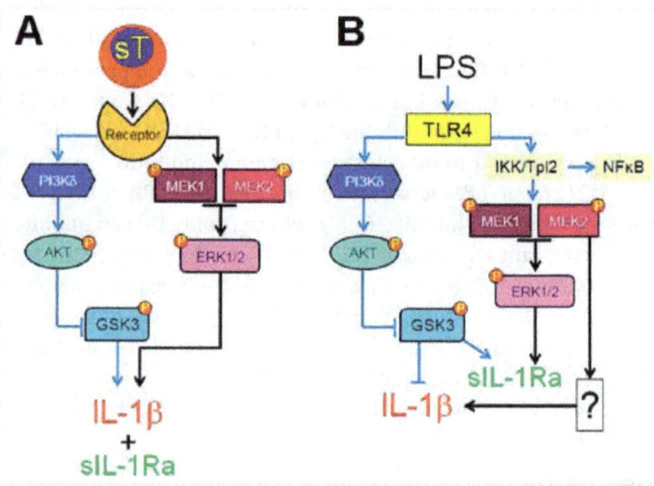

Figure 4: Control of IL-1β and sIL-1Ra production upon chronic/sterile and infectious conditions.

A) Upon chronic/sterile condition (contact with stimulated T cells), MEK1 and MEK2 are activated and contribute to the induction of IL-1β and sIL-1Ra production in human monocytes, and both MEK-ERK and PI3Kδ-AKT-GSK3 pathways trigger cytokine expression.

B) Upon infectious conditions (LPS), MEK1 and MEK2 are activated. Both MEK1 and MEK2 contribute to the induction of sIL-1Ra expression whereas only MEK2 is needed to the induction of IL-1β production. The PI3Kδ-AKT-GSK3 pathway dampens IL-1β production but triggers sIL-1Ra expression. Black lines and arrows represent results described in the present study, blue lines and arrows were previously established by us and others elsewhere (24,40). IKK, IκBα kinase; Tpl2, tumor progression locus 2; PI3Kδ, phosphatidylinositide-3 kinase δ; AKT, protein kinase B; GSK3, glycogen synthase kinase 3; TLR4, toll-like receptor 4.

MEK1/2 inhibitors are currently assessed in clinical trials mainly in cancer patients [54]. However, as recently stated by P. Cohen [40], whether safe drugs that modulate protein kinase activities can also be developed for the treatment of chronic diseases, where they may need to be taken for decades, is an issue that is still unresolved.

Acknowledgements

The authors are grateful to Dr. S. Pagano for endotoxin measurements and to Dr. N. Molnarfi who initiated this study. This work was supported by the Swiss National Science Foundation (grant #310030-141071), the Swiss Society for Multiple Sclerosis, and the Hans Wilsdorf Foundation.

References

1. Dinarello CA (2010) IL-1: discoveries, controversies and future directions. Eur J Immunol 40(3): 599-606.
2. Dinarello CA (2005) Blocking IL-1 in systemic inflammation. J Exp Med 201(9): 1355-1359.
3. Dinarello CA (2005) The many worlds of reducing interleukin-1. Arthritis Rheum 52(7): 1960-1967.
4. Burger D, Dayer JM (2000) IL-1Ra. In Cytokine Reference. Oppenheim JJ, Feldmann M, eds. Academic Press, London, 319-336.
5. Dinarello C A, Simon A, van der Meer JW (2012) Treating inflammation by blocking interleukin-1 in a broad spectrum of diseases. Nat Rev Drug Discov 11(8): 633-652.
6. Huang J, Xie ZK, Lu RB, Xie ZF (2013) Association of interleukin-1 gene polymorphisms with multiple sclerosis: a meta-analysis. Inflamm Res 62(1): 97-106.
7. Burger D, Dayer JM, Palmer G, Gabay C (2006) Is IL-1 a good therapeutic target in the treatment of arthritis? Best Pract Res Clin Rheumatol 20(5): 879-896.
8. Burger D (2000). Cell contact-mediated signaling of monocytes by stimulated T cells: a major pathway for cytokine induction. Eur Cytokine Netw 11(3): 346-353.
9. Burger D, Dayer JM, Molnarfi N (2007) Cell contact dependence of inflammatory events. In Contemporary Targeted Therapies in Rheumatology. Smolen JS, Lipsky PE, eds. Taylor & Francis Books Ltd, Abingdon/UK, 85-103.
10. Stout RD, Suttles J (1995) T-Cell Signaling of Macrophage Activation: Cell Contact-dependent and Cytokine Signals. Springer, R.G. Landes Company, New York.
11. McInnes IB, Leung BP, Sturrock RD, Field M, Liew FY (1997) Interleukin-15 mediates T cell-dependent regulation of tumor necrosis factor-alpha production in rheumatoid arthritis. Nat Med 3(2): 189-195.
12. Parry SL, Sebbag M, Feldmann M, Brennan FM (1997). Contact with T cells modulates monocyte IL-10 production. Role of T cell membrane TNF-alpha. J Immunol 158(8): 3673-3681.
13. Sebbag M, Parry SL, Brennan FM, Feldmann M (1997) Cytokine stimulation of T lymphocytes regulates their capacity to induce monocyte production of tumor necrosis factor-alpha, but not interleukin-10: Possible relevance to pathophysiology of rheumatoid arthritis. Eur J Immunol 27(3): 624-632.
14. Avice MN, Demeure CE, Delespesse G, Rubio M, Armant M, et al. (1998) IL-15 promotes IL-12 production by human monocytes via T cell-dependent contact and may contribute to IL-12-mediated IFN-gamma secretion by CD4+ T cells in the absence of TCR ligation. J Immunol 161(7): 3408-3415.
15. Avice MN, Sarfati M, Triebel F, Delespesse G, Demeure CE (1999) Lymphocyte activation gene-3, a MHC class II ligand expressed on activated T cells, stimulates TNF-alpha and IL-alpha production by monocytes and dendritic cells. J Immunol 162(5): 2748-2753.
16. Chabot S, Charlet D, Wilson TL, Yong VW (2001) Cytokine production consequent to T cell--microglia interaction: the PMA/IFN gamma-treated U937 cells display similarities to human microglia. J Neurosci Methods 105(2): 111-120.
17. Burger D, Roux-Lombard P, Chizzolini C, Dayer JM (2004). Cell-cell contact in chronic inflammation: the importance to cytokine regulation in tissue destruction and repair. In Cytokines and Joint Injury. van den Berg WB Miossec P, eds. Birkhäuser Verlag, Basel. 165-188.
18. Dasgupta S, Jana M, Liu X, Pahan K (2003). Role of very-late antigen-4 (VLA-4) in myelin basic protein-primed T cell contact-induced expression of proinflammatory cytokines in microglial cells. J Biol Chem 278(25): 22424-22431.
19. Dasgupta S, Jana M, Liu X, Pahan K (2005) Myelin basic protein-primed T cells of female but not male mice induce nitric oxide synthase and proinflammatory cytokines in microglia: Implications for gender bias in multiple sclerosis. J Biol Chem 280(38): 32609-32617
20. Brennan FM, Foey AD (2002) Cytokine regulation in RA synovial tissue: role of T cell/macrophage contact-dependent interactions. Arthritis Res 4 suppl 3: S177-S182.
21. Molnarfi N, Gruaz L, Dayer JM, Burger D (2004) Opposite effects of IFNbeta on cytokine homeostasis in LPS- and T cell contact-activated human monocytes. J Neuroimmunol 146(1-2): 76-83.
22. Burger D, Dayer JM (2007) Assays of T-cell contact dependent monocyte-macrophage functions. Methods Mol Med 136: 139-148.
23. Molnarfi N, Gruaz L, Dayer JM, Burger D (2007) Opposite regulation of IL-1beta and secreted IL-1 receptor antagonist production by phosphatidylinositide-3 kinases in human monocytes activated by lipopolysaccharides or contact with T cells. J Immunol 178(1): 446-454.
24. Molnarfi N, Brandt KJ, Gruaz L, Dayer JM, Burger D (2008) Differential regulation of cytokine production by PI3Kdelta in human monocytes upon acute and chronic inflammatory conditions. Mol Immunol 45(12): 3419-3427.
25. Roskoski R Jr (2012) MEK1/2 dual-specificity protein kinases: structure and regulation. Biochem Biophys Res Commun 417(1): 5-10.
26. Zheng CF, Guan KL (1993) Cloning and characterization of two distinct human extracellular signal-regulated kinase activator kinases, MEK1 and MEK2. J Biol Chem 268(15): 11435-11439.
27. Belanger LF, Roy S, Tremblay M, Brott B, Steff AM, et al. (2003) Mek2 is dispensable for mouse growth and development. Mol Cell Biol 23(14): 4778-4787.
28. Giroux S, Tremblay M, Bernard D, Cardin-Girard JF, Aubry S, et al. (1999). Embryonic death of Mek1-deficient mice reveals a role for this kinase in angiogenesis in the labyrinthine region of the placenta. Curr Biol 9(7): 369-372.
29. Pan CQ, Liou YC, Low BC (2010) Active Mek2 as a regulatory scaffold that promotes Pin1 binding to BPGAP1 to suppress BPGAP1-induced acute Erk activation and cell migration. J Cell Sci 123(Pt 6): 903-916.

30. Brandt KJ, Carpintero R, Gruaz L, Molnarfi N, Burger D (2010) A novel MEK2/PI3Kd pathway controls the expression of IL-1 receptor antagonist in IFN-b-activated human monocytes. J Leukoc Biol 88(6): 1191-1200.

31. Davies SP, Reddy H, Caivano M, Cohen P (2000) Specificity and mechanism of action of some commonly used protein kinase inhibitors. Biochem J 351(Pt 1): 95-105.

32. Cohen P (1999) The development and therapeutic potential of protein kinase inhibitors. Curr Opin Chem Biol 3(4): 459-465.

33. Molnarfi N, Hyka-Nouspikel N, Gruaz L, Dayer JM, Burger D (2005) The production of IL-1 receptor antagonist in IFN-{beta}-stimulated human monocytes depends on the activation of phosphatidylinositol 3-kinase but not of STAT1. J Immunol 174(5): 2974-2980.

34. Burger D, Molnarfi N, Gruaz L, Dayer JM (2004) Differential induction of IL-1beta and TNF by CD40 ligand or cellular contact with stimulated T cells depends on the maturation stage of human monocytes. J Immunol 173(2): 1292-1297.

35. Jungo F, Dayer JM, Modoux C, Hyka N, Burger D (2001) IFN-beta inhibits the ability of T lymphocytes to induce TNF-alpha and IL-1beta production in monocytes upon direct cell-cell contact. Cytokine 14(5): 272-282.

36. Hyka N, Dayer JM, Modoux C, Kohno T, Edwards CK 3, et al. (2001) Apolipoprotein A-I inhibits the production of interleukin-1beta and tumor necrosis factor-alpha by blocking contact-mediated activation of monocytes by T lymphocytes. Blood 97(8): 2381-2389.

37. Bradford M (1976) A rapid and sensitive method for the quantitation of microgram quantities of protein utilizing the principle of protein-dye binding. Anal Biochem 72: 248-254.

38. Carpintero R, Brandt KJ, Gruaz L, Molnarfi N, Lalive PH, et al. (2010). Glatiramer acetate triggers PI3Kd/Akt and MEK/ERK pathways to induce IL-1 receptor antagonist in human monocytes. Proc Natl Acad Sci U S A 107(41): 17692-17697.

39. O'Neill LA (2006) How Toll-like receptors signal: what we know and what we don't know. Curr Opin Immunol 18(1): 3-9.

40. Cohen P (2009) Targeting protein kinases for the development of anti-inflammatory drugs. Curr Opin Cell Biol 21(2): 317-324.

41. Su SC, Hua KF, Lee H, Chao LK, Tan SK, et al. (2006) LTA and LPS mediated activation of protein kinases in the regulation of inflammatory cytokines expression in macrophages. Clin Chim Acta 374(1-2): 106-115.

42. Winston BW, Remigio LK, Riches DW (1995) Preferential involvement of MEK1 in the tumor necrosis factor-alpha-induced activation of p42(mapk/erk2) in mouse macrophages. J Biol Chem 270(46): 27391-27394.

43. Wu X, Noh SJ, Zhou G, Dixon JE, Guan KL (1996) Selective activation of MEK1 but not MEK2 by A-Raf from epidermal growth factor-stimulated Hela cells. J Biol Chem 271(6): 3265-3271.

44. Xu S, Khoo S, Dang A, Witt S, Do V, et al. (1997). Differential regulation of mitogen-activated protein/ERK kinase (MEK)1 and MEK2 and activation by a Ras-independent mechanism. Mol Endocrinol 11(11): 1618-1625.

45. Eblen ST, Slack JK, Weber MJ, Catling AD (2002) Rac-PAK signaling stimulates extracellular signal-regulated kinase (ERK) activation by regulating formation of MEK1-ERK complexes. Mol Cell Biol 22(17): 6023-6033.

46. Liu X, Yan S, Zhou T, Terada Y, Erikson RL (2004) The MAP kinase pathway is required for entry into mitosis and cell survival. Oncogene 23(3): 763-776.

47. Battcock SM, Collier TW, Zu D, Hirasawa K (2006) Negative regulation of the alpha interferon-induced antiviral response by the Ras/Raf/MEK pathway. J Virol 80(9): 4422-4430.

48. Voisin L, Julien C, Duhamel S, Gopalbhai K, Claveau I, et al. (2008). Activation of MEK1 or MEK2 isoform is sufficient to fully transform intestinal epithelial cells and induce the formation of metastatic tumors. BMC Cancer 8: 337.

49. Skarpen E, Flinder LI, Rosseland CM, Orstavik S, Wierod L, et al. (2008) MEK1 and MEK2 regulate distinct functions by sorting ERK2 to different intracellular compartments. FASEB J 22(2): 466-476.

50. Catalanotti F, Reyes G, Jesenberger V, Galabova-Kovacs G, de Matos Simoes R, et al. (2009) A Mek1-Mek2 heterodimer determines the strength and duration of the Erk signal. Nat. Struct. Mol. Biol. 16(3): 294-303.

51. Zhang H, Feng H, Luo L, Zhou Q, Luo Z, et al. (2010) Distinct effects of knocking down MEK1 and MEK2 on replication of herpes simplex virus type 2. Virus Res 150(1-2): 22-27.

52. Sturgill TW (2008) MAP kinase: it's been longer than fifteen minutes. Biochem Biophys Res Commun 371(1): 1-4.

53. Thangavelu K, Pan CQ, Karlberg T, Balaji G, Uttamchandani M, (2012) Structural basis for the allosteric inhibitory mechanism of human kidney-type glutaminase (KGA) and its regulation by Raf-Mek-Erk signaling in cancer cell metabolism. Proc Natl Acad Sci U S A 109: 7705-7710.

54. Akinleye A, Furqan M, Mukhi N, Ravella P, Liu D (2013) MEK and the inhibitors: from bench to bedside. J Hematol Oncol 6: 27.

24

Autoimmune disease: Pathogenesis, Genetics, Immunotherapy, Microbial triggers, Prophylaxis

Duncan D. Adams*

Faculty of Medicine, University of Otago, Dunedin, New Zealand

***Corresponding author:** Duncan D. Adams, Faculty of Medicine, University of Otago, Dunedin, New Zealand, Email: duncan. adams@xtra.co.nz*

Abstract

Pasteur's correct Germ Theory of Disease led to discovery of the Immunity System that exists for defence against diseases caused by bacterial or viral infections. In order to cope with the huge diversity of microbial infections, the immune system needs a flexibility of immune response. This is provided by use of semi-random somatic gene mutations in lymphocytes, the cells that recognise invading bacteria and viruses.

Burnet realised that this flexibility leads to development of the Forbidden Clones of lymphocytes that cause the autoimmune diseases by accidentally being reactive with a host antigen instead of a microbial one. Ebringer has recently discovered the microbial triggers of rheumatoid arthritis and ankylosing spondylitis and has determined the amino acid sequences of the antigens on them that trigger the related autoimmune diseases. This has explained how histocompatibility antigens can predispose to autoimmune diseases. The Salk and Sabin anterior poliomyelitis vaccines have prevented the leg paralyses of the polio epidemics. We postulate that these paralyses were rare autoimmune complications of virtually universal poliovirus infection. As autoimmune diseases are triggered by microbial infections, we suggest that the triggering bacteria or viruses be sought, so that prophylaxis of the autoimmune diseases can be achieved by vaccination against their triggering microbes.

Keywords: Forbidden clone theory; H gene theory; Microbial triggers of autoimmune diseases; Poliomyelitis vaccines; Prophylaxis of poliomyelitis paralyses; Prophylaxis of autoimmune diseases

Introduction

Endemic goitre and its conquest

In mountainous parts of the world, such as Switzerland, the Himalayas, the Andes and New Zealand, Goiter (an enlargement of the thyroid gland) is common. This was discovered to be due to shortage in the soil of the trace element iodine [1]. Surprisingly, Kelly et al. [2], London scientists recruited by the Chilean government discovered that iodine in soil does not come from the weathering of rock. Iodine is dissolved in the sea from which it vaporizes by oxidation to be deposited in soils by rain. From the soil, iodine is taken up by plants from where it comes to man, directly in vegetables and fruit or indirectly through meat animals. Iodine takes a long time to build up in soils to the levels we need for manufacture of sufficient thyroid hormone. Geologically old soils, like those of England and France and the Eastern United States, contain adequate iodine for human need but the new soils of Switzerland, the Himalayas, Chile and New Zealand, where there has been geologically recent up-thrust of mountains have not had time to accumulate adequate amounts of iodine to meet the needs for humans and other animals. This causes goitre, an enlargement of the thyroid gland, caused by increased secretion of thyroid stimulating hormone from the pituitary gland aimed at enabling an enlarged thyroid to obtain more iodine from the blood. In New Zealand, HD Purves [3], with brilliant studies, found that the amount of iodine needed to be added to the domestic salt was 1 part of potassium iodide per 20,000 parts of sodium chloride, 10 times more than the ineffectual level previously used. This abolished New Zealand's endemic goitre [4].

As Professor of Public Health and Preventive Medicine at Otago Medical School, Charles Hercus followed classic academic practice in requiring his 5th year medical students to write a thesis, on any topic they chose, giving scope for expression of originality. Duncan Adams seized this opportunity to write a thesis on the aetiology of asthma, from which he himself had suffered severely, leading to an invitation by Sir Charles Hercus for him to join the research world via a Medical Research Council Fellowship. This led on to years of professional research and Directorship of the MRC Autoimmunity Research Unit (Figure 1).

Pathogenesis

The long mysterious pathogenesis and genetics of the autoimmune diseases is now solved [4,5]. The key to the pathogenesis is Jerne's selection theory of antibody formation [6], which led to Burnet's clonal selection theory of acquired immunity [7] and his forbidden clone theory of autoimmune disease [8]. This states that somatic gene mutations in multiplying lymphocytes produce the appropriately-named Forbidden Clones that cause the autoimmune diseases by reacting with a host antigen instead of a microbial one. Figure 2 illustrates clonal selection by antigenic stimulation and clonal diversification by somatic mutations in lymphocyte V genes. After discovery that the thyroid gland over-activity of Graves' disease is caused by auto-antibodies that react with the thyroid's receptor for thyroid-stimulating hormone from the pituitary gland (Figure

***Corresponding author email:** duncan.adams@xtra.co.nz*

2) study of the properties of these auto-antibodies confirmed the **Forbidden Clone theory** by showing that they originate from single lymphocytes and show fine variation from patient to patient indicative of the random element in the mutations that produced them [9].

Classification of immune reactions causing disease

This is shown in (Table 1), where allergy and anaphylaxis and serum sickness are distinguished from the two types of autoimmunity, that caused by B cell forbidden clones and that caused by T cell forbidden clones.

Graves' disease, a paradigm for autoimmune disease

In 1986, at the University of Pisa, Professor Aldo Pinchera et al. led an International Symposium on Thyroid Autoimmunity [23], at which **Graves' disease** was described as a paradigm of autoimmune disease [15] and Duncan Adams (Figure 4) was awarded a Medal for Fundamental Contributions to Biomedical Science, namely discovery of the thyroid-stimulating auto-antibodies, first called the long-acting thyroid stimulator (LATS).

Figure 1: Dr Duncan Adams, MD DSc FRACP. Selected by Sir Charles Hercus for professional research with a Medical Research Council Fellowship which led on to discovery of the thyroid stimulating autoantibodies with HD Purves, Directorship of the MRC Autoimmunity Research Unit, The H Gene Theory of Autoimmune disease and the autoimmune model of schizophrenia with JG Knight.

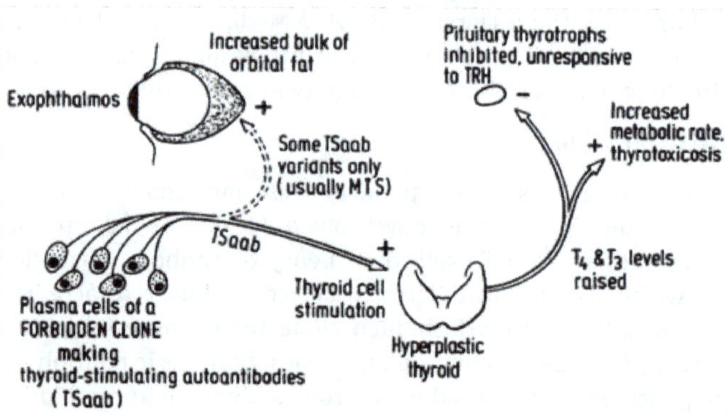

Figure 2: The pathogenesis of Graves' disease. Thyroid-stimulating autoantibodies (TSaab) from forbidden clones of immunocytes stimulate the thyroid cells, causing overproduction of thyroid hormones T4 and T3 and the manifestations of thyrotoxicosis. The thyrotroph cells of the anterior pituitary are inhibited by the high blood thyroid hormone level, so TSH secretion ceases and response to TRH is absent. Some variants of TSaab react with receptors on fat cells in the orbit to cause exophthalmos from orbital adipocyte proliferation, demonstrated by Rundle and Pochin [9,57].

Table 1: Classification of immune reactions causing disease.

Type I. **Allergy and Anaphylaxis.** Gut worm-defence mechanism reacting to non-worm antigens [10,11]. Fault: a B lymphocyte IgE clone reactive with an allergen. e.g. hay fever, anaphylaxis, gut allergy, skin allergy.
Type II. **Serum Sickness and Immune Complex Disease.** Fault: excessive quantity of antigen. This swamps complement-neutralising mechanisms, leading to complement-mediated damage. Anti-microbial immune defense is designed to cope with picogram quantities of antigen, not milligrams of horse serum protein nor micrograms of released intra-cellular protein, such as nuclei [12,13]. e.g. serum sickness following passive immunization against diphtheria toxin with horse serum, systemic lupus erythaematosus, lupus nephritis.
Type III. **Autoimmunity.** Fault: forbidden clones, which are anti-microbial lymphocyte clones with accidental host-antigen specificity, arising from unlucky somatic mutations in their lymphocyte V genes [9-14].
Type III B. Diseases caused by B lymphocyte forbidden clones: e.g. Graves' disease[15], myasthenia gravis [16], rheumatoid arthritis [17].
Type III T. Diseases caused by T lymphocyte forbidden clones: e.g. Diabetes Type 1 [18,19], diabetic retinopathy [20,21], experimental autoimmune encephalomyelitis[22] and presumptively Addison's disease, hypoparathyroidism, and other autoimmune diseases with specific parenchymal cell destruction[14].

This came from the experimental advantages provided by the presence of iodine in thyroid hormone, the hormone receptor nature of the auto-antigen and the control of thyroid activity by the pituitary gland.

Genetics

The familial aggregation

Studies of families, including twins, show that autoimmune diseases are weakly inherited, with disease specificity. The Mendelian pattern is that of multiple co-dominant genes with incomplete penetrance [14]. What genes are these? Before discussing this we need to describe the histocompatibility System.

The histocompatibility system

This system is essential for defense against virus infection, prevents allo-transplantation, and influences risk of autoimmune disease. Unlike the blood group antigens, A, B, O, on red blood cells, important for blood transfusions, the histocompatibility antigens are on the surface of all nucleated cells, including the white blood cells, the leucocytes, where they were discovered, and named by Dausset and Svejgard, the human leukocyte antigens (HLA) [24].

Figure 3: Dr. Duncan Adams receiving a prize from the Chancellor of the University of Pisa for discovering the thyroid-stimulating autoantibodies that cause Graves' disease.

Involvement of the MHC in autoimmune disease

Vladutiu and Rose [25] discovered involvement in autoimmune disease of the major histocompatibility gene complex (MHC), which had been discovered in the field of surgical transplantation [26]. In man, a collection of genes on chromosome 6 code for peptides expressed on the surface of all nucleated cells. In the mouse a similar collection is on chromosome 17. Why does the MHC exist?

Functions of the MHC

First function: defense against virus disease: One of the classical experiments of recent times is that of Zinkernagel, et al. [27], who found that virus-infected cells extrude peptides from the virus into the Bjorkman Groove of their major histocompatibility antigens, this combined viral-histocompatibility antigen on the cell surface being the target for attack by the defensive cytotoxic T-cells. Adams [28] realized that the explosive speed of viral replication [29] necessitates this histocompatibility antigen involvement, which directs the cytotoxic T-cell attack on to the surface of the infected cell, destroying the virus factory, rather than ineffectively being muffled by the myriad numbers of free virions, as shown in (Table 2 [A,B,C]). This explains the Simonsen phenomenon [30], our having huge clones of cytotoxic T-cells reactive with allo-histocompatibility antigens, which our immune system mistakes for viral peptides on host histocompatibility antigens. The explosive speed of viral replication makes this mechanism necessary to prevent swift death of the virus-infected animal [28]. Polly Matzinger's "Danger Theory" [31] is in full accord with Simonsen's discovery that the immune system is far less concerned with things that are foreign than those that are dangerous.

Second function: Imposition of polymorphism on the immune repertoire. This diversity usually enables some members of a population to survive an epidemic infection, preventing whole populations from being wiped out.

Third function: Imperfect defense against autoimmune disease. [4].

Fourth function: Provision of a gene haven for MHC Class-III gene products [4].

The H gene Theory of Inheritance of Autoimmune disease.

Building on the Bielschowsky's discovery [32] of autoimmune anaemia in their NZB/BL strain of inbred mice, Howie and Helyer discovered that F^1 hybrids of this strain with the healthy NZW strain unexpectedly develop the autoimmune kidney disease, lupus nephritis [33]. The occurrence of this disease in a hybrid, both of whose parents lack it, shows that at least one gene from each of the two parental strains causes the lupus nephritis. Back-cross and linkage studies showed that three genes contribute to the lupus nephritis [34], one from the NZB mice and two from the NZW mice. This has been confirmed, corrected and extended by Drake et al. [35] and Kono, et al. [36], using the wonderfully detailed microsatellite gene markers.

In all, Knight and Adams found four genes, with linkage information coding for autoimmune disease in mice [37]. None were the expected V genes, one was in the MHC and two appeared to be in the neighbourhood of the minor histocompatibility antigens, Hh and H-18 [37]. The linkage studies show that these are not the immunoglobulin V genes that were expected. Einstein [38] stated.

"The history of scientific and technical discovery teaches us that the human race is poor in independent thinking and creative imagination. Even when the external and scientific requirements for the birth of an idea have long been present, it generally needs an external stimulus to make it actually happen; man has, so to speak, to stumble right up against the thing before the idea comes."

In an example of creative imagination, Adams and Knight put together the research fields of autoimmunity and transplantation, arriving at the **H-Gene Theory of inheritance of autoimmune disease** [39]. This states that histocompatibility antigen genes, major, minor and HY (the male sex antigen), together with the V (variable region) genes coding for antigen receptors on B and T-lymphocytes, are the germ-line immune response genes, the genes that influence the risk of autoimmune disease. The H-Gene Theory received general acceptance and wide admiration when delivered by John Knight to a distinguished audience at a Ciba Foundation Symposium in London in 1982 [40].

Immune Response Genes [14].

Table 3 lists the genes that govern the specificity of immune responses, indicating how they work. Section-A shows the germline variable region (V) genes, it provide the repertoire by coding for antigen receptors on B-cell and T-cell lymphocytes. For B-cell antigen receptors, the heavy chains are coded on chromosome14q, the κ light chains on chromosome 2p and the λ light chains on chromosome 22q. There are separate genes for the T-cell antigen receptors, which are of two types, αβ and γδ, coded for by genes on chromosomes 14q and 7q as shown in the Table 3. Section-B shows that new clones are added to the repertoire by somatic mutations in the V-genes of multiplying lymphocytes. Section-C shows that clones are subtracted from the repertoire by the H (histocompatibility antigen) genes which delete nascent complementary clones.

The Class-I major histocompatibility antigens are on all nucleated cells but the Class II are only on B-lymphocytes and professional antigen presenting cells such as macrophage and dendritic cells.

Females have two X-chromosomes; males have an X-chromosome and a Y-chromosome which confers masculinity.

Aire (Autoimmune regulator) [41]

Mutations in the transcriptional regulator, Aire, cause APECED, a poly-glandular autoimmune disease. Animal models of APECED have revealed that Aire plays an important role in T cell tolerance induction in the thymus, mainly by promoting ectopic expression of a large repertoire of transcripts encoding proteins normally restricted to differentiated organs residing in the periphery. The absence of Aire results in impaired clonal deletion of self-reactive thymocytes, which escape into the periphery and attack a variety of organs.

The significance of this phenomenon for human autoimmune disease is most intriguing.

Microbial triggers

Autoimmune diseases are caused by malfunction of the immunity system, being consequences of infection by bacteria or viruses [41,42].

A. **Rheumatic Carditis and *Streptococci*.**

Before the advent of penicillin, rheumatic fever, with crippling or fatal lesions of the heart, was a frequent consequence of infection by *β-haemolytic streptococci* of Lancefield Group A. This was because of an antigenic similarity between a component of these *streptococci* and heart tissue, discovered by Kaplan and Meyeserian [43]. Today, with such infection therapeutically aborted by penicillin, rheumatic heart disease, once common, has become rare.

B. **Glomerulonephritis and *Streptococci*.**

Post-infective glomerulonephritis follows infection by Group-A *streptococci* of multiple M types. This disease is also less frequent due to use of antibiotics.

Table 2: The Histocompatibility System exists for defence against virus infection.
(Adams DD. Lancet 1987; **ii**: 245-9.)

1. The race between virus and cytotoxic T cell.

The contestants	Replication time	Progeny
Influenza virus	10 hours [29]	1,000 virions
Cytotoxic T- cell	18 hours	2 T cells

The race	Virions	T cells	Virion/T cell ratio
Day 1	1	10^6	$1/10^6$
Day 2	$1 \times 1{,}000^{2.4}$	$106 \times 2^{1.3}$	6/1
Day 3	2.5×10^{14}	6.3×10^6	$10^7/1$
Day 4	4×10^{21}	1.6×10^7	$10^{14}/1$

The result: the virus wins, the patient dies

Table 3: Immune response genes [14]. The Germline H and V genes provide the germline predispositions to the autoimmune diseases and the random element of the somatic mutations in the lymphocyte V genes causes incomplete penetrance.

	Chain	Chromo-some
A. Providing the repertoire, the germline V genes		
a) Genes for B cell antigen receptors	Heavy (V,D,J)	14q32.3
	k light (V,J)	2p12
	l light (V,J)	22q11
b) Genes for T cell antigen receptors	a (V,J)	14q11-12
	b (V,D,J)	7q32-33
	g (V,J)	7p15
	d (V,D,J)	14q11-12
B. Adding new clones to the repertoire, somatic mutations in the V genes of multiplying lymphocytes.		
C. Subtracting from the repertoire, the H (histocompatibility Antigen) genes which delete nascent complementary clones.		
Major		
Class I (A, B, C)	a (very polymorphic)	6p Class II (DP, DQ, DR)
b-2 microglobulin	15	
	b (very polymorphic)	6p
	A (less polymorphic)	6p
Erythrocyte alloantigens	A, B, O	9
	Others	various
The H-Y antigen, expressed in Bjorkman	Grooves	Y
Other minor H antigens expressed in Bjorkman	Grooves.	various

C. **Reactive Arthritis.**

This has been observed after enteric infection with *Shigella, Salmonella, Yersinia, Compylobacter* and genital infection with *Neisseria gonorrhea.*

D. **Rheumatoid Arthritis (RA) and *Proteus mirabilis*** [44].

Multiple studies over three decades have found high titres of antibodies against this bacterium in a total of 1375 RA patients, but not in other diseases or healthy controls in studies by independent groups in 15 different countries. There was no such elevation in antibodies against 27 other microbial agents. There is evidence that the upper urinary tract is the main site of *Proteus* infection in RA.

E. **Ankylosing Spondylitis (AS) and *Klebsiella*** [45].

In worldwide studies involving 1330 AS patients and 1191 healthy controls, the AS patients showed significantly increased antibody titres to *Klebsiella*. There is evidence that the gut is the

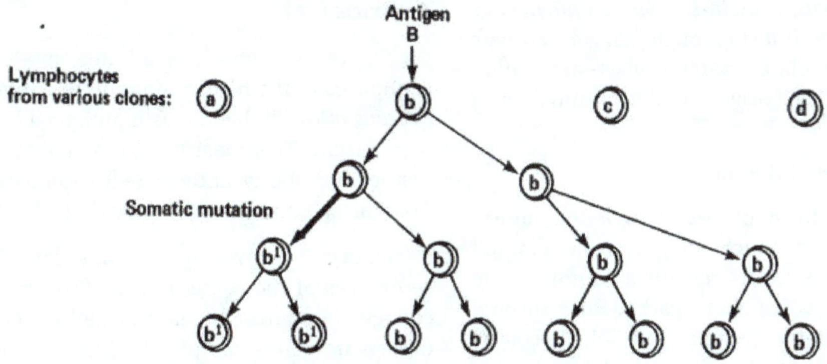

Figure 4: Clonal selection by antigenic stimulation and clonal diversification by somatic mutation. Concept of Jerne [6] and Burnet [7].

Figure 5: Professor Alan Ebringer, BSc, MD, FRCP, FRACP.

1. Discovered that *Proteus mirabilis* in the upper urinary tract triggers rheumatoid arthritis
2. Discovered that *Klebsiella pneumoniae* in the gut triggers ankylosing spondylitis
3. Confirmed the H Gene Theory by finding two antigens on Proteus, one resembling HLA-DR1/4 the predisposing HLA antigen, one resembling the autoantigen attacked.
4. Showed how HLA-B27 predisposes, 69-fold, to ankylosing spondylitis.

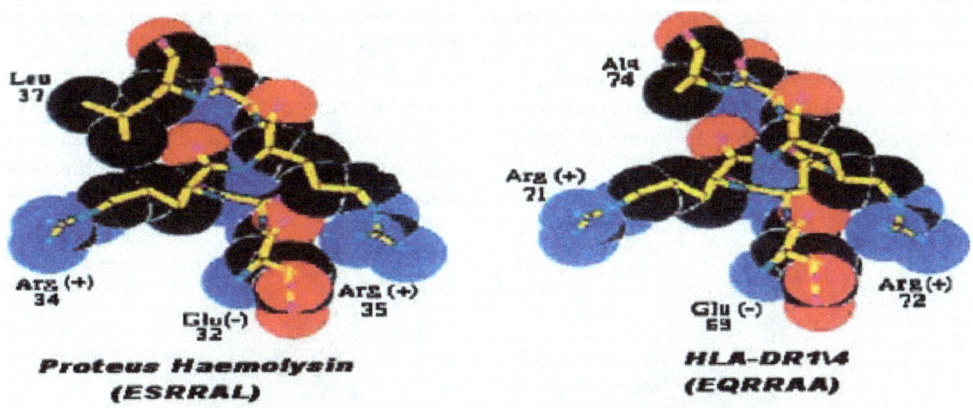

Figure 6: Molecular similarity between histocompatibility antigen HLA-DR1/4 and *Proteus* haemolysin, preventing immune reaction against this bacterial antigen.

main site of *Klebsiella* infection in AS.

F. **Type 1 Diabetes and *Coxsackie virus*.**

Richter and Horowitz [46] present the considerable evidence that Coxsackie viruses, especially B4, may trigger type1 diabetes, including the discovery that there are shared regions of homology between the Coxsackie virus protein PC-2 and the islet antigen GAD65.

G. **Systemic Sclerosis and Infections.**

Randone, et al. [47] describe the evidence suggesting that *Parvovirus B19, Cytomegalovirus, Ebstein-Bar virus, Endogenous retrovirus*, or *Helicobacter pylori* infection might trigger systemic sclerosis, where it has been postulated that fibroblast-stimulating forbidden clones, probably of B cell origin, are the immunological agent [8].

H. **Schizophrenia and virus infection.**

Acute schizophrenia has been observed to follow upper respiratory tract virus infection. Knight et al have assembled much evidence indicating that schizophrenia is an autoimmune disease caused by auto-antibodies that react with neuronal receptors influencing the limbic system [48-50]. Recently, Fabienne Brilot [51], a Paediatrician, has discovered dopamine-2 receptor antibodies in children with encephalitis. If these auto-antibodies are found in cases of acute schizophrenia, it will confirm the autoimmune basis of this disease.

Information from Sequencing Antigens on Triggering Bacteria

a. **Basic books**

Details of the development of the methods used for successful determination of the amino acid sequences of antigens on the rheumatoid arthritis-triggering bacteria, *Proteus mirabilis* are described in the book, "Rheumatoid arthritis and *Proteus*" by Ebringer [52] (Figure 5).

Similarly, the book "Ankylosing spondylitis and *Klebsiella*", also by Ebringer [53], describes how the amino acid sequences of antigens on *Klebsiella pneumoniae,* that enormously increase the risk of ankylosing spondylitis, were determined, and how they exert their effect.

b. **Confirmation of the H Gene theory**

This research provides experimental confirmation, at the molecular level, of the H-Gene Theory of the inheritance of the autoimmune diseases, described above, in confirming the speculated presence of multiple antigens on triggering bacteria and alternative clonal development causing development of the forbidden clones that cause the associated autoimmune diseases.

c. **How HLA-DR1/4 predisposes to Rheumatoid Arthritis** [54]

Figure 6 shows space-filling models of the amino acid sequences of the histocompatibility antigen HLA-DR1/4 and the *Proteus mirabilis* haemolysin antigen. Close structural similarity is apparent. This means the immune tolerance imposed by the histocompatibility antigen will extend to this *Proteus* antigen, preventing immune reaction with it.

Figure 7 shows space-filling models of the amino acid sequences of the *Proteus mirabilis* urease antigen and Type-11 collagen, an auto-antigen attacked in rheumatoid arthritis. The urease antigen is completely different from HLA-DR1-4, so will not be protected from immune reaction, being free to stimulate development of a forbidden clone reacting with the closely similar Type-11 collagen molecule, an auto-antigen attacked in rheumatoid arthritis.

d. **How HLA-B27 predisposes to Ankylosing Spondylitis** [55]

Figure 8 shows space-filling models of the amino acid sequences of the histocompatibility antigen HLA-B27 and two antigenic peptides on the bacterium *Klebsiella pneumoniae.* The *Klebsiella* nitrogenase antigen closely resembles HLA-B27, so will be covered by the tolerance induced by HLA-B27, but the bacterium's pullanase peptide is different and able to stimulate development of a forbidden clone attacking the spine to cause ankylosing spondylitis.

Prophylaxis of autoimmune diseases

a. The *Poliomyelitis virus* epidemics

A New Zealand example occurred in 1938, reported in the press and observed by Adams trapped in a boarding school in Masterton. An epidemic of leg paralyses occurred in Christchurch and spread progressively north, from town to town, to Picton, Wellington, Featherston, then Carterton, the town next to Masterton, engendering great fear. Then the boy in the bed next to Adams complained of a stiff neck, was taken away and reported to have polio. Six months later he returned with a paralyzed leg. Adams and his schoolmates were all unaffected.

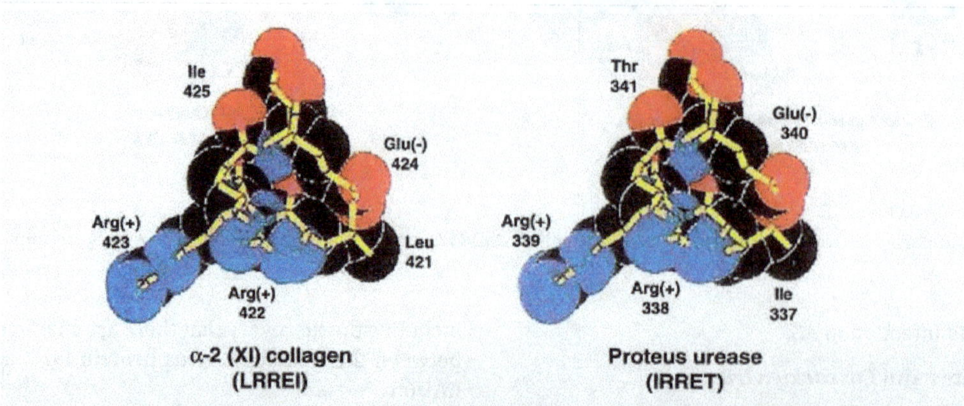

Figure 7: Molecular similarity of *Proteus* urease with Type X1 collagen, an autoantigen attacked in rheumatoid arthritis.

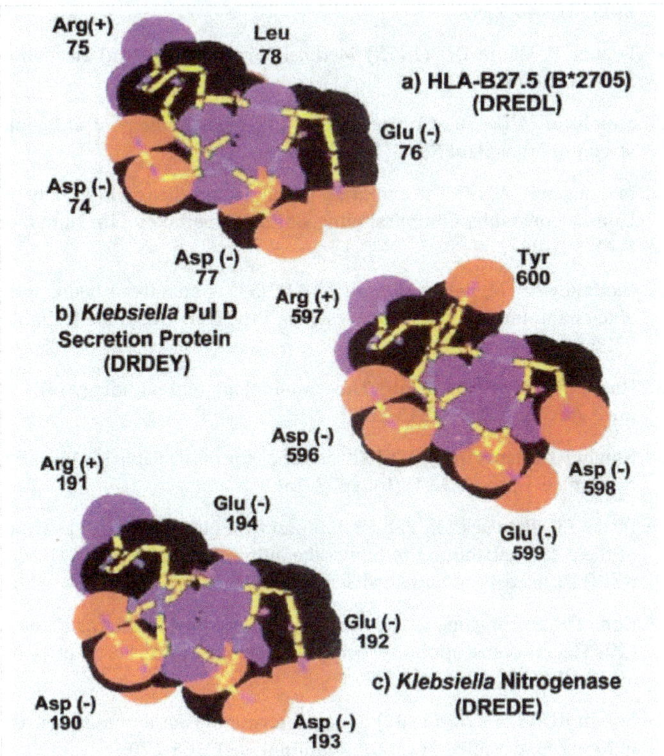

Figure 8: Molecular similarity between HLA-B27 and the nitrogenase reductase peptide of *Klebsiella pneumoniae,* preventing immune reaction, and dissimilarity with the pullanase peptide (Pul D), of *Klebsiella pneumoniae,* allowing immune reactivity that can lead to development of the forbidden clones that cause ankylosing spondylitis, explaining the strong the strong genetic predisposition (69-fold) by HLA-B27 discovered by Schlosstein, Terasaki (1973).

b. Paralysis a rare autoimmune complication of universal virus infection?

I postulate that the leg paralyses of poliomyelitis were a rare autoimmune complication of virtually universal virus infection, the paralyses probably caused by forbidden clones of cytotoxic T-cells which attacked anterior horn neurons in the spine, hence the Pathologists' appropriate name, "acute anterior poliomyelitis."

c. The lead in prophylaxis given by the polio vaccines

The Salk (killed) and Sabin (attenuated) polio vaccines have been brilliantly successful in preventing the polio leg paralyses. This exemplifies how autoimmune diseases in general, can be prevented by finding and vaccinating against their microbial triggers.

d. Finding microbial triggers

Ebringer has succeeded in this with rheumatoid arthritis and ankylosing spondylitis. He has pioneered this new field of medical research, developing a whole new technology that needs to be copied in other diseases, especially schizophrenia. Systematic studies of autoimmune diseases, with collaboration between clinicians and microbiologists are needed. The American Academy of Microbiology would be an ideal organization for doing the research needed to provide this urgently-needed knowledge.

Discussion

Some disease associations are cross-tissue autoimmunity, for example the eye proptosis of Graves' disease, caused by variants of the thyroid-stimulating auto-antibodies that react with receptors on orbital fat cells [56], and diabetic retinopathy [57], probably caused by destruction of retinal pericyte cells by antigenic variants of the T-cell forbidden clones that destroy the pancreatic islet β cells to cause Type-1 diabetes. Many autoimmune diseases, such as Graves' disease, already have satisfactory therapy. Immunotherapy, by radiological or chemical immune ablation, with immune reconstitution by autologous bone marrow cells, pioneered by Tyndall [58], can be used to save the lives of patients with dangerous autoimmune diseases, such as systemic scleroderma [59].

Selective destruction of forbidden clones could be achieved by isolating their auto-antigen (such as the TSH receptor of Graves' disease, cloned by Vassart and Dumont) [60] and attaching it to a cytotoxic moiety, such as bungarotoxin or ^{131}I iodine (emitting short-range β particles), then administering the molecular complex intravenously to destroy the pathogenic clones of plasma cells.

When monoclonal antibody technology was discovered, it was mistakenly assumed that this would provide cures for the autoimmune diseases, a notion greatly encouraged by the drug companies. Struggling to help his patients get benefit from use of Rituximab, Dreyfus [61] envisages major progress from anti-viral therapies and ultimately virus vaccines. Prevention is better than cure, so finding and countering antigenic triggers of autoimmune diseases is the ideal. Recognition of the universality of microbial triggers of autoimmune diseases is a major advance, in showing how the diseases can be prevented by finding and vaccinating against their triggers. Ebringer has led the way by discovering two triggers and developing the technology for finding others.

References

1. Hercus CE, Benson WN, Carter CL (1925) Endemic goitre in New Zealand and its relation to the soil iodine. J Hyg 24(3-4): 321-402.
2. Kelly FC, Snedden WW (1960) Prevalence and geographical distribution of endemic goitre. WHO, Palais des Nations, Geneva, Switzerland, UK.
3. Purves HD (1974-1975) Symposium. N Z Med J 80-81(529-532): 475, 548, 15, 61.
4. Adams DD, Adams CD (2013) Autoimmune Disease. Springer, London.
5. Adams DD, Knight JG, Ebringer A (2010) Autoimmune diseases: Solution of the environmental, immunological and genetic components with principles for immunotherapy and transplantation. Autoimmun Rev 9(8): 525-530.
6. Jerne NK (1955) The natural-selection theory of antibody formation. Proc Natl Acad Sci USA 41(11): 849-857.
7. Burnet FM (1959) The clonal selection theory of acquired immunity. Vanderbilt University Press, Cambridge, Nashville, USA.

8. Burnet FM (1959) Auto-immune disease. BMJ 2(5153,5154): 645-650, 720-725.

9. Adams DD (1980) Thyroid-stimulating autoantibodies. Vitam Horm 38: 119-203.

10. Jarrett EE, Miller HR (1982) Production and activities of IgE in helminth infection. Prog Allergy 31: 178-233.

11. Bell RG (1996) IgE, allergies and helminth parasites: a new perspective on an old conundrum. Immunol Cell Biol 74(4): 337-345.

12. Adams DD (1983) Autoimmune mechanisms In: Davies TF (Ed.) Autoimmune Endocrine Disease. John Wiley & Sons, New York, USA, pp. 1-39.

13. Adams DD (1982) Systemic lupus erythaematosus: a simple concept of the pathogenesis and its genetic basis. In: Dawkins RL, Christiansen FT, Zilko PJ (Ed.) Immunogenics in Rheumatology. Excerpta Medica Amsterdam, Netherlands, pp. 242-243.

14. Adams DD, Knight JG (2003) Principles of autoimmune disease: pathogenesis, genetics and specific immunotherapy. J Clin Lab Immunol 52: 1-22.

15. Adams DD, Knight A, Knight JG, Laing P (1987) Graves' disease: a paradigm for autoimmunity. In: Pinchera A, Ingbar SH, McKenzie JM, Fenzi GF (Ed.) Thyroid Autoimmunity. Plenum, New York, USA, pp. 1-10.

16. Lindstrom J, Shelton D, Fujii Y (1988) Myasthenia gravis. Adv Immunol 42: 233-284.

17. Ebringer A, Rashid T, Wilson C (2010) Rheumatoid arthritis, *Proteus*, anti-CCP antibodies and karl popper. Autoimmune Rev 9(4): 216-223.

18. Sherwin RS (2000) Diabetes mellitus. In: Goldman L, Bennet JC (Ed.) Cecil Textbook of Medicine. (21stedn), Saunders, Philadelphia, USA, pp. 1263-1285.

19. Sutherland DE, Goetz FC, Sibley RE (1989) Recurrence of disease in pancreas transplants. Diabetes 38(suppl 1): 85-87.

20. Klintworth GK (1988) The eye. In: Rubin E, Farber JL (Ed.) Pathology (2ndedn.) Lippincott, Philadelphia, USA pp. 1456-1483.

21. Adams DD (2008) Autoimmune destruction of pericytes as the cause of diabetic retinopathy. Clin Ophthalmol 2(2): 295-298.

22. Ben-Nun A, Wekerle H, Cohen IR (1981) The rapid isolation of clonable antigen-specific T cell lines capable of mediating autoimmune encephalomyelitis. Eur J Immunol 11(3): 195-199.

23. Pinchera A, Ingbar SH, McKenzie JM, Fenzi GF (1987) Thyroid Autoimmunity. Plenum Press, New York, USA.

24. Dausset J, Svejgaard A (1977) HLA and disease. Williams & Wilkins Co, Copenhagen, Munksgaard, North and South America.

25. Vladutiu AO, Rose NR (1971) Autoimmune murine thyroiditis: relation to histocompatibility (H-2) type. Science 174(4014): 1137-1139.

26. Auchincloss H, Sykes M, Sachs DH (1999) Transplantation Immunology. In: Paul WE (Ed.) Fundamental Immunology (4th edn) Lippincott-Raven, Philadelphia, USA, pp. 1175-1235.

27. Zinkernagel RM, Doherty PC (1974) Restriction of in vitro T cell-mediated cytotoxicity in lymphocytic choriomeningitis within a syngeneic or semi-allogeneic system. Nature 248(5450): 701-702.

28. Adams DD (1987) Protection from autoimmune disease as the third function of the major histocompatibility gene complex. Lancet 2(8553): 245-249.

29. Fenner F, White DO (1975) Medical Virology. (4th edn) Academic Press, London.

30. Simonsen M (1970) On the nature and measurement of antigenic strength. Transplant Rev 3(1): 22-35.

31. Matzinger P (2012) The evolution of the danger theory. Interview by Lauren Constable, Commissioning Editor. Expert Rev Clin Immunol 8(4): 311-317.

32. Bielschowsky M, Helyer BJ, Howie JB (1953) Spontaneous haemolytic anaemia in mice of the NZB/BL strain. Proc Univ Otago Med Sch 37: 9-11.

33. Howie JB, Helyer BJ (1968) The immunology and pathology of NZB mice. Adv Immunol 9: 215-266.

34. Knight JG, Adams DD (1978) Three genes for lupus nephritis in NZB x NZW mice. J Exp Med 147(6): 1653-1660.

35. Drake CG, Babcock SK, Palmer E, Kotzin BL (1994) Genetic analysis of the NZB contribution to lupus-like autoimmune disease in (NZB x NZW) F1 mice. Proc Natl Acad Sci USA 91(9): 4062-4066.

36. Kono DH, Burlingame RW, Owen DG, Kuramochi A, Balderas RS et al (1994). Lupus susceptibility loci in New Zealand mice. Proc Natl Acad Sci USA 91(21): 10168-10172.

37. Knight JG, Adams DD (1981) Genes determining autoimmune disease in New Zealand mice. J Clin Lab Immunol 5(3): 165-170.

38. Einstein A (1934) Essays in Science. (1st edn) Philosophical Library, New York, USA, pp. 92.

39. Adams DD, Knight JG (1980) H gene theory of inherited autoimmune disease. Lancet 1(8165): 396-398.

40. Knight JG, Adams DD (1982) The genetic basis of autoimmune disease. In: David Evered, Julie Whelan (Ed.) Ciba Foundation Symposium 90 - Receptors, Antibodies and Disease, John Wiley & Sons, USA, pp. 35-56.

41. Mathis D, Benoist C (2009) Aire. Annu Rev Immunol 27: 287-312.

42. McKusick VA (1978) Mendelian Inheritance in Man. (5th edn), Johns Hopkins University Press, Baltimore, USA.

43. Doria A, Sarzi-Puttini P, Shoenfeld Y (2008) Infections, rheumatism and autoimmunity: the conflicting relationship between humans and their environment. Autoimmun Rev 8(1): 1-4.

44. Doria A, Sampieri S, Sarzi-Puttini P (2008) Exploring the complex relationship between infection and autoimmunity. Autoimmun Rev 8(2): 89-91.

45. Kaplan MH, Meyeserian M (1962) An immunological cross-reaction between Group A streptococcal cells and human heart tissue. The Lancet 279(7232): 706-710.

46. Ebringer A, Rashid T, Wilson C, Ptazynska T, Fielder M (2006) Ankylosing spondylitis, HLA-B27 and *Klebsiella pneumoniae* – An overview proposed for early diagnosis and treatment. Curr Rheumatol Rev 2: 55-68.

47. Richter MJ, Horowitz MS (2009) Coxsackievirus infection as an environmental factor in the etiology of type1 diabetes. Autoimmun Rev 8(7): 611-615.

48. Randone SB, Guiducci S, Cerinic MM (2008) Systemic sclerosis and infection. Autoimmun Rev 8(1): 36-40.

49. Knight JG, Knight A, Pert CB (1987) Is schizophrenia a virally triggered

antireceptor autoimmune disease? In: Helmchen H, Henn FA (Ed.) Biological Perspectives of Schizophrenia. John Wiley & Sons, New York, pp. 107–127.

50. Laing P, Knight JG, Hill JM, Harris AG, Webster RG, et al. (1989) Influenza viruses induce autoantibodies to a brain-specific 37-kDa protein in rabbit. Proc Natl Acad Sci USA 86(6): 1998-2002.

51. Knight JG, Menkies DB, Highton J, Adams DD (2007) Rationale for a trial of immuno-suppressive therapy in acute schizophrenia. Mol Psychiatry 12(5): 424-431.

52. Dale RC, Merhev V, Pillai S, Wang D, Cantrill L, et al. (2012) Antibodies to surface dopamine-2 receptor in autoimmune movement and psychiatric disorders. Brain 135(11): 3453-3468.

53. Ebringer A (2012) Rheumatoid arthritis and *Proteus*. Springer, London, UK, pp. 233.

54. Ebringer A (2013) Ankylosing spondylitis and *Klebsiella*. Springer, Lonon, UK, pp 256.

55. Wilson C, Ebringer A, Ahmadi K, Wrigglesworth J, Tiwana H, et al. (1995) Shared amino acid sequences between major histocompatibility complex class II glycoproteins, type XI collagen and *Proteus mirabilis* in rheumatoid arthritis. Ann Rheum Dis 54(3): 216-220.

56. Ebringer A, Rashid T, Wilson C, Ptazynska T, Fielder M (2006) Ankylosing spondylitis, HLA-B27 and *Klebsiella pneumoniae* – An overview proposed for early diagnosis and treatment. Curr Rheumatol Rev 2: 55-68.

57. Rundle F, Pochin E (1944) The orbital tissues in thyrotoxicosis: a quantitative analysis relating to exophthalmos. Clin Sci 5: 51-74.

58. Forge D, Passweg J, van Laar JM, Marjanovic Z, Besenthal C, et al. (2004) Autologous stem cell transplantation in the treatment of systemic sclerosis: report from the EBMT/EULAR registry. Ann Rheum Dis 63(8): 974-981.

59. Englert H, Katelaris C, McGill N, Schrieber L, Moore J (2005) Grape-sultana sign represents a favourable response to aggressive treatment of early diffuse systemic scleroderma. Intern Med J 35(7): 436-437.

60. Peret J, Ludgate M, Libert F, Gerard C, Dumont JE, et al. (1990) Stable expression of the human TSH receptor in CHO cells and characterization of differentially expressing clones. Biochem Biophys Res Commun 171(3): 1044-1050.

61. Deyfus DH (2011) Autoimmune disease: A role for new anti-viral therapies. Autoimmun Rev 11(2): 88-97.

25

Promoter Hypomethylation and Expression of *PLS3* in Human Sezary Lymphoma Cells

Elodie Begue[1*], Laurence Michel[1*], Francette Jean-Louis[1], Martine Bagot[1,2] and Armand Bensussan[1]

[1]*UMR976, Skin Research Center and Universite Paris Diderot, Sorbonne Paris Cite, Hopital Saint-Louis, Paris, France*
[2]*Assistance Publique-Hopitaux de Paris, Hopital Saint Louis, Paris, France*

***Corresponding authors:** Laurence Michel, Skin Research Center and Universite Paris Diderot, Sorbonne Paris Cite, Hopital Saint-Louis, Paris, F-75475 Paris, France, E-mail: laurence.michel@inserm.fr*

Elodie Begue, Skin Research Center and Universite Paris Diderot, Sorbonne Paris Cite, Hopital Saint-Louis, Paris, F-75475 Paris, France, E-mail: begueelodie@gmail.com

Abstract

Previous studies about the mechanism responsible for the regulation of *PLS3* gene expression in cutaneous T cell lymphoma suggested a possible role of DNA methylation in the control of tissue-specific expression of *PLS3*. Using methylation-specific polymerase chain reaction adapted for quantitative real-time PCR, the present study investigated the methylation status in two human lymphocyte cell lines derived from tumor lymphocytes of patients with Sezary syndrome, HuT-78 cells which express *PLS3* mRNA and SeAx cells that do not, and used well-established inhibitor of DNA methylation, the nucleoside analog 5-aza-2-deoxycytidine (5-aza-dC), to further determine its effect on *PLS3* gene expression in both cell lines. Our results clearly revealed a quantitative relationship between the methylation status of the *PLS3* CpG region and *PLS3* expression in HuT-78 and SeAx Sezary cell lines.

Keywords: T-plastin; *PLS3*; Methylation; Sezary syndrome (SS); CTCL

Introduction

A molecular feature of Sezary syndrome (SS), the leukemic variant of cutaneous T-Cell Lymphomas (CTCL), is the abnormal gene expression of *PLS3* in 3 out of 4 patients [1-9]. Currently, *PLS3* is usually not expressed in normal T lymphocytes, although a recent study observed *PLS3* transcription in PBLs from less than 5% healthy individuals [10] in contrast to our own data [9]. Plastins, also known as fimbrins, are a family of actin-bundling proteins that are evolutionarily conserved from yeast to humans. In humans, three distinct isoforms have been identified and are expressed in a tissue-specific manner: I-plastin (*PLS1*) in intestinal and renal brush borders, L-plastin (*LCP1*) in hematopoïetic cells, and T-plastin (*PLS3*) in all other non hematopoietic tissues [11].

The mechanism responsible for the regulation of *PLS3* gene expression remains to be investigated. Previous studies suggested a possible role of DNA methylation in the control of tissue-specific expression of *PLS3* [12]. Actually, an inverse correlation between transcriptional activation and hypermethylation of local CpG sites has been reported for a number of genes in association with cell-type-specific expression and repression during tumorigenesis [13]. In CTCL, a recent work found no evidence for PLS3 mutations within coding or promoter regions, but showed significant hypomethylation of CpG dinucleotides 95–99 within the PLS3 CpG island restricted to the PLS3$^+$ population [8]. Using Methylation-Specific Polymerase (MSP) chain reaction adapted for quantitative real-time PCR (qRT-PCR), we investigated herein the methylation status in CTCL. We studied two human lymphocyte cell lines derived from tumor lymphocytes of patients with Sezary syndrome, HuT-78 cells which express PLS3 mRNA and SeAx cells that do not, for methylation status and used a well-established inhibitor of DNA methylation, the nucleoside analog 5-aza-2-deoxycytidine (5-aza-dC), to further determine its effect on *PLS3* gene expression in both cell lines.

Material and Methods

Cell cultures

The Sezary cell line HuT-78 was purchased from European Collection of Animal Cell Cultures (ECACC, Salisbury, England). SeAx cell line established from a SS patient was kindly provided by Dr Keld Kaltoft (University of Aarhus, Denmark). The CTCL cell lines were grown at 37°C with 5 % CO_2 at the concentration 5 x 10^4 cells/ml in RPMI 1640 Glutamax™-I medium, supplemented with 10% fetal bovine serum (FBS) and 100 U/mL penicillin, 100 μg/mL streptomycin, (GIBCO).

Multiplex and quantitative Real-time PCR (qRT-PCR)

RNA was isolated from cell lines using RNeasy Mini Kit according to the manufacturer's instructions with one step DNAse I (QIAGEN). Genomic DNA-free RNA was then converted into cDNA using the ThermoScript qRT-PCR system (INVITROGEN). cDNAs were used for multiplex PCR according to the manufacturer's recommendations (QIAGEN) and quantitative

***Corresponding author email:** s.laurence.michel@inserm.fr*

real-time PCR using a Power SYBR Green mixture on an AB7300 apparatus (APPLIED BIOSYSTEMS). PCR primer sequences and specific PCR conditions are available upon request.

Sodium bisulfite conversion

Genomic DNA was isolated from cell lines using the QIAmpDNA Mini Kit (QIAGEN). Extracted DNA was measured using a Nanodrop® Spectrophotometer and 1µg DNA was modified by sodium bisulfite using an EZ DNA Methylation™ Kit (ZYMO RESEARCH) according to the manufacturer's recommendations.

Methylation-Specific PCR (MSP) and quantitative MSP (qMSP) amplification

After sodium bisulfate conversion, the recovered DNA template was quantified using a Nanodrop® Spectrophotometer and 50ng DNA were used for fluorescence-based, qRT-PCR amplification as introduced above. The PCR primers were specifically designed for bisulfite converted DNA sequence, with sets of primers distinguishing fully methylated, fully unmethylated and wild-type DNA for the PLS3 gene and an internal reference set for the MYOD1 gene control for input DNA as previously described [14,15]. The methylated and unmethylated primers were designed with Methyl Primer Express® Software (APPLIED BIOSYSTEMS) in order to overlap four potential CpG dinucleotide sites. Quantitative SyBR Green PCR reactions were simultaneously performed with primers designed for the bisulfite-converted methylated, unmethylated and wild-type sequences of PLS3 and with MYOD1 reference primers. The designed pairs of primer sequences were as follows: PLS3 Wild-type sense 5'-AACTTCCCTCTGTCGTCCTTCTC-3', anti-sense 5'-AGGAAATC-CCGGAGCCAG-3'; PLS3 Methylated sense 5'-GTTCGGATT TAG-GAAATTTC-3', anti-sense 5'-TCGTATCCTTCCCTCGAC-3', PLS3 Unmethylated sense 5'-TTTTGTTTGGATTTAGGAAATTTT-3', anti-sense 5'-CCTCATATCCTTCCCTC A-ACC-3'; MYOD1 sense 5'-CCAACTCCAAATCCCCTCTCTAT3', anti-sense 5'-TGATT AATT-TAGATT-GGGTTTAGAGAAGGA-3'. Positive controls were generated by bisulfite conversion of CpGenome™ Universal Methylated DNA (CHEMICON), a commercially available enzymatically methylated human male genomic DNA used as a methylation-positive control for gene methylation studies. The thermocycling program included: an initial denaturation at 95°C for 10min, 40 cycles of denaturation at 95°C for 15s, annealing at 60°C for 60s and elongation at 72°C for 30s. Each assay was run in duplicate. Reference gene MYOD1 was quantified to normalize mRNA level between cell line samples.

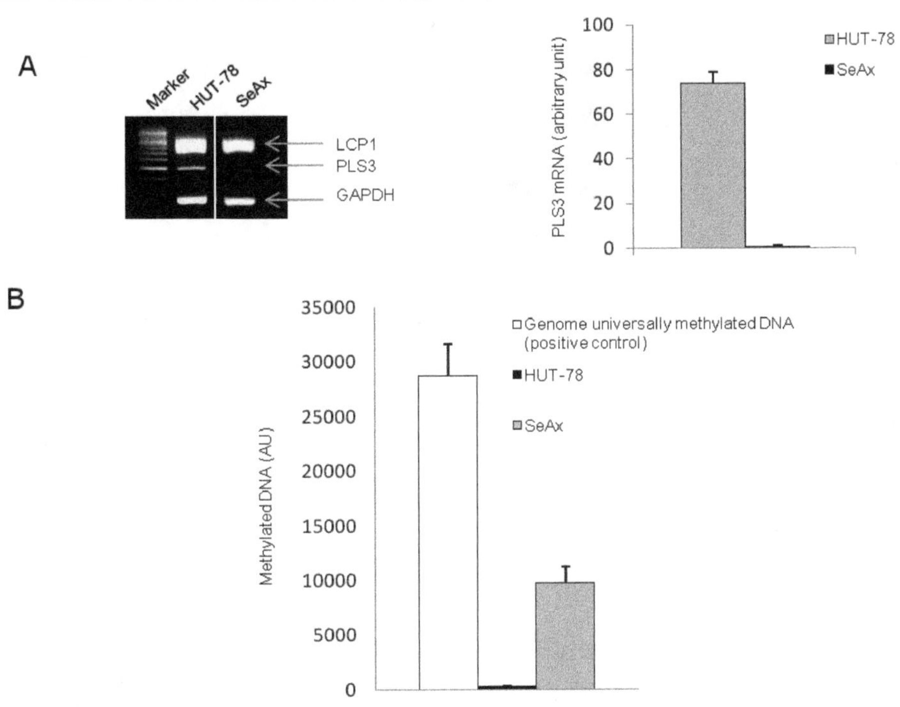

Figure 1: Expression of *PLS3* gene and methylation status of a portion of CpG island in HuT-78 and SeAx cell lines.

(A) Expression of *PLS3* mRNA levels in HuT-78 and SeAx cells. Basal expression levels of *PLS3* and *LCP1* transcripts were determined by multiplex PCR and normalized to GAPDH (left panel). *PLS3* expression levels were assessed by qRT-PCR (right panel): quantification of the target gene expression was done using the comparative cycle threshold (*Ct*) method according to the manufacturer's instructions (Applied Biosystems). An average *Ct* was calculated for the duplicate reactions and normalized to GAPDH (_*Ct* = *Ct* sample – *Ct* GAPDH). The obtained D*C*ts values are expressed as arbitrary units (A.U.) and presented as mean ± SD (n=3).

(B) Methylation status of HuT-78 and SeAx cell lines. qMSP analysis Genomic purifed DNA was treated with sodium bisulfite as briefly described in material and methods. Modified bisulfite DNA was amplified using PLS3-methylated primers and wild-type to ensure complete bisulfite conversion. Positive controls were generated by bisulfite modifying CpG universally methylated DNA. The results were normalized with *MYOD1* and presented as mean ± SD (n=3) in arbitrary units (A.U.).

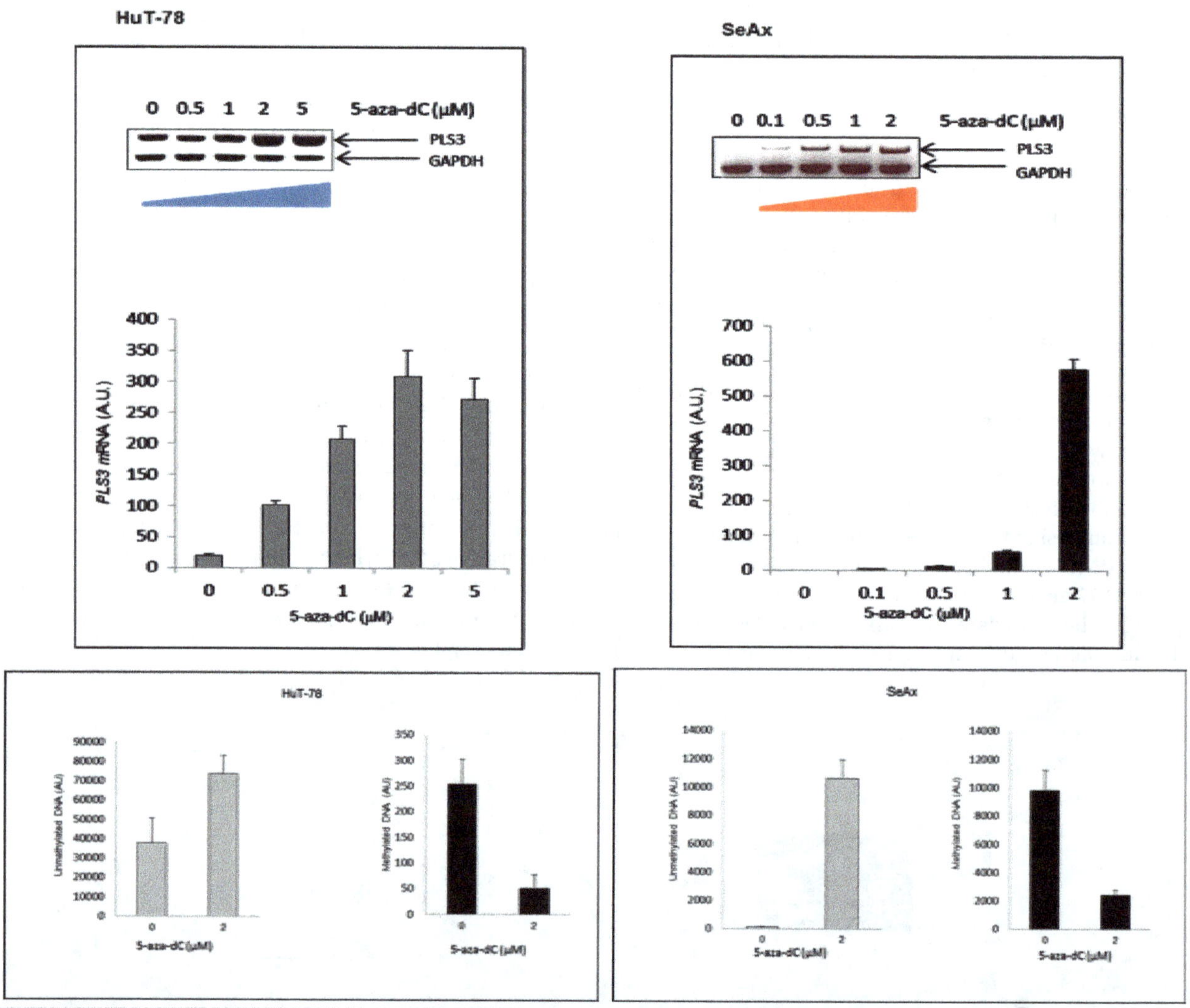

Figure 2: *PLS3* mRNA expression (Upper panel) and Methylation status of a portion of CpG island (4 CpG dimers) in the human *PLS3* gene (Lower panels) in HuT-78 (A) and SeAx(B) cell lines after treatment with the DNA methyltranferase inhibitor 5-aza-dC.
Upper panels: Cells were exposed or not to increasing concentrations of 5-aza-dC for 6 days. Transcripted PLS3 normalized to GADPH mRNA was determined by multiplex PCR or qRTPCR.
Lower panels: Genomic DNA was isolated from cell lines using the QIAmpDNA Mini Kit (QIAGEN). Extracted DNA was measured using a Nanodrop® Spectrophotometer and 1μg DNA was modified by sodium bisulfite using an EZ DNA Methylation™ Kit (ZYMO RESEARCH) according to the manufacturer's recommendations.
Modified bisulfite DNA was amplified with PLS3-Unmethylated and PLS3-methylated primers and wild-type to ensure complete bisulfite conversion. The results were normalized with MYOD1.
Results from qMSP analysis are expressed as levels of unmethylated DNA (left) and methylated DNA (right) detected in HuT-78 cells (A) or SeAx cells (B) either unexposed (0) or exposed to 5 aza-dC 2μM for 6 days and presented as mean ± SD (n=3) in arbitrary units (A.U.).

Treatment with 5-aza-2'-deoxycytidine (5-aza-dC)

HuT-78 and SeAx cells were seeded at the concentration of 5×10^4 cells/mL and incubated in a culture medium with or without 5-aza-dC (SIGMA) at a concentration of 0.1, 0.5, 1, 2 or 5μM for 6 days with daily medium changes and fresh 5-aza-dC addition every day. Total RNA or DNA from control and treated cells was analyzed by multiplex PCR and qRT-PCR.

Results and Discussion

Methylation status

Our present results obtained by quantitative MSP analysis revealed a quantitative relationship between the methylation status of the *PLS3* CpG region and *PLS3* expression in HuT-78 and SeAx Sezary cell lines. As shown in figure 1, hypomethylation of *PLS3* promoter was detected in HuT-78 cell line and related to the significant expression of *PLS3* mRNA, which was detected by multiplex PCR and qRT-PCR. In contrast, hypermethylation of *PLS3* promoter was observed in SeAx cell line and this could be associated with undetectable *PLS3* mRNA.

Rescue of *PLS3* expression by epigenetic 5-Aza-dC exposure

To further examine whether the silencing of *PLS3* expression in SeAx cells was attributed to the methylation status of the promoter and whether PLS3 synthesis could be restored by treatment with DNA methylation inhibitors, SeAx cells were exposed to 5-aza-dC for 6 days and then analyzed for their methylation status by qMSP and for *PLS3* expression by qRT-PCR. As control, we similarly treated HuT-78 cells that constitutively expressed PLS3. As shown in figure 2, *PLS3* promoter hypomethylation and gene expression was induced in SeAx cells after the 5-aza-dC treatment according to a dose-dependent manner, with a maximal expression for 2μM 5-aza-dC. Of interest, *PLS3* promoter hypomethylation and gene expression in HuT-78 cells was increased by 5-aza-dC treatment, with a maximal expression with 2μM inhibitor. These data indicate that hypomethylation of the *PLS3* promoter is linked to constitutive *PLS3* gene expression in HuT-78 cell line and that *PLS3* promoter methylation controls gene expression in SeAx cells.

During the last decade, substantial evidence has demonstrated the importance of epigenetic mechanisms in the transcriptional regulation of genes that play critical roles in the process of cancer progression [16]. Some cancers display aberrant methylation profiles in multiple genes and global genomic hypomethylation have been described in several malignant cancers.

In the present study, *PLS3* mRNA expression was confirmed to be associated with aberrant promoter hypomethylation in HuT-78 CTCL cell line and was increased by treatment with DNA methylation inhibitors such as 5-aza-dC. Significant *PLS3* expression in SeAx cells that do not constitutively express PLS3 was also induced by treatment with 5-aza-dC.

A hallmark of cancer is a paradoxical aberration of DNA methylation patterns, with a global loss of DNA methylation that coexists with regional hypermethylation of certain genes [17]. It has been proposed that hypermethylation and hypomethylation in cancers are independent processes, which target different programs at different stages in tumorigenesis. Hypermethylation and silencing of genes that regulate proliferation were proposed to be critical for deregulation growth early in carcinogenesis, while hypomethylation and activation of other genes may be more important for metastasis [13]. Activation of PLS3 expression in hematopoietic cells through alteration of DNA methylation may play a role in cutaneous T-cell lymphoma oncogenesis and tissue invasion by modulating actin cytoskeleton, as suggested by our recent data demonstrating that PLS3 expression favors tumor cell migration [9].

In conclusion, our results confirmed that an epigenetic mechanism regulates PLS3 expression in CTCL as shown by *PLS3* promoter hypomethylation status in CTCL lines. As PLS3 expression has potential as molecular biomarker in Sezary syndrome, it might be suggested that methylation status may serve as a novel biomarker in malignant T-cells from SS patients.

References

1. Kari L, Loboda A, Nebozhyn M, Rook AH, Vonderheid EC, et al. (2003) Classification and prediction of survival in patients with the leukemic phase of cutaneous T cell lymphoma. J Exp Med 197: 1477-1488.
2. Nebozhyn M, Loboda A, Kari L , Rook AH, Vonderheid EC, et al. (2006) Quantitative PCR on 5 genes reliably identifies CTCL patients with 5% to 99% circulating tumor cells with 90% accuracy. Blood 107: 3189-3196.
3. van Doorn R, Dijkman R, Vermeer MH, Out-Luiting JJ, van der Raaij-Helmer EM, et al. (2004) Aberrant expression of the tyrosine kinase receptor EphA4 and the transcription factor twist in Sezary syndrome identified by gene expression analysis. Cancer Res 64: 5578-5586.
4. Su MW, Dorocicz I, Dragowska WH, Ho V, Li G, et al. (2003) Aberrant expression of T-plastin in Sezary cells. Cancer Res 63: 7122-7127.
5. Booken N, Gratchev A, Utikal J, Weiss C, Yu X, et al. (2008) Sezary syndrome is a unique cutaneous T-cell lymphoma as identified by an expanded gene signature including diagnostic marker molecules CDO1 and DNM3. Leukemia 22: 393-399.
6. Capriotti E, Vonderheid EC, Thoburn CJ, Wasik MA, Bahler DW, et al. (2008) Expression of T-plastin, FoxP3 and other tumor-associated markers by leukemic T-cells of cutaneous T-cell lymphoma. Leuk Lymphoma 49: 1190-1201.
7. Tang N, Gibson H, Germeroth T, Porcu P, Lim HW, et al. (2010) T-plastin (PLS3) gene expression differentiates Sezary syndrome from mycosis fungoides and inflammatory skin diseases and can serve as a biomarker to monitor disease progression. Br J Dermatol 162: 463-466.
8. Jones CL, Ferreira S, McKenzie RC, Tosi I, Caesar JA, et al. (2012) Regulation of T-plastin expression by promoter hypomethylation in primary cutaneous T-cell lymphoma. J Invest Dermatol 132: 2042-2049.
9. Begue E, Jean-Louis F, Bagot M, Jauliac S, Cayuela JM, et al. (2012) Inducible expression and pathophysiologic functions of T-plastin in cutaneous T-cell lymphoma. Blood 120: 143-154.
10. Oprea GE, Krober S, McWhorter ML, Rossoll W, Muller S, et al. (2010) Plastin 3 is a protective modifier of autosomal recessive spinal muscular atrophy. Science 320: 524-527.
11. Lin CS, Park T, Chen ZP, Leavitt J(1993) Human plastin genes. Comparative gene structure, chromosome location, and differential expression in normal and neoplastic cells. J Biol Chem 268: 2781-2792.
12. Lin CS, Lau A, Huynh T, Lue TF (1999) Differential regulation of human T-plastin gene in leukocytes and non-leukocytes: identification of the promoter, enhancer, and CpG island. DNA Cell Biol 18: 27-37.
13. Szyf M, Pakneshan P, Rabbani SA (2004) DNA demethylation and cancer: therapeutic implications. Cancer Lett 211: 133-143.
14. Eads CA, Danenberg KD, Kawakami K, Saltz LB, Blake C, et al. (2000) MethyLight: a high-throughput assay to measure DNA methylation. Nucleic Acids Res 28: E32.
15. Trinh BN, Long TI, Laird PW (2001) DNA methylation analysis by MethyLight technology. Methods 25: 456-462.
16. Rodriguez-Paredes M, Esteller M (2011) Cancer epigenetics reaches mainstream oncology. Nat Med 17: 330-339.
17. Ehrlich M (2002) DNA methylation in cancer: too much, but also too little. Oncogene 21: 5400-5413.

Plasma Membrane-derived Vesicles (PMVs) in G6PD Deficient Patients

Samuel Antwi-Baffour*, Andrew Owusu Boafo, Ransford Kyeremeh and Seidu Abdulai Mahmood

Department of Medical Laboratory Sciences, School of Allied Health Sciences, College of Health Sciences, University of Ghana, Korle-Bu, Accra

***Corresponding author:** Samuel Antwi-Baffour, Department of Medical Laboratory Sciences, School of Allied Health Sciences, College of Health Sciences, University of Ghana, Accra; Email: s.antwi-baffour@chs.edu.gh*

Abstract

Background: Plasma membrane-derived vesicles (PMVs) are described as sub-cellular vesicles released upon shear stress, cell activation, injury or apoptosis. They are considered as universal biomarkers in a wide range of physiological and pathological processes. Glusoe-6-phosphate dehydrogenase (G6PD) deficiency on the other hand is an x-linked enzymopathy characterized by the inability of red blood cells (RBCs) to withstand oxidative stress. The presence and role of PMVs in G6PD deficiency has not been fully investigated.

Objectives: The aim of this study was to quantify and compare the levels of PMVs in G6PD deficient and non-deficient patients.

Methods: G6PD deficiency test using the methaemoglobin reduction method was performed on the blood samples. The levels of PMVs in those samples were quantified using quantitative flow cytometry. This was done by centrifuging the blood samples to obtain platelet-free plasma. Annexin V was then added to label the PMVs and quantified by flow cytometry.

Results: The mean level of PMVs among G6PD full defects subjects was 26×10^4 PMVs/ml and that among the G6PD normal subjects was 10×10^4 PMVs/ml. Statistically, the difference between the PMV levels of the G6PD normal samples and G6PD deficient samples was significant ($p<0.05$).

Conclusion: There is an increased level of PMVs in the circulating blood of G6PD deficient individuals as compared to G6PD normal individuals.

Keywords: G6PD, Plasma membrane; Deficiency; Apoptosis.

Abbreviations

ATP: Adenosine Triphosphate; EDTA: Ethylene Diamine Tetracetic Acid; FACS: Fluorescence Activated Cell Sorting; G6PD: Glucose-6-phosphate Dehydrogenase; NADP: Nicotinamide Adenine Dinucleotide Phosphate; NADPH: Reduced Nicotinamide Adenine Dinucleotide Phosphate; PBS: Phosphate Buffered Saline; PFP: Platelet Free Plasma; PMVs: Plasma Membrane-derived Vesicles; RBC: Red Blood Cells; SLE: Systemic Lupus Erythematosus

Introduction

Red blood cells (RBCs) require constant energy to maintain their biconcave disc shape and haemoglobin in reduced form [1]. Without adequate energy, RBCs will lyse and/or deform. RBCs gain energy in the form of adenosine triphosphate (ATP) from glucose which is derived from metabolism via anaerobic glycolysis (Embden-Myerhof pathway)[1,2]. There is also an alternate aerobic pathway (pentose-phosphate shunt) which starts with Glucose-6-Phosphate under the catalysis of Glucose-6-phosphate dehydrogenase (G6PD) [2]. Low G6PD activity results in low levels of nicotinamide adenine diphosphate hydrogenase (NADPH) and reduced glutathione, which are required to protect hemoglobin from oxidative damage [3]. In the absence of adequate reducing ability provided by G6PD, oxidizing agents convert haemoglobin to methemoglobin, and then denature it, causing it to precipitate as Heinz bodies within the cell [3]. The spleen pinches off the Heinz body and the overlying membrane, leaving a "bite cell" or "blister cell" which eventually haemolysis.

G6PD deficiency is a X-chromosomally transmitted disorder of RBCs that affects 400 million people Worldwide. The deficiency is mainly found in malaria endemic areas and it has been suggested that it offers protection against malaria infection [4]. While the molecular biology of G6PD deficiency has been extensively studied, the molecular mechanisms leading to the haemolytic crisis are still unclear [5]. However, some membrane alterations have been described in G6PD deficient red blood cells, such as the oxidation and clustering of membrane proteins, the binding of hemichromes to the internal face of the membrane, the destabilization of the membrane and the release of micro-vesicles [6]. There is therefore the suspicion of increased levels of Plasma Membrane-derived Vesicles (PMVs) in G6PD deficient patients.

PMVs are membrane-coated vesicles of diameter 0.1 to 1 µm. They are released when cells undergo activation/apoptosis via blebbing and shedding and have a function in intercellular communication. PMVs are released from the membranes of cells and therefore contain cell surface or membrane proteins and cytoplasmic components of their parent cells [7]. The presence of basal levels of PMVs is common in healthy individuals and an increase in their release although a controlled event, is a hallmark of cellular alteration [8]. Increased numbers of circulating PMVs have been identified in individuals with diseases such as hypertension, thrombotic thrombocytopenic purpura, systemic lupus erythematosus (SLE), multiple sclerosis, cerebral malaria and many others associated with inflammation [9]. G6PD deficiency disease is also associated with inflammation and hence the suspicion that PMVs levels may be high in those patients. Although the prognostic potential of circulating PMVs is still in its infancy, the different studies being carried out clearly demonstrate that their detection and quantification is an interesting and potentially valuable tool to appreciate in disease states and also may indicate the severity of that disease state [10].

Materials And Methods

This was a prospective study conducted from the month of May to June, 2013.

Study population

The study population was made up sixty six (66) patients who were suspected of G6PD deficiency and were requested to do the test. Out of the total number, 47 were males and 19 females. Their ages ranged from 5 to 40 years which were categorized into 5 – 10yrs; 11 – 20yrs; 21 – 30 yrs and 31 – 40yrs. Those who tested positive for G6PD full defect (33) became the study group and those who tested negative (33) for the defect were the control group.

Ethical consideration

Ethical clearance for this research was sought from the Ethics and Protocol Review Committee at the School of Allied Health Sciences, University of Ghana, Legon. All the participants gave their informed consent before their samples were collected. No G6PD partial defects were obtained.

Materials

Some of the materials needed for the work include: FACS-Flow Cytometer, sonicating water-bath (Townson and Mercer Ltd, Croydon), Centrifuge – Eppendorf 5810R, Rotor number A-4-62, Micro-centrifuge – Eppendorf 5417R, Rotor number FA 45-24-11, Ethylene diamine tetracetic acid (EDTA) tubes, Pasteur pipettes, sodium nitrite (0.7g%) and methylene blue (10 mg/ml) reagents and phosphate buffered saline (PBS- pH 7.2).

Sample collection and processing: 10 ml of whole blood from each individual was collected into EDTA anticoagulant tubes. 5 ml of the blood was used to perform a G6PD deficient test to know or confirm participant's G6PD status. The remaining 5 ml was spun at 2 000 g for 10 minutes and the plasma separated for the quantification of the PMVs. For the G6PD screening the methaemoglobin reduction method as in Cheesbrough, 2010 was used.

With the flow cytometry analysis of PMVs, Frozen plasma was allowed to thaw, and centrifuged at a higher speed (4 000 g, 60 minutes) to obtain platelet free plasma (PFP) and to remove any cell debris. The resultant supernatant was sonicated in a sonicating water-bath for 5 x 1 minute prior to further centrifugation, in order to disperse aggregated exosomes. The supernatant was then centrifuged at 25 000 g for 90 minutes to pellet the PMVs. The supernatant was discarded, and the pellets containing the PMVs resuspended in 200 µl phosphate buffered saline (PBS-pH 7.2). An aliquot of this was diluted in the ratio of 1:40 in PBS (pH 7.2). A 100 µl of the diluted sample was put in a 96 well plate, equal volume of annexin V added and quantified using Guava Express plus (FACS- Flow Cytometer). Three (3) different measurements (technical repeats) were done for each patient.

Data analysis: Data from the study was analysed using the Statistical Package for Social Sciences (SPSS) version 20.0 and summary was presented as a descriptive statistics of mean. The student's t-test for unpaired data was used to compare the differences in PMVs levels between G6PD deficient and non-deficient samples. A p-value < 0.05 was considered significant. Graph pad prism version 6.0 and Microsoft Excel 2007 was used to construct the graph and bar chart respectively.

Results

The study population was made up of 47 males and 19 females who were between the ages of 5 to 40 years. Out of this number, 27 of the males and 6 of the females showed G6PD full defect whilst 20 of the males and 13 of the females showed no defect (the control group). The age category of 21 – 30 yrs showed the highest number of full defect followed by the 31 – 40yrs. Generally, 4 patients in age category 5 – 10yrs were full defect as against 6 who were no defect, then 6 against 7 in age 11-20yrs; 13 against 11 in category 21 -30yrs and 10 against 9 in age category 31 – 40yrs (Table 1).

All the samples were analysed for their G6PD status. 50% of the 66 samples obtained were G6PD full defect and 50% No defect. No partial defects were obtained. There was clear difference between the levels of PMVs in the G6PD full defect patients as against the No defect ones (Figure 1). The table below (Table 2) shows the descriptive statistics of the PMV levels in the G6PD normal and full defect samples. The minimum, maximum and mean values of the full defect and no defect samples clearly emphasize the difference in PMVs levels of the two G6PD statuses.

The highest PMV level among the full defect subject was 32

Age categories (yrs)	Males		Females		Percentage
	Full Defect	No Defect	Full Defect	No Defect	
5 – 10	3	4	1	2	**13.6**
11 – 20	6	4	0	3	**21.2**
21 – 30	10	9	3	2	**36.4**
31 – 40	8	3	2	6	**28.8**
Total	**27**	**20**	**6**	**13**	**100**

Table 1: A table representing participants' age, sex and G6PD status.

× 10^4 PMVs/ml and the lowest was 22 × 10^4 PMVs/ml whilst in the normal the highest PMV level was 22 ×10^4 PMVs/ml and the lowest 7 × 10^4 PMVs/ml. The mean level of PMVs among the full defects subjects was 26 × 10^4 PMVs / ml whilst the mean level among the normal subjects was 10 × 10^4 PMVs/ml (Figure 2).

Discussion

Plasma membrane-derived vesicles (PMVs) have been identified at basal levels (2 – 10 x 10^4 PMVs /ml) in blood of normal individuals whilst in diseases such as hypertension, thrombotic thrombocytopenic purpura, multiple sclerosis, cerebral malaria and many others associated with inflammation the circulating PMV levels are increased (10 – 30 x10^4 PMVs/ml) [9,10]. In this study, there was a significant difference ($p<0.005$) between the level of the PMVs in the full defect sample and the normal samples. Even the minimum level of PMVs recorded in the G6PD full defect sample (22 × 10^4 PMVs/ml) was higher than the maximum level (12 × 10^4 PMVs /ml) recorded in the G6PD normal samples. This clearly demonstrates that there are increased levels of circulating PMVs in the blood of the G6PD full defect individuals who participated in the study.

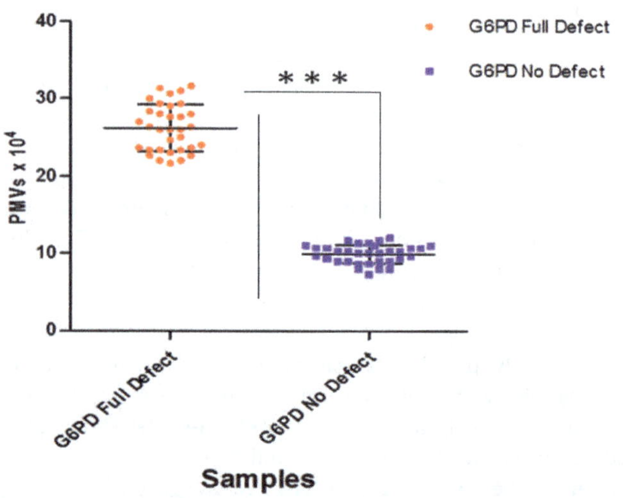

Figure 1: G6PG full defect plasma levels of PMVs are higher than those of G6PG No defect plasma.

	Num-ber	Mini-mum value	Maxi-mum value	Mean value	Std. Error	Std. Deviation
Full Defect	33	22	32	26.20	0.528	3.033
No Defect	33	7	12	9.93	0.209	1.198
Valid N (listwise)	33					

Table 2: Descriptive statistics of PMVs levels.

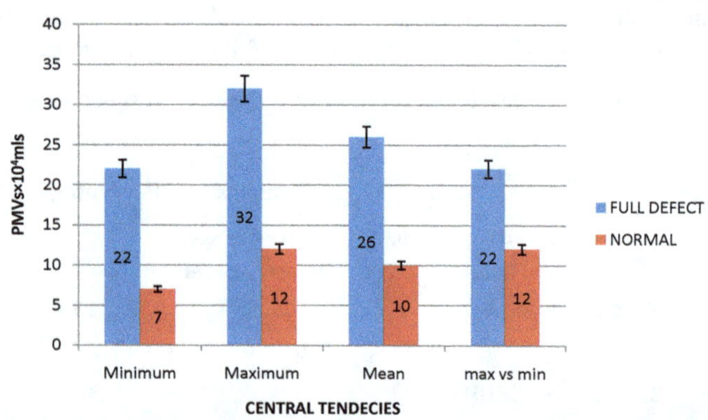

Figure 2: Shows the central tendencies of the PMV level of the G6PD full defect and normal samples.

A study conducted by Pantaleo showed that the treatment of G6PD deficient red cells with diamide or divicine causes an increase in the oxidation and tyrosine phosphorylation of AE1. This leads to subsequent vesiculation of the red cells thus production of PMVs [11]. Diamide and divicine are compounds extracted from fava beans which can cause severe haemolytic crisis in G6PD deficient subjects [12]. In the study, Pantaleo identified and characterized the PMVs as likely to have been produced from G6PD deficient red blood cells by illustrating hemichromes in the PMVs produced [11].

The accumulation of hemichromes leads to the formation of Heinz Bodies which are additional hallmark of the haemolytic crisis in G6PD deficient individuals [11]. Based on Pantaleo's illustration, the higher levels of PMVs recorded in the G6PD full defect samples can be said to be as a result of increased activated red blood cells in the full defect subjects. This implies that the level of PMVs in a G6PD deficient individual can be used as a marker of the level of severity of erythrocyte stress and therefore crisis [13,14]. There were some limitations during the research work. These include not obtaining G6PD partial defect participants as well as low sample population.

Conclusion

There are increased levels of PMVs in the circulating blood of G6PD deficient individuals as compared to G6PD normal individuals. The increased PMVs levels may be attributed to the stress the G6PD deficient red blood cells may be going through.

Acknowledgements

We are grateful to the Central Laboratory of the Korle-Bu Teaching Hospital for providing us with the samples and the Haematology Laboratory of the School of Allied Health Sciences for their help in analysing them.

References

1. Beutler E, Duparc S (2007) Glucose-6-phosphate dehydrogenase deficiency and antimalarial drug development. The American journal of tropical medicine and hygiene 77: 779–789.

2. Bertil E, Glader A (2002) A review on genetics and Pathophysiology of Glucose-6-Phosphate Dehydrogenate deficiency. Blood 15: 150.

3. Baker MA, Bosia A, Pescarmona G, Turrini F, Arese P. (1984) Mechanism of action of divicine in a cell-free system and in glucose-6-phosphate dehydrogenase-deficient red cells. Toxicologic pathology 12(4): 331-336.

4. Tishkoff SA, Varkonyi R, Cahinhinan N, Abbes S, Argyropoulos G, et al. (2001) Haplotype diversity and linkage disequilibrium at human G6PD: recent origin of alleles that confer malarial resistance. Science 293: 455–462.

5. Beutler E. (2008). Glucose-6-phosphate dehydrogenase deficiency: a historical perspective. Blood 111: 16–24.

6. Caprari PP, Bozzi AA, Ferroni LL, Giuliani AA, La Chiusa BF BF et al. Membrane alterations in G6PD- and PK-deficient erythrocytes exposed to oxidizing agents. Biochemical medicine and metabolic biology 45: 16–27.

7. Bevers EM, Wiedmer T, Comfurius P, Zhao J, Smeets EF, et al. (1995) The complex of phosphatidylinositol 4,5-bisphosphate and calcium ions are not responsible for Ca2+-induced loss of phospholipid asymmetry in the human erythrocyte: a study in Scott syndrome, a disorder of calcium-induced phospholipid scrambling. Blood 86: 1983–1991.

8. Berckmans RJ, Nieuwland R, Boing AN, Romijn FP, Hack CE,et al. (2001) Cell-derived microparticles circulate in healthy humans and support low grade thrombin generation.Thromb Haemost 85: 639–646.

9. Antwi-Baffour S, Kholia S, Aryee YK, Ansa-Addo EA, Stratton D, et al. (2010) Human Plasma Membrane-derived Vesicles inhibits the phagocytosis of apoptotic cells possible role in SLE. Biochem. Biophys Res Commun 398: 278-283.

10. Boulanger CM, Scoazec A, Ebrahimian T, Henry P, Mathieu E, et al. (2001) Circulating microparticles from patients with myocardial infarction cause endothelial dysfunction. Circulation 68: 87-89.

11. Pantaleo A, Ferru E, Carta F, Mannu F, Simula LF, et al. (2011) Irreversible AE1 Tyrosine Phosphorylation Leads to Membrane Vesiculation in G6PD Deficient Red Cells. PLoS ONE 6(1):15847

12. Chevion M, Navok T, Glaser G, Mager J. (1982) The chemistry of favism-inducing compounds. The properties of isouramil and divicine and their reaction with glutathione. European journal of biochemistry/ FEBS 127: 405-409.

13. Ansa-Addo EA, Lange S, Stratton D, Antwi-Baffour S, Cestari I, et al. (2010) Human plasma membrane-derived vesicles halt proliferation and induce differentiation of THP-1 acute monocytic leukaemia cell. Jour. Immunol 185:5 236-5246.

14. Cheesbrough M (2010) District Laboratory Practice in Tropical Countries Part 2. (2ndedn), Cambridge: Tropical Health Technology and Cambridge University Press.

27

Heterogeneous Nuclear Ribonucleoprotein K Autoantibodies in Patients who Suffered Severe Traumatic Brain Injury

María J. Acevedo-Calado[1,4], Marco A. Montes-Cano[1,4]*, Francisco Murillo-Cabezas[1,3], Ainara Madrazo-Atutxa[2,4], Salvador Toro- Llamas[1,4], Ángel Luis Castaño Núñez[1,4], José L. Caro Oleas[1,4], Alfonso Leal-Cerro[2,4] and Antonio Núñez-Roldán[1,4]

[1]*Servicio de Inmunología, Hospital Universitario Virgen del Rocío, Sevilla, Spain*

[2]*Unidad de Gestión de Endocrinología, Hospital Universitario Virgen del Rocío, Sevilla, Spain*

[3]*División de Cuidados Críticos, Hospital Universitario Virgen del Rocío, Sevilla, Spain*

[4]*Instituto de Biomedicina de Sevilla (IBiS), Sevilla, Spain*

***Corresponding author:** Marco Antonio Montes-Cano, Servicio de Inmunología, Instituto de Biomedicina de Sevilla (IBiS), Hospital Universitario Virgen del Rocío, 41013-Seville, Spain; E-mail: marcoa.montes.sspa@juntadeandalucia.es*

Abstract

The immune-inflammatory response as well as cerebral endothelial activation has been described after Traumatic Brain Injury (TBI). Although inflammation can have both beneficial and detrimental effects in TBI, the mechanisms underlying this dichotomy are mostly unknown. Moreover, emerging data indicates that chronic alterations produced after TBI are probably the result of systemic and persistent inflammation that activates immune response, culminating in the production of different specific auto antibodies against central nervous system antigens. In the previous study we demonstrated the production of anti-hnRNPK antibodies, belonging to Anti-Endothelial Cell Antibodies (AECA), in heart transplanted patients who developed Cardiac Allograft Vasculopathy (CAV). These antibodies could be produced in response to the vascular endothelial damage that is also present in TBI patients. In the current study we analyse the presence of AECA by Indirect Immunofluorescence assay (IFI) and anti-hnRNPK IgG antibodies by Enzyme-Linked Immunosorbent Assay (ELISA) in a cohort of 19 patients who suffered TBI. We detected a significant increase in the number of patients with AECA and anti-hnRNPK antibodies after TBI. Moreover, anti-hnRNPK antibody levels were also higher during follow-up, being the values obtained after TBI significantly higher than those detected at the time of trauma ($p = 0.001$). We also found that AECA and anti-hnRNPK antibodies were mostly present in the sera of patients with a worse outcome ($p = 0.018$ and $p = 0.04$ respectively). Taking into account, measuring these auto antibodies could be useful for evaluating endothelial injury and/or the outcome after TBI.

Keywords: Traumatic brain injury; Anti-endothelial cell antibodies; Heterogeneous nuclear ribonucleoprotein K

Introduction

In recent year's the knowledge about the alterations occurred after Traumatic Brain Injury (TBI) has undergone significant developments [1]. These advances have contributed to clarify the relationship between post-traumatic edema and neuropathological sequelae that are largely responsible for adverse outcome. Several investigations have been conducted to clarify the role of immunological and inflammatory response produced after TBI. Thus, it has been described two types of brain damage: an immediate and irreversible primary lesion, followed by a secondary lesion, which begins at the time of the injury and continues in the ensuing days to weeks. Primary and secondary events lead to a variety of physiological, cellular and molecular responses aimed at restoring the homeostasis of the damaged tissue which, if not controlled, could increase the brain lesion [2]. Therefore, TBI initiates a series of related events including edema, cytotoxicity and an intense inflammatory response which affects the injured cerebral tissue as well as the healthy one. The immuno-inflammatory response that occurs after TBI persists and auto amplifies by complement activation, proinflammatory cytokine production, increased expression of endothelial cell adhesion molecules, as well as the production of processes of necrosis and apoptosis; these processes could be a potential source for presentation and generation of autoimmune responses [3]. In this way, Anti-Pituitary Antibodies (APA), Neuron-Specific Enolase (NSE), S100β, heat shock protein-70, among other possible biomarkers, has been studied in an effort to reach a more accurate of prognosis [4,5], and more recently it has been described different molecules as biomarkers that are thought to play a part in secondary injury following severe TBI [6-8].

In the previous study, we identify a new antigen that seems to be associated with endothelial damage, related auto-antigens with a significative increase in the production of antibodies against heterogeneous nuclear Ribonucleoprotein K (hnRNPK), and a type of AECA, in heart transplanted patients who developed Cardiac Allograft Vasculopathy (CAV) [9]. This protein, is a member of the hnRNP family which has several different cellular roles including transcription, mRNA shuttling, RNA editing and translation. These cellular functions might be related to the involvement of this protein in apoptosis, tumors development, angiogenesis, cell invasion [10-12], and also altered gene expression patterns of hnRNPK have been found in many human cancers [13,14]. Additionally it has been described, as a high expression of this protein during smooth muscle cell proliferation in both, aortas from animal models of atherosclerosis and in human occluded veins [15]. We cannot rule out that hnRNPK auto antibodies could be also over expressed after endothelial damage and inflammation produced after TBI. However, at present there is no information on the involvement of anti-hnRNPK antibodies generation and the outcome of TBI. Therefore, the aim of this work was to analyze the presence of hnRNPK auto antibodies in patients who suffered different severity of TBI and its possible role in the outcome.

Materials and Methods

This study included patients with severe TBI admitted to neurosurgical ICU at the Virgen Del Rocio University Hospital, Seville, Spain, between July 2004 and Jan 2006. The study was approved by the hospital ethics committee and informed consent was obtained from a next of kin, given that all eligible patients were in coma. Inclusion criteria were: (1) male or female over 14 years of age with severe TBI (GCS score ≤ 8 after resuscitation) diagnosed by history and clinical examination with at least one reactive pupil; (2) to obtain at least two serum samples from each patient, one of them done within 24 hours after accident and a second serum sample at least six months post-TBI and (3) haemodynamically stable (mean arterial pressure > 75 mmHg with no or low-dose vasoactive drugs). Patients were excluded based on the following criteria: (1) presence of two reactive pupils; (2) suffered cardiac arrest after TBI; (3) any spinal cord injury, pregnancy, or coma suspected to be primarily due to other causes (E.g: alcohol); (4) suffered multiple injury as measured through the Abbreviated Injury Score (AIS) > 2; (5) no possibility of follow-up during one year and (6) presence of chronic or auto-immune diseases.

Data collection included demographic and clinical variables (age, sex, cause of injury, GCS and pupil reaction after resuscitation, occurrence of prehospital hypotension (systolic blood pressure < 90 mmHg), hypoxia (peripheral oxygen saturation (SpO_2)) < 90% and occurrence of sepsis (< 4 days). Sepsis was defined according to the Sepsis Consensus Conference criteria. Assessment of overall injury severity was based on the Injury Severity Score (ISS). Patients underwent an initial CT scan after resuscitation. Neuroradiological findings were classified according to the Traumatic Coma Data Bank (TCDB). This classification is divided into six groups: the first four indicate the presence and severity of the diffuse injury and the rest of the categories indicate the presence of an evacuated or non-evacuated mass lesion. A neuroradiologist, blind to the study goals and data reviewed and completed this classification.

Intra-Parenchymal Intra-Cranial Pressure (VENTRIX, INTEGRA Neuroscience, and Plainsboro, NJ), mean arterial blood pressure (obtained with a radial artery fluid-coupled system), Cerebral Perfusion Pressure (CPP), brain tissue oxygen pressure (LICOX, IMC System, GMS Kiel-Mielkendorf, and Germany), end-tidal carbon dioxide and SaO_2 were continuously monitored in the ICU. All patients were managed according to Brain Trauma Foundation guidelines and local protocols. Treatment was targeted at maintaining intracranial pressure at < 25 mmHg, CPP at > 60 mmHg, and brain tissue oxygen pressure at > 15 mmHg. The outcome assessment was carried out 12 months after patient discharge using the Glasgow Outcome Score (GOS). To relate our findings with the auto antibodies we dichotomized our results according to the GOS into two groups, severe (2-3) and mild (4-5) sequelae.

Two serum samples were obtained and analyzed from each patient; one at the time of TBI (basal sample, obtained within 24 hours after TBI), and the second within six months after trauma. As healthy controls, we included a total of 124 non related individuals with a range of age 26-53 (39.9 ± 13.9 years). Samples were immediately centrifuged, frozen at - 80°C and were stored for later analysis. Anti-hnRNPK IgG antibodies were detected by ELISA using hnRNPK recombinant protein previously purified in our laboratory [9]. Recombinant hnRNP-K protein, once affinity-purified by polyhistidine-tag, was plated at 2 μg/ml (in 0.1 mol/liter carbonate-bicarbonate buffer, p^H 9.5) onto a polystyrene flat-bottom ELISA plate (Nunc, Roskilde, Denmark) and incubated for 16 hours at 4°C. The non-specific binding of Igs was prevented by adding Tris-buffered saline with TBS-TM for 2 hours. The same serum-positive sample used in the library screening was used to construct a standard curve (dilutions of 1:100 to 1:1,600) to rule out non-specific antibody activities. For each ELISA, standards and samples (diluted 1:100 in TBS-TM) were added to duplicate wells and incubated for 1 hour at room temperature. Plates were washed, horseradish peroxidase conjugated rabbit anti-human IgG (Phadia AB, Uppsala, Sweden) was added, and the plates were for 1 additional hour at room temperature. The enzyme reaction was started by adding 100 μl of 3,3', 5,5'- tetramethylbenzidine and stopped 45 minutes later with 50 μl of 0.5 mol/liter H_2SO_4. Finally, the optical density was read at 450 nm in a micro titer plate reader (Bio-Tek Instruments, Winooski, VT). To calculate results, background reactivity of the reference mixture was subtracted to establish the frequency of newly identified antibodies in healthy individuals, a healthy control group of 124 individuals (72 men and 52 women; mean age, 40 ± 1.24 years) was also included. Patient serum samples with optical density values higher than the 95th percentile for control subjects (0.5966) were considered positive.

The study of Anti-Endothelial Cell Antibodies (AECA) was performed using the Indirect Immunofluorescence (IIF) method on commercially available slides of Human Umbilical Vein Endothelial Cells (HUVEC; Euroimmun, Lübeck, Germany). Antibodies present in patients' sera were detected with a fluorescein-conjugated secondary antibody against human IgG (Euroimmun, Lübeck, Germany) and hnRNP-K was localized using a rabbit anti-human hnRNP-K antibody (AbCam, Cambridge, UK) with a fluorescein conjugated secondary antibody against rabbit IgG (Jackson Immuno Research Laboratories Inc, West Grove, PA). After washing with phosphate-buffered saline-Tween, slides were read on an epifluorescence microscope. Cutoff was set at 1:80 dilutions, at this dilution; all sera from 124 healthy individuals gave negative results. Positive samples were classified according to their IIF pattern.

Statistical analysis was performed with SPSS software version 18.0 (SPSS©, Chicago, IL). Qualitative variables were compared for statistically significant using the χ^2 test and Fisher's exact test. The Wilcoxon signed-rank test was used to evaluate the statistical significance between the levels of hnRNPK antibodies at basal time and after TBI. The p values below 0.05 were considered statistically significant.

Results

During the present study 32 TBI patients were enrolled. Only 19 patients aged 19-64 (28.48 ± 10.3) met the inclusion criteria. None of the patients were excluded due to cardiac arrest. Twelve patients were excluded due to extra cerebral injury (AIS > 2), and 1- year follow-up could not be done on one patient due to administrative issues. Data concerning age, sex, initial GCS, GOS, mechanism of injury and other clinical information were displayed in Table 1. All patients showed pathological CT findings (none within TCDB category I). Six patients underwent neurosurgery, 4 of them at the admission and two require neurosurgery while in the ICU. No decompression craniectomy (duraplasty plus bone removal) was found in the sample. After discharge from the ICU, patients were admitted to the hospital´s Neuro-Rehabilitation Unit. In this series we did not have any death and the final outcome was evaluated one year after TBI by GOS and was performed by a Neuro radiologist not involved in the study.

AECA auto antibodies were present in 2 patients at basal time and also during post - TBI evolution, while 5 more patients developed de novo auto antibodies after trauma (Table 2). Additionally, after the screening of hnRNPK by ELISA assay, we observed that only 3 (15.8%) out of 19 patients had antibodies at the time of TBI, nevertheless 10/19 (52.6%) developed the antibodies during follow-up, showing a significant increase of anti-hnRNPK antibodies after TBI (Table 3). A statistically significant difference for the AECA factor (*p = 0.1*) was not reached, although in the case of hnRNPK the differences were statistically significant despite the small number of patients included (*p = 0.01*). Furthermore, when we analyze the concentration of anti-hnRNPK at the time and after TBI, we detected a significant increase of IgG hnRNPK, that were higher after TBI than the antibody levels found at the time of trauma *p = 0.001*, (Figure 1).

From our study cohort, eleven patients developed permanent injuries (ranked by GOS) after TBI and also, in all the cases it shows the presence of AECA auto antibodies in the serum samples obtained after trauma, and only one patient with sequelae was negative for hnRNPK antibodies. We found a significant correlation between the appearance of AECA and hnRNPK auto antibodies and the development of permanent injuries (*p = 0.018* and *p = 0.040* respectively).

Discussion

To our knowledge, this study first demonstrates the

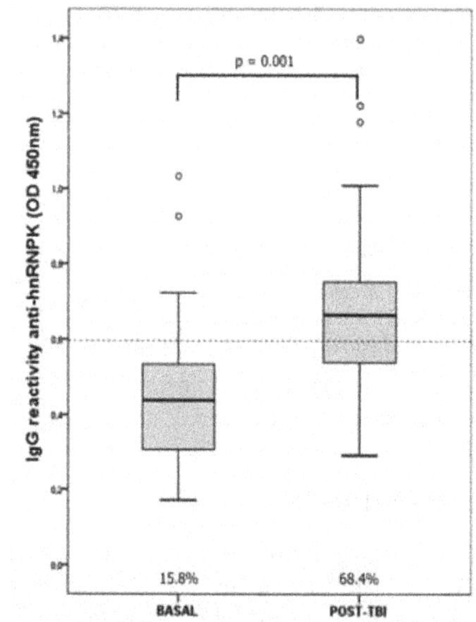

Figure 1: Immunoglobulin (IgG) titers against human recombinant hnRNP-K in TBI patients at the time of trauma (basal) and post-TBI. The box plots show the median and 25th-75th percentile range as well as the minimum and maximum values. Lines inside the boxes represent the mean. Circles indicate outliers. Values under the boxes are the percentage of patients with IgG titers above the cut-off value (0.5966), calculated as the 95th percentile from the control population.

Table 1: Demographic and clinical data for 19 patients with severe TBI.

Variable	n (%)
Age, years, median (range)	28 (19.0-64.0)
Sex n (%)	
Men	16 (84.2)
Women	3 (15.8)
Type of accident	
Traffic accident	12 (63.2)
Fall	12 (63.2)
Other	5 (26.3)
Hypoxia	
Absence	15 (78.9)
Presence	4 (21.1)
Hypotension	
Absence	15 (79.9)
Presence	4 (21.1)
GCS at admission	
3-4	2 (10.5)
5-6	3 (15.8)
7-8	14 (73.7)
Pupillary reaction	
Unilateral absence	4 (21.1)
Bilateral absence	1 (5.3)
Presence	14 (73.6)

CT-TCDB classification at admission	
II	5 (26.3)
III	9 (47.3)
IV	0 (0.0)
V	5 (26.3)
VI	0 (0.0)
Intracranial hypertension	
Yes	9 (47.4)
No	10 (52.6)
Neurosurgery	
Yes	6 (31.6)
No	13 (68.4)
Early sepsis	
Yes	12 (63.1)
No	7 (36.9)
GOS 1 year	
1	0 (0.0)
2	0 (0.0)
3	5 (26.3)
4	6 (31.5)
5	8 (42.1)
Sequelae	
Yes	11 (57.8)
No	8 (42.2)

Table 2: Evolution of AECA autoantibodies analyzed in basal and post-TBI sera.

Patient	Sex/Age (yr)	AECA (basal)	AECA (post - TBI)
1	M/32	-	-
2	M/32	-	-
3	F/28	-	-
4	M/64	-	+
5	M/19	-	-
6	M/26	-	-
7	M/24	-	+
8	M/31	+	+
9	M/20	-	-
10	F/21	-	-
11	M/27	-	+
12	M/34	-	-
13	M/24	-	-
14	M/24	-	-
15	M/27	+	+
16	M/25	-	-
17	F/22	-	-
18	M/40	-	+
19	M/27	-	+

Table 3: Evolution of hnRNPK autoantibodies analyzed in basal and post-TBI sera.

Patient	Sex/Age (yr)	hnRNPK (basal)	hnRNPK (post-TBI)
1	M/32	-	+
2	M/32	-	+
3	F/28	+	+
4	M/64	-	+
5	M/19	-	+
6	M/26	-	-
7	M/24	-	-
8	M/31	-	+
9	M/20	-	+
10	F/21	-	-
11	M/27	-	+
12	M/34	-	-
13	M/24	-	-
14	M/24	+	+
15	M/27	-	-
16	M/25	+	+
17	F/22	-	+
18	M/40	-	+
19	M/27	-	+

development of AECA and anti-hnRNPK IgG antibodies after TBI and its likely relationship with the outcome. It is well established that the persistent inflammation produced after TBI activate the immune response and culminating in the production of different auto antibodies against endothelial and central nervous system antigens. In fact, the presence of AECA after TBI could be explained by an ulterior exposition of antigens produced by endothelial lesions. Similarly, in human and experimental models, the alterations produced in the blood-brain barrier after TBI have been related to neuronal damage caused by the activation of immune system cells that also impairs neurological recovery after TBI [16]. On the other hand, there are evidences of cell-mediated immune response within the brain and the systemic circulation after TBI, were antibodies and B cells are also pivotal players [1]. More specifically, the pathological sequelae that accompanies to CNS trauma has characteristics of a self-directed immunological disease, and the production of specific auto antibodies such as gangliosides, phospholipids, beta-tubulin III , nuclear antigens or anti pituitary antibodies has been related with neurological/neuroendocrinological diseases [17-22]. Although some of these antibodies have been found in healthy individuals, several findings suggest a pathogenic role of AECA auto antibodies in autoimmune rheumatic diseases such as systemic lupus erythematosus, scleroderma and vasculitis. These antibodies are able to induce proinflammatory and procoagulant effects on endothelium (increased expression of adhesion molecules and tissue factors, increased release of cytokines). In addition, in systemic vasculitis such as Takayasu arteritis and antineutrophil cytoplasm antibody-positive vasculitis, AECA have been reported to correlate with disease activity [23]. These antibodies have been found after TBI in 15 out of 19 patients and only in five of them were preformed (Table 2). Thus, after trauma a specific immune response against endothelium cell tissue has been occurred. With respect to hnRNPK antibodies, the present study shows that 3 out of the 19 patients (15.8%) had antibodies against hnRNPK at the basal time sera, while after trauma these antibodies could be detected in serum sample from 13 of the patients (68.4%). Among the patients who were hnRNPK antibodies positive before the TBI, 2 suffered some type of trauma prior TBI (one suffered a serious stab wound and the other a motorcycle accident). These patients remained anti-hnRNPK positive antibodies during follow-up. Surprisingly, when patients were stratified by the presence/absence of any type of sequels, 100% and 91% of the patients with permanent injuries were AECA and hnRNPK antibodies positive respectively.

In the previous study we describe that antibodies against hnRNPK were associated with a vascular endothelial damage in heart transplanted patients [9]. Similarly, in the current study, TBI patients who developed autoantibodies after trauma are probably due to vascular endothelium damage. Accordingly, a significant association between the presence and, in particular, higher levels of anti-hnRNPK antibodies were found. These differences could be explained by the endothelial damage/activation that occurs after of trauma triggering immune mechanisms. Our results can be added to other published studies that demonstrate the importance of the definition of a good marker and/or the combination of different markers in outcome prediction after TBI [5].

A major limitation of this study is the small number of patients included. However, the study included a very homogeneous cohort, with clinical follow-up of at least one year. Furthermore, this study addresses a pathway which is not sufficiently explored, and it is the role of the endothelium in the genesis of some accompanying phenomena to the pathology of TBI.

Summary

Measuring anti-hnRNPK antibodies in patients who suffered a severe TBI could be a helpful biomarker for evaluating endothelial injury as well as the pathophysiological events accompanying these traumas.

More studies including larger populations are needed to clarify whether these antibodies can predict the occurrence of secondary damage produced by the trauma.

Acknowledgement

This work was supported by grants: PI10-2332 from Fondo de Investigaciones Sanitarias, Ministerio de Ciencia e Innovación of Spain and P07-CTS-02886 from Incentivos a Proyectos de Investigación de Excelencia, Consejería de Economía, Innovación y Ciencia, Junta de Andalucía.

Declaration of interest

None of the authors has a financial relationship with a commercial entity that has an interest in the subject of the submitted manuscript or other conflict of interest to disclose.

References

1. Ankeny DP, Popovich PG. B cells and autoantibodies: complex roles in CNS injury. Trends Immunol. 2010; 31(9): 332-338. doi: 10.1016/j.it.2010.06.006.
2. Ziebell JM, Morganti-Kossmann MC. Involvement of pro- and anti-inflammatory cytokines and chemokines in the pathophysiology of traumatic brain injury. Neurotherapeutics. 2010; 7(1): 22-30. doi: 10.1016/j.nurt.2009.10.016.
3. Gopcevic A, Mazul-Sunko B, Marout J, Sekulic A, Antoljak N, Siranovic M, et al. Plasma interleukin-8 as a potential predictor of mortality in adult patients with severe traumatic brain injury. Tohoku J Exp Med. 2007; 211(4): 387-393.
4. De Oliveira CO, Ikuta N, Regner A. Outcome biomarkers following severe traumatic brain injury. Rev bras ter Intensiva. 2008; 20(4): 411-421.
5. Murillo-Cabezas F, Muñoz-Sánchez MA, Rincón-Ferrari MD, Martín-Rodríguez JF, Amaya-Villar R, García-Gómez S, et al. The prognostic value of the temporal course of S100beta protein in post-acute severe brain injury: A prospective and observational study. Brain Injury. 2010; 24(4): 609-619. doi: 10.3109/02699051003652823.
6. Stein DM, Lindell AL, Murdock KR, Kufera JA, Menaker J, Bochicchio GV, et al. Use of Serum Biomarkers to Predict Cerebral Hypoxia after Severe Traumatic Brain Injury. J Neurotrauma. 2012; 29(6): 1140-1149. doi: 10.1089/neu.2011.2149.
7. Lange RT, Iverson GL, Brubacher JR. Clinical Utility of the Protein S100B to Evaluate Traumatic Brain Injury in the Presence of Acute Alcohol Intoxication. J Head Trauma Rehabil. 2012; 27(9): 123-134. doi: 10.1097/HTR.0b013e31820e6840.
8. Czeiter E, Mondello S, Kovacs N, Sandor J, Gabrielli A, Schmid K, et al. Brain injury biomarkers may improve the predictive power of the IMPACT outcome calculador. J Neurotrauma. 2012; 29(9): 1770-8. doi: 10.1089/neu.2011.2127.
9. Acevedo MJ, Caro-Oleas JL, Álvarez-Márquez AJ, Sobrino JM, Lage-Gallé E, Aguilera I, et al. Antibodies against heterogeneous nuclear ribonucleoprotein K in patients with cardiac allograft vasculopathy. J Heart Lung Transplant. 2011; 30(9): 1051-1059. doi: 10.1016/j.healun.2011.02.014.
10. Lynch M, Chen L, Ravitz MJ, Mehtani S, Korenblat K, Pazin MJ, et al. hnRNPK binds a core polypyrimidine element in the eukaryotic translation initiation factor 4E (eIF4E) promoter, and its regulation of eIF4E contributes to neoplastic transformation. Mol Cell Biol. 2005; 25(15): 6436-6453.
11. Carpenter B, McKay M, Dundas SR, Lawrie LC, Telfer C, Murray GI, et al. Heterogeneous nuclear ribonucleoprotein K is over expressed, aberrantly localised and is associated with poor prognosis in colorectal cancer. Br J Cancer. 2006; 95(7): 921-927.
12. Klimek-Tomczak K, Mikula M, Dzwonek A, Paziewska A, Karczmarski J, Hennig E, et al. Editing of hnRNPK protein mRNA in colorectal adenocarcinoma and surrounding mucosa. Br J Cancer. 2006; 94(4): 586-592.
13. Zhou R, Shanas R, Nelson MA, Bhattacharyya A, Shi J. Increased expression of the heterogeneous nuclear ribonucleoprotein K in pancreatic cancer and its association with the mutant p53. Int J Cancer. 2010; 126: 395-404. doi: 10.1002/ijc.24744.

14. Barboro P, Repaci E, Rubagotti A, Salvi S, Boccardo S, Spina B, et al. Heterogeneous nuclear ribonucleoprotein K: altered pattern of expression associated with diagnosis and prognosis of prostate cancer. Br J Cancer. 2009; 100(10): 1608-1616. doi: 10.1038/sj.bjc.6605057.

15. Laury-Kleintop LD, Tresini M, Hammond O. Compartmentalization of hnRNP-K during cell cycle progression and its interaction with calponin in the cytoplasm. J Cell Biochem. 2005; 95(5): 1042-1056.

16. Lotocki G, de Rivero Vaccari JP, Pérez ER, Sánchez-Molano J, Furones-Alonso O, Bramlett HM, et al. Alterations in blood-brain barrier permeability to large and small molecules and leukocyte accumulation after traumatic brain injury: effects of post-traumatic hypothermia. J Neurotrauma. 2009; 26(7): 1123-1134. doi: 10.1089/neu.2008.0802.

17. Zhang Y, Popovich P. Roles of autoantibodies in central nervous system injury. Discov Med. 2011; 11(60): 395-402.

18. Davies AL, Hayes KC, Dekaban GA. Clinical correlates of elevated serum concentrations of cytokines and autoantibodies in patients with spinal cord injury. Arch Phys Med Rehabil. 2007; 88(11): 1384-1393.

19. Hayes KC, Hull TC, Delaney GA, Potter PJ, Sequeira KA, Campbell K, et al. Elevated serum titers of proinflammatory cytokines and CNS autoantibodies in patients with chronic spinal cord injury. J Neurotrauma. 2002; 19(6): 753-761.

20. Skoda D, Kranda K, Bojar M, Glosová L, Bäurle J, Kenney J, et al. Antibody formation against beta-tubulin class III in response to brain trauma. Brain Res Bull. 2006; 68(4): 213-216.

21. López-Escribano H, Miñambres E, Labrador M, Bartolomé MJ, López-Hoyos M. Induction of cell death by sera from patients with acute brain injury as a mechanism of production of autoantibodies. Arthritis Rheum. 2002; 46(2): 3290-3300.

22. Tanriverdi F, De Bellis A, Bizzarro A, Sinisi AA, Bellastella G, Pane E, et al. Antipituitary antibodies after traumatic brain injury: is head trauma-induced pituitary dysfunction associated with autoimmunity? Eur J Endocrinol. 2008; 159(1): 7-13. doi: 10.1530/EJE-08-0050.

23. Guilpain P, Mouthon L. Antiendothelial cells autoantibodies in vasculitis-associated systemic diseases. Clin Rev Allergy Immunol. 2008; 35(1,2): 59-65. doi: 10.1007/s12016-007-8069-3.

Sensitivity of Cytokine and Cytokine Mediator Detection aiding in Diagnosis of Premature Coronary Artery Disease Patients

Wafa Munir Ansari[1], Dilshad Ahmad Khan[1], Omer Jamshed Khan[2] and Farooq Ahmad Khan[1]

[1]*Department of Chemical Pathology, Army Medical College, National University of Science and Technology (NUST), Islamabad, Pakistan*

[2]*Armed Forces Institute of Rehabilitation Medicine/Cardiac Rehab, Rawalpindi, Pakistan*

***Corresponding author:** Wafa Munir Ansari, Department of Chemical Pathology, Army Medical College, National University of Science and technology (NUST), Islamabad, Pakistan; E-mail: wafamuniransari@yahoo.com*

Abstract

Introduction: The objective was to evaluate diagnostic accuracy of high sensitivity C-reactive protein (hS CRP), Interleukin-18 (IL-18), Tumor Necrosis Factor-alpha (TNF-alpha), Interleukin-10 (IL- 10), IL-18/IL- 10 ratio and their serum cut off values for identification of Premature Coronary Artery Disease (PCAD) patients.

Method: It was a diagnostic validation case-control study carried out at the Clinical Pathology Laboratories of the Army Medical College, Rawalpindi, Pakistan, from May, 2013 to Jan, 2014. Two hundred and fifty subjects aged < 45 years scheduled to undergo coronary angiography were consecutively screened. Out of these ninety-eight patients with > 70% stenosis, in at least one coronary vessel on angiography, were labeled to have PCAD. An equal number of angio-negative subjects were taken as controls from the study group. Serum IL- 10, IL- 18 and TNF alpha were measured using Enzyme Linked Immuno-sorbent Assay (ELISA), hS CRP on Immulite 1000 and serum cholesterol, Triglycerides and High Density Lipoprotein (HDL) by colorimetric methods. Statistical analysis was done using SPSS-17 and MedCalc software.

Results: Total 196 subjects consisting of 98 PCAD patients and 98 angio-negative controls participated in the study. Mean ± SD age of PCAD patients was 40.7 ± 4.23yrs (89 males and 9 females) while in controls it was 35.1 ± 7.55yrs (93 males and 5 females). Serum hS CRP had highest area under curve AUC (95% confidence interval) of 0.936 (0.89- 0.97) while AUC (95% CI) of IL-18, TNF alpha and IL-10 were 0.853 (0.79-0.90), 0.731 (0.659-0.796) and 0.574 (0.497-0.649) respectively. Sensitivity-specificity of hS CRP and IL-18 at cut off value of 3.18 pg/ml and 200pg/ml were 86%-91% and 77%-81% respectively while that of IL-18/IL-10 ratio and IL-10 at cut off values > 138.9 and > 0.3pg/ml were 68-64% and 20-100% respectively. Significant correlation was observed between hS CRP ($p < 0.01$), IL-18 ($p < 0.01$) and TNF alpha ($p < 0.05$).

Conclusion: hS CRP and IL-18 have the best diagnostic potential among the cytokines for detection of PCAD with high sensitivity and specificity. IL-18/IL-10 ratio had moderate significance in the diagnosis of PCAD although it correlated well with the disease burden. IL-10 had 100% specificity highlighting its role in the diagnosis of the disease but its low sensitivity is a limiting factor. Thus, hS CRP and IL-18 are promising potential biochemical parameters aiding in the clinical diagnosis of Premature Coronary Artery Disease.

Keywords: Diagnosis; Heart disease; hS CRP; Interleukin- 18; Interleukin- 10; TNF alpha

Abbreviations

TNF alpha: Tumor Necrosis Factor-alpha; IL-18: Interleukin-18; IL-10: Interleukin-10; hS CRP: High Sensitivity C- Reactive Protein; PCAD: Premature Coronary Artery Disease; AUC: Area under Curve; ROC: Receiver Operator Curves; CV: Coefficient of Variation

Introduction

Premature Coronary Artery Disease (PCAD) is one of the earliest manifestations of coronary atherosclerosis. Effective management of PCAD therefore requires timely detection and accurate evaluation of these patients. Inflammation is proposed to be the main event in the pathogenesis of atherosclerotic plaque formation and progression in CAD [1]. The inflammatory cascade has counter balancing factors that maintain a delicate balance of pro- and anti-inflammatory molecules that regulate vascular homeostasis and maintain integrity of the vessel wall [2]. Multiple pro-inflammatory and anti-inflammatory cytokines are involved in the pathogenesis of CAD having overlapping, antagonistic and synergistic effects. Interleukin-18 (IL-18), tumor necrosis factor-alpha (TNF alpha), High-sensitivity C-reactive protein (hS CRP) and interleukin-10 (IL-10) are being considered to play an essential role in the modulation of immunological and inflammatory processes during various stages of premature coronary artery disease.

IL-18 induces the expression of pro-inflammatory cytokines and Chemokines such as Interleukin-6 (IL-6), IL-8 and Monocyte Chemo attractant Protein-1 (MCP-1) [3]. TNF alpha is also a pro-inflammatory cytokine which is up regulated in ischemia inducing activation of xanthine oxidase and production of $O_2^{\bullet-}$, leading to dysfunctional coronary endothelium [4]. *Samnegard et al.* [5] demonstrated that the systemic concentrations of IL-18 and TNF alpha were higher in post myocardial patients as compared to the controls. Moreover, individuals with elevated levels of TNF alpha were at increased risk for acute myocardial infarction and CAD [6]. High sensitive C-reactive protein (hS CRP) is an established pro-inflammatory biomarker for the detection of individuals at a risk of coronary artery disease [7]. Several studies have also shown that increased levels of fibrinogen, CRP and IL-6 are associated not only with the risk of coronary artery disease but also with its clinical course, progression and severity [8]. Interleukin-10 (IL-10) on the other hand is an anti-inflammatory cytokine which is associated with a humoral immune response that acts by limiting the local inflammatory response which provides stability to the atherosclerotic lesion [9]. We sought to assess the role of cytokine mediators IL-18, TNF alpha, hS CRP, IL-10, IL-18/IL-10 ratio and TNF alpha/IL-10 ratio in the diagnosis of PCAD patients.

Methodology

The diagnostic validation case-control study was conducted at the Clinical Pathology Laboratory (CPL), Army Medical College, Rawalpindi, in collaboration with the National Institute of Heart Diseases (NIHD), Rawalpindi, Pakistan. Ethical approval was duly sought from the Institutional Review Committee of Army Medical College (AM College) under the auspices of the National University of Science and Technology (NUST) Islamabad, Pakistan. Duration of the study was 8 months.

Subjects

Two hundred and fifty subjects aged < 45 years scheduled to undergo coronary angiography were consecutively screened. Out of these 196 subjects who gave consent and fulfilled our inclusion criteria participated in the study. These included ninety-eight patients with > 70% stenosis in at least one coronary vessel on angiography which is the gold standard, were labeled to have PCAD. Those who were angiographically proven to be disease free were labeled as controls (n = 98) after informed consent. Only those patients who had > 70% stenosis in at least 1 coronary vessel on coronary angiography which is the current gold standard were taken as PCAD patients. Patients with a history of hematological, neoplastic, renal, liver or thyroid disease were excluded. Furthermore, patients with infectious or autoimmune diseases, familial hyperlipidemia, congenital heart disease, hepatitis, valvular heart disease, asthma, rheumatoid arthritis or life expectance of less than 12 months and those unable to give informed consent were also excluded. In the controls subjects with any acute or chronic illness or those on anti- inflammatory drugs were excluded. Pregnant women and women on contraceptives were also excluded from the study. Demographic characteristics including cardiovascular risk factors such as hypertension; diabetes mellitus, dyslipidaemia and smoking were noted. Medical examination by a general physician was conducted.Written informed consent was obtained from subjects.

Biochemical Analysis

All laboratory investigations were carried out in the Clinical Pathology Laboratories of AM College, NUST Rawalpindi, Pakistan. Blood samples were taken in the morning on the day of angiography of the respective patient. 5 ml blood sample was obtained by venipuncture and transferred to a plain vacutainer tube for serum analysis. Serum was separated by centrifugation at 1500x g for 15 minutes and stored at -70°C until biochemical analysis.

Enzyme Linked Immuno-sorbent Assay (ELISA) technique was used for measuring the concentrations of serum IL-10, IL-18, and TNF alpha using human IL-10, IL-18, and TNF alpha, (Bender med Systems, Austria) commercial kits with monoclonal antibodies. The calculated overall intra assay Coefficient of Variation (CV) for IL-18, TNF alpha and IL-10 was 6.5%, 6.0% and 3.2% respectively while the limit of detection was 9pg/ml, 2.3pg/ml and 0.1pg/ml respectively. The inter assay CV was 3.4%, 3.8% and 3.0% for IL-18, TNF alpha and IL-10 respectively. Analysis of hS CRP was done by a chemiluminescent immunoassay on Immulite 1000 using kit provided by Siemens (UK). The inter-assay coefficient of variation (CV) was 3.5%. Serum cholesterol was measured by cholesterol oxidase method (CHOD-POD) and serum triglyceride was measured by GPO-POD Colorimetric method. All the analytes were run on Selectra E (Vital Scientific, Netherland) using kits provided by Pioneer Diagnostics (USA). CV of the method was < 1%.

Coronary angiography was performed by trained cardiologists by Jut kin technique using a quantitative coronary angiographic system. Degree of atherosclerosis was calculated using the Gensini score [10]. All coronary angiograms were evaluated by the cardiologists who are unaware of the biochemical analysis results to avoid bias.

Statistical Analysis

Statistical analysis was performed using SPSS- 17 (SPSS Inc, Chicago) and MedCalc software version 9.6.4.0. Kolmogorov-Smirnov test was applied on data which revealed non-Gaussian distribution for cytokines ($P < 0.05$). Mean, SD, median was calculated for descriptive statistics. Median and Interquartile Range (IQR) were calculated for cytokine levels and ratios. Mann-Whitney U test was applied for comparison of PCAD patients and controls. Receiver operating characteristic curves (ROC) was constructed using MedCalc software in order to evaluate the diagnostic values including sensitivity, specificity, likelihood ratios and diagnostic odds ratios of cytokine mediators. Spearman Correlation test was applied to see the association between hS CRP and IL-18 and TNF alpha. A p-value of < 0.05 was considered to be significant.

Results

A total of 196 subjects participated in this study comprising of 98 cases and 98 controls. The mean age of the cases and controls was 40.7 ± 4.23 and 35.1 ± 7.55 respectively. Baseline characteristic data of the subjects is shown in Table 1. The PCAD patients were more frequently hypertensive, diabetic, and hyperlipidemia with a positive family history of PCAD.

ROC analysis of biomarkers in PCAD patients and controls revealed that Area under curve (AUC) and 95% (CI) of hS CRP 0.936 (0.89-0.97) was the highest closely followed by IL-18: 0.853 (0.79-0.90). Area under curve (AUC) and 95% (CI) of TNF alpha, IL-18/IL-10 ratio and IL-10 were 0.731 (0.659-0.796), 0.667 (0.592-0.737) and 0.574 (0.497-0.649) respectively [Figure 1]. The diagnostic odds ratio of hS CRP was highest at a cutoff level of 3.18 mg/L. The sensitivity and specificity of hS CRP at this cutoff for the diagnosis of PCAD was 86% and 91% respectively.

Table 1: Demographic Characteristics of Patients and Controls.

Parameters	Cases (n = 98)	Controls (n = 98)	*p- value*
Age (years)	40 ± 4.23	35 ± 7.55	0.76
Gender(M/F)	89 / 9	93 / 5	0.63
Weight (Kg)	74.1 ± 11.7**	67.7 ± 10.1	0.0009
Height (m)	1.68 ± 0.06	1.69 ± 0.08	0.85
BMI (Kg/ m^2)	26.32 ± 3.7**	23.6 ± 3.5	< 0.002
Positive Diabetes n (%)	36 (37)**	3(4)	< 0.01
Positive Premature CAD Family History n (%)	43(44)**	2 (3)	< 0.01
Positive DM family history n (%)	36(37)**	11(15)	< 0.01
Smoker's n (%)	60 (61)*	38 (39)	< 0.05
Total Cholesterol (mmol/ l)	4.9 ± 3.77	4.5 ± 1.38	0.56

PCAD: Premature Coronary Artery Disease; BMI: Body Mass Index; CAD: Coronary Artery Disease: DM: Diabetes Mellitus; SD=Standard Deviation Categorical variables were compared using x^2 test while continuous variables were compared using Welch's t-tests. *$*p < 0.05$; $**p < 0.01$.*

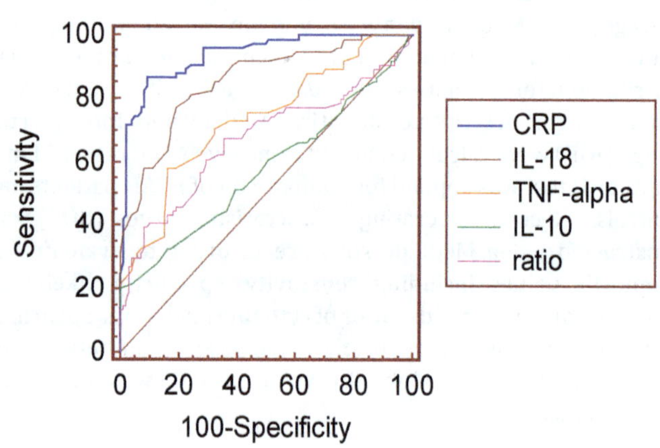

Figure 1: Receiver-Operator Characteristic (ROC) curve analyses of Interleukin-18 (IL- 18), Interleukin-10 (IL- 10), Tumor Necrosis Factor-alpha (TNF alpha), IL- 18/ IL- 10 ratio and hS C-reactive protein (hS CRP) for diagnosis of premature coronary artery disease (PCAD) (n = 98).

The maximum sensitivity and specificity of IL-18, TNF alpha and IL-10 were seen at levels of 200 pg/ml, 4.6 pg/ml and 0.3 pg/ml respectively, their DOR's were however lower than that of hS CRP as shown in [Table 2]. Significant correlation was observed between hS CRP and IL-18 (r = 0.436, *$p < 0.01$*) and hS CRP and TNF alpha (r = 0.287, $p < 0.05$). The serum cytokine levels were significantly higher in cases as compared to the controls [Supplementary Table 1]. On performing logistic regression analysis after adjusting for the confounding variables IL-18, TNF-alpha and hS-CRP remained highly significant *($p < 0.01$)* while IL-10 was moderately significant *($p < 0.05$)* [Supplementary Table 2].

Discussion

The results of our study show that hS CRP have the best diagnostic ability out of all the cytokine mediators for the detection of PCAD. Patients with CAD had higher CRP levels than those without CAD and correlated with the severity of coronary atherosclerosis [11]. CRP is being quoted as the marker of disease activity as well as disease severity in cases of CAD [12]. This is because in the early stages of atherosclerosis it correlates extremely well with the rate of progression of atherosclerosis. hS CRP is the most predictive inflammatory marker for coronary artery disease [7]. Role of hS CRP in the risk stratification of CAD has also been shown by *Peer et al.* [13] more recently, *Koc et al.* [14] demonstrated that the levels of hS CRP were elevated in CAD when compared to controls regardless of the time of sampling and also had good diagnostic power for detection of stable CAD. Our study highlights the diagnostic performance of hS CRP in identifying premature CAD patients. *Elzahwy et al.* [15] demonstrated a strong correlation between hS CRP and coronary intima-media thickness which is an independent predictor of multi-level atherosclerosis and has high sensitivity and specificity. However, *Hung et al.* [16] state that CRP provides only modest predictive value for future CAD Inflammatory biomarkers have immense potential in detection of coronary disease [17]. The importance of hS CRP in the diagnosis of PCAD emphasizes the need to determine its optimal cutoff value with the best sensitivity and specificity. The definitive cutoff values have still not been determined as the cut off value in our study was slightly higher and with much greater sensitivity and specificity as compared to a previous study [18].

Similarly, we observed that the sensitivity and specificity of IL-18 was also very high for the diagnosis of PCAD. Positive correlation has been shown between IL-18 and carotid intima-media thickness and coronary plaque area [19]. This is probably because IL-18 over expression induces atheroma formation while the endogenous inhibitor of IL-18 slows plaque development [20]. *Li et al.* [21] demonstrated that IL-18 had AUC of 0.86 for detecting and predicting thin cap fibro-atheromas as confirmed on intravascular optical coherence tomography. On the other hand *Martins et al.* [22] found only marginal significance of IL-18 for prediction of CAD with or without adverse events.

TNF alpha and IL-18/IL-10 ratio showed moderate diagnostic performance for the diagnosis of PCAD. According to *Branen et al.* [23] TNF alpha is actively involved in the progression of atherosclerosis and its inhibition reduces atherosclerosis in mice. TNF alpha is being considered to be a sensitive and specific biomarker in the early diagnosis of metabolic syndrome which increases the risk to develop PCAD [24]. TNF alpha levels have shown significant association with the atherosclerotic burden consistently [25]. *Chalikias et al.* [26] demonstrated that an imbalance between the pro-and anti-inflammatory forces leads to plaque disruption and recurrent cardiovascular accidents making IL-18/IL-10 an independent predictor of adverse events

Table 2: Diagnostic performance of hS CRP, IL-18, TNF alpha, IL-10 and IL-18/IL-10 ratio at different cut offs for diagnosis of PCAD.

Biomarkers	SN (%)	95%CI	SP (%)	95%CI	LR+	LR-	DOR
hS- CRP (mg/ L) ≥ 3.14	85.14	75-92.3	90.82	83.3-95.7	9.27	0.16	58
≥ 3.18*	86.49	76.5-93.3	90.82	83.3-95.7	9.42	0.15	62
≥ 3.2	86.49	76.5-93.3	89.80	82.0-95.0	8.48	0.15	57
IL-18 (pg/ ml) ≥ 195	70.27	58.5-80.3	82.65	73.7-89.6	4.05	0.36	11
≥ 200*	77.03	65.8-86.0	80.61	71.4-87.9	3.97	0.28	14
≥ 210	78.38	67.3-87.1	78.57	69.1-86.2	3.66	0.28	13
TNF alpha(pg/ ml) ≥ 4.5	56.76	44.7-68.2	83.67	74.8-90.4	3.48	0.52	6.7
≥ 4.6*	58.11	46.1-69.5	83.67	74.8-90.4	3.56	0.50	7.1
≥4.8	58.11	46.1-69.5	80.61	71.4-87.9	3.00	0.52	5.8
IL-18/IL-10 ratio ≥135.7	66.22	54.3-76.8	64.29	54.0-73.7	1.85	0.53	3.5
≥138.9*	67.57	55.7-78.0	64.29	54.0-73.7	1.89	0.50	3.8
≥140	67.57	55.7-78.0	63.27	52.9-72.8	1.84	0.51	3.6
IL-10 (pg/ml) ≥0.25	13.51	6.7-23.5	100.0	96.3-100	5.5	0.86	6.4
≥0.3*	20.27	11.8-31.2	100.0	96.3-100	5.35	0.80	6.7
≥0.4	21.62	12.9-32.7	95.92	89.9-98.9	5.3	0.82	6.5

hS- CRP: high sensitive C reactive protein, IL-18: Interleukin-18; IL-10: Interleukin-10; TNF alpha: Tumor Necrosis Factor-alpha; SN: Sensitivity; SP: Specificity; LR+: Positive likelihood ratio; LR-: Negative likelihood ratio; DOR: Diagnostic Odds Ratio; CI: Confidence Interval

in hospitalized coronary syndrome patients. The reason for their moderate diagnostic efficacy in our study is the possibility that acute cardiovascular event in our study population occurred over 2 months before the day of their angiography. So as the protective response comes into play after the acute event in the form of rise in IL-10 the IL-18/IL-10 ratio starts decreasing and TNF alpha degrades due to its short half life. IL-10 turned out to be highly specific at 0.3 pg/ml and may be used for correctly identifying the subjects not having PCAD. It can be employed as a rule out biochemical marker.

Significant positive correlation was observed between hS CRP and IL-18 highlighting the role and probable synergism of pro-inflammatory cytokine mediators in the diagnostic evaluation and pathogenesis of PCAD. *Yamaoka-Tojo et al.* [26] suggests that CRP induces the production of IL-18 further enhancing the pro-inflammatory component of the cytokine cascade. Moderate correlation was observed between hS CRP and TNF alpha also. This is in agreement with certain studies which suggest close correlation between hS CRP and TNF alpha [27]. One of the major strengths of our study is that we have independently established the reference interval of novel cytokine mediators in our study and have included angio-negative individuals as healthy controls thus allowing for a better comparison between patients and controls. It is also the first kind of its study to the best of our knowledge which has studied the diagnostic accuracy of pro- and anti-inflammatory cytokines in Premature Coronary artery disease patients who are young (< 45 years) and whose number is increasing alarmingly in Pakistan and South Asia. Moreover, we have seen that while IL-18 and hS CRP have high diagnostic performance for diagnosis of CAD and may be used in combination with clinical assessment for better therapeutic management. The major limitation of our study is its small sample size and the fact that we have included stable PCAD patients. We therefore recommend that future studies be carried out in multiple centers on larger patient populations and should also include acute coronary syndrome cases of PCAD for a better assessment of the diagnostic efficacy of cytokines in PCAD patients.

Conclusion

The study demonstrated that serum hS-CRP and IL-18 had more than 85% and 70% sensitivity respectively among the cytokines mediators for diagnosis of PCAD in the high risk group of patients. IL-18/IL-10 ratio revealed moderate diagnostic significance for PCAD. IL-10 had 100% specificity highlighting its role in the exclusion of PCAD diagnosis in the cardiac clinic setup. These serum cytokines correlated well with the atherosclerosis burden. Thus, these potential cytokines can be used for triage of patients reported with chest pain and avoid the un-necessary costly angiography procedures in the healthy individual.

Acknowledgement

We wish to acknowledge the efforts of all the laboratory technicians who participated in the study.

References

1. Hansson GK, Robertson AK, Soderberg-Naucler C. Inflammation and atherosclerosis. Annu Rev Pathol. 2006; 1: 297-329.
2. Mills R, Bhatt DL. The Yin and Yang of arterial inflammation. J Am Coll Cardiol. 2004; 44(1): 50-52.
3. Sahar S, Dwarakanath RS, Reddy MA, Lanting L, Todorov I, Natarajan R. Angiotensin II Enhances Interleukin-18 Mediated Inflammatory Gene Expression In Vascular Smooth Muscle Cells -A Novel Cross-Talk in the Pathogenesis of Atherosclerosis. Circ Res. 2005; 96(10): 1064-1071.
4. Zhang C, Xu X, Potter BJ, Wang W, Kuo L, Michael L, et al. TNF alpha contributes to endothelial dysfunction in ischemia/reperfusion injury. Arterioscler Thromb VascBiol. 2006; 26(3): 475-480.

5. Samnegård A, Hulthe J, Silveira A, Ericsson CG, Hamsten A, Eriksson P. Gender specific associations between matrix metalloproteinases and inflammatory markers in post myocardial infarction patients. Atherosclerosis. 2009; 202: 550-556. doi: 10.1016/j.atherosclerosis.2008.05.050.

6. Biswas S, Ghoshal PK, Mandal SC, Mandal N. Relation of anti- to Pro-inflammatory cytokine ratios with acute myocardial infarction. Korean J Intern Med. 2010; 25(1): 44-50. doi: 10.3904/kjim.2010.25.1.44.

7. Pearson TA, Mensah GA, Alexander RW, Anderson JL, Cannon RO 3rd, Criqui M, et al. Markers of inflammation and cardiovascular disease: application to clinical and public health practice: a statement for healthcare professionals from the Centers for Disease Control and Prevention and the American Heart Association. Circulation. 2003; 107(3): 499-511.

8. Sukhija R, Fahdi I, Garza L, Fink L, Scott M, Aude W, et al. Inflammatory markers, angiographic severity of coronary artery disease, and patient outcome. Am J Cardiol. 2007; 99(7): 879-884.

9. Pérez Fernández R, Kaski JC. [Interleukin-10 and Coronary Disease]. Rev Esp Cardiol. 2002; 55(7): 738-750.

10. Gensini GG. A more meaningful scoring system for determining the severity of coronary heart disease. Am J Cardiol 1983; 51(3): 606-607.

11. Momiyama Y, Ohmori R, Fayad ZA, Kihara T, Tanaka N, Kato R, et al. Associations between plasma C-reactive protein levels and the severity of coronary and aortic atherosclerosis. J Atheroscler Thromb. 2010; 17(5): 460-467.

12. Arroyo-Espliguero R, Avanzas P, Cosín-Sales J, Aldama G, Pizzi C, Kaski JC. C- protein elevation and disease activity in patients with coronary artery disease. Eur Heart J. 2004; 25(5): 401-408.

13. Peer A, Falkensammer G, Alber H, Kroiss A, Griesmacher A, Ulmer H, et al. Limited utilities of N-terminal pro B-type natriuretic peptide and other newer risk markers compared with traditional risk factors for prediction of significant angiographic lesions in stable coronary artery disease. Heart. 2004; 95(4): 297-303. doi: 10.1136/hrt.2008.145425.

14. Koc M, Karaarsalan O, Abali G, Batur MK. Variation in high sensitivity C reactive protein levels over 24 hours in patients with stable coronary artery disease. Tex Heart Inst J. 2010; 37(1): 42-48.

15. Elzahwy SS, Ewida II, Ismaeel MG, Hakeem SK. Value of carotid intima-media thickness in estimating the spread of atherosclerosis and its relation to C-Reactive Protein. Heart Mirror J.2010; 4(1): 80-85.

16. Hung J, Knuiman MW, Divitini ML, Langton PE, Chapman CL, Beilby J. C-reactive protein and interleukin-18 levels in relation to coronary heart disease: prospective cohort study from Busselton Western Australia. Heart Lung Circ. 2008; 17(2): 90-95.

17. Yayan J. Emerging families of biomarkers for coronary artery disease: inflammatory mediators. Vasc Health Risk Manag. 2013; 9: 435-456. doi: 10.2147/VHRM.S45704.

18. Yilmaz N, Cicek HK, Celik A, Meram I, Kocabas R, Davutoglu V. Diagnostic value of homocysteine, C- reactive protein and bilirubin for coronary artery disease. Eastern Mediterr Health Journal. 2007; 13(3): 522-535.

19. Li C, Zhang XL, Liu H, Wang ZG, Yin KS. Association among plasma interleukin-18 levels, carotid intima- media thickness and severity of obstructive sleep apnea. Chin Medical J. 2009; 122(1): 24-29.

20. De Nooijer R, Von der Thusen JH, Verkleij CJN, Kuiper J, Jukema JW, van der Wall EE, et al. Over expression of IL-18 decreases intimal collagen content and promotes a vulnerable plaque phenotype in apolipoprotein-E-deficient mice. Arterioscler Thromb Vasc Biol. 2004; 24(12): 2313-2319.

21. Li QX, Fu QQ, Shi SW, Wang YF, XieJJ, Yu X, et al. Relationship between plasma inflammatory markers and plaque fibrous cap thickness determined by intravascular optical coherence tomography. Heart. 2010; 96: 196-201. doi: 10.1136/hrt.2009.175455.

22. Martins TB, Anderson JL, Muhlestein JB, Horne BD, Carlquist JF, Roberts WL, et al. Risk Factor Analysis of Plasma Cytokines in Patients with coronary artery disease by a multiplexed fluorescent immunoassay. Am J Clin Pathol. 2006; 125(6): 906-913.

23. Branen L, Hovgaard L, Nitulescu M, Bengtsson E, Nilsson J, Jovinge S. Inhibition of tumor necrosis factor-α reduces atherosclerosis in apolipoprotein E knockout mice. Arterioscler Thromb Vasc Biol. 2004; 24 (11): 2137-2142.

24. Gurrola-Díaz CM, Sánchez-Enriquez S, Oregon-Romero E, García-López PM, Garzón de la Mora P, Bastidas-Ramírez BE, et al. Establishment of a cut-point value of serum TNF-α levels in the metabolic syndrome. J Clin Lab Anal. 2009; 23(1): 1-69. doi: 10.1002/jcla.20289.

25. Battes LC, Cheng JM, Oemrawsingh RM, Boersma E, Garcia-Garcia HM, de Boer SP, et al. Circulating cytokines in relation to the extent and composition of coronary atherosclerosis: results from the ATHEROREMO-IVUS study. Atherosclerosis. 2014; 236(1): 18-24. doi: 10.1016/j.atherosclerosis.2014.06.010.

26. Chalikias GK, Tziakas DN, Kaski JC, Hatzinikolaou EI, Stakos DA, Tentes IK, et al. Interleukin-18: interleukin-10 ratio and in-hospital adverse events in patients with acute coronary syndrome. Atherosclerosis. 2005; 182(1): 135-143.

27. Yamaoka-Tojo M, Tojo T, Masuda T, Machida Y, Kitano Y, Kurosawa T, et al. C-reactive protein-induced production of interleukin-18 in human endothelial cells: a mechanism of orchestrating cytokine cascade in acute coronary syndrome. Heart Vessels. 2003; 18(4): 183-187.

Supplementary Table 1: Comparison of serum cytokine levels in Premature Coronary Artery Disease Patients and Controls.

Parameters	B	SE	Exp(B) (95%CI)	Sig.
IL- 18 (pg/ ml)	0.020	0.003	1.020 (1.004- 1.019)**	0.005
IL- 10 (pg/ ml)	0.310	0.182	1.062 (0.842- 2.037)*	0.04
TNF- alpha (pg/ ml)	0.609	0.174	1.723 (1.274- 2.523)**	0.001
hS- CRP (mg/ dl)	0.450	0.124	1.561(1.253-2 .145)**	0.000

IL: Interleukin; TNF-alpha: tumor necrosis factor-alpha; hS-CRP: High-sensitivity C - reactive protein; IQR: Interquartile range
$*p < 0.05$; $**p < 0.01$ applying Mann-Whitney *U* test.

Supplementary Table 2: Multivariate Logistic Regression Analysis for Prediction of Premature Coronary artery disease.

Parameters	B	SE	Exp(B) (95%CI)	Sig.
IL- 18 (pg/ ml)	0.020	0.003	1.020(1.004 - 1.019)**	0.005
IL- 10 (pg/ ml)	0.310	0.182	1.062(0.842 - 2.037)*	0.04
TNF- alpha (pg/ ml)	0.609	0.174	1.723(1.274 - 2.523)**	0.001
hS- CRP (mg/ dl)	0.450	0.124	1.561(1.253 - 2.145)**	0.000

Logistic regression, adjusted for age, sex, smoking, BMI and diabetes, was performed for each group. SE: standard error; Exp (B): odds ratio; CI: Confidence interval; Sig: significance.
$**p < 0.01$; $*p < 0.05$

CD36 in Atherosclerotic Coronary Artery Disease: Correlation with Angiogenesis and Inflammation

Enas A. El-Hussieny[1], Amira M. Gamal Eldeen[2,3]*, Roba M. Talaat[4], Ahmed F. Tamara[5], Mohamed S. Salama[6] and Nadia M. El-Beih[1]

[1]*Department of Zoology, Faculty of Science, Ain Shams University, Cairo, Egypt*

[2]*Department of Cancer Biology Laboratory, Center of Excellence for Advanced Sciences, National Research Center (NRC), Dokki, 12622, Cairo, Egypt*

[3]*Department of Biochemistry, National Research Center, Dokki, 12622, Cairo, Egypt*

[4]*Department of Molecular Biology, Genetic Engineering and Biotechnology Research Institute (GEBRI), University of Sadat City, Egypt*

[5]*Department of Cardiology, Faculty of Medicine, Ain Shams University, Cairo, Egypt*

[6]*Department of Entomology, Faculty of Science, Ain Shams University, Cairo, Egypt*

***Corresponding author:** *Roba M. Talaat, Department of Molecular Biology, Genetic Engineering and Biotechnology Research Institute (GEBRI), University of Sadat City, Egypt, E- mail: robamtalaat@yahoo.com*

Abstract

Background: CD36 is one of the macrophage scavenger receptor that has been implicated as a key player in the pathogenesis of atherosclerosis.

Aim: This study focused on the role of CD36 in coronary artery disease (CAD) and its association with proangiogenic and inflammatory mediators.

Patients and Methods: We studied 100 atherosclerotic CAD patients and 100 healthy controls. All patients had angiographic evidence of stenosis ≥ 50% in at least one major coronary artery Soluble CD36 (sCD36), peripheral blood mononuclear cells (PBMC-cCD36), vascular endothelial and platelet- derived growth factors (VEGF & PDGF), Tumor necrosis factor (TNF-α) and cyclooxygenase-2 (COX-2) were measured using enzyme-linked immunosorbent assay (ELISA). CD36 mRNA expression was measured using reverse transcriptase polymerase chain reaction (RT-PCR).

Results: A significant reduction in cCD36 expression (protein and mRNA) and plasma levels of sCD36, VEGF, PDGF in CAD patients was demonstrated. In contrast, a marked elevation in the plasma level of TNF-α was observed. There was a strong inverse correlation between severity of CAD and sCD36 ($r = -0.413$; $P < 0.001$), cCD36 ($r = -0.4663$; $P < 0.001$), CD36 mRNA ($r = -0.328$; $P < 0.01$). Furthermore, CD36 was found to be positively correlated with VEGF ($r = 0.446$; $P < 0.001$) and PDGF ($r = 0.287$; $P < 0.05$), while it was negatively correlated with TNF-α ($r= - 0.352$; $P < 0.01$).

Conclusion: CD36 deficiency might represent a good expression marker for the development of CAD in Egyptian patients in association with inflammatory and proangiogenic mediators. Targeting the underling mechanism of the direct and inverse association of CD36 with angiogenic and inflammatory mediators could be used in the treatment of athero-inflammatory disorders.

Keywords: Coronary artery disease; CD36; Angiogenesis; Inflammation.

Abbreviations

CAD: Coronary Artery Disease; VEGF: Vascular Endothelial Growth Factor; PDGF: Platelet - Derived Growth Factor; TNF-α: Tumor Necrosis Factor-α; COX- 2: Cycloxygenase-2; HC: Hypercholesterolemia; HT: Hypertension; DM: Diabetes Mellitus; RT-PCR: Reverse Transcription Polymerase Chain Reaction.

Introduction

Cardiovascular Diseases (CVDs) accounting for ~ 38% of all deaths worldwide. Without doubt, the largest contributor to CVD is Coronary Artery Disease (CAD), the leading cause of the death in developed countries [1]. By 2015, almost 20 million people will die from CVDs [2]. In Egypt CVDs are now the main causes of death among Egyptians. In 1970, CVDs were accounted for 12.4% of all deaths, whereas two decades later they were responsible for 42.5% of the mortality [3]. Major risk factors of CVDs include Hypercholesterolemia (HC), Hypertension (HT), Diabetes Mellitus (DM) and smoking [4].

Atherosclerosis, the main cause of CAD, is a progressive and chronic inflammatory disease in which lipids, immune cells, vascular smooth muscle cells, and extracellular matrix accumulate in the subendothelial space to form the growing atherosclerotic lesion. The risk for of cardiovascular events is related to the composition and stability of the plaque rather than to the degree of arterial stenosis, and many studies suggest that inflammation is a critical determinant of plaque stability [5]. More than 80% of sudden cardiac death is caused by atherosclerosis [6]. The progression of the atherosclerotic disease and the increasing severity of atherosclerosis relate not only to the presence and extent of cardiovascular risk factors but also to the persistence of risk factors over time [4]. The study by Allah, et al. [7] found that atherosclerotic carotid artery disease (intima-media thickness and/ or plaques) was present in 41% of the study population.

CD36 is one of the monocyte-derived macrophage scavenger receptors that have been implicated as a key player in the pathogenesis of atherosclerosis [8]. It is believed to play a critical role in the initiation and progression of atherosclerosis through its ability to bind and internalize modified low-density lipoprotein, facilitating the formation of lipid-engorged macrophage foam cells [9]. A variety of studies have been shown that CD36 deficiency may be related to the metabolic syndrome, which is strongly associated with atherosclerotic CVD [10].

Inadequate blood supply to the heart and other tissues resulting from partially lose of the functional blood flow and insufficient new blood vessel growth is a hallmark feature of atherosclerosis. The formation of new blood vessels out of pre-existing capillaries (angiogenesis) is a consequence of events that is important in atherosclerosis. The most potent pro angiogenic factors are Vascular Endothelial Growth Factor (VEGF) and platelet- derived growth factor (PDGF). They played a significant role in angiogenesis and regulation of vascular endothelial cell growth and maintenance and development of new blood vessels [11].

In atherosclerotic plaques, monocytes/macrophages are significant producer of inflammatory cytokines that have a central role in atherogenesis, and cause the clinical manifestations and acute clinical complications of atherosclerosis [12]. Tumor necrosis factor (TNF-α) is involved in the pathogenesis of atherosclerosis and it can initiate many signaling pathways resulting in production of factors influencing angiogenesis [13]. Substantial evidence indicates that unregulated cycloxygenase-2 (COX-2) expression and prostaglandin synthesis influence chronic inflammatory condition, including atherosclerosis and its complications. COX-2 is rapidly induced by various stimuli, including proinflammatory cytokines, such as TNF-α, growth factors, resulting in prostaglandin synthesis associated with inflammation [14].

Risk factors for the development of atherosclerotic CAD have been identified, but several biomarkers may also be important in identifying patients at risk. Therefore, the aim of this study was to evaluate the potential predictive role of CD36 expression in atherosclerosis severity in Egyptian patients. The study was also focused on the relationship between the inflammatory (TNF-α, and COX-2), angiogenic (VEGF and PDGF) mediators and the scavenger receptor (CD36) in the development of accelerated atherosclerotic CAD in Egyptian patients. To the best of our knowledge, this study is the first to be done in Egyptian patients with CAD.

Patients and Methods

Study population

The study population was consisted of 100 consecutive atherosclerotic patients (69 males and 31 females) with mean age 54 ± 7 years (range 35-77 years), who had been referred to Nasser Institute Hospital, Egypt International Hospital, Specialized Hospital of Ain Shams Universities, El-Demerdash Hospital and Cardiovascular Cath-Lab Center, Egypt. All investigations were done in accordance with the Ministry of Health, Health and Human Ethical Clearance Committee guidelines for Clinical Researches. Ain-Shams University local Ethics Committee approved the study protocol. All patients and healthy controls agreed to be enrolled in this study and written consent was obtained from all participants.

All patients had angiographic evidence of stenosis ≥ 50% in at least one major coronary artery or their main branches. Stenosis after a careful history was taken; the patients underwent a physical examination, chest X-ray, 12-lead Echocardiography (ECG) was performed to detect segmental wall motion abnormalities (SWMA) as a result of myocardial ischemia, and ultrasound examination of the heart. Blood was drawn from fasting patients on the morning of catheterization for assay of glucose and lipid concentrations by standard laboratory methods. Risk factors for cardiovascular disease, carotid artery ultra sonography, and coronary artery angiography-left ventriculography were performed. According to the number of diseased vessels, the patients were classified into three groups: Patients with one- (SV), two- (DV), or Multiple-Vessels Disease (MV). Clinical features of CAD cases and controls were shown in Table 1.

The study was also included 100 normal healthy individuals to serve as a control group, including 25 females and 75 males.

Table 1: Demographics and clinical status of Egyptian CAD patients.

Characteristic	Patients (%) N = 100
Age (yr) (mean ± SE)	58 ± 10
Gender (%)	
Male	69%
Females	31%
Coronary risk factors	
Family history	25.60%
Diabetes Mellitus (DM)	13.30%
Hypertension (HT)	29.30%
Hypercholesterolemia (HC)	13.30%
DM+HT	21.20%
DM+HC	8%
HT+HC	24%
DM+HT+HC	45.30%
Coronary angiography	
Severe	60%
Moderate	30%
Mild	10%
Number of stenosed vessels	
Single vessel (1)	38.30%
Double vessels (2)	31.60%
Multi-vessels (3 or more)	28.30%
Smoking status	
Mild	10%
Moderate	11.60%
Severe	30%

Controls included persons who had no major risk factors such as family history of early cardiovascular disease, diabetes, hypertension, hypercholesterolemia, and smoking. They had normal resting ECG or with no blood vessel stenosis or disease (at angiography). Blood samples were drawn from patients prior to angiography. Plasma were separated by centrifugation at 1500 rpm for 15 min at 4°C, aliquotted, and stored at -80°C until analysis. The Peripheral Blood Mononucleated Cells (PBMCs) were obtained from heparinized blood by centrifugation through a Ficoll-Hypaque separating media (Bio Basic Canada Inc).

Evaluation of risk factor variables

The recorded risk factors included smoking status, HC, DM, HT and family history of CAD. Current smokers were categorized according to the number of cigarette smoked per day as reported at baseline into mild smoker (1-10 cigarette/ day), moderate (10-20 cigarette/ day), and severe smokers (20 or more cigarette/ day). Non-smokers group had never smoked any cigarette. Hypertension was defined as blood pressure over 140/90 mm Hg, as measured on several occasions, or the use of antihypertensive treatment; DM as a fasting plasma glucose over 126 mg/ dl or the use of glucose-lowering treatment; hyperlipidemia as low density lipoprotein levels over 130 mg/ dl or high density lipoprotein < 35 mg/ dl and triglycerides > 200 mg/ dl or the use of lipid-lowering therapy. Accordingly, our CAD patients were sub-classified based on the presence of one or more risk factors into: DM, HT, HC, DM+HC, DM+HT, HT+HC, and DM+HT+HC.

Estimation of CD36

Soluble and PBMCs CD36 (sCD36 and cCD36) were quantified by indirect ELISA as previously described [15]. Plasma CD36 levels were expressed as milli absorpance (mA).

Analysis of CD36 mRNA expression

Expression of CD36 by PBMCs was performed using reverse transcription polymerase chain reaction (RT-PCR). Total RNA was isolated from PBMCs samples using BIOZOL reagent (Bioflux) following manufacturer's instructions. Purity and concentration of total RNA was quantified using UV-microplate reader (FLUOstar OPTIMA), on the basis of the A260/ A280 ratio. RT–PCR was performed using total RNA template, where 1 µg of total RNA was reversely transcribed to produce cDNA by sequential incubation with anchored 0.6 µm oligo-dT primer and RT enzyme (SibEnzyme, Ltd, Russia) at 37°C for 10 min, 42°C for 1 h and 70°C for 10 min. The final volume for each PCR reaction was 25 µl that was consisted of DreamTaq Green PCR Master Mix (2X) (Fermentas, Thermo Fisher Scientific Inc., USA), 50 pmoles of specific primers for human CD36: Sense: 5'-TCCTCGAAGAAGGTACAATTGC-3' and antisense: 5'-5'-CAATACAATGACATTTGCCAAG-3'. PCR cycling conditions were consisted of 95°C for 2 min followed by 40 cycles (94°C for 40 seconds, 60°C for 60 seconds and 72°C for 60 seconds) and then a final extension cycle at 72°C for 10 min [16]. PCR products (250 bp) were visualized by 2.5% agarose gel electrophoresis. The sizes of PCR products were determined relatively to the migration of a 50 bp step ladder (Fermentas). The signed intensity of the bands was quantified by UV gel documentation system (Biometra® goettingen, Germany, Bioanalyze 1.0).

Estimation of plasma pro-angiogenic factors

Plasma levels of VEGF, PDGF and TNF-α were quantified by ELISA as previously described [16], using matched paired antibodies (R&D Systems). The ELISA reader-controlling software (Softmax) readily processes the digital data of raw absorbance values into a standard curve from which unknown concentrations can be derived directly. Results were expressed as pg/ ml of plasma.

Estimation of COX-2 by Dot Blot assay

Plasma COX-2 was quantified by Dot Blot assay. A polyvinylidene difluoride (PVDF) membrane (Amersham Pharmacia Inc., CA, USA) was pre-wet in 100% methanol for 2 min, and then soaked in Tris-buffered saline (TBS) (50 mM Tris-HCl and 150 mM NaCl; pH 7.5) for rehydration. 50 µl of plasma were added to each well, and incubated for 45 min. After washing with TTBS (TBS containing Tween 20), 200 µl of blocking buffer (TTBS with 7% FBS) were added for 60 min. Goat anti-human COX-2 (diluted 1:1000; Abcam, MA, USA) and rabbit anti-goat polyclonal antibody- horseradish-peroxidase conjugated (diluted 1:1000; Sigma) were used. After 60 min incubation, the membrane was washed with TBS then soaked in 3,3' Diaminobenzidine (DAB) solution (5 mg DAB + 10 ml TTBS + 30% H_2O_2) (Biobasic). After color development, the membrane was soaked in distilled water, dried for imaging and analyzed by BioDocAnalyze (BDA) soft ware (Biometra®).

Statistical analysis

All of the statistical analysis was performed using the Statistical Package for Social Science (SPSS) version 10. Data are presented as means with corresponding standard error. Comparisons among different groups of patients were performed by one-way analysis of variance (ANOVA). Tukey's post-hoc test and Dunnett's test for multiple comparisons were used. Correlation between variables was determined using Spearman's correlation test. In all of the tests the level of significance was set at $P < 0.05$. For those variables that were significantly influencing serum CD36 levels ($P < 0.05$), a standard linear multiple regression analysis with CD36 levels as the dependent variable was performed. For the regression model, adjusted R^2, β, b-value and standard error b were recorded.

Results

Soluble CD36, PMNCs CD36, and Expression of CD36 mRNA in PMNCs in CAD patients versus control subjects

sCD36 level in plasma and cCD36 expression (cCD36 and CD36 mRNA) of CAD patients classified according to risk factors is presented in Figures 1 & 2. sCD36 level and cCD36 expression were significantly reduced in all groups of CAD patients suffered from risk factors [DM ($P < 0.05$, $P < 0.001$), HT($P < 0.001$, $P < 0.05$), HC ($P < 0.001$, $P < 0.01$), DM+HT ($P < 0.001$, $P < 0.001$), DM+HC ($P < 0.05$, $P < 0.01$), HT+HC ($P < 0.01$, $P < 0.001$) and DM+HT+HC

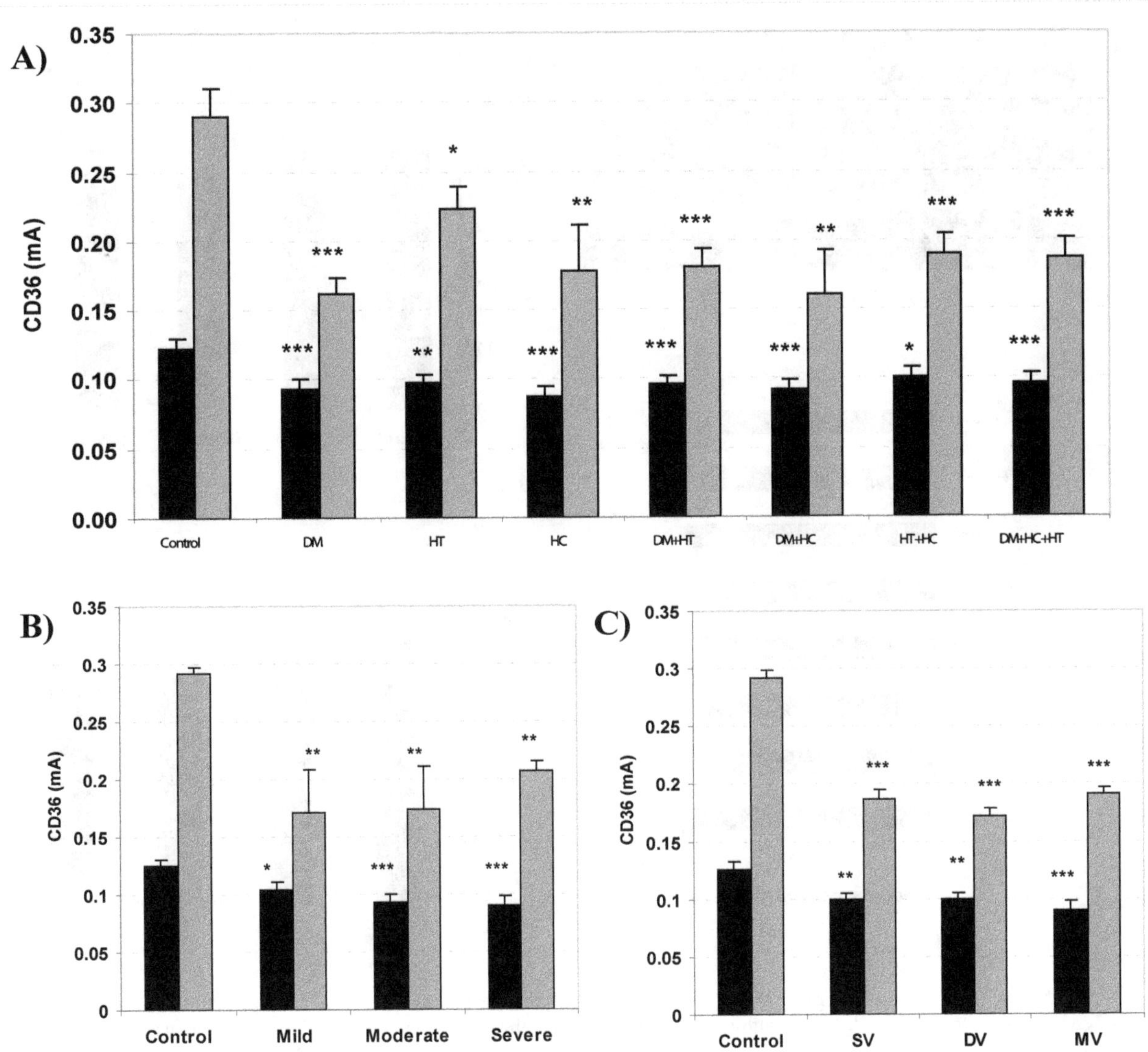

Figure 1: Plasma (black bars) and PMNCs surface (grey bars) levels of CD36 in CAD patients classified according to risk factors (a), smoking status (b), and number of stenoic arteries (c). Results are expressed as mean ± standard error. *Denotes groups significantly different from controls. *$P < 0.05$, **$P < 0.01$, ***$P < 0.001$.
DM: Diabetes Mellitus; HT: Hypertension; HC: Hypercholesterolemia; SV: Single Vessel Stenosis; DV: Double Vessel Stenosis; MV: Multi-Vessel Stenosis

($P < 0.001$, $P < 0.001$)], as shown in figure 1a, where's a reduced cCD36 mRNA expression was noticed in hypertensive groups of CAD patients [HT *($P < 0.05$)*, DM+HT *($P < 0.01$)* and DM+HT+HC *($P < 0.01$)*] in comparison to the healthy control (Figure 2c). When considering smoking status groups, a significant reduction in sCD36 level ($P < 0.05$ and $P < 0.001$) and cCD36 expression *($P < 0.01$)* were demonstrated in all groups of different smoking degrees in relation to normal subjects (figure1b). cCD36 mRNA expression was significantly reduced *($P < 0.01$)* only in severe smokers versus control subjects (figure 2b). The analysis according to the number of stenotic arteries indicated that SV, DV and MV groups had a significant reduction in sCD36 level *($P < 0.001$, $P < 0.01$, $P < 0.001$* for SV, DV and MV; respectively) and cCD36 expression *($P < 0.001$)* versus control subjects (figure 1c). A significant reduction in mRNA expression of cCD36 was found in SV *($P < 0.05$)*, DV *($P < 0.01$)* and MV *($P < 0.01$)* when compared to the healthy controls (Figure 2d).

Evaluation of plasma VEGF plasma PDGF

Compared with normal controls, a significant reduction in VEGF was observed in CAD patients who suffered from hypercholesterolemia [HC *($P < 0.05$)*] and hypertensive groups [DM+HT *($P < 0.001$)*; HT+HC *($P < 0.01$)* and DM+HT+HC *($P < 0.001$)*], as shown in figure 3a. It was noticed that there was a significant decrease in plasma PDGF concentration in diabetic *($P < 0.001$)*, hypercholesterolemic patient *($P < 0.01$)* and the patients suffered from combined DM+HT, HT+HC and DM+HC *($P < 0.001$)* as shown in figure 3a.

Our findings revealed that there was a significant reduction in plasma VEGF in mild, moderate *($P < 0.05$)* and severe smokers *(P*

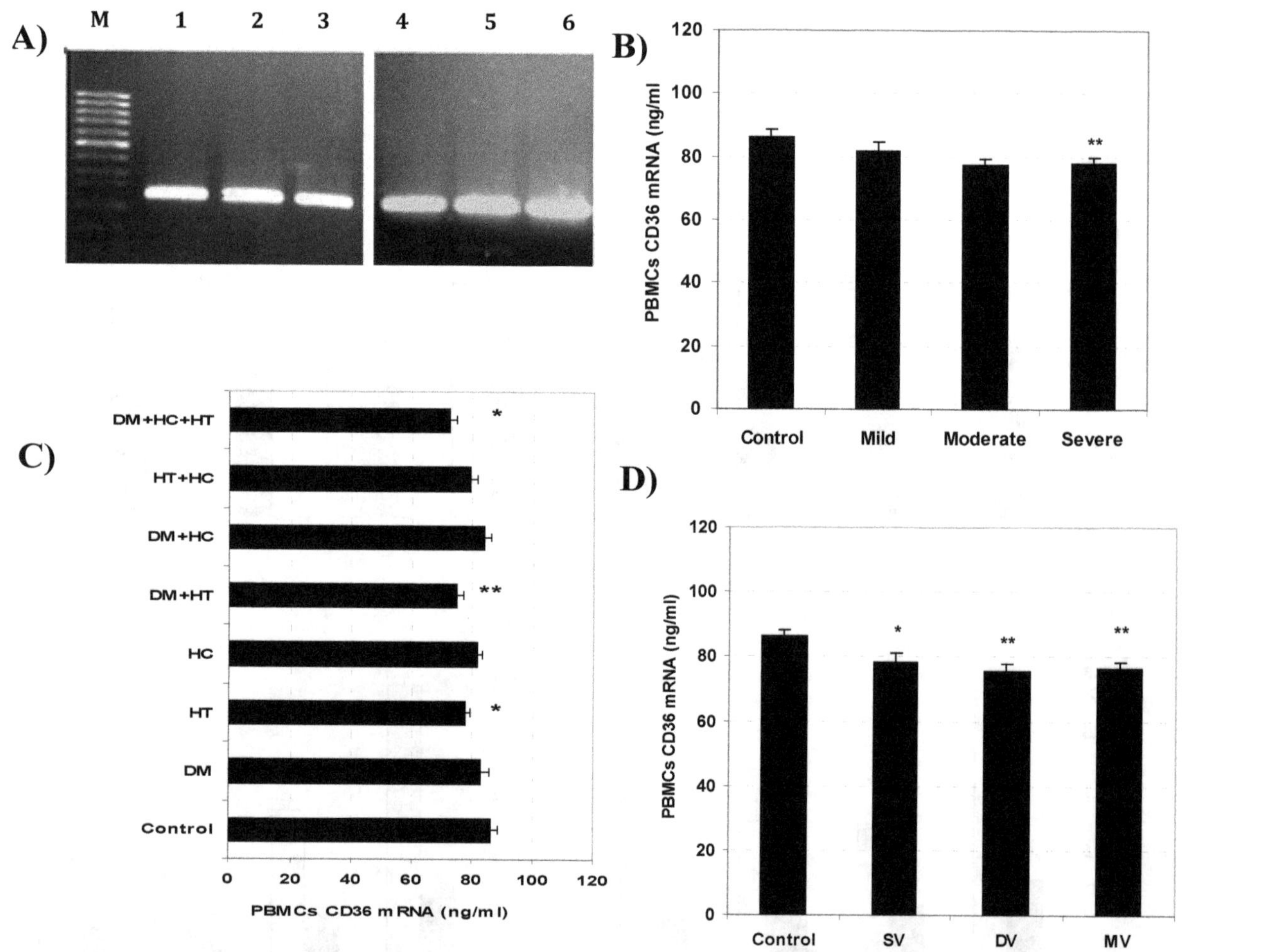

Figure 2: CD36 mRNA expression quantified by RT-PCR. (a) 2% agarose gel for the amplified fragment (250 bp) Lane 1: DNA ladder (50 bp); lane (2-4): CD36 of CAD patients; lane (5-7): CD36 of healthy subjects. CD36 mRNA in CAD patients classified according to according to risk factors (b), smoking status (c), and number of stenoic arteries (d). Results are expressed as mean ± standard error. *Denotes groups significantly different from controls *$P < 0.05$, $^{**}P < 0.01$.*
DM: Diabetes Mellitus; HT: Hypertension; HC: Hypercholesterolemia; SV: Single Vessel Stenosis; DV: Double Vessel Stenosis; MV: Multi-Vessel Stenosis.

< 0.01) compared to the healthy controls (figure 3b). Additionally, a significant reduction *($P < 0.001$)* was recorded in plasma PDGF in mild, moderate, and severe smokers (figure 3b). Concerning stenosis, diminution in plasma VEGF and PDGF levels was also demonstrated in CAD patients with SV, DV and MV stenosis compared to the healthy controls *($P < 0.001$)* (figure 3c and 3c).

Evaluation of plasma TNF- α and COX-2

In comparison to the healthy controls, patients had a significant increase in plasma TNF-α secretion level in DM *($P < 0.05$)* and HT *($P < 0.001$)*. CAD patients who suffered from HC had insignificant reduction in TNF-α level. In groups of combined risk factors, DM+HT, DM+HC, and HT+HC groups have a significant increase *($P < 0.01$)* in TNF-α level. A significant *($P < 0.001$)* elevation in plasma TNF-α level was demonstrated in DM+HT+HC group (figure 4a). Degrees of smoking frequency showed a significant increase in plasma TNF-α in mild *($P < 0.001$)*, moderate and severe smoking *($P < 0.01$)* as compared to the healthy controls. However, compared with mild smoking status, a significant reduction in TNF-α was observed in moderate *($P < 0.05$)* and severe *($P < 0.05$)* smokers (figure 4b). According to number of stenotic arteries, plasma TNF-α showed a significant increase in SV *($P < 0.001$)*, DV *($P < 0.05$)* and MV *($P < 0.001$)* stenosis groups, as compared to the healthy controls (figure 4c). The maximum production was observed in MV patients. There was a non-significant change *($P > 0.05$)* in plasma COX-2 levels in CAD patients classified according to risk factors, smoking status or stenosis, as compared to the healthy controls.

Statistical correlations

CAD progression in Egyptian patients showed statistically significant direct correlation with plasma levels of TNF-α *($r = 0.480, P < 0.001$)* and indirect significant correlation with VEGF *($r = -0.499, P < 0.001$)*, sCD36 *($r = -0.413, P < 0.001$)*, cCD36 *($r = -0.466$,*

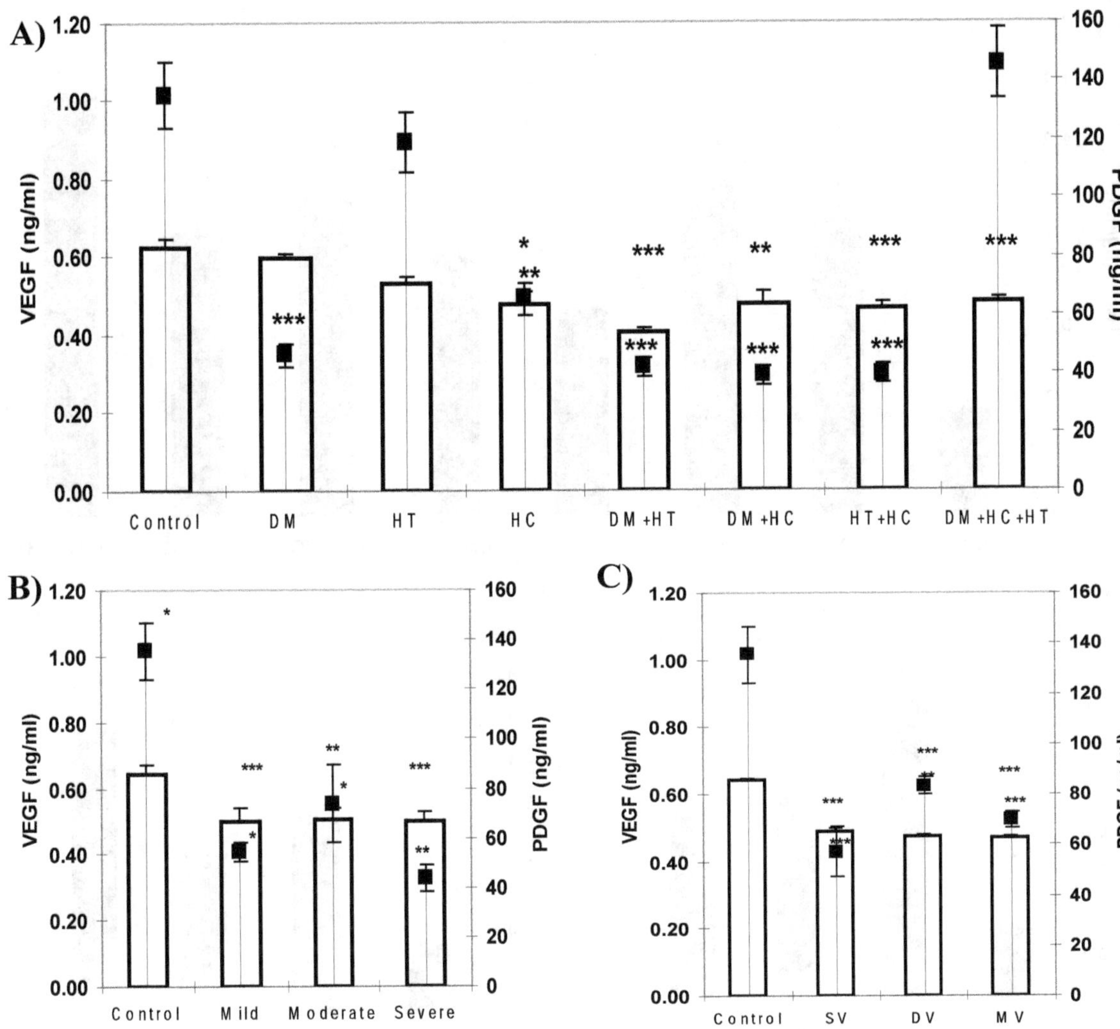

Figure 3: Plasma VEGF (white columns) and PDGF (black lines) in CAD patients classified according to risk factors (a), smoking status (b), and number of stenoic arteries (c). Results are expressed as mean ± standard error. *Denotes groups significantly different from controls *$P < 0.05$, $^{**}P < 0.01$, $^{***}P < 0.001$.*
DM: Diabetes Mellitus; HT: Hypertension; HC: Hypercholesterolemia; SV: Single Vessel Stenosis; DV: Double Vessel Stenosis; MV: Multi-Vessel Stenosis

$P < 0.001$) and cCD36 mRNA *($r = -0.328$, $P < =0.01$).* Statistically significant negative correlation was observed between TNF-α and VEGF production in plasma *($r = -0.323$, $P < 0.01$).* sCD36 showed a significant positive correlation with cCD36 *($r=0.234$, $P < 0.05$)*, and both of them showed a direct correlation with plasma VEGF *($r = 0.446$, $P < 0.001$* and *$r = 0.321$, $P < 0.01$*; respectively), PDGF *($r = 0.287$, $P < 0.05$* and *$r = 0.401$, $P < 0.001$*; respectively) and a negative correlation with plasma TNF-α *($r = -0.352$, $P < 0.01$* and *$r = -0.380$, $P < 0.001$*; respectively).

Finally, using multiple linear regression, disease association remained significant *(β = -0.239, $P < 0.05$)* between serum levels of sCD36, taken as the dependent variable ($R^2 = 0.082$), for the presence of down regulated TNF-α. On the other hand, other independent variables (VEGF, PDGF) were not significant in the multivariate analysis.

Discussion

Atherosclerotic CAD progression, whether clinically silent or associated with acute coronary events, has been shown to be a powerful predictor of cardiovascular risk [17]. It is now widely accepted that atherosclerosis is a chronic inflammatory disease [18] and that CD36 as a scavenger receptor for oxidized LDL played an important role in the pathogenesis of atherosclerosis [19]. This study focused on the role of CD36 in the development of atherosclerotic CAD in Egyptian patients and the influence of various risk factors on CD36, with particular attention to its association with inflammatory and angiogenic mediators.

The findings showed that Egyptian CAD patients had reduced serum sCD36 levels compared to healthy controls. This observation indicated that sCD36 levels might be affected

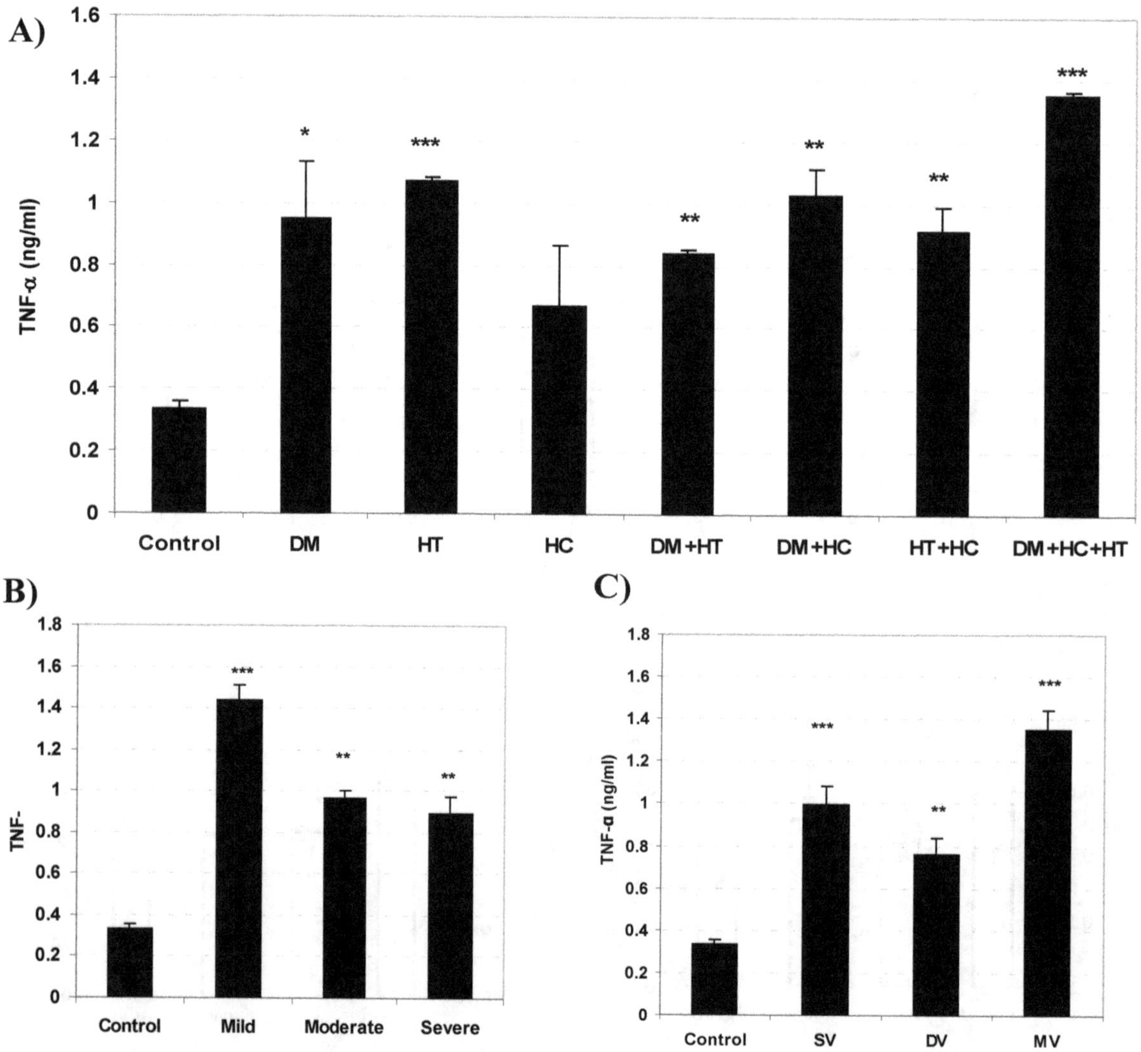

Figure 4: Plasma TNF-α in CAD patients classified according to risk factors (a), smoking status (b), and number of stenoic arteries (c). Results are expressed as mean ± standard error. (*)Denotes groups significantly different from controls. **P < 0.05, **P < 0.01, ***P < 0.001.*
DM: Diabetes Mellitus; HT: Hypertension; HC: Hypercholesterolemia; SV: Single Vessel Stenosis; DV: Double Vessel Stenosis; MV: Multi-Vessel Stenosis

by the degree of smoking status in CAD patients and indicated that a reduction in plasma level of sCD36 could be considered as a potential marker and a strong predictor of CAD. These data are in agreement with previously reported data that suggested that sCD36 may be derived from a proteolytic cleavage of the extracellular part of CD36 protein from CD36- expressing tissues or from micro particles of activated or apoptotic monocytes/macrophages [20]. A previous explanatory studies provided that soluble form of CD36 could serve as a biomarker that is associated with altered PBMCs CD36 expression, thus, the levels of sCD36 seem to be parallel CD36 expression on intact monocytes. Based on this study, the down regulated CD36 in our patients' plasma might be used as an indicator to a PBMCs CD36 down regulation. This hypothesis is confirmed through the detection of CD36 either in PBMCs surface and mRNA expression. A comparable remarkable result was observed, where a significant reduction in cCD36 expression was documented in our work in all types of risk factors.

Our data confirmed the previously reported studies [21-22]; that revealed the association of CD36 deficiency with HC, DM and arterial HT, and agreed with [23] who reported a severe atherosclerosis in subjects whom naturally deficient in CD36, which suggests that CD36 has an anti-atherogenic role. Elevation in LDL-Cholesterol was previously documented [24], modified LDL, and triglycerides [25] levels in CD36 deficient patients, which were considered as risk factors for CAD.

Furthermore, patients with severe stenosis (DV and MV) showed a marked reduction in CD36 mRNA than those with SV. These data suggest that CD36 mRNA expression is associated with the severity of CAD. However, the molecular mechanism by which the reduction of CD36 RNA expression occurs in PMMCs is still unclear. One possible explanation is that the decrease in CD36 mRNA might be linked to a mutation in CD36 gene that affects transcriptional efficiency in macrophages. A significant indirect correlation between CAD and CD36 profile (sCD36, cCD36 and CD36 mRNA) was found. Thus CD36 may turn out to be a good expression marker for the severity of atherosclerosis in Egyptian CAD patients.

CD36 not only implicated in atherosclerosis, but also has an angiostatic function, as demonstrated by the inability of thrombospondin-1 (TSP-1) to inhibit angiogenesis in CD36 null mice [26]. In atherosclerosis, the role of angiogenesis remains a highly contentious issue. No consensus exists whether angiogenesis is a key causative factor in the pathogenesis of atherosclerotic plaque formation or it is a way to treat coronary artery disease [27]. For further investigation of this issue, we extended our work to study the exact relationship between CAD and both of CD36 and angiogenic mediators in CAD patients.

Growth factors, such as VEGF and PDGF play an important role in angiogenesis. Dysfunction of endothelial cells may promote abnormal vascular growth, such as that in atherosclerosis and arteriosclerosis [28]. In the current study, we demonstrated that plasma VEGF was reduced markedly in all HC CAD groups. In addition, decreased plasma VEGF level was also found in DM CAD patients. These findings are consistence with previous work that reported a reduction in VEGF in diabetic rats, that followed by ischemic cardiovascular disease [29]. Our results strengthen previous reports that document the reduction in VEGF in diabetic rats, that followed by ischemic cardiovascular disease [30]. In agreement with the presented results, it was demonstrated that the reduction of VEGF levels lead to elevation of peripheral resistance and HT [31].

When considering smoking status, a marked reduction of VEGF and PDGF levels were observed in all groups. These results are in agreement with previous studies that reported a reduced level of VEGF as a result of cigarette smoking [31-32]. The observed reduction of VEGF among smokers of CAD patients might be related to the fact that cigarette smoke disrupts components of the VEGF165-VEGFR-2 tertiary signaling complex by decreasing neuropilin-1 (NRP-1) expression together with reducing expression of VEGFR-2 and VEGF [33]. Whereas, the current data demonstrated a significant decrease of VEGF and PDGF levels in plasma of CAD patients with SV, DVs, and MVs. A significant indirect correlation between CAD and angiogenic factors (VEGF and PDGF) is best expressed by the Pearson correlation coefficient analysis indicated that the VEGF has a prognostic importance in atherosclerotic CAD in Egyptian patients. Decreased VEGF and PDGF levels in CAD patients may be contributed to the reduction of CD36 expression. This explanation is supported by a direct correlation between CD36 and angiogenic mediators observed in our study. This is consistent with Howell, et al. [34] who reported a diminished expression VEGF in CD36 deficient mice.

Inflammatory mediators can; either directly or indirectly, promote angiogenesis, which in turn contributes to inflammatory pathology. New blood vessels can maintain the chronic inflammatory state by transporting inflammatory cells to the site of inflammation and supplying nutrients and oxygen to the proliferating inflamed tissue. The increased endothelial surface area also creates an enormous capacity for the production of cytokines, adhesion molecules, and other inflammatory stimuli. Generally speaking, pro-inflammatory mediators promote angiogenesis and the proangiogenic effects mediated by IL-1 and TNF-α support such hypothesis [35]. The consequences of inflammation in atherosclerosis are difficult to predict. Although it may be beneficial at early stage, with prosperities reversing atherogenesis, it may be detrimental to the individual with more aggressive disease progression and plaque rupture at later stage.

Mediators of inflammation such as TNF-α have been associated with an increased risk for cardiovascular events in several clinical studies [36]. TNF-α is a central pro-inflammatory cytokine involved in the propagation of atherosclerosis. TNF-α is secreted in the vascular wall by endothelial smooth muscle cells and monocytes/ macrophages and it is a powerful inducer of local inflammation [37]. It promotes the expression of leukocyte adhesion molecules [38] and increases the uptake of macrophages in atherosclerotic lesions [39] thus directly promoting atherosclerosis.

The current study showed a significant elevation in TNF-α level in DM, HT and multi-risk factors CAD patients. In addition, a marked elevation in TNF-α, plasma levels were reported in mild, moderate, and severe smokers, while the elevation is more significant in severe smokers than other groups. These results are in agreement with an earlier report that recorded elevated TNF-α in congestive heart failure that was associated with atherosclerotic risk factors, including smoking, hyperglycemia, HT, LDL cholesterol [40].

Inflammation plays a pivotal role in the development of metabolic syndrome features, including dyslipidemia and altered glucose tolerance. These metabolic changes constitute the substrate for the subsequent development of atherosclerotic plaque [41]. Chronic inflammatory conditions have been shown to be associated with pro-atherogenic lipid pattern and altered glucose tolerance. TNF-α has been demonstrated to directly interfere with metabolic pathways of triacylglyceride and cholesterol [42].

In addition to the modifications that occur in lipids metabolism, TNF-α may interfere with glucose metabolism pathways [43], where it is likely to increase hepatic glucose production and decreases glucose uptake and catabolism in the muscle. While, in adipocytes, TNF-α down regulate the expression of several proteins implicated in the insulin receptor pathway [44]. The lipid and glucose changes induced by TNF-α are pro-atherogenic in terms of both quality and quantity. Therefore, the persistence of these modified lipids in the circulation will promote the development of atherosclerotic lesions [41]. These data strongly support the hypothesis that the inflammatory cytokines (such as TNF-α) are surrogate biomarker of grade inflammation burden present in patients with atherosclerotic CAD who suffered from DM or HT or both of them [45]. According to the number of stenotic coronary vessels, we reported an elevation in plasma TNF-α in SV, DVs, and MVs CAD patients, where TNF-α is more significantly elevated in MVs group than others. These data are supported by statistical analysis that showed a direct correlation between CAD and TNF-α level, which indicates the pivotal role played by TNF-α in atherosclerosis. In accordance with our data, it was previously found that the elevated plasma TNF-α concentration is associated with severity of atherosclerosis [46].

With regard to angiogenesis, our study showed an indirect correlation between TNF-α with VEGF and PDGF. Consistent with our data, it was reported that TNF-α directly inhibits VEGF cellular effects and that TNF-mediated inhibition of angiogenesis results from down regulation of receptors for proangiogenic factors and activation of angiogenesis inhibitors [47]. Furthermore, CD36 showed indirect correlation with TNF-α in CAD patients, which agreed with Boyer, et al. [48], who reported that TNF-α inhibits both CD36 membrane and mRNA expression and that this inhibition of CD36 expression involves a reduction in Peroxisome proliferator-activated receptor gamma (PPAR-γ) activation in human monocytes.

In conclusion, we found significant reduction of sCD36 and PBMCs CD36 surface protein and CD36 mRNA expression, for the first time in Egyptian CAD patients. CD36 was correlated with the severity of the disease. Our data approved that CD36 deficiency may turn out to be a good biomarker and strong predictor for the severity of atherosclerosis in Egyptian CAD patients. Furthermore, the marked elevation of inflammatory cytokine TNF-α support the hypothesis that TNF-α could be used as a surrogate marker of grade inflammation present in patients with atherosclerosis and a pivotal role of TNF-α in elevation of atherosclerosis. On the other hand, the marked reduction in proangiogenic mediators (VEGF & PDGF) and the strong indirect correlation between VEGF and CAD indicating that VEGF has a prognostic importance in atherosclerosis in CAD patients.

Taken together CD36 deficiency might represent a new risk factor for the development of CAD in association with inflammatory and angiogenic mediators. Therefore, targeting the underlying mechanism of the direct and inverse association of CD36 with angiogenic mediators (VEGF & PDGF) and inflammatory cytokine (TNF-α) in Egyptian CAD patients could be a strategy in the treatment of athero-inflammatory disorders and deserves exploration in future prospective study. Our assessment of this evidence leads us to conclude that, there is a need for early-stage prediction of those populations that have the risk of developing atherosclerotic CAD. This need could be met by analysis of CD36 genetic polymorphisms related to CD36 deficiency. Thus, our future prospective study will be directed towards studying the association between change in CD36 genotype and the incidence of CAD in a large, prospective cohort of Egyptian patients.

Acknowledgement

This work was financially supported by the National Research Center, Cairo, Egypt.

Conflict of Interest

We declare no conflict of interest.

Author Contribution

Enas, Amira, Nadia and Mohamed participated in the experimental design. Enas and Roba conducted the experiments. Enas, Amira and Roba interrelated and analyzed the data. Ahmed Tamara participated in clinical work. All Authors participated in writing the manuscript and approved its final version.

References

1. Wang Q. Molecular genetics of coronary artery disease. Curr Opin Cardiol. 2005; 20(3): 182-188.
2. World Health Organization (WHO): The annual health report of the year 2009. Cairo, WHO, 2009.
3. Central Agency for Public Mobilization and Statistics (CAPMAS): The annual health report of the year 1990. Cairo, CAPMAS, 1990.
4. Wilson J M. Diagnosis and treatment of acquired coronary artery disease in adults. Postgrad Med J. 2009; 85(1005): 364-365. doi: 10.1136/pgmj.2008.068791.
5. Libby P. Inflammation and cardiovascular disease mechanisms. Am J Clin Nutr. 2006; 83(2): 456S-460S.
6. Ouguerram K, Nguyen P, Krempf M, Pouteau E, Briand F, Bailhache E, et al. Selective uptake of high-density lipoproteins cholesteryl ester in the dog, a species lacking in cholesteryl ester transfer protein activity; an in vivo approach using stable isotopes. Comp Biochem Physiol B Biochem Mol Biol. 2004; 138(4): 339-345.
7. Abd Allah F, Baligh E, Ibrahim M. Clinical relevance of carotid atherosclerosis among Egyptians: a 5-year retrospective analysis of 4,733 subjects. Neuroepidemiology. 2010; 35(4): 275-279. doi: 10.1159/000319899.
8. Yamashita S, Hirano K, Kuwasako T, Janabi M, Toyama Y, Ishigami M, et al. Physiological and pathological roles of a multi-ligand receptor CD36 in atherogenesis; insights from CD36-deficient patients. Mol Cell Biochem. 2007; 299(1-2): 19-22.
9. Collot-Teixeira S, Martin J, McDermott-Roe C, Poston R, McGregor JL. CD36 and macrophages in atherosclerosis. Cardiovasc Res. 2007; 1; 75(3): 468-477.
10. Febbraio M, Podrez EA, Smith JD, Hajjar DP, Hazen SL, Hoff HF, et al. Targeted disruption of the class B scavenger receptor CD36 protects against atherosclerotic lesion development in mice. J Clin Invest. 2000; 105(8): 1049-1056.
11. Connolly DT, Heuvelman DM, Nelson R, Olander JV, Eppley BL, Delfino JJ, et al. Tumor vascular permeability factor stimulates endothelial cell growth and angiogenesis. J Clin Invest. 1989; 84(5): 1470–1478.
12. Uchida K. A Lipid-derived Endogenous Inducer of COX-2: a Bridge between Inflammation and Oxidative Stress. Mol Cells. 2008; 25(3): 347-351.
13. Bigda J, Okrój M, Onkol W. The role of tumor necrosis factor (TNF) in angiogenesis. Often has. 2002; 6(2): 57-59.
14. FitzGerald GA. COX-2 and beyond: approaches to prostaglandin inhibition in human disease. Nat Rev Drug Discov. 2003; 2(11): 879-890.
15. Bassyouni IH, Gheita TA, Talaat RM. Clinical significance of serum levels of sCD36 in patients with systemic sclerosis: preliminary data. Rheumatology (Oxford). 2011; 50(11): 2108-2112. doi: 10.1093/rheumatology/ker297.
16. Talaat RM. Soluble angiogenesis factors in sera of Egyptian patients

with hepatitis C virus infection: correlation with disease severity. Viral Immunol. 2010; 23(2): 151-157. doi: 10.1089/vim.2009.0089.

17. Ross R. Atherosclerosis—an inflammatory disease. N Engl J Med. 1999; 14; 340(2): 115-126.

18. Tiong A Y, Brieger D. Inflammation and coronary artery disease. Am Heart J. 2005; 150(1): 11–18.

19. Handberg A, Levin K, Hojlund K, Beck- Nielsen H. Identification of the oxidized low-density lipoprotein scavenger receptor CD36 in plasma: a novel marker of insulin resistance. Circulation. 2006; 12; 114(11): 1169-1176. Epub 2006 Sep 4.

20. Wang L, Bao Y, Yang Y, Wu Y, Chen X, Si S, et al. Discovery of antagonists for human scavenger receptor CD36 via an ELISA like high-throughput screening assay. J Biomol Screen. 2010; 15(3): 239-250. doi: 10.1177/1087057109359686.

21. Hajri T, Han XX, Bonen A, Abumrad NA. Defective fatty acid uptake modulates insulin responsiveness and metabolic responses to diet in CD36-null mice. J Clin Invest 2002, 109(10):138-1389.

22. Miyaoka K, Kuwasako T, Hirano K, Nozaki S, Yamashita S, Matsuzawa Y. CD36 deficiency associated with insulin resistance. Lancet. 2001; 357(9257):686-687.

23. Masuda D, Hirano K, Oku H, Sandoval JC, Kawase R, Yuasa-Kawase M, et al. Chylomicron Remnants Are Increased in the Postprandial State in CD36 Deficiency. J Lipid Res. 2009; 50: 999-1011. doi: 10.1194/jlr. P700032-JLR200.

24. Sato O, Kuriki C, Fukui Y, Motojima K. Dual promoter structure of mouse and human fatty acid translocase/CD36 genes and unique transcriptional activation by peroxisome proliferator-activated receptor alpha and gamma ligands. J Biol Chem. 2002; 277: 15703-15711.

25. Khurana R, Simons M, Martin JF, Zachary IC. Role of Angiogenesis in Cardiovascular Disease: A Critical Appraisal. Circulation. 2005; 112(12):1813-1824.

26. Dzau VJ. The role of mechanical and humoral factors in growth regulation of vascular smooth muscle cells and cardiac myocytes. Curr Opin Nephrol Hypertens. 1993;2(1): 27-32.

27. Yoon YS, Uchida S, Masuo O, Cejna M, Park JS, Gwon HC, et al. Progressive Attenuation of Myocardial Vascular Endothelial Growth Factor Expression Is a Seminal Event in Diabetic Cardiomyopathy. Circulation. 2005; 111(16): 2073-2085.

28. Koyama S, Sato E, Haniuda M, Numanami H, Nagai S, Izumi T. Decreased level of vascular endothelial growth factor in bronchoalveolar lavage fluid of normal smokers and patients with pulmonary fibrosis. Am J Respir Crit Care Med. 2002; 166(3): 382–385.

29. Khorram O, Khorram N, Momeni M, Han G, Halem J, Desai M, et al. Maternal undernutrition inhibits angiogenesis in the offspring: a potential mechanism of programmed hypertension. Am J Physiol Regul Integr Comp Physiol. 2007; 293(2): R745-R753.

30. Felmeden DC, Spencer CGC, Chung NA, Belgore FM, Blann A D, Beevers DG, et al. Relation of thrombogenesis in systemic hypertension to angiogenesis and endothelial damage/ dysfunction (a Sub study of the Anglo-Scandinavian Cardiac Outcomes Trial [ASCOT]. Am J Cardiol. 2003; 92: 400-405.

31. Melkonian G, Cheung L, Marr R, Tong C, Talbot P. Mainstream and Sidestream Cigarette Smoke Inhibit Growth and Angiogenesis in the Day 5 Chick Chorioallantoic Membrane. Toxicolo Sci. 2002; 68: 237-248.

32. Cheng SL, Wang HC, Yu CJ, Yang PC. Increased expression of placenta growth factor in COPD. Thorax. 2008; 63: 500-506. doi: 10.1136/thx.2007.087155.

33. Houssier M, Raoul W, Lavalette S, Keller N, Guillonneau X, Baragatti B, et al. CD36 Deficiency Leads To Choroidal Involution Via COX2 Down-Regulation In Rodents. PLos Med. 2008; 5(2): E39. doi: 10.1371/journal.pmed.0050039.

34. Howell WM, Ali S, Rose-Zerilli MJ, Ye S. VEGF polymorphisms and severity of atherosclerosis. J Med Genet. 2005; 42(6): 485-490.

35. Leung, DW, Cachianes G, Kuang WJ, Goeddel DV, Ferrara N. Vascular endothelial growth factor is a secreted angiogenic mitogen. Science. 1989; 246(4935): 1306-1309.

36. Ridker PM, Rifai N, Pfeffer M, Sacks F, Lepage S, Braunwald E. Elevation of tumor necrosis factor-and increased risk of recurrent coronary events after myocardial infarction. Circulation. 2000; 101(18): 2149-2153.

37. Libby P. Inflammation in atherosclerosis. Nature. 2002; 420(6917): 868-874.

38. Kim CJ, Khoo JC, Gillotte-Taylor K, Li A, Palinski W, Glass CK, et al. Polymerase chain reaction-based method for quantifying recruitment of monocytes to mouse atherosclerotic lesions in vivo: enhancement by tumor necrosis factor-alpha and interleukin-1 beta. Arterioscler Thromb Vasc Biol. 2000; 20(8): 1976-82.

39. Skoog T, Dichtl W, Boquist S, Skoglund-Andersson C, Karpe F, Tang R et al. Plasma tumor necrosis factor-alpha and early carotid atherosclerosis in healthy middle-aged men. Eur Heart J. 2002; 23(5):376-383.

40. Popa C, Netea MG, van Riel PL, van der Meer J W, Stalenhoef AF. The role of TNF-α in chronic inflammatory conditions, intermediary metabolism, and cardiovascular risk. J Lipid Res. 2007; 48(4): 751-762.

41. Feingold KR, Pollock AS, Moser AH, Shigenaga JK, Grunfeld C. Discordant regulation of proteins of cholesterol metabolism during the acute phase response. J Lipid Res. 1995; 36(7): 1474-1482.

42. Memon RA, Grunfeld C, Moser AH, Feingold KR. Tumor necrosis factor mediates the effects of endotoxin on cholesterol and triglyceride metabolism in mice. Endocrinology. 1993; 132(5): 2246-2253.

43. Feingold KR, Hardardottir I, Grunfeld C. Beneficial effects of cytokine induced hyperlipidemia. Z Ernahrungswiss. 1998; 37: 66-74.

44. Stephens JM, Pekala PH. Transcriptional repression of the GLUT4 and C/EBP genes in 3T3-L1 adipocytes by tumor necrosis factor-alpha. J Biol Chem. 1991; 266(32): 21839-21845.

45. Elneihoum AM, Falke P, Hedblad B, Lindgarde F, Ohlsson K. Leukocyte activation in atherosclerosis: correlation with risk factors. Atherosclerosis. 1997; 131(1): 79-84.

46. Gotsman I, Stabholz A, Planer D, Pugatsch T, Lapidus L, Novikov Y, et al. Serum Cytokine Tumor Necrosis Factor-Alpha and Interleukin 6 Associated with the severity of Coronary Artery Disease: Indicators of an Active Inflammatory Burden? Isr Med Assoc J. 2008; 10(7): 494-498.

47. Nakagami H, Cui T-X, Iwai M, Shiuchi T, Takeda-Matsubara Y, Wu L, et al. Tumor Necrosis Factor- Inhibits Growth Factor–Mediated Cell Proliferation Through SHP-1 Activation in Endothelial Cells. Arterioscler Thromb Vasc Biol. 2002; 22(2): 238-242.

48. Boyer JF, Balard P, Authier H, Faucon B, Bernad J, Mazières B, et al. Tumor necrosis factor alpha and adalimumab differentially regulate CD36 expression in human monocytes. Arthritis Research & Therapy. 2007; 9(2): R22.

30

The Incidence and Prevalence of Crohn's Disease in Global Scale

Payam Behzadi[1], Elham Behzadi[1], and Reza Ranjbar[2]*

[1]Department of Microbiology, College of Basic Sciences, Shahr-e-Qods Branch, Islamic Azad University, Tehran, Iran
[2]Molecular Biology Research Center, Baqiyatallah University of Medical Sciences, Tehran, Iran

***Corresponding author:** Reza Ranjbar, Molecular Biology Research Center, Baqiyatallah University of Medical Sciences, Tehran, Iran*
E-mail: ranjbarre@gmail.com

Abstract

The chronic and lifelong gastrointestinal tract disorder of Crohn's Disease (CD) is a type of inflammatory bowel disease with unclear causative agent. CD is directly correlated with a triad group of predisposing factors including genetic problems, immune system malfunctions and environmental factors. The incidence and prevalence of CD in industrialized countries are high and in developing countries are low. However, industrial progressions have risen up the occurrence of CD in some developing countries in Asia and Africa. Because of some significant changes regarding to the incidence and prevalence of CD in global scale, the aim of this review article is to show the incidence and prevalence of CD worldwide.

For preparing this review, Google Scholar search engine and PubMed database were used to find the most recent and valuable original and review articles. The searched articles were then studied and sifted for writing the present review literature.

The results of different investigations show a significant increase among industrialized developing countries within Asia and Africa continents. Besides, the rate of CD is stable within western countries with no significant changes.

Continuous studies relating to the incidence and prevalence of CD will lead to discover the real causative agent of the disease. Moreover, detecting a diversity of the predisposing factors in different geographic areas may help to control and prevent the progression of CD in future.

Keywords: Crohn's Disease; Autoimmune disease; Anti *Saccharomyces cerevisiae* antibodies

Introduction

Inflammatory Bowel Disease (IBD) is known as a chronic sickness that involves CD and Ulcerative Colitis (UC). CD is a lifelong IBD which may affect any part of alimentary tract. The predominant age range of patients with CD is reported between 20 and 30; but it can happen at any age [1-3].

CD is usually recognized via clinical symptoms of diarrhea (with blood or/and mucus), fever, weight loss and abdominal pain. The ileocolonic lesions start with mucosal ulcers and change into fistulas by the time. About 80% of patients with CD have experienced a type of gut surgeries. The foci of CD ulcers and lesions in patients are localized in determined parts of the gut including only small intestine (25%), only large intestine (20%), small and large intestines (50%), and perianal zone (5%) [1,2,4].

The clear etiopathogenesis of CD is not recognized yet; but, there is a triangle of the main predisposing factors such as genetic, immunologic and peripheral parameters which triggers the disease. In other word, genetic malfunctions regarding to immune system in parallel with environmental factors like human gut microbiota, foods, smoke, chemical contamination induce the occurrence of CD in predisposed people. The environmental risk factors increase the appearance of CD in immune dysregulated individuals up to 40 times [1,2,4,5,6].

According to epidemiological investigations, the prevalence of CD in industrialized countries is higher than developing countries. However, in recent years the rate of CD in industrialized countries has been stabilized while the number of patients with CD is rising up in developing countries. The progression of CD in developing countries is relating to changes in people's lifestyle. This feature has paled the idea that genetic characteristics rank in first [4,6,7].

Because of some significant changes regarding to the incidence and prevalence of CD in global scale, the aim of this review article is to show the incidence and prevalence of CD worldwide.

Environmental Risk Factors

The achievement of a wide range of scientific investigations has proved the strong influence of environmental factors on the occurrence of CD among different races and populations. The environmental factors are able to control the efficacy of genetic parameters. Thus, a known race or population with low risk of CD disease in their homelands (such as Asians and Hispanics) may be triggered for CD in an industrialized country and vice versa (e.g. Jews). Today, the role of environmental factors ranks first. Table 1 shows the most important environmental factors regarding to CD induction [4,6-9].

In parallel with environmental factors, genetic characteristics have been studied. Developments in Genetics and Bioinformatics resulted in identification of 163 susceptible loci for IBD in which 110 loci contribute in CD and UC [4,6].

Epidemiology

Several studies show a high incidence and prevalence of CD in Western countries including USA, Canada, New Zealand, UK, Scandinavian area and Western Europe with stable increasing rate while CD is not a common autoimmune disease among African, Asian and Eastern European countries. Recent studies indicate a considerable increase of CD among industrialized developing countries such as a number of Asian countries [3,7,16,17].

The results of previous surveys in Western countries confirm a higher rate of CD among female patients in comparison with male populations; while this ratio is reversed in developing countries with a low incidence rate. Moreover there are two identified peaks for occurrence of CD among patients. The first peak appears in the age range of 15 to 35 and the second peak is happened in 60-80 year-old patients. The race and ethnical characteristics have no correlation with the appearance of CD; because the recorded reports show that the disease occurs among immigrants in destination countries not in their homelands. This fact is detected within Asians and Hispanics who are living out of their homelands as immigrants [3,4,6,18].

Until today, many statistical investigations in association with the prevalence of CD have been done around the world. The newest results are indicated in table 2.

In epidemiologic studies, the rates of disease incidence and prevalence have their interpretations. Generally, the high incidence rate is in association with the presence of predisposing factors while the prevalence of a disease is correlated with geographical variations [7,16].

In this review, we have studied the ranges of incidence and prevalence of CD in global scale. The results are presented as cases per 10^5 (100,000) individuals regarding to annual incidence [table 2].

Global Incidence and Prevalence of CD

The Persian Gulf region countries

According to table 2, no complete data belonging to the all 8 countries of the Persian Gulf Region is available. The results of different studies in Iran and Saudi Arabia show a low incidence rate of CD. However, the number of patients with CD is growing in Iran and Saudi Arabia. In addition to Iran and Saudi Arabia, the increase of CD is observed in the six left countries of the Persian Gulf Region too. Saudi Arabia owns the highest rates of incidence and prevalence of CD among the Persian Gulf Region countries [2,8,9,19,21,29].

Asia

The rate of incidence and prevalence of CD in 12 large and crowded Asian countries is low. But, previous investigations indicate a significant increase of CD within Asian countries. The first three countries with highest incidence of CD involve Turkey, Lebanon and South Korea, respectively. On the other hand, Lebanon, Japan and South Korea rank as the first three countries with highest prevalence of CD, respectively. Previous studies show that countries of South Korea, Japan and Hong Kong have experienced a significant increase of CD within last three decades [2,7,16,17,20,22,23].

Europe

Although, the incidence and prevalence of CD in Europe are high, they vary in a vast range [Table 2] [7,16,20,23-25].

Among Western European countries Scotland has the highest incidence (11.7 per 10^5 cases) and Switzerland ranks first with 100.7 per 10^5 cases regarding to the prevalence of CD. Finland, Denmark and Sweden own the highest incidence of CD with 9.2, 8.6, and 8.3 per 10^5 cases; respectively while Sweden (213 per 10^5 cases) and Denmark (151 per 10^5 cases) rank first and second in the prevalence of CD in Northern Europe and even in whole Europe, respectively. The rate of CD (the incidence and prevalence) has been stabilized in Western Europe and especially in Scandinavian countries [7,16,20,23-25].

The incidence and prevalence of CD are considerably much lower in Eastern Europe. Hungary with the incidence of 8.9 per 10^5 cases ranks first in association with CD and Bosnia Herzegovina owns the highest ranking for the prevalence of CD with 28.2 per 10^5 cases among Eastern European countries. Last investigations reveal that, those countries from Eastern Europe which are trying to be modernized and industrialized are experiencing a high increase of the incidence and prevalence of CD. This feature

Table 1: The association of environmental factors with CD.

Environmental factor	Effect
Smoking [6,8]	The most important environmental factor for inducing CD. It increases the occurrence of CD two times. An obvious decrease in secretion of interleukin-8 (IL-8), IL-10 and IL-23 from mononuclear cells is recognized.
Vitamin D [6]	The loss of vitamin D leads to CD. Furthermore, sunlight is an indirect important factor for production of vitamin D in people. The lack of sunlight leads to the absence of vitamin D and presence of CD.
Antibiotics [6,10-12]	Some studies show that long time consumption of antibiotics increases the rate of CD [6]. While some other surveys indicate that a long period consumption of a combination of two or more antibiotics has no clear positive effect on CD. However, some studies show positive effects of antibiotic consumption on CD for a short time [10-12].
Chemical pollution [4,6]	Chemical agent like NO_2 rises up the risk of CD.
Food Diet [4,7,13,14]	Vegetable and fruits decrease the rate of CD while fast food increases the risk of CD.
Season of birth [9,15]	Male patients born in April to June are predisposed for CD [15]. Another survey indicates no correlation between the season of birth and CD [9].

Table 2: The incidence and prevalence of Crohn's Disease in different countries.

Region	Country	Crohn's Disese per 10^5 cases	
		Incidence	Prevalence
The Persian Gulf Region Countries (Asia)	Bahrain	NR*	NR
	Iran [19]	0.8	-
	Iraq	NR	NR
	Kuwait [20]	-	-
	Oman	NR	NR
	Qatar	NR	NR
	Saudi Arabia [21]	1.91	6.72
	United Arab Emirate (UAE)	NR	NR
Asia	China [17]	1.22	-
	Hong Kong [17]	1.31	-
	India [17]	-	-
	Indonesia [17]	0.33	-
	Japan [17]	-	21.2
	Lebanon [20]	1.4	53.1
	Malaysia [17]	0.24	-
	Singapore [20]	-	7.2
	South Korea [20]	1.34	11.2
	Sri Lanka [17]	0.59	-
	Thailand [17]	0.3	-
	Turkey [22]	2.2	-
Europe	Bosnia and Herzegovina [23]	-	28.2
	Croatia [23]	6.5	-
	Czech [16]	1.5	-
	Denmark [16]	8.6	151
	Estonia [23]	1.4	-
	Finland [24]	9.2	-
	France [23]	6.7	-
	Germany [23]	6.6	-
	Greece [23]	2.7	-
	Holland [23]	6.2	-
	Hungary [23]	8.9	-
	Iceland [23]	5.5	-
	Ireland [25]	6	-
	Italy [16]	2.3	40
	Norway [25]	5.8	-
	Poland [25]	Insignificant	Insignificant
	Portugal [25]	4.2	-
	Romania [16]	0.5	8.3
	Russia [25]	NR	NR
	Scotland [25]	11.7	-
	Slovakia [25]	Insignificant	Insignificant
	Spain [23]	7.3	-
	Sweden [23]	8.3	213
	Switzerland [23]	-	100.7
	UK [23]	6.6	-
North America	Canada [16]	20.2	319
	USA [26]	-	241.3
Latin America	Argentina [27]	Insignificant	Insignificant
	Brazil [25]	14.6	-
	Chile [25]	Insignificant	Insignificant
	Panama [27]	-	-
Oceania	Australia [28]	17.4	-
	New Zealand [28]	15.2	145
Africa	South Africa [16]	1.57 (Average)	-

NR*: Not Recorded

is completely noticeable for Croatia; a considerable increase in the incidence of CD from 0.7 per 10^5 (in 1989) to 6.5 (in 2004) [7,16,20,23-25].

North America

The incidence and prevalence of CD in North America are highest worldwide. There are several studies in different states of Canada and the USA. Canada ranks first relating to the incidence and prevalence of CD around the world. USA ranks second in association with the prevalence of CD both in North America and worldwide, previous reports from different states show a high increase of incidence and prevalence for CD. In recent decade, the rate of incidence and prevalence of CD has been stabilized [7,16,20,23-25].

Latin America

Latin America is consisted of Central and South America. Several researches indicate low levels for the incidence and prevalence of CD in Latin America. According to table 2, only Brazil encompasses high ranking in the incidence of CD. The incidence and prevalence of CD in Panama (Central America), Argentina and Chile (South America) have been recorded insignificant and low [7,16,17,20,23,25,27].

Oceania

Australia and New Zealand rank first and second for the incidence of CD within Oceania continent, respectively. The Oceania continent shows a remarkable high ranking for the incidence and prevalence of CD. Previous reports from 1980s relating to New Zealand show a very low incidence of 1.75 per 10^5 for CD [7,16,23,28].

Africa

There are few recorded studies from Africa. Some reports from South Africa show a low rank for the incidence and prevalence of CD. The results from African studies are classified into white, black and colored; However, we have indicated an average from three different records [7,16,25].

There are different types of diseases including bacterial infections, UC, celiac disease, food intoxication and CD which have similar clinical demonstrations. Therefore, laboratory and imaging tests must be performed. Microbiological assays involve stool culture, ELISA, microscopy and PCR for detecting pathogenic parasites (such as helminthes and their eggs, protozoaires and their ova), pathogenic bacteria (like *Campylobacter*, *Escherichia coli*, *Salmonella* spp. and *Clostridium difficile*) and viral agents, serologic tests for tracing anti-*Escherichia coli* antibodies (AECA) (against outer membrane), anti-*Saccharomyces cerevisiae* antibodies (ASCA), C-reactive protein (CRP) and anti-neurophil cytoplasmic antibodies (ANCA) [1-4,18].

Among different items relating to serologic tests, a positive result for ASCA represents a high possibility for CD suspended cases. The prevalence of ASCA+ patients with CD is reported within a range of 45% to 60% in global scale. The profile of CRP+/ASCA+/ANCA- shows an active CD. A CRP+ serologic test devotes the active form of CD. The CRP+/ASCA-/ANCA+ profile belong to a patient with UC [1-4].

In parallel with microbiological and serological tests, histological biopsies and observations including colonoscopy, computed tomography (CT), endoscopy, fluoroscopy, magnetic resonance imaging (MRI), radiology, and ultrasonic radiations must be done [1-4].

Conclusion

CD is an autoimmune disease which is resulted from a multiple predisposing factors with unknown causative agent. Dysregulated immune system, genetic characterizations and environmental factors are considerable risk factors in patients with CD.

Although, genetic (ethnical/racial) characterizations are important items for occurring CD, this is not a predominant agent. There are a vast range of environmental factors which effectively influence the occurrence of CD. Smoking, vitamin D, pollutants and food diet are remarkable environmental factors which directly stimulate genetic properties for the appearance of CD. The importance of sex, age, antibiotic consumption and seasons as other environmental parameters rank in second. According to table 2, Canada and the USA are the biggest countries that have a high ranking of fast foods. Simultaneously, in some states of Canada lacks sunlight. Besides, the range of smoking and level of chemical pollutions is high. These items are determined environmental factors that rising up the increase of CD in Canada and the USA. However in recent years, the rate of CD has been stabilized in both aforementioned countries.

The rate of CD in Latin American countries is low. However, some countries like Brasilia own high ranking of CD. Insignificant rates of CD are reported from Argentina, Chile and Panama. It seems that these countries suffer from some inaccuracies in their diagnostic techniques.

Western European countries have huge similarities with Canada and the USA in their lifestyle. Fast foods, lack of sunlight, chemical pollutions and smocking are predominant environmental items in a big part of Western Europe. Scotland, Switzerland, Scandinavian countries are suffering from these environmental parameters. Despite the high occurrence of CD in Western European countries, the rate of CD has been stabilized.

Eastern Europe countries possess a low ranking level in association with CD. However, those countries which are modernizing for being industrialized as well as Western European countries are experiencing significant increase in the rate of CD occurrence. This is completely obvious in some countries such as Hungary, Croatia, and Bosnia Herzegovina.

Australia and New Zealand encompass high rates of CD. The high rate of CD in these countries is related to their modernized lifestyles. Smoking, chemical pollutions and fast food are predominated environmental factors in Oceania continent.

Africa experiences a low rate of CD. There are not enough studies and reports in this continent. Besides, the majority of

African countries suffer from hunger and poor diagnostic tools.

The Persian Gulf Region countries are experiencing a significant increase in association with the rate of CD. Modernization and industrializing are the most important factors. Moreover, the use of modern diagnostic tools may help to detect accurate cases in these countries. However, there is a considerable lack of data from the most countries located in the Persian Gulf Region, exclusive Iran and Saudi Arabia. The increase of smoking, chemical pollutions and western food diet is responsible for the increase of CD in the Persian Gulf Region countries.

Asia involves a wide range of countries with different characterizations. Japan, Hong Kong and South Korea are pioneers in high ranking rate of CD. These aforementioned countries are experiencing industrialized societies from old pasts; so the high rate of CD is recognized from the past. Lebanon, Turkey and China are countries that are trying to be modernized; hence they are experiencing a new increase in the rate of CD. The most important environmental factors in these countries are smoking, chemical pollutions and fast food consumption.

Long time consumption of antibiotics and the season as environmental factors rank in second place among; because according to table 1 there is paradox information in association with the recent factors. The rate of mortality among patients with CD is high, worldwide. However, it has decreased in recent decades.

Finally we can conclude that environmental factors are dominated factors for changing the rate of CD in different countries. Besides, diagnostic procedures that are employed in enable specialists to detect and identify the disease with high accuracy and reliability.

Because of the importance of CD, we suggest our colleagues to investigate and present the incidence and prevalence of CD in their countries; in particular, colleagues in the Persian Gulf Region countries.

References

1. Laass MW, Roggenbuck D, Conrad K. Diagnosis and classification of Crohn's disease. Autoimmunity reviews. 2014; 13(4-5): 467-471.
2. Behzadi P, Behzadi E, Ranjbar R. Autoimmunity, Crohn's disease and anti-Saccharomyces cerevisiae autoantibodies. farmacist.ro. 2014; 156(1): 50-52.
3. Pinzon MCM, Hayden DM. Crohn's Disease. Common Surgical Diseases: Springer. 2015; 161-163.
4. Baumgart DC, Sandborn WJ. Crohn's disease. The Lancet. 2012; 380(9853): 1590-1605. doi: 10.1016/S0140-6736(12)60026-9.
5. Khor B, Gardet A, Xavier RJ. Genetics and pathogenesis of inflammatory bowel disease. Nature. 2011; 474(7351): 307-317. doi: 10.1038/nature10209.
6. Ananthakrishnan AN. Environmental triggers for inflammatory bowel disease. Current gastroenterology reports. 2013; 15(1): 302. doi: 10.1007/s11894-012-0302-4.
7. Logan I, Bowlus CL. The geoepidemiology of autoimmune intestinal diseases. Autoimmun rev. 2013; 9(5): A372-A8. doi: 10.1016/j.autrev.2009.11.008.
8. Safarpour AR, Hosseini SV, Mehrabani D. Epidemiology of inflammatory bowel diseases in iran and Asia; a mini review. Iran J Med Sci. 2013; 38(2 Suppl): 140-149.
9. Taghavi SA, Safarpour AR, Hosseini SV, Noroozi H, Safarpour M, Rahimikazerooni S. Epidemiology of Inflammatory Bowel Diseases (IBD) in Iran: A review of 740 patients in Fars province, Southern Iran. Ann Colorectal Res. 2013; 1(1): 17-22.
10. Selby W, Pavli P, Crotty B, Florin T, Radford-Smith G, Gibson P, et al. Two-year combination antibiotic therapy with clarithromycin, rifabutin, and clofazimine for Crohn's disease. Gastroenterology. 2007; 132(7): 2313-2319.
11. Thia KT, Mahadevan U, Feagan BG, Wong C, Cockeram A, Bitton A, et al. Ciprofloxacin or metronidazole for the treatment of perianal fistulas in patients with Crohn's disease: a randomized, double-blind, placebo-controlled pilot study. Inflamm Bowel Dis. 2009; 15(1):17-24. doi: 10.1002/ibd.20608.
12. Feller M, Huwiler K, Schoepfer A, Shang A, Furrer H, Egger M. Long-term antibiotic treatment for Crohn's disease: systematic review and meta-analysis of placebo-controlled trials. Clinical infectious diseases. 2010; 50(4): 473-80. doi: 10.1086/649923.
13. Chassaing B, Koren O, Goodrich JK, Poole AC, Srinivasan S, Ley RE, et al. Dietary emulsifiers impact the mouse gut microbiota promoting colitis and metabolic syndrome. Nature. 2015; 519(7541):92-96. doi: 10.1038/nature14232.
14. Devkota S, Wang Y, Musch MW, Leone V, Fehlner-Peach H, Nadimpalli A, et al. Dietary-fat-induced taurocholic acid promotes pathobiont expansion and colitis in Il10-/-mice. Nature. 2012; 487(7405):104-108. doi: 10.1038/nature11225.
15. Shaw SY, Nugent Z, Targownik LE, Singh H, Blanchard JF, Bernstein CN. Association between Spring Season of Birth and Crohn's Disease. Clin Gastroenterol Hepatol. 2014; 12(2):277-282. doi: 10.1016/j.cgh.2013.07.028.
16. Cosnes J, Gower–Rousseau C, Seksik P, Cortot A. Epidemiology and natural history of inflammatory bowel diseases. Gastroenterology. 2011; 140(6):1785-1794. e4. doi: 10.1053/j.gastro.2011.01.055.
17. Ng SC. Epidemiology of inflammatory bowel disease: Focus on Asia. Best Pract Res Clin Gastroenterol. 2014; 28(3):363-372. doi: 10.1016/j.bpg.2014.04.003.
18. Chan SS, Hart AR. Aetiology and Clinical Features of Crohn's Disease. Crohn's Disease: Springer. 2015; p. 1-9.
19. Zahedi MJ, Darvish Moghadam S, Hayat Bakhsh Abbasi M, Dehghani M, Shafiei Pour S, Zydabady Nejad H. Middle East J Dig Dis; 2014; 6(1):32-36.
20. Thia KT, Loftus EV, Sandborn WJ, Yang S-K. An update on the epidemiology of inflammatory bowel disease in Asia. Am J Gastroenterol. 2008; 103(12): 3167-3182. doi: 10.1111/j.1572-0241.2008.02158.x.
21. Al-Mofleh IA, Azzam NA. Crohn's disease. Increasing trend in Saudi Arabia. Saudi Med J. 2013; 34(11):1105-13.
22. Tozun N, Atug O, Imeryuz N, Hamzaoglu HO, Tiftikci A, Parlak E, et al. Clinical characteristics of inflammatory bowel disease in Turkey: a multicenter epidemiologic survey. J clin Gastroenterol. 2009; 43(1):51-7. doi: 10.1097/MCG.0b013e3181574636.

23. Burisch J, Jess T, Martinato M, Lakatos PL. The burden of inflammatory bowel disease in Europe. J Crohn's Colitis. 2013; 7(4):322-37. doi: 10.1016/j.crohns.2013.01.010.

24. Jussila A, Virta LJ, Kautiainen H, Rekiaro M, Nieminen U, Färkkilä MA. Increasing incidence of inflammatory bowel diseases between 2000 and 2007: a nationwide register study in Finland. Inflamm Bowel Dis. 2012; 18(3):555-561. doi: 10.1002/ibd.21695.

25. Economou M, Zambeli E, Michopoulos S. Incidence and prevalence of Crohn's disease and its etiological influences. Annals of gastroenterology. 2009; 22(3):158-67.

26. Kappelman MD, Moore KR, Allen JK, Cook SF. Recent trends in the prevalence of Crohn's disease and ulcerative colitis in a commercially insured US population. Digestive diseases and sciences. 2013; 58(2): 519-525. doi: 10.1007/s10620-012-2371-5.

27. Linares dlCJ, Canton C, Hermida C, Perez-Miranda M, Mate-Jimenez J. Estimated incidence of inflammatory bowel disease in Argentina and Panama (1987-1993). Rev Esp Enferm Dig. 1999; 91(4):277-286.

28. Day AS, Lemberg DA, Gearry RB. Inflammatory Bowel Disease in Australasian Children and Adolescents. Gastroenterol Res Pract. 2014; 2014:703890. doi: 10.1155/2014/703890.

29. Mansour-Ghanaei F, Haghkerdar M, Joukar F, Aminian K, Mashhour MY, Shafaghi A, et al. Epidemiologic Features of Inflammatory Bowel Disease in Guilan Province, North of Iran, During 2002-2012. Middle East Journal of Digestive Diseases (MEJDD). 2015; 7(2): 69-74.

Permissions

All chapters in this book were first published in SOJ BIOCHEMISTRY, by Symbiosis; hereby published with permission under the Creative Commons Attribution License or equivalent. Every chapter published in this book has been scrutinized by our experts. Their significance has been extensively debated. The topics covered herein carry significant findings which will fuel the growth of the discipline. They may even be implemented as practical applications or may be referred to as a beginning point for another development.

The contributors of this book come from diverse backgrounds, making this book a truly international effort. This book will bring forth new frontiers with its revolutionizing research information and detailed analysis of the nascent developments around the world.

We would like to thank all the contributing authors for lending their expertise to make the book truly unique. They have played a crucial role in the development of this book. Without their invaluable contributions this book wouldn't have been possible. They have made vital efforts to compile up to date information on the varied aspects of this subject to make this book a valuable addition to the collection of many professionals and students.

This book was conceptualized with the vision of imparting up-to-date information and advanced data in this field. To ensure the same, a matchless editorial board was set up. Every individual on the board went through rigorous rounds of assessment to prove their worth. After which they invested a large part of their time researching and compiling the most relevant data for our readers.

The editorial board has been involved in producing this book since its inception. They have spent rigorous hours researching and exploring the diverse topics which have resulted in the successful publishing of this book. They have passed on their knowledge of decades through this book. To expedite this challenging task, the publisher supported the team at every step. A small team of assistant editors was also appointed to further simplify the editing procedure and attain best results for the readers.

Apart from the editorial board, the designing team has also invested a significant amount of their time in understanding the subject and creating the most relevant covers. They scrutinized every image to scout for the most suitable representation of the subject and create an appropriate cover for the book.

The publishing team has been an ardent support to the editorial, designing and production team. Their endless efforts to recruit the best for this project, has resulted in the accomplishment of this book. They are a veteran in the field of academics and their pool of knowledge is as vast as their experience in printing. Their expertise and guidance has proved useful at every step. Their uncompromising quality standards have made this book an exceptional effort. Their encouragement from time to time has been an inspiration for everyone.

The publisher and the editorial board hope that this book will prove to be a valuable piece of knowledge for researchers, students, practitioners and scholars across the globe.

List of Contributors

Cheng-Yuan Huang and Suefay Liu
UBC Center for Reproductive Health, Vancouver, Canada

Gregory Lee and Hao Zhang
UBC Center for Reproductive Health, Vancouver, Canada
Department of Pathology, Shantou University Medical College, Shantou, China

Gabriela Mellado-Sánchez, Héctor Vivanco-Cid and Adriana Sumoza-Toledo
Multidisciplinary Laboratory of Biomedical Sciences, Institute of Medico-Biological, Research Campus Veracruz Universidad Veracruzana, Iturbide S/N. CP. 91700, Veracruz, Mexico

Patricia R. Wardwell, David R. Wilson, Dewayne L. Garner and Rebecca A. Bader
Department of Biomedical & Chemical Engineering, Syracuse Biomaterials Institute, Syracuse University, Syracuse, NY, USA
Onondaga Community College, Syracuse, NY, USA

Cristin Coquillard, Valery Vilchez, Francesc Marti and Roberto Gedaly
Department of Surgery, Division of Transplantation, University of Kentucky - College of Medicine, 800, Rose Street, C453 Lexington, KY, 40536, USA

Markus Gäbel and Michael E. Breimer
Department of Surgery, Institute of Clinical Sciences, Sahlgrenska University Hospital, Bruna stråket 5, S-413 45 Gothenburg, Sweden

Lennart Rydberg and Jan Holgersson
Department of Clinical Chemistry and Transfusion Medicine, Institute of Biomedicine, Sahlgrenska Academy at the University of Gothenburg, Sahlgrenska University Hospital, Gothenburg, Sweden

Ayeda AlMahri
Department of Clinical Chemistry and Transfusion Medicine, Institute of Biomedicine, Sahlgrenska Academy at the University of Gothenburg, Sahlgrenska University Hospital, Gothenburg, Sweden
Sheikh Khalifa Medical City (SKMC), Abu Dhabi, United Arab Emirates

Diego Saporta
Associates in ENY & Allergy PA, 470 North Ave, Elizabeth NJ 07208, USA

Brian J. Hayes and Jean S. Campbell
Department of Pathology, University of Washington, Seattle, WA 98195, USA

Kimberly J. Riehle
Department of Pathology, University of Washington, Seattle, WA 98195, USA
Department of Surgery, University of Washington, Seattle, WA 98195, USA

Debra G. Gilbertson and Wendy R. Curtis
Zymogenetics Inc, a Bristol-Myers Squibb Company, Seattle, WA 98102, USA

Edward J. Kelly
Department of Pharmaceutics, University of Washington, Seattle, WA 98195, USA

Piper M. Treuting
Department of Comparative Medicine, University of Washington, Seattle, WA 98195, USA

Mepur H. Ravindranath, Vadim Jucaud and Paul I. Terasaki
Terasaki Foundation Laboratory, 11570 W, Olympic Blvd, Los Angeles, CA 90064, California, USA

Hirota Fujiki, Tatsuro Watanabe and Eisaburo Sueoka
Department of Clinical Laboratory Medicine, Faculty of Medicine, Saga University, Saga, Japan

Masami Suganuma
Graduate School of Science and Engineering, Saitama University, Saitama, Japan

Judy F. Glass and Joshua Dziba
VA Medical Center, Lexington Kentucky 40502

Debra L. Hardin
Department of Internal Medicine, College of Medicine, Lexington, Kentucky 40536

Jeffrey Campion
College of Pharmacy Lexington Kentucky 40536

John S. Thompson and Stephen A. Brown
VA Medical Center, Lexington Kentucky 40502
Department of Internal Medicine, College of Medicine, Lexington, Kentucky 40536

Tatsuaki Tsuruyama
Department of Pathology, Graduate School of Medicine, Kyoto University, Yoshida-konoe-cho, 606-8501 Kyoto, Japan
Center for Anatomical, Pathological, and Forensic Medical Research, Graduate School of Medicine, Kyoto University, Yoshida-konoe-cho, Kyoto 606-8501, Japan

David Andrés Bernal-Estévez and Diana Rocio Tovar Murillo
Immunology and Clinical Oncology Research Group (GIIOC). Fundación Salud de los Andes, Bogotá-Colombia South-America
Immunology and Translational Medicine Research Group, Universidad Nacional de Colombia, Graduated School in Biomedical Sciences. Department of Microbiology. Carrera 30 Calle 45, Bogotá-Colombia South-America

Carlos Alberto Parra-López
Immunology and Translational Medicine Research Group, Universidad Nacional de Colombia, Graduated School in Biomedical Sciences. Department of Microbiology. Carrera 30 Calle 45, Bogotá-Colombia South-America

Ilker Ates, Asuman Karakaya and Sinan H. Suzen
Department of Toxicology, Faculty of Pharmacy, Ankara University, 06100 Tandogan, Ankara, Turkey

Berran Yucesoy
Health Effects Laboratory Division, CDC/NIOSH, Morgantown, WV, USA

Haidong Dong
Department of Immunology, Mayo Clinic, Rochester, MN, USA
Department of Urology, Mayo Clinic, Rochester, MN, USA

Roxana S. Dronca
Division of Medical Oncology, Mayo Clinic, Rochester, MN, USA.

Yiyi Yan
Division of Medical Oncology, Mayo Clinic, Rochester, MN, USA
Division of Hematology, Mayo Clinic, Rochester, MN, USA

Svetomir N. Markovic
Department of Immunology, Mayo Clinic, Rochester, MN, USA
Division of Hematology, Mayo Clinic, Rochester, MN, USA

Najdat Ali Al-Kadhi
Kirkuk Technical College, Northern Technical University, Iraq

Fatima Mustafa Mohammed and Chateen I Ali Pambuk
College of Dentistry, Tikrit University, Iraq

Ektifa Abdulhameed
College of Education for Women, Tikrit University, Iraq

Shahin Asadi
Student of Molecular Genetics, Director of the Division of Medical Genetics and Molecular Research, Molecular Genetics-IRAN-TABRIZ

Vida Vahdani Kia and Rana Bagheri
Master of Molecular Biology-Genetics, Director of the Division of Medical Genetics and Molecular Research, Molecular Genetics-IRAN-TABRIZ

Mahsa Jamali and Samaneh Sadeh Dell
Master of Molecular Genetics, Director of the Division of Medical Genetics and Molecular Research, Molecular Genetics-IRAN-TABRIZ

S Daniel Chukwuemaka Ogbuabor and Alphosus O Ogbabor
Department of Health Management, Faculty of Health Sciences and Technology, University of Nigeria, Nigeria

Peter U Achukwu
Department of Medical Laboratory Sciences, Faculty of Health Sciences and Technology, University of Nigeria, Nigeria

Nosa Terry Omorodion
Department of Medical Laboratory Sciences, Faculty of Health Sciences and Technology, University of Nigeria, Nigeria
Health Services Department, University of Benin, Nigeria

Kaze Paul Davou and Dogo Goni Abraham
Department of Veterinary Parasitology and Entomology, Faculty of Veterinary Medicine, University of Jos, Nigeria

Idris Lawal and Ajanusi Joseph
Department of Veterinary Parasitology and Entomology, Faculty of Veterinary Medicine, Ahmadu Bello University, Zaria, Nigeria

Saidu Lawal
Department of Veterinary medicine, Faculty of Veterinary Medicine, Ahmadu Bello University, Zaria, Nigeria

Chidozie E Agu
Prime Health Response Initiative (PHRI)–Sub-recipients of Global Fund Impact HIV Project, Ilorin, Kwara State, Nigeria

Ikenna K Uchendu, Chigozie N Okwuosa and Peter U Achukwu
Department of Medical Laboratory Science, Health Sciences and Technology, College of Medicine, University of Nigeria Enugu Campus

Jena R Wirth, Osama Naga, Jackie Eudaly and Gary S Gilkeson
Division of Rheumatology and Immunology, Medical University of South Carolina, USA

Melissa A Cunningham
Division of Rheumatology and Immunology, Medical University of South Carolina, USA
Ralph H. Johnson Veterans Affairs Hospital, Charleston, South Carolina, USA

Rakel Carpintero, Karim J. Brandt, Lyssia Gruaz and Danielle Burger
Division of Immunology and Allergy, Inflammation and Allergy research group, Hans Wilsdorf Laboratory, Department of Internal Medicine, University Hospital and Faculty of Medicine, University of Geneva, Switzerland

Duncan D. Adams
Faculty of Medicine, University of Otago, Dunedin, New Zealand

Elodie Begue, Laurence Michel, Francette Jean-Louis and Armand Bensussan
UMR976, Skin Research Center and Universite Paris Diderot, Sorbonne Paris Cite, Hopital Saint-Louis, Paris, France

Martine Bagot
UMR976, Skin Research Center and Universite Paris Diderot, Sorbonne Paris Cite, Hopital Saint-Louis, Paris, France
Assistance Publique-Hopitaux de Paris, Hopital Saint Louis, Paris, France

Samuel Antwi-Baffour, Andrew Owusu Boafo, Ransford Kyeremeh and Seidu Abdulai Mahmood
Department of Medical Laboratory Sciences, School of Allied Health Sciences, College of Health Sciences, University of Ghana, Korle-Bu, Accra

Salvador Toro- Llamas and Ángel Luis Castaño Núñez
Servicio de Inmunología, Hospital Universitario Virgen del Rocío, Sevilla, Spain
Instituto de Biomedicina de Sevilla (IBiS), Sevilla, Spain

Wafa Munir Ansari, Dilshad Ahmad Khan and Farooq Ahmad Khan
Department of Chemical Pathology, Army Medical College, National University of Science and Technology (NUST), Islamabad, Pakistan

Omer Jamshed Khan
Armed Forces Institute of Rehabilitation Medicine/ Cardiac Rehab, Rawalpindi, Pakistan

Amira M. Gamal Eldeen
Department of Cancer Biology Laboratory, Center of Excellence for Advanced Sciences, National Research Center (NRC), Dokki, 12622, Cairo, Egypt
Department of Biochemistry, National Research Center, Dokki, 12622, Cairo, Egypt

Payam Behzadi and Elham Behzadi
Department of Microbiology, College of Basic Sciences, Shahr-e-Qods Branch, Islamic Azad University, Tehran, Iran

Reza Ranjbar
Molecular Biology Research Center, Baqiyatallah University of Medical Sciences, Tehran, Iran

Yakup Cevik and Ugur Acar
Department of Ophthalmology, Ardahan State Hospital, Ardahan, Turkey

Bekir Kucuk and Gungor Sobaci
Department of Ophthalmology, Hacettepe University, Faculty of Medicine, Ankara, Turkey

Kalyan Srivastava
New York Blood Center, New York, USA

Kamna Srivastava
Dr. Ambedkar Center for Biomedical Research, University of Delhi, India

Elaine Y. Cheng
Terasaki Foundation Laboratory, Los Angeles, California, USA
Department of Surgery, David Geffen School of Medicine at University of California, Los Angeles

Paul I. Terasaki
Terasaki Foundation Laboratory, Los Angeles, California, USA

Index

www.ingramcontent.com/pod-product-compliance
Lightning Source LLC
LaVergne TN
LVHW060541210926
848485LV00005B/1239